POWER ELECTRONICS:
Converters, Applications, and Design

About the Authors

Ned Mohan is an Associate Professor in the Department of Electrical Engineering at the University of Minnesota. He has worked on several power electronics projects sponsored by industry and the electric power utilities including the Electric Power Research Institute. He has numerous publications and patents in this field.

Tore M. Undeland is a Professor in Power Electronics in the Department of Electrical Engineering and Computer Science at the Norwegian Institute of Technology. He is also Scientific Advisor to the Norwegian Research Institute of Electricity Supply. He has been a Visiting Scientific Worker in the Power Electronics Converter Department of ASEA in Vaasteras, Sweden and a Visiting Professor in the Department of Electrical Engineering at the University of Minnesota. He has worked on many industrial research and development projects in the power electronics field and has numerous publications.

William P. Robbins is an Associate Professor in the Department of Electrical Engineering at the University of Minnesota. Prior to joining the University of Minnesota, he was a research engineer at the Boeing Company. He has taught numerous courses in electronics and semiconductor device fabrication. His research interests are in acoustic microscopy and ultrasonics, and he has numerous publications in this field.

POWER ELECTRONICS:
Converters, Applications, and Design

———————
———————
———————

NED MOHAN

Department of Electrical Engineering
University of Minnesota
Minneapolis, Minnesota

TORE M. UNDELAND

Department of Electrical Engineering and Computer Science
Norwegian Institute of Technology
Trondheim, Norway

WILLIAM P. ROBBINS

Department of Electrical Engineering
University of Minnesota
Minneapolis, Minnesota

WILEY

JOHN WILEY & SONS
New York   Chichester   Brisbane   Toronto   Singapore

Library of Congress Cataloging in Publication Data:

Mohan, Ned
  Power electronics: converters, applications, and design/Ned
Mohan, Tore M. Undeland, William P. Robbins
      p.   cm.
  Includes bibliographies and index.
    1. Power semiconductors.   2. Electric current converters.
I. Undeland, Tore M.   II. Robbins, William P.   III. Title.
TK7871.85.M57345   1989
621.381'52—dc19   89-5707
                                                        CIP

ISBN 0-471-61342-8

Printed in the United States of America

10 9 8 7 6 5 4

To Our Families ...

Mary, Michael, and Tara
Mona, Hilde, and Arne
Joanne and Jon

PREFACE

This book describes how to design, specify, and apply power semiconductor converters. The reader learns to analyze converter characteristics in order to understand how these converters interact with the electric utility grid and with the load in various applications. The physics of power semiconductor devices and their terminal characteristics are described so that these devices can be used appropriately in the design of converters.

Why offer another book on power electronics? In the rapidly evolving field of power electronics, new power semiconductor devices and new circuit topologies for improved converter performance have opened up new applications. The majority of these applications are in the power range of 100 kW or less, where a huge market exists and where the demand for power electronic engineers is likely to be. At present, there are many topics of intense research within the power electronic community. An example is resonant circuits to minimize switching losses in the power semiconductor devices. As a consequence, the converter efficiency is improved and the converter can be operated at a higher frequency, which reduces equipment size. Another example is the interface of power electronic systems with the utility grid. Proliferation of power electronic systems raises concerns about the distortion of the utility voltage waveform and the electromagnetic interference with other loads that are connected on the same grid. An up-to-date discussion of these topics is included in this book.

This book is written as a textbook, but should be equally useful to practicing engineers both as a comprehensive reference and as a text for self-study. While undertaking the task of writing this book, we recognized the multidisciplinary nature of power electronics and each of us contributes a different area of expertise. By working together, we have prepared a cohesive presentation of the fundamentals.

Organization of the Book

This book is divided into six parts. In Part 1, Chapter 1 introduces the scope and applications of power electronics. Chapter 2 presents an overview of power semiconductor devices in terms of their terminal characteristics. The discussion in Chapter 2 justifies the assumption of replacing the actual semiconductor device (e.g., a transistor) by an ideal switch. This ideal switch assumption makes it easier to discuss basic converter topologies and their applications.

Part 2 of the book, which includes Chapters 3 to 7, discusses the generic converter topologies that are used in most applications. This approach avoids the repetition of converter topology with each new application in Parts 3 to 5. In Chapter 3, diode rectifiers are described separately because of the trend to use them in place of thyristor converters for ac mains rectification. In Chapter 4, phase-controlled thyristor converters for three-phase rectification at high power levels are described. Chapters 5 and 6 deal with switch-mode dc–dc converters and dc-to-ac inverters, respectively. Chapter 7 presents the recent developments in resonant-mode converter topologies for zero-loss switching.

Part 3 of the book, which includes Chapters 8 and 9, discusses switch-mode dc and uninterruptible power supplies. Power supplies represent one of the two major applications of power electronic converters.

Part 4, which includes Chapters 10 through 14, considers the other major applications area of power electronics—motor drives. This includes dc motors in Chapter 11, induction motor drives in Chapter 12, synchronous motor drives in Chapter 13, and step-motor drives in Chapter 14. Although we assume that the reader has a basic knowledge of electric motors, the appropriate theory is discussed in order to interface these motors with the power electronic converters.

Part 5, which includes Chapters 15 through 17, discusses other applications of power electronic converters. Chapter 15 describes several industrial and residential applications. High power electric utility applications are discussed in Chapter 16. Chapter 17 examines the concerns and remedies for interfacing power electronic systems with the electric utility. The topics of harmonic current injection, distortion of the utility waveform, and EMI have lately become of great concern.

Part 6 (Chapters 18 to 26) discusses the semiconductor power devices used in power electronic converters and many practical considerations in designing power converters. Chapter 18 briefly reviews the basic semiconductor physics that undergirds the operation of power semiconductor devices. Each major power device is described in a separate chapter. The discussion includes basic device construction, terminal characteristics, a qualitative description of the physical basis of the device operation, and the operational limitations. In the same chapter, the design of drive and protective (snubber) circuits for the device are also considered. This close linking of device physics and drive and snubber circuits ensures that the unique electrical requirements of each device are met by the circuit design. In Chapter 26 many other practical converter design considerations are explored, including the design of inductors and transformers, capacitor selection, and circuit layout for minimizing stray inductance.

To the Instructor

We offer our own experiences in teaching courses on this subject at the University of Minnesota and the Norwegian Institute of Technology for the last 10 years in the hope that they may be of benefit. Core courses in circuits, fields, and electronics are prerequisites to these power electronic courses.

At the University of Minnesota, there is a two quarter (10 weeks per quarter) sequence in power electronics at the first year graduate level which is also open to seniors. In the first quarter, we consider the basic converter topologies and power supply applications (Chapters 1–6 and Chapters 8–9). In the second quarter, we discuss semiconductor power devices and drive and snubber circuits (Chapters 18–26). In this arrangement, each single quarter course is relatively independent of the other and one is not a prerequisite for the other. A one-quarter course at the senior elective level on electromechanics in robotics utilizes Chapters 10 to 14. An advanced graduate level course on special topics in power electronics utilizes Chapters 7 and 17.

At the Norwegian Institute of Technology, several one-semester (one semester equals 14 weeks) courses in power electronics at the senior level or first year graduate level use this material. A one-semester course covers practical converter design considerations including semiconductor power devices; this course utilizes Chapters 18 to 26. A second course covers power supplies and motor drives and utilizes Chapters 6 to 17 with the exception of Chapter 7. Another one-semester course, at the senior elective level, utilizes Chapters 1 to 5.

It is possible to combine Chapters 1 to 6 and Chapters 18 to 26 in several different ways in the formulation of a two-quarter sequence or a one-semester course.

A number of problems and examples are included that emphasize both the concepts and practical aspects of power electronics. An instructor's manual with completely worked-out solutions to all of the problems at the end of each chapter is available. The problems in this book are designed such that they do not require computer simulation tools. However, we recognize their usefulness.

Computer-Aided Analysis and Design

As a companion to this book, a large number of computer examples have been prepared to aid in teaching and in the design of power electronic systems. The data files for these examples are compatible with the widely used, general-purpose, royalty-free computer program named EMTP (Electro-Magnetics Transients Program) which executes on many computers including MS-DOS personal computers. Further information can be obtained from the University of Minnesota Media Distribution, Box 734 Mayo Building, 420 Delaware Street SE, Minneapolis, Minnesota 55455.

Acknowledgments

We are grateful to the following individuals who reviewed the manuscript and whose comments and advice improved the text: Professor Peter Lauritzen of the University of Washington, Professor Bimal K. Bose of the University of Tennessee—Knoxville, Professor Deepak Divan of the University of Wisconsin—Madison, Professor William Sayle III of Georgia Institute of Technology, Dr. Mirka Mikes Lindback and her colleagues at ABB in Vasteras, Sweden, Professor Jimmie Cathey of the University of Kentucky, Professor Alexander Emanuel of Worcester Polytechnic Institute, and Dr. Heinrich Boenig of Los Alamos National Laboratory.

We thank Mr. Owen Schott, president of the Schott Foundation, and the University of Minnesota Center for Electric Energy for their financial support of our power electronics program.

We thank Ms. Mary English for patiently typing the original version of the manuscript which was used as lecture notes, our editor Christina Kamra, and the rest of the editorial and production staff of John Wiley and Sons. Finally we are grateful to our colleagues at the University of Minnesota and the Norwegian Institute of Technology for their helpful comments and advice.

Ned Mohan
Tore M. Undeland
William P. Robbins

CONTENTS

PART 2   GENERIC POWER ELECTRONIC CONVERTERS

Chapter 3   Line-Frequency Diode Rectifiers: 60 Hz ac → Uncontrolled dc

Chapter 4   Line-Frequency Phase-Controlled Rectifiers and Inverters: 60 Hz ac ⇔ Controlled dc

Chapter 5   DC-to-DC Switch-Mode Converters

Chapter 6   Switch-Mode DC-to-AC Inverters: dc ⇔ Sinusoidal ac

Chapter 7   Resonant Converters: Zero-Voltage and / or Zero-Current Switchings

PART 3   POWER SUPPLY APPLICATIONS

Chapter 8   Switching DC Power Supplies

Chapter 9   Power Conditioners and Uninterruptible Power Supplies

PART 4 MOTOR DRIVE APPLICATIONS

Chapter 10 Introduction to Motor Drives

Chapter 11 DC-Motor Drives

Chapter 12 Induction Motor Drives

Chapter 13 Synchronous-Motor Drives

Chapter 14   Step-Motor Drives

PART 5   OTHER APPLICATIONS

Chapter 15   Residential and Industrial Applications

Chapter 16   Electric Utility Applications

Chapter 17   Optimizing the Utility Interface with Power Electronic System

PART 6   SEMICONDUCTOR DEVICES AND CONVERTER DESIGN

Chapter 18   Basic Semiconductor Physics

Chapter 19   Power Diodes

Chapter 20   BJTs with Drive and Snubber Circuits

Chapter 21 Power MOSFETs

Chapter 22 Thyristors

Chapter 23 Gate Turn-Off Thyristors

Chapter 24 Insulated Gate Bipolar Transistors

Chapter 25   Emerging Devices and Circuits

Chapter 26   Passive Components and Practical Converter Design Considerations

PART

1

Introduction

1

Power Electronic Systems

1-1 INTRODUCTION

In broad terms, the task of power electronics is to control the flow of power by shaping the utility-supplied voltages by means of power semiconductor devices. In recent years, the field of power electronics has experienced a large growth due to confluence of several factors. There have been revolutionary advances in microelectronics methods, which have led to the development of linear integrated circuits and digital signal processors as controllers in power electronic systems. Moreover, these advances in fabrication technology have made it possible to significantly improve the voltage and current ratings of power semiconductor devices and to increase their switching speeds. There has also been a significant expansion in the market for power electronics.

This expanded market demand has several dimensions. There is an increasing demand for variable-speed motor drives for compressors and pumps in process control. Robots in automated factories are powered by servo drives. It should be noted that the availability of process computers is a significant factor in making process control and factory automation feasible. Advances in microelectronic fabrication technology has led to the development of computers, communication systems, and consumer electronics, all of which require regulated power supplies and often uninterruptible power supplies. The increasing cost of energy has made it mandatory that the energy in all these systems be utilized efficiently. Power electronic systems offer the most cost-effective means of achieving efficient energy utilization.

In linear electronic systems, the semiconductor devices are used in their linear (active) regions of operation where they act as adjustable resistors. They have a low energy efficiency, which can be tolerated because the power levels are usually low, being on the order of a few tens of watts.

In power electronic applications, the power to be converted in a controlled manner ranges from a few watts to several hundred megawatts. Therefore, in contrast to linear electronic systems, semiconductor devices in power electronic systems operate as switches being either fully on or fully off. This results in a substantially higher

TABLE 1-1
Power Electronic Applications

a. *Residential*
   Refrigeration and freezer
   Space heating
   Air conditioning
   Cooking
   Lighting
   Electronics (personal computers, other entertainment equipment)
b. *Commercial*
   Heating, ventilating, and air conditioning
   Central refrigeration
   Lighting
   Computers and office equipment
   Uninterruptible power supplies (UPS)
   Elevators
c. *Industrial*
   Pumps
   Compressors
   Blowers and fans
   Machine tools (robots)
   Arc furnaces, induction furnaces
   Lighting
   Industrial lasers
   Induction heating
   Welding
d. *Transportation*
   Traction control of electric vehicles
   Battery chargers for electric vehicles
   Electric locomotives
   Street cars, trolley buses
   Subways
   Automotive electronics including engine controls
e. *Utility systems*
   High-voltage dc transmission (HVDC)
   Static var generation (SVG)
   Supplemental energy sources (wind, photovoltaic)
   Energy storage systems
   Induced-draft fans and boiler feed-water pumps
f. *Aerospace*
   Space shuttle power supply system
   Satellite power systems
   Aircraft power systems
g. *Telecommunications*
   Battery chargers
   Power Supplies (dc and UPS)

energy efficiency. This increased efficiency is extremely important because of the cost of wasted energy and the difficulty of removing the heat generated by wasted energy.

1-2 SCOPE AND APPLICATIONS OF POWER ELECTRONICS

The importance of power electronics can be appreciated by considering Table 1-1, which lists various applications of power electronics. These systems cover a wide power range from a few watts to several hundred megawatts. As power semiconductor devices improve in performance and fall in price, more applications will undoubtedly make use of power electronic converters. For example automotive electronics is a rapidly growing area of power electronic applications.

1-3 CLASSIFICATION OF POWER ELECTRONIC CONVERTERS

Power electronic systems consist of one or more power electronic converters which utilize power semiconductor devices controlled by integrated circuits. The converter is the basic module of power electronic systems. In general, a power electronic converter controls and shapes an electrical input of magnitude V_i, frequency f_i, and number of phases m_i into an electrical output of magnitude V_o, frequency f_o, and number of phases m_o. The power flow through these converters may be reversible, thus interchanging the roles of input and output. Specifically, a dc–dc converter converts one dc voltage level into another. In ac–dc conversion, a rectifier refers to a converter if the power flow is from the ac to the dc side, otherwise the converter is called an inverter.

There are many ways to classify converters used in power electronics. These ways include classification by type of device used, function of the converter, how the devices in the converter are switched, and so on. Unfortunately no well-defined categories based on these criteria are possible because there are always exceptions.

Some insight can be gained by classifying converters according to how the devices within the converter are switched. There are two possibilities including:

1. *Line frequency converters* where the utility line voltages present at one side of the converter facilitate the turn-off of the power semiconductor devices. Similarly, the devices are turned on, phaselocked to the line-voltage waveform. Therefore the devices switch on and off at the line frequency of 50 or 60 Hz.

2. *Switching converters* where the controllable switches in the converter are turned on and off at frequencies that are high compared to the line frequency. In spite of the high switching frequency internal to the converter, the converter output may be either dc or at a frequency comparable to the line frequency.

These two classifications can be further divided into subcategories according to the type of conversion (dc–dc conversion, rectification, inversion, etc.), as done in later chapters of this book.

1-4 ABOUT THE TEXT

The purpose of this book is to facilitate the study of practical and emerging power electronic converters made feasible by the new generation of power semiconductor devices. The topics are arranged such that the basic converter topologies used in more than one application can be studied in a generic manner, based on the assumption that the actual power semiconductor switches can be treated as ideal switches; an overview of power semiconductor devices and the justification for assuming them as ideal switches are provided in Chapter 2. Interactions of power electronic converters with loads and the utility system in a broad spectrum of applications are discussed. Finally, the design of converters based on the physical explanation of power semiconductor device characteristics and trade-offs are presented.

The reader is urged to read the overview of the text presented in the preface.

2

Overview of Power
Semiconductor Switches

2-1 INTRODUCTION

The increased power capabilities, ease of control, and reduced costs of modern power semiconductor devices compared to those of just a few years ago have made converters affordable in a large number of applications and have opened up a host of new converter topologies for power electronic applications. In order to clearly understand the feasibility of these new topologies and applications, it is essential that the characteristics of available power devices be put in perspective. To do this, a brief summary of the terminal characteristics and the voltage, current, and switching speed capabilities of currently available power devices are presented in this chapter.

If the power semiconductor devices can be considered as ideal switches, the analysis of converter topologies becomes much easier. This approach has the advantage that the details of device operation will not obscure the basic operation of the circuit. Therefore, the important converter characteristics can be more clearly understood. The summary of device characteristics will enable us to determine how much the device characteristics can be idealized.

Presently available power semiconductor devices can be classified into three groups according to their degree of controllability. The groupings are:

1. Diodes—on and off states controlled by the power circuit.
2. Thyristors—latched on by a control signal but must be turned off by the power circuit.
3. Controllable switches—turned on and off by control signals.

The controllable switch category includes several device types including bipolar junction transistors, MOS field effect transistors, gate-turn-off thyristors, and insulated gate bipolar transistors. There have been major advances in recent years in this category of devices.

7

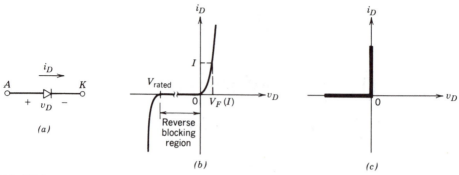

FIGURE 2-1: Diode: (*a*) symbol, (*b*) *i-v* characteristics, (*c*) idealized characteristics.

2-2 DIODES

Figures 2-1*a* and 2-1*b* show the circuit symbol for the diode and its steady-state *i-v* characteristic. When the diode is forward biased, it begins to conduct with only a small forward voltage across it, which is on the order of one volt. When the diode is reverse biased, only a negligibly small leakage current flows through the device until the reverse breakdown voltage is reached. In normal operation, the reverse bias voltage should not reach the breakdown rating.

In view of the very small leakage currents in the blocking (reverse bias) state and the small voltage in the conducting (forward bias) state as compared to the operating voltages and currents of the circuit in which the diode is used, the *i-v* characteristics for the diode can be idealized as shown in Fig. 2-1*c*. This idealized characteristic can be used for analyzing the converter topology but should not be used for the actual design, when, for example, heat sink requirements for the device are being estimated.

At turn-on, the diode can be considered an ideal switch because it turns on rapidly compared to the transients in the power circuit. However at turn-off, the diode current reverses for a reverse recovery time t_{rr}, as is indicated in Fig. 2-2 before falling to zero. This can lead to overvoltages in inductive circuits. In many circuits, this reverse current does not affect the converter characteristic and so the diode can also be considered as ideal during the turn-off transient.

Depending on the application requirements, various types of diodes are available:

1. *Schottky diodes.* These diodes are used where a low forward voltage drop (typically 0.3 V) is needed in very low output voltage circuits. These diodes are limited in their blocking voltage capabilities to 50–100 V.

2. *Fast-recovery diodes.* These are designed to be used in high-frequency circuits in combination with controllable switches where a small reverse recovery time is needed. At power levels of several hundred volts and several hundred amperes, such diodes have t_{rr} ratings of less than a few microseconds.

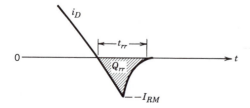

FIGURE 2-2: Diode turn-off.

3. *Line-frequency diodes*. The on-state voltage of these diodes is designed to be as low as possible and as a consequence have larger t_{rr}, which are acceptable for line-frequency applications. These diodes are available with blocking voltage ratings of several kilovolts and current ratings of several kiloamperes. Moreover, they can be connected in series and parallel to satisfy any voltage and current requirement.

2-3 THYRISTORS

The circuit symbol for the thyristor and its *i-v* characteristic are shown in Figs. 2-3a and 2-3b. The main current flows from the anode (A) to the cathode (K). In its off-state, the thyristor can block a forward polarity voltage and not conduct as is shown in Fig. 2-3b by the off-state portion of the *i-v* characteristic.

The thyristor can be triggered into the on-state by applying a pulse of positive gate current for a short duration provided that the device is in its forward blocking state. The resulting *i-v* relationship is shown by the on-state portion of the characteristics shown in Fig. 2-3b. The forward voltage drop in the on-state is only a few volts (typically 1 to 3 V depending on the device blocking voltage rating).

Once the device begins to conduct, it is latched on and the gate current can be removed. The thyristor cannot be turned off by the gate, and the thyristor conducts as a diode. Only when the anode current tries to go negative under the influence of the circuit in which the thyristor is connected does the thyristor turn off and the current go

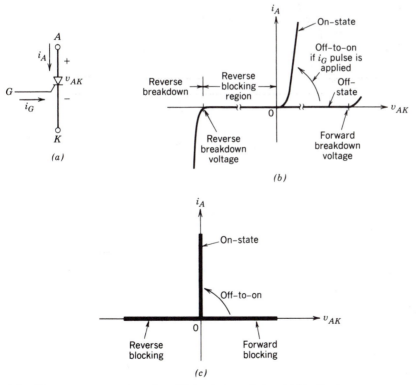

FIGURE 2-3: Thyristor: (*a*) symbol, (*b*) *i-v* characteristics, (*c*) idealized characteristics.

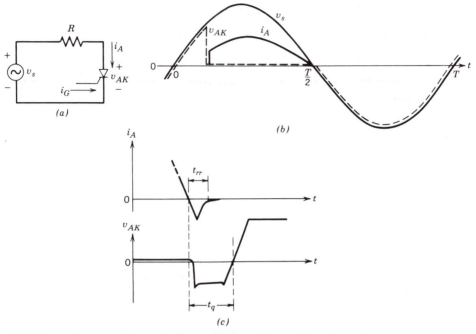

FIGURE 2-4: Thyristor: (*a*) circuit, (*b*) waveforms, (*c*) turn-off time interval t_q.

to zero. This allows the gate to regain control in order to turn the device on at some controllable time after it has again entered the forward blocking state.

In reverse bias at voltages below the reverse breakdown voltage, only a negligibly small leakage current flows in the thyristor as is shown in Fig. 2-3*b*. Usually the thyristor voltage ratings for forward and reverse blocking voltages are the same. The thyristor current ratings are specified in terms of maximum rms and average currents that it is capable of conducting.

Using the same arguments as for diodes, the thyristor can be represented by the idealized characteristics shown in Fig. 2-3*c* in analyzing converter topologies.

In an application such as the simple circuit shown in Fig. 2-4*a*, control can be exercised over the instant of current conduction during the positive half cycle of source voltage. When the thyristor current tries to reverse itself when the source voltage goes negative, the idealized thyristor would have its current become zero immediately after $t = T/2$ as is shown in the waveform in Fig. 2-4*b*.

However, as specified in the thyristor data sheets and illustrated by the waveforms in Fig. 2-4*c*, the thyristor current reverses itself before becoming zero. The important parameter is not the time it takes for the current to become zero from its negative value, but rather the turn-off time interval t_q defined in Fig. 2-4*c* from the zero crossover of the current to the zero crossover of the voltage across the thyristor. During t_q a reverse voltage must be maintained across the thyristor and only after this time is the device capable of blocking a forward voltage without going into its on-state. If a forward voltage is applied to the thyristor before this interval has passed, the device may prematurely turn on and damage to the device and/or circuit could result. Thyristor data sheets specify t_q with a specified reverse voltage applied during this interval as well as a specified rate-of-rise of voltage beyond this interval. This interval t_q is sometimes called the circuit-commutated-recovery time of the thyristor.

Depending on the application requirements, various types of thyristors are available. In addition to voltage and current ratings, turn-off time t_q, and the forward voltage drop, other characteristics that must be considered include the rate-of-rise of the current (di/dt) at turn-on and the rate-of-rise of voltage (dv/dt) at turn-off.

1. *Phase-control thyristors*. Sometimes termed converter thyristors, these are used primarily for rectifying line-frequency voltages and currents in applications such as phase-controlled rectifiers for dc and ac motor drives and in high-voltage dc power transmission. The main device requirements are large voltage and current handling capabilities and a low on-state voltage drop. This type of thyristor has been produced in wafer diameters of up to 10 cm, where the average current is about 4000 A with blocking voltages of 5–7 kV. On-state voltages range from 1.5 V for 1000-V devices to 3.0 V for the 5–7-kV devices.

2. *Inverter-grade thyristors*. These are designed to have small turn-off times t_q in addition to low on-state voltages, although on-state voltages are larger in devices with shorter values of t_q. These devices are available with ratings up to 2500 V and 1500 A. Their turn-off times are usually in the range of a few microseconds to 100 μs depending on their blocking voltage ratings and on-state voltage drops.

3. *Light-activated thyristors*. These can be triggered on by a pulse of light guided by optical fibers to a special sensitive region of the thyristor. The light-activated triggering of the thyristor uses the ability of light of appropriate wavelengths to generate excess electron–hole pairs in the silicon. The primary use of these thyristors are in high-voltage applications such as high-voltage dc transmission where many thyristors are connected in series to make up a converter valve. The differing high potentials that each device sees with respect to ground poses significant difficulties in providing triggering pulses. Light-activated thyristors have been reported with ratings of 4 kV and 3 kA, on-state voltages of about 2 V, and light trigger power requirements of 5 mW.

There are other variations of these thyristors such as gate-assisted-turn-off thyristors (GATT), asymmetrical silicon-controlled-rectifiers (ASCR), and reverse-conducting-thyristors (RCT). These are utilized based on the application.

2-4   DESIRED CHARACTERISTICS IN CONTROLLABLE SWITCHES

As mentioned in the introduction, several types of semiconductor power devices including BJTs, MOSFETs, GTOs, and IGBTs can be turned on and off by control signals applied to the control terminal of the device. These devices we term controllable switches which are represented in a generic manner by the circuit symbol shown in Fig. 2-5. No current flows when the switch is off, and when it is on, current can flow in the direction of the arrow only. The ideal controllable switch has the following

FIGURE 2-5: Generic controllable switch.

characteristics:

1. Block arbitrarily large forward and reverse voltages with zero current flow when off.

2. Conduct arbitrarily large currents with zero voltage drop when on.

3. Switch from on to off or vice versa instantaneously when triggered.

4. Vanishingly small power required from control source to trigger the switch.

Real devices, as we intuitively expect, do not have these ideal characteristics, and hence will dissipate power when they are used in the numerous applications already mentioned. If they dissipate too much power, the devices can fail and in doing so will not only destroy themselves, but may also damage the other system components.

Power dissipation in semiconductor power devices is fairly generic in nature; that is, the same basic factors governing power dissipation apply to all devices in the same manner. The converter designer must understand what these factors are and how to minimize the power dissipation in the devices.

In order to consider power dissipation in a semiconductor device, a controllable switch is connected in the simple circuit shown in Fig. 2-6a. This circuit models a very commonly encountered situation in power electronics; the current flowing through a switch also must flow through some series inductance(s). The dc current source approximates the current that would actually flow due to inductive energy storage. The diode is assumed to be ideal because our focus is on the switch characteristics, though in practice the diode reverse-recovery current can significantly affect the stresses on the switch.

When the switch is on, the entire current I_o flows through the switch and the diode is reverse biased. When the switch is turned off, I_o flows through the diode and a voltage equal to the input voltage V_d appears across the switch, assuming a zero voltage drop across the ideal diode. Figure 2-6b shows the waveforms for the current through the switch and the voltage across the switch when it is being operated at a repetition rate or switching frequency of $f_s = 1/T_s$ with T_s being the switching time period. The switching waveforms are represented by linear approximations to the actual waveforms in order to simplify the discussion.

When the switch has been off for a while, it is turned on by applying a positive control signal to the switch, as is shown in Fig. 2-6b. During the turn-on transition of this generic switch, the current buildup consists of a short delay time $t_{d(on)}$ followed by the current rise time t_{ri}. Only after the current I_o flows entirely through the switch can the diode become reverse biased and the switch voltage fall to a small on-state value of V_{on} with a voltage fall time of t_{fv}. The waveforms in Fig. 2-6b indicate that large values of switch voltage and current are present simultaneously during the turn-on crossover interval $t_{c(on)}$ where

$$t_{c(on)} = t_{ri} + t_{fv} \tag{2-1}$$

The energy dissipated in the device during this turn-on transition can be approximated from Fig. 2-6b as

$$W_{c(on)} = \tfrac{1}{2} V_d I_o t_{c(on)} \tag{2-2}$$

where it is recognized that no energy dissipation occurs during the turn-on delay interval $t_{d(on)}$.

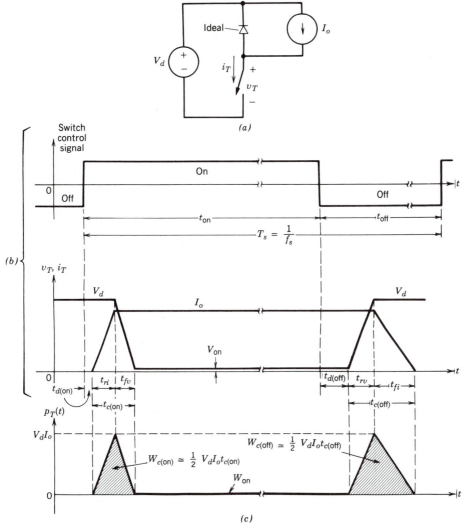

FIGURE 2-6: Generic-switch switching characteristics (linearized): (*a*) simplified inductive-switching circuit, (*b*) switch waveforms, (*c*) instantaneous switch power loss.

Once the switch is fully on, the on-state voltage V_{on} will be on the order of a volt or so depending on the device, and it will be conducting a current I_o. The switch remains in conduction during the on interval t_{on}, which in general is much larger than the turn-on and turn-off transition times. The energy dissipation W_{on} in the switch during this on-state interval can be approximated as

$$W_{on} = V_{on}I_o t_{on} \qquad (2\text{-}3)$$

where $t_{on} \gg t_{c(on)}, t_{c(off)}$.

In order to turn the switch off, a negative control signal is applied to the control terminal of the switch. During the turn-off transition period of the generic switch, the voltage buildup consists of a turn-off delay time $t_{d(off)}$ and a voltage rise time t_{rv}. Once the voltage reaches its final value of V_d (see Fig. 2-6*a*), the diode can become forward biased and begin to conduct current. The current in the switch falls to zero with a current fall time t_{fi} as the current I_o commutates from the switch to the diode. Large values of switch voltage and switch current occur simultaneously during the crossover interval $t_{c(\text{off})}$ where

$$t_{c(\text{off})} = t_{rv} + t_{fi} \tag{2-4}$$

The energy dissipated in the switch during this turn-off transition can be written, using Fig. 2-6b, as

$$W_{c(\text{off})} = \tfrac{1}{2} V_d I_o t_{c(\text{off})} \tag{2-5}$$

where any energy dissipation during the turn-off delay interval $t_{d(\text{off})}$ is ignored since it is small compared to $W_{c(\text{off})}$.

The instantaneous power dissipation $p_T(t) = v_T i_T$ plotted in Fig. 2-6c makes it clear that a large instantaneous power dissipation occurs in the switch during the turn-on and turn-off intervals. There are f_s such turn-on and turn-off transitions per second. Hence the average switching power loss P_s in the switch due to these transitions can be approximated from Eqs. 2-2 and 2-5 as

$$P_s = \tfrac{1}{2} V_d I_o f_s \left[t_{c(\text{on})} + t_{c(\text{off})} \right] \tag{2-6}$$

This is an important result because it shows that the switching power loss in a semiconductor switch varies linearly with the switching frequency and the switching times. Therefore if devices with short switching times are available, it is possible to operate at high switching frequencies in order to reduce filtering requirements and at the same time keep the switching power loss in the device from being excessive.

The other major contribution to the power loss in the switch is the average power dissipated during the on-state P_{on}, which varies in proportion to the on-state voltage. From Eq. 2-3 P_{on} is given by

$$P_{\text{on}} = V_{\text{on}} I_o \frac{t_{\text{on}}}{T_s} \tag{2-7}$$

which shows that the on-state voltage in a switch should be as small as possible.

The leakage current during the off-state (switch open) of controllable switches is negligibly small and therefore the power loss during the off-state can be neglected in practice. Therefore, the total average power dissipation P_T in a switch equals the sum of P_s and P_{on}.

From the considerations discussed in the preceding paragraphs, the following characteristics in a controllable switch are desirable:

1. Small leakage current in the off-state.

2. Small on-state voltage V_{on} to minimize on-state power losses.

3. Short turn-on and turn-off times. This will permit the device to be used at high switching frequencies.

4. Large forward and reverse voltage blocking capability. This will minimize the need for series connection of several devices, which complicates the control and protection of the switches. Moreover, most of the device types have a minimum on-state voltage regardless of their blocking voltage rating. A series connection of several such devices would lead to a higher total on-state voltage and hence

higher conduction losses. In most (but not all) converter circuits, a diode is placed across the controllable switch to allow the current to flow in the reverse direction. In those circuits, controllable switches are not required to have any significant reverse voltage blocking capability.

5. **High on-state current rating.** In high current applications, this would minimize the need to connect several devices in parallel, thereby avoiding the problem of current sharing.

6. **Positive temperature coefficient of on-state resistance.** This ensures that paralleled devices will share the total current equally.

7. **Small control power required to switch the device.** This will simplify the control circuit design.

8. **Capability to withstand rated voltage and rated current simultaneously while switching.** This will eliminate the need for external protection (snubber) circuits across the device.

9. **Large dv/dt and di/dt ratings.** This will minimize the need for external circuits otherwise needed to limit dv/dt and di/dt in the device so that it is not damaged.

We now will briefly consider the steady-state i-v characteristics and switching times of the commonly used semiconductor power devices that can be used as controllable switches. As mentioned previously these devices include BJTs, MOSFETs, GTOs, and IGBTs. The details of the physical operation of these devices, their detailed switching characteristics, commonly used drive circuits, and needed snubber circuits are discussed in Chapters 18 through 24.

2-5   BIPOLAR JUNCTION TRANSISTORS (BJTs) AND MONOLITHIC DARLINGTONS (MDs)

The circuit symbol for a NPN bipolar junction transistor is shown in Fig. 2-7a and its steady-state i-v characteristics are shown in Fig. 2-7b. As shown in the i-v characteristics, a sufficiently large base current (dependent on the collector current) results in the device being fully on. This requires that the control circuit provide a

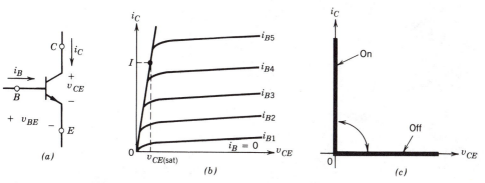

FIGURE 2-7: Bipolar junction transistor: (a) symbol, (b) i-v characteristics, (c) idealized characteristics.

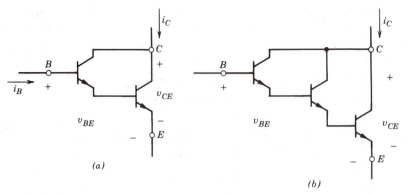

FIGURE 2-8: Darlington configurations: (*a*) darlington, (*b*) triple darlington.

base current that is sufficiently large so that

$$I_B > \frac{I_C}{h_{FE}} \tag{2-8}$$

where h_{FE} is the dc current gain of the device.

The on-state voltage $V_{CE(\text{sat})}$ of the power transistors is usually in the 1–2 V range, so that the conduction power loss in the BJT is quite small. The idealized *i-v* characteristics of the BJT operating as a switch are shown in Fig. 2-7*c*.

BJTs are current-controlled devices and base current must be supplied continuously to keep them in the on-state. The dc current gain h_{FE} is usually only 5–10 in high-power transistors and so these devices are sometimes connected in a Darlington or triple Darlington configuration as is shown in Fig. 2-8 to achieve a larger current gain. Some disadvantages accrue in this configuration including slightly higher overall $V_{CE(\text{sat})}$ values and slower switching speeds.

BJTs, whether in single units or made as a Darlington configuration on a single chip [a monolithic Darlington (MD)], have significant storage time during the turn-off transition. Typical switching times are in the range of a few hundred nanoseconds to a few microseconds.

BJTs including MDs are available in voltage ratings up to 1400 V and current ratings of a few hundred amperes. In spite of a negative temperature coefficient of on-state resistance, modern BJTs fabricated with good quality control can be paralleled provided that care is taken in the circuit layout and that some extra current margin is provided, that is, where theoretically four transistors in parallel would suffice based on equal current sharing, five may be used to tolerate a slight current imbalance.

2-6   METAL-OXIDE-SEMICONDUCTOR FIELD EFFECT TRANSISTORS (MOSFETs)

The circuit symbol of an *n*-channel MOSFET is shown in Fig. 2-9*a*. It is a voltage-controlled device, as is indicated by the *i-v* characteristics shown in Fig. 2-9*b*. The device is fully on and approximates a closed switch when the gate-source voltage is sufficiently large. The MOSFET is off when the gate-source voltage is below the threshold value, $V_{GS(\text{th})}$. The idealized characteristics of the device operating as a switch are shown in Fig. 2-9*c*.

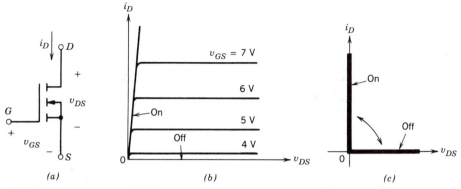

FIGURE 2-9: N-channel MOSFET: (*a*) symbol, (*b*) *i-v* characteristics, (*c*) idealized characteristics.

MOSFETs require the continuous application of a gate-source voltage of appropriate magnitude in order to be in the on-state. No gate current flows except during the transitions from on to off or vice versa when the gate capacitance is being charged or discharged. The switching times are very short, being in the range of a few tens of nanoseconds to a few hundred nanoseconds depending on the device type.

The on-state resistance $r_{DS(on)}$ of the MOSFET between the drain and source increases rapidly with the device blocking voltage rating. On a per unit area basis, the on-state resistance as a function of blocking voltage rating BV_{DSS} can be expressed as

$$r_{DS(on)} = k \quad BV_{DSS}^{2.5 \text{ to } 2.7} \tag{2-9}$$

where k is a constant that depends on the device geometry. Because of this, only devices with small voltage ratings are available that have low on-state resistance and hence small conduction losses.

However, because of their fast switching speed, the switching losses can be small in accordance with Eq. 2-6. From a total power loss standpoint, 300–400-V MOSFETs compete with bipolar transistors only if the switching frequency is in excess of 30–100 kHz. However, no definite statement can be made about the crossover frequency because it depends on the operating voltages, with low voltages favoring the MOSFET.

MOSFETs are available in voltage ratings in excess of 1000 V but with small current ratings and with up to 100 A at small voltage ratings. The maximum gate-source voltage is ± 20 V, although MOSFETs that can be controlled by 5-V signals are becoming available.

MOSFETs are easily paralleled because their on-state resistance has a positive temperature coefficient. This causes the device conducting the higher current to heat up and thus forces it to equitably share its current with the other MOSFETs in parallel.

2-7   GATE-TURN-OFF THYRISTORS (GTOs)

The circuit symbol for the GTO is shown in Fig. 2-10*a* and its steady-state *i-v* characteristic is shown in Fig. 2-10*b*.

Like the thyristor, the GTO can be turned on by a short-duration gate current pulse, and once in the on-state, the GTO may stay on without any further gate

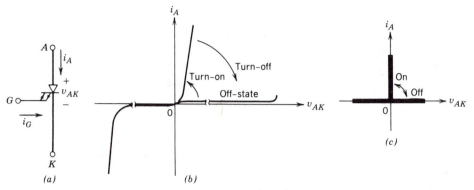

FIGURE 2-10: Gate-turn-off thyristor (GTO): (*a*) symbol, (*b*) *i-v* characteristics, (*c*) idealized characteristics.

current. However, unlike the thyristor, the GTO can be turned off by applying a negative gate-cathode voltage and therefore causing a sufficiently large negative gate current to flow. This negative gate current need only flow for a few microseconds (during the turn-off time) but it must have a very large magnitude, typically as large as one-third the anode current being turned off. GTOs can block negative voltages whose magnitude depends on the details of the GTO design. Idealized characteristics of the device operating as a switch are shown Fig. 2-10*c*.

Even though the GTO is a controllable switch in the same category as MOSFETs and BJTs, its turn-off switching transient is different from that shown in Fig. 2-6*b*. This is because presently available GTOs cannot be used for inductive turn-off such as is illustrated in Fig. 2-6 unless a snubber circuit is connected across the GTO (see Fig. 2-11*a*). This is a consequence of the fact that a large dv/dt that accompanies inductive turn-off cannot be tolerated by present-day GTOs. Therefore a circuit to reduce dv/dt at turn-off which consists of R, C, and D, as shown in Fig. 2-11*a*, must be used across the GTO. The resulting waveforms are shown in Fig. 2-11*b*, where dv/dt is significantly reduced compared to the dv/dt that would result without the turn-off snubber circuit. The details of designing a snubber circuit to shape the switching waveforms of GTOs are discussed in Chapter 23.

The on-state voltage (2–3 V) of a GTO is slightly higher than those of thyristors. The GTO switching speeds are in the range of a few microseconds to 25 μs.

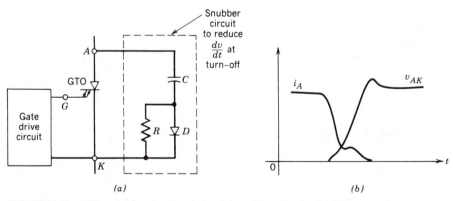

FIGURE 2-11: GTO transient characteristics: (*a*) snubber circuit, (*b*) GTO turn-off characteristic.

Because of their capability to handle large voltages (up to 4.5 kV) and large currents (up to a few kiloamps), the GTO is used when a switch is needed for high voltages and large currents in a switching frequency range of a few hundred hertz to 10 kHz.

2-8   INSULATED GATE BIPOLAR TRANSISTORS (IGBTs)

The circuit symbol for an IGBT is shown in Fig. 2-12a and its i-v characteristics are shown in Fig. 2-12b. IGBTs have some of the advantages of the MOSFET, the BJT, and the GTO combined. Similar to the MOSFET, the IGBT has a high impedance gate, which requires only a small amount of energy to switch the device. Like the BJT, the IGBT has a small on-state voltage even in devices with large blocking voltage ratings (for example $V_{on} = 2$–3 V in a 1000-V device). Similar to the GTO, IGBTs can be designed to block negative voltages as their idealized switch characteristics shown in Fig. 2-12c indicate.

IGBTs have turn-on and turn-off times on the order of 1 μs and are available in ratings as large as 1200 V and 100 A. Voltage ratings of up to 2000 V and current ratings of several hundred amperes are projected.

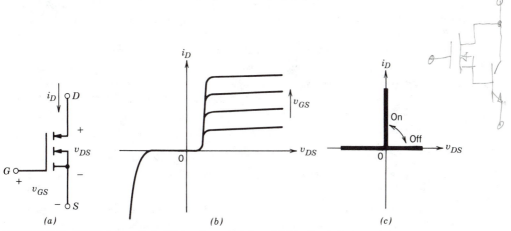

FIGURE 2-12: IGBT: (*a*) symbol, (*b*) *i-v* characteristics, (*c*) idealized characteristics.

2-9   COMPARISON OF CONTROLLABLE SWITCHES

Only a few definite statements can be made in comparing these devices since a number of properties must be considered simultaneously and because the devices are still evolving at a rapid pace. However, the qualitative observations given in Table 2-1 can be made.

TABLE 2-1
Relative Properties of Controllable Switches

Device	Power Capability	Switching Speed
BJT/MD	Medium	Medium
MOSFET	Low	Fast
GTO	High	Slow
IGBT	Medium	Medium

It should be noted that in addition to the improvements in these devices, there are other types of devices such as MOS-controlled thyristors (MCT) and static induction thyristors that are presently being investigated. The progress in semiconductor technology will undoubtedly lead to higher power ratings, faster switching speeds, and lower costs.

On the other hand, the forced-commutated thyristor, which was once widely used in circuits for controllable switch applications, is no longer being used in new converter designs with the possible exception of power converters in multi-MVA ratings. This is a pertinent example of how the advances in semiconductor power devices have modified converter design.

2-10   DRIVE AND SNUBBER CIRCUITS

In a given controllable power semiconductor switch, its switching speeds and on-state losses depend on how it is controlled. Therefore, for a proper converter design, it is important to design the proper control circuit for driving the base of a BJT or the gate of a MOSFET, GTO, or IGBT. The future trend is to integrate a large portion of the control circuitry along with the power switch within the device package, with the intention that the logic signals, for example from a microprocessor, can be used to control the switch directly. These topics are discussed in Chapters 18 through 25. In Chapters 3 through 17, where idealized switch characteristics are used in analyzing converter circuits, it is not necessary to consider these control circuits.

Snubber circuits, which were mentioned briefly in conjunction with GTO, are used to modify the switching waveforms of controllable switches. In general, snubbers can be divided into three categories:

1. Turn-on snubbers to minimize large overcurrents through the device at turn-on.
2. Turn-off snubbers to minimize large overvoltages across the device during turn-off.
3. Stress-reduction snubbers that shape the device switching waveforms such that the voltage and current associated with a device are not high simultaneously.

In practice, some combination of snubbers mentioned before are used, depending on the type of device and converter topology. The snubber circuits are discussed in Chapters 18 through 25. Since ideal switches are assumed in the analysis of converters, snubber circuits are neglected in Chapters 3 through 17.

The future trend is to design devices that can withstand high voltage and current simultaneously during the short switching interval, and thus minimize the stress-reduction requirement. However, for a device with a given characteristic, an alternative to the use of snubbers is to alter the converter topology such that large voltages and currents do not occur at the same time. These converter topologies, called resonant converters, are discussed in Chapter 7.

2-11   JUSTIFICATION FOR USING IDEALIZED DEVICE CHARACTERISTICS

In designing a power electronic converter, it is extremely important to consider the available power semiconductor devices and their characteristics. The choice of devices depends on the application. Some of the device properties and how they influence the selection process are listed here:

1. On-state voltage or on-state resistance dictates the conduction losses in the device.

2. Switching times dictate the energy loss per transition and determine how high the operating frequency can be.

3. Voltage and current ratings determine the device power handling capability.

4. The power required by the control circuit determines the ease of controlling the device.

5. The temperature coefficient of the device on-state resistance determines the ease of connecting them in parallel to handle large currents.

6. Device cost is a factor in its selection.

In designing a converter from the system viewpoint, the voltage and current requirements must be considered. Other important considerations include acceptable energy efficiency, the minimum switching frequency to reduce the filter and the equipment size, cost, and the like. Hence the device selection must ensure a proper match between the device capabilities and the requirements on the converter.

These observations help to justify the use of idealized device characteristics in analyzing converter topologies and their operation in various applications as follows: (1) since the energy efficiency is usually desired to be high, the on-state voltage must be small compared to the operating voltages and hence it can be ignored in analyzing converter characteristics; (2) the device switching times must be short compared to the period of the operating frequency, and thus the switchings can be assumed to be instantaneous; and (3) similarly the other device properties can be idealized. The assumption of idealized characteristics greatly simplifies the converter analysis, with no significant loss of accuracy. However, in designing the converters, not only must the device properties be considered and compared, but the converter topologies must also be carefully compared based on the properties of the available devices and the intended application.

2-12 SUMMARY

Characteristics and capabilities of various power semiconductor devices are presented. A justification is provided for assuming ideal devices, unless stated explicitly, in Chapters 3 through 17. The benefits of this approach are the ease of analysis and a clear explanation of the converter characteristics, unobscured by the details of device operation.

PROBLEM

2-1. The data sheets of a switching device specify the following switching times corresponding to the linearized characteristics shown in Fig. 2-6b for inductive switchings:

$$t_{ri} = 100 \text{ ns}$$

$$t_{fv} = 50 \text{ ns}$$

$$t_{rv} = 100 \text{ ns}$$

$$t_{fi} = 200 \text{ ns}$$

Calculate and plot the switching power loss as a function of frequency in a range of 25 kHz to 100 kHz, assuming $V_d = 300$ V and $I_o = 4$ A.

2-13 REFERENCES

1. R. Sittig and P. Roggwiller, (Eds.), *Semiconductor Devices for Power Conditioning,* Plenum Press, New York, 1982.
2. M. S. Adler, S. W. Westbrook and A. J. Yerman, "Power Semiconductor Devices—An Assessment," IAS Conference Record, pp. 723–728, 1980.

P A R T

2

Generic Power Electronic Converters

3

Line-frequency Diode Rectifiers:
60-Hz ac → Uncontrolled dc

3-1 INTRODUCTION

In most power electronic applications, the power input is in the form of a 50- or 60-Hz sinewave ac voltage provided by the electric utility, which is first converted to a dc voltage. Increasingly, the trend is to convert the input ac into dc in an uncontrolled manner, using rectifiers with diodes. In such diode rectifiers, the power flow can only be from the utility ac side to the dc side. A majority of the power electronics applications such as switching dc power supplies, ac-motor drives, dc-servo drives, and so on use such uncontrolled rectifiers. In most of these applications, these rectifiers are supplied directly from the utility source without a 60-Hz transformer. The avoidance of this costly and bulky 60-Hz transformer is important in most modern power electronic systems.

The dc output voltage of a rectifier should be as ripple-free as possible. Therefore, a large capacitor is connected as a filter on the dc side. This capacitor gets charged to a value close to the peak of the ac input voltage. As a consequence, the current through these rectifiers does not flow continuously, that is, it becomes zero for finite durations during each half-cycle of the line frequency. Such rectifiers, with single-phase and three-phase inputs are discussed. As discussed in Chapter 2, the diodes are assumed to be ideal in the analysis of rectifiers. In a similar manner, the electromagnetic interference (EMI) filter at the ac input to the rectifier is ignored, since it does not influence the basic operation of the rectifier. EMI and EMI filters are discussed in Chapter 17.

3-2   BASIC RECTIFIER CONCEPTS

To illustrate the basic concepts, consider the circuit shown in Fig. 3-1a where v_s is a sinusoidal source voltage in series with a diode and a load resistance. The current through the diode flows only during the positive half-cycle of v_s.

With both R and L present, the voltage v_R (which is linearly proportional to the diode current) is plotted in Fig. 3-1b instead of the current. The inductor voltage v_L, which equals $v_s - v_R$ during $i > 0$, shows that the current continues to flow for an interval even after v_s has become negative. Once the current goes to zero, $v_L = 0$, and v_s appears as reverse voltage across the diode. The time interval during which the current flows is governed by the fact that the inductor voltage v_L, integrated over one time period T must be zero in steady state as shown below: During the time interval between any arbitrary time t_1 and $t_1 + T$ (in Fig. 3-1b, t_1 is chosen as the zero-crossing instant of the source voltage v_s)

$$di = \frac{1}{L}v_L \cdot dt \tag{3-1}$$

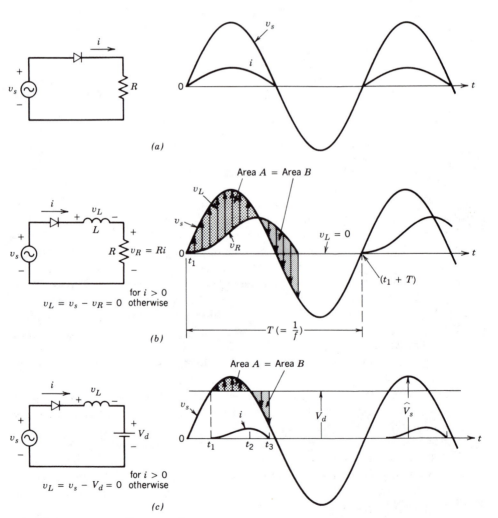

FIGURE 3-1: Basic rectifier concepts.

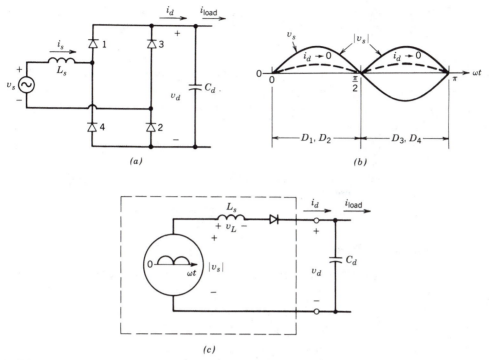

FIGURE 3-2: Single-phase diode bridge rectifier: (*a*) circuit, (*b*) open-circuit waveform without C_d and $i_d \to 0$, (*c*) equivalent circuit.

Integrating both sides of the foregoing equation between t_1 and $t_1 + T$

$$I(t_1 + T) - I(t_1) = \frac{1}{L} \int_{t_1}^{t_1 + T} v_L \cdot dt \qquad (3\text{-}2)$$

In steady state, $I(t_1 + T)$ will always be equal to $I(t_1)$. Therefore,

$$\int_{t_1}^{t_1 + T} v_L \cdot dt = 0 \qquad (3\text{-}3)$$

which implies that area A equals area B in Fig. 3-1b.

Next, L and v_d are included as shown in Fig. 3-1c, where $v_d(t) = V_d$. The diode begins to conduct at t_1 when v_s exceeds V_d. The current reaches its peak at t_2 and decays to zero at t_3, with t_3 determined by the requirement that area A be equal to area B. The waveforms in Fig. 3-1c are drawn where V_d is chosen to be a large fraction of $\hat{V}_s$, which is usually the case. This causes the current to become zero prior to the next half-cycle of v_s.

3-3 SINGLE-PHASE DIODE BRIDGE RECTIFIERS

3-3-1 Equivalent Circuit

A single-phase diode bridge rectifier is shown in Fig. 3-2a. A large filter capacitor is connected on the dc side. The utility supply is modeled as a voltage source v_s in series with its internal impedance, which in practice is primarily inductive. Therefore, it is represented by L_s in Fig. 3-2a. To improve the line-current waveform, a filter-inductor may be added on the ac side, which will result in a higher value of L_s.

In order to obtain v_d and i_d waveforms, it is desirable to obtain a simple equivalent circuit of the diode bridge rectifier, as seen from the dc side. This can be achieved by first obtaining its dc output voltage under an open-circuit condition, which is approximated by assuming a resistor R_d connected across the rectifier output without C_d, where i_d is approximately zero for a large value of R_d. Figure 3-2b shows the rectifier waveforms. During the positive half-cycle of v_s, D_1 and D_2 conduct the current and $v_d = v_s$. During the negative half-cycle of v_s, D_3 and D_4 conduct and $v_d = -v_s$. This implies that

$$v_d|_{\text{open-circuit}} = |v_s| \tag{3-4}$$

which results in the equivalent circuit of Fig. 3-2c. The diode in the equivalent circuit ensures that i_d cannot reverse. This equivalent circuit is valid only for a highly discontinuous i_d, which is usually the case.

3-3-2 Rectifier Characteristic

In analyzing the circuit of Fig. 3-2a, the following assumptions and observations are made:

1. Since L_s and C_d in Fig. 3-2a form a low-pass filter and if C_d is large, v_d can be assumed to be ripple-free, that is, $v_d(t) \simeq V_d$.

2. Assuming a large value of C_d, i_d is highly discontinuous and goes to zero prior to v_s going to zero every half-cycle. This allows the equivalent circuit of Fig. 3-2c to be used.

3. C_d, being large, appears as a short-circuit to the ripple in the load-current i_{load}. The average value I_d of the rectifier output current equals the average value I_{load} of the load current, since the average current through C_d in steady state is zero.

For design purposes, it is useful to know V_d as a function of I_{load} for given values of source voltage V_s and the inductance L_s. Instead of deriving a complicated analytical expression, a graphical method to calculate V_d is presented. In this method, a value is assumed for V_d and a corresponding I_d is calculated. From the plot of V_d versus I_d, the operating value of V_d is obtained corresponding to a specified $I_d(= I_{\text{load}})$. The steps involved are as follows:

1. Choose a value for V_d to be slightly less than the peak of the ac input. The current i_d begins to flow at θ_b in Fig. 3-3a, where θ_b can be calculated:

$$V_d = \sqrt{2}\, V_s \sin \theta_b \tag{3-5}$$

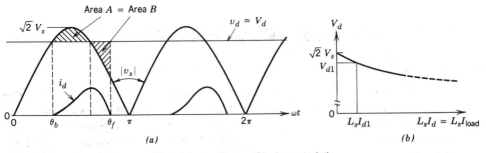

FIGURE 3-3: Single phase rectifier: (a) waveforms, (b) characteristic.

2. The voltage across L_s in Fig. 3-2c is

$$v_L = \sqrt{2}\, V_s \sin \omega t - V_d \tag{3-6}$$

which is valid as long as i_d keeps flowing. Using Eq. 3-1 in Fig. 3-3a

$$L_s i_d(\omega t) = \frac{1}{\omega} \int_{\theta_b}^{\omega t} v_L d(\omega t) \qquad \theta_b < \omega t < \theta_f \tag{3-7}$$

Equations 3-5 through 3-7 allow θ_f to be calculated as the value of ωt at which i_d becomes zero in Fig. 3-3a, satisfying that area A equals area B. If θ_f turns out to be greater than π, a larger value of V_d should be chosen in step (1).

3. Knowing θ_f, $L_s I_d$ can be calculated by averaging $L_s i_d(\omega t)$ given by Eq. 3-7 between $\omega t = 0$ and π:

$$L_s I_d = \frac{1}{\pi} \int_{\theta_b}^{\theta_f} L_s i_d(\omega t) \cdot d(\omega t) \tag{3-8}$$

By using this procedure, V_d can be plotted as a function of $L_s I_d$ as shown in Fig. 3-4b. The curve in Fig. 3-3b for large $L_s I_d$ with $\theta_f > \pi$ is shown as dotted. For given V_s, L_s and $I_{\text{load}}(= I_{d1})$, the actual voltage V_{d1} can be obtained as shown in Fig. 3-3b. By knowing the range over which I_{load} may vary in a given application, it is possible for us to obtain the variation in V_d from Fig. 3-3b, if V_s and L_s are given. In practice, circuit simulation programs are used to implement the foregoing steps to obtain the rectifier characteristic. This characteristic shows that for a given load current, a higher L_s results in a lower V_d.

Having calculated the instantaneous waveform i_d and its average value I_d using Eqs. 3-7 and 3-8, the assumption of zero ripple in v_d can be relaxed and the ripple can be calculated.

EXAMPLE 3-1

In the circuit of Fig. 3-2a, the input ac voltage is 230 V rms at 60 Hz and $L_s = 5$ mH. By circuit simulation, the rectifier characteristic is obtained, as shown in Fig. 3-4. Obtain the average dc voltage for a load current of 3 A.

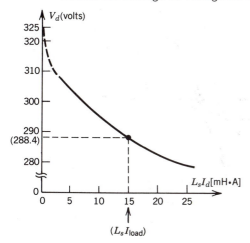

FIGURE 3-4: Example 3-1 (V_s = 230 V rms at 60 Hz).

SOLUTION

L_sI_d equals 15 mH-A. From Fig. 3-4, V_d equals 288.4 V.   ∎

3-3-3 Line-Current i_s

The line-current i_s at the input to the rectifier deviates significantly from a sinusoidal waveform, as is shown in Fig. 3-5. The distortion in the line-current waveform can be quantified as described in the following section.

3-3-3-1 BASIC DEFINITIONS

By Fourier analysis, the line-current can be expressed in terms of its fundamental frequency component i_{s1} (shown as dotted in Fig. 3-5) plus other harmonic components. If v_s is assumed to be purely sinusoidal, then only i_{s1} contributes to the average power flow, because the frequencies of voltage and current must be the same for average power flow. In terms of the rms voltage V_s and the rms value I_{s1} of the fundamental frequency component of i_s, the average power P flowing through the rectifier is

$$P = V_sI_{s1} \cos \phi_1 \qquad (3\text{-}9)$$

where ϕ_1 is the angle by which i_{s1} lags v_s, as shown in Fig. 3-5. The magnitude of the apparent power S is the product of rms voltage V_s and the rms value I_s of the line current i_s in Fig. 3-5

$$S = V_sI_s \qquad (3\text{-}10)$$

The power factor is defined as

$$\text{power factor } PF = \frac{P}{S} \qquad (3\text{-}11)$$

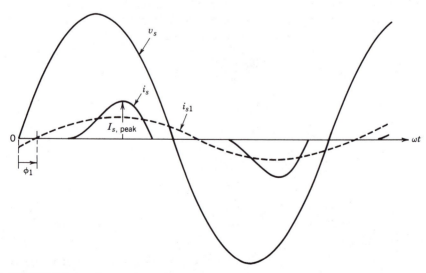

FIGURE 3-5: Input current waveform.

From Eqs. 3-9 through 3-11

$$PF = \frac{V_s I_{s1} \cos \phi_1}{V_s I_s} = \frac{I_{s1}}{I_s} \cdot \cos \phi_1 \qquad (3\text{-}12)$$

The displacement power factor (which is the same as the power factor in circuits with sinusoidal voltages and currents) is defined as the cosine of the angle ϕ_1

$$\text{displacement power factor } DPF = \cos \phi_1 \qquad (3\text{-}13)$$

Therefore

$$PF = \frac{I_{s1}}{I_s} \cdot DPF \qquad (3\text{-}14)$$

From Eq. 3-14, it can be noted that a large distortion in the line current will result in a small value of the current ratio I_{s1}/I_s, and hence a small value of PF, even if DPF is close to unity.

The rms value I_s of the line current can be calculated by root-mean-squaring the i_s waveform as in the following Eq. 3-15 (where $T = 2\pi/\omega$) or in terms of its rms Fourier components, I_{s1} and I_{sh}, as in Eq. 3-16:

$$I_s = \left[\frac{1}{T} \int_0^T i_s^2(t) \cdot dt \right]^{1/2} \qquad (3\text{-}15)$$

or

$$I_s = \left[I_{s1}^2 + \sum_{h=2}^{\infty} I_{sh}^2 \right]^{1/2} \qquad (3\text{-}16)$$

From Eq. 3-16, the rms value of the distortion component in the line current can be defined as

$$I_{\text{dis}} = \left[I_s^2 - I_{s1}^2 \right]^{1/2} = \left[\sum_{h=2}^{\infty} I_{sh}^2 \right]^{1/2} \qquad (3\text{-}17)$$

To quantify the distortion in the current waveform, a quantity called the total harmonic distortion THD is defined as

$$\%THD = 100 \times \frac{I_{\text{dis}}}{I_{s1}} \qquad (3\text{-}18)$$

In many applications, it is important to know the peak value $I_{s,\text{peak}}$ of the i_s waveform in Fig. 3-5, as a ratio of the total rms current I_s, which is defined as follows:

$$\text{Crest factor} = \frac{I_{s,\text{peak}}}{I_s} \qquad (3\text{-}19)$$

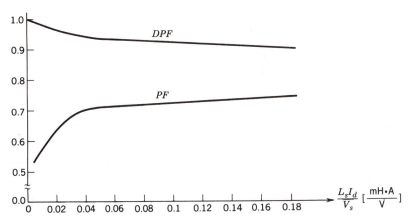

FIGURE 3-6: Effect of L_s on the ac current waveform (L_s is in mH, I_d is the average current and V_s is the rms voltage).

Another quantity called the form factor is defined as

$$\text{Form factor} = \frac{I_s}{I_d} \tag{3-20}$$

which is used in the following section.

3-3-3-2   EFFECT OF L_s

Based on circuit simulation and verified by measurements, the effect of L_s on the displacement power factor and the power factor is shown in Fig. 3-6, which can be used for any combination of L_s, I_d, and V_s. Figure 3-6 shows that in diode rectifiers, the displacement power factor is better than 0.9. However, for given I_d and V_d, the power factor is very poor at small L_s and improves to approximately 0.75 as L_s increases.

In Fig. 3-7, the ratio V_d/V_s, the form factor and the crest factor for the line current are plotted. This figure shows that a small L_s results in a large crest factor and

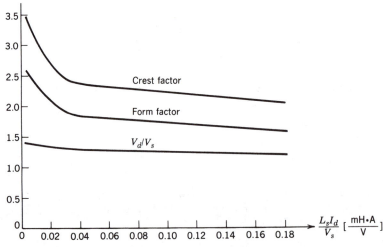

FIGURE 3-7: The dc-voltage to line rms voltage (V_d/V_s) ratio, form factor and crest factor as a function of $\dfrac{L_s I_d}{V_s} \left[\dfrac{\text{mH.A}}{\text{V}} \right]$

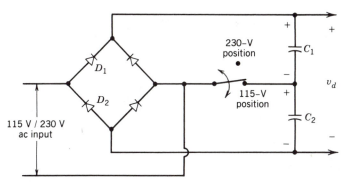

FIGURE 3-8: 115/230 V Rectifier.

a large form factor. This implies that the $I_{s,\text{peak}}$ will be several times the average current I_d (for example, at $L_s I_d / V_s$ of 0.04, the ratio $I_{s,\text{peak}}/I_d$ equals approximately $2.4 \times 1.87 \simeq 4.5$). It should be noted that in the actual circuit, the line resistance in series with L_s may not be negligible. As a consequence, the peak current will be smaller than the value calculated from Fig. 3-7.

3-4   VOLTAGE-DOUBLER (SINGLE-PHASE) RECTIFIERS

In many applications, the input line-voltage magnitude may be insufficient to meet the dc output voltage requirement. More importantly, the equipment may be required to operate with a line voltage of 115 V as well as 230 V. Therefore, a voltage-doubler rectifier, as shown in Fig. 3-8 may be used to avoid a voltage step-up transformer.

When the switch in Fig. 3-8 is in the 230-V position with a line voltage of 230 V, the circuit is similar to a full-bridge rectifier discussed earlier. With the switch in the 115-V position and the line voltage of 115 V, each capacitor gets charged to approximately the peak of the ac input voltage, and therefore V_d (which is the sum of voltages across C_1 and C_2) is approximately the same as in the 230-V operation. The capacitor C_1 is charged through the diode D_1 during the positive half-cycle of the input ac voltage, and C_2 is charged through D_2 during the negative half-cycle. In this mode, the circuit operates as a voltage-doubler rectifier.

3-5   THREE-PHASE FULL-BRIDGE RECTIFIERS

In industrial applications where three-phase ac voltages are available, it is preferable to use three-phase rectifier circuits compared to single-phase rectifiers, because of their lower ripple content in the waveforms and a higher power-handling capability. A three-phase, six-pulse, full-bridge diode rectifier shown in Fig. 3-9 is a very commonly used circuit arrangement. A filter capacitor is connected at the dc side of the rectifier.

Similar to the analysis of the single-phase full-bridge rectifier, it is desirable to obtain a simple equivalent circuit. Again, to obtain the open-circuit waveform of v_d, we will assume a resistance R_d connected at the dc side without C_d, where $R_d \rightarrow \infty$

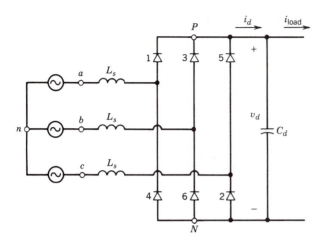

FIGURE 3-9: Three-phase diode-bridge rectifier.

and therefore $i_d \rightarrow 0$. Out of the diodes 1, 3, and 5 with common cathode connections, the diode connected to the highest positive phase voltage would conduct. Similarly, out of the diodes 2, 4, and 6 with common anode connections, the diode connected to the most negative phase voltage would conduct. v_{Pn} is the voltage of the top terminal P on the dc side of the rectifier, with respect to the ac source neutral n. Similarly, v_{Nn} is the voltage of bottom terminal N on the dc side of the rectifier, with respect to n. Both v_{Pn} and v_{Nn} are shown in Fig. 3-10a by means of darker curves. The dc side rectifier voltage $v_d (= v_{Pn} - v_{Nn})$ at open circuit is also plotted.

By looking at the waveforms in Fig. 3-10a, it is clear that at any instant of time, two devices conduct: one from the top diode group 1, 3, and 5 and one from the bottom diode group 2, 4, and 6. Also, by numbering the diodes as in Fig. 3-9, the diodes conduct in a sequence 1, 2, 3, and so on. Each diode conducts for 120° per cycle and a new diode begins to conduct after a 60° interval. In fact, v_d waveform, which consists of portions of the line-to-line ac voltage waveforms, repeats with a 60° duration, making this a six-pulse rectifier.

This analysis results in a simple equivalent circuit shown in Fig. 3-10b, where a diode in series with v_d at open-circuit is used to make sure that i_d through the circuit cannot reverse. Since two phases in Fig. 3-9 would conduct when the current is flowing, the series inductance in Fig. 3-10b is $2L_s$. This equivalent circuit can be used, provided the current i_d through the rectifier becomes zero prior to each 60° interval as shown in Fig. 3-10c. Because of a large C_d, the dc-side voltage is approximated as $v_d(t) \simeq V_d$.

For the conditions depicted in Fig. 3-10c, a graphical analysis similar to that in Section 3-3-2 for single-phase rectifiers can be used to obtain the three-phase rectifier characteristics: V_d as a function of I_d. Otherwise, a computer modeling of the rectifier will be necessary to determine its characteristic.

Often in three-phase rectifiers, if an inductor is used to improve the current waveforms and the ripple in the dc voltage output, it is placed on the dc side between the rectifier and the filter capacitor, as is shown by L_d in Fig. 3-11a. In this circuit, even if i_d flows continuously, the equivalent circuit of Fig. 3-10b can be modified and used as shown in Fig. 3-11b, assuming L_s to be negligibly small.

For the same value of I_d, the effect of the inductance L_d on the waveforms is illustrated by means of Figs. 3-11c and 3-11d resulting in a discontinuous and a continuous i_d, respectively.

It is possible to calculate the minimum value of inductance L_d required to make the current i_d continuous, for given values of I_d and the input voltage at the line

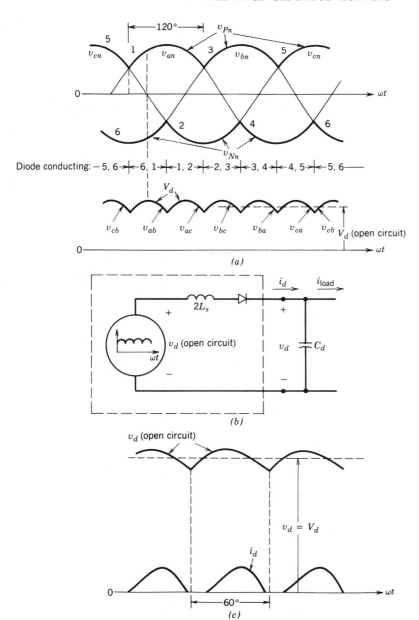

FIGURE 3-10: Three-phase rectifier: (*a*) Open circuit waveforms without C_d and with $i_d \to 0$, (*b*) Equivalent circuit, (*c*) i_d waveform.

frequency. It can be derived as (see Problem 3-3)

$$L_{d,\min} = \frac{0.013V_{LL}}{\omega I_d} \tag{3-21}$$

where V_{LL} is the rms line-to-line voltage and ω is the line frequency in radians per second (rad/s).

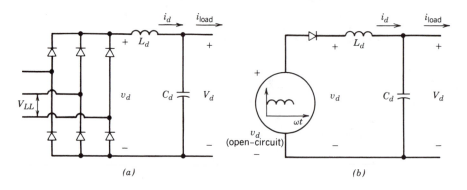

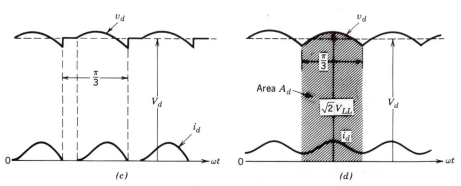

FIGURE 3-11: Three-phase rectifier; finite L_d: (a) circuit, (b) equivalent circuit, (c) discontinuous i_d, (d) continuous i_d.

With a continuously flowing i_d, the average value of the output voltage can be calculated from Fig. 3-11d. The area under the v_d waveform during a 60° interval is

$$A_d = \int_{-\pi/6}^{\pi/6} \sqrt{2}\, V_{LL} \cos \omega t \cdot d(\omega t) = \sqrt{2}\, V_{LL} \qquad (3\text{-}22)$$

Therefore

$$V_d = \frac{A_d}{\pi/3} = \frac{3\sqrt{2}}{\pi} V_{LL} = 1.35 V_{LL} \qquad (3\text{-}23)$$

The effect of L_d on the average voltage V_d is illustrated in Fig. 3-12 for a given I_{load}. With small L_d, i_d becomes highly discontinuous and V_d approaches $\sqrt{2}\, V_{LL}$. For $L_d \geq L_{d,\min}$, i_d is continuous and V_d equals 1.35 V_{LL} as given by Eq. 3-23. From Eq. 3-21 and Fig. 3-12, it can also be concluded that for a given L_d, as I_{load} increases (with i_d going from being highly discontinuous to continuous), there is only a small drop of approximately 4.5% in V_d.

3-5-1 Line-Current i_s

Figures 3-13a and 3-13b show the line-current waveforms for one of the three phases with a discontinuous i_d and a continuous i_d (with $L_s = 0$), respectively. In the waveform of Fig. 3-13b with a continuous i_d, the rectifier current on the ac side shifts

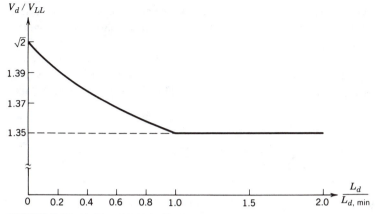

FIGURE 3-12: V_d/V_{LL} versus $L_d/L_{d,min}$.

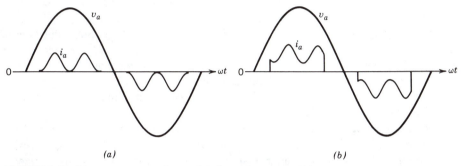

FIGURE 3-13: Line-current waveforms: (a) discontinuous i_d, (b) continuous i_d ($L_s = 0$).

instantaneously from one pair of phases to the next, for example, from (a, b) to (b, c). This instantaneous shift of current is possible since the ac-side inductance L_s has been assumed to be zero. If i_d flows continuously and L_s is finite, then the current on the ac side cannot shift or commutate instantaneously from one phase pair to the next. This current-commutation process is described in detail in Chapter 4, which deals with line-frequency controlled rectifiers and inverters. Since the uncontrolled diode recti-

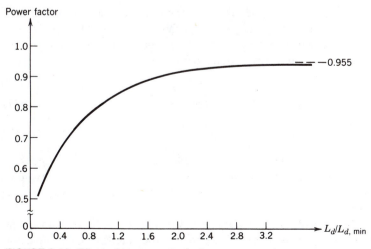

FIGURE 3-14: Effect of L_d on power factor.

fiers are a special case of the controlled rectifiers of Chapter 4, this current commutation process is not described here to avoid repetition, and will be discussed in a general manner in Chapter 4.

Assuming $L_s = 0$, Fig. 3-14 shows that increasing L_d improves the input power factor. If L_d is very large such that i_d is essentially a constant $i_d(t) \simeq I_d$, then the power factor approaches 0.955 (see Problem 3-4). The displacement power factor remains better than 0.98.

3-6   COMPARISON OF SINGLE-PHASE AND THREE-PHASE RECTIFIERS

Comparison of the line-current waveforms in Figs. 3-5 and 3-13 shows that the line-current in a single-phase rectifier contains significantly more distortion compared to a three-phase rectifier. This results in a much poorer power factor in a single-phase rectifier compared to a three-phase rectifier. The displacement power factor (cos ϕ_1) is high in both rectifiers.

Comparison of the i_d waveforms in Figs. 3-3 and 3-11 shows that the ripple in the dc current is smaller in a three-phase rectifier in comparison to a single-phase rectifier. The ripple current, which flows through the filter capacitor, dictates the capacitance and the current-handling capability required of the filter capacitor. Therefore, in some applications, the filter-capacitance required may be much smaller in a three-phase rectifier compared to a single-phase rectifier.

In a three-phase rectifier, the maximum regulation in the dc voltage V_d from no-load to a full-load condition will generally be less than 4.5%, as discussed earlier. This regulation is often much larger in single-phase rectifiers. Based on the foregoing discussion, it is always preferable to use a three-phase rectifier over a single-phase rectifier.

3-7   INRUSH CURRENT AND OVERVOLTAGES AT TURN-ON

In the previous sections, we have considered only the steady-state rectifier operation. However, considerable overvoltages and large inrush currents can result at turn-on if the ac voltage is suddenly applied to the circuit by means of a contactor.

For the worst case analysis, the filter capacitor is assumed to be initially completely discharged. Furthermore, it is assumed that at turn-on ($\omega t = 0$), the ac source input is at its peak value ($\sqrt{2} V_{LL}$ in the three-phase circuit). Therefore, the theoretical maximum voltage across the capacitor due to this series L-C connection approaches

$$V_{d, \max} = 2\sqrt{2} V_s \quad \text{(single-phase)} \tag{3-24}$$

$$V_{d, \max} = 2\sqrt{2} V_{LL} \quad \text{(three-phase)} \tag{3-25}$$

Normally, the load across the filter capacitor is a voltage-sensitive electronic circuit such as a switch-mode inverter in an ac-motor drive, and this large overvoltage can cause serious damage. Moreover, large inrush currents at turn-on may destroy the diodes in the rectifier.

To overcome these problems, one possible solution is to use a current-limiting resistor on the dc side in series between the rectifier output and the filter capacitor. This resistor is shorted out, either by a mechanical contactor or by a thyristor after a few cycles subsequent to turn-on, in order to avoid power dissipation and a substantial loss of efficiency due to its presence. A better alternative is presented in Reference 3.

3-8   CONCERNS AND REMEDIES FOR LINE-CURRENT HARMONICS AND POOR POWER FACTOR

Typical ac current waveforms in single-phase and three-phase diode rectifier circuits are far from being sinusoidal. The power factor is also very poor because of the harmonic contents in the line current.

As the power electronic systems proliferate, ac to dc rectifiers will undoubtedly play an increasingly important role. A large number of systems injecting harmonic currents into the utility grid can have significant impact on the quality of the ac voltage waveform (i.e., it will become distorted), thus causing problems with other sensitive loads connected to the same supply. Moreover, these harmonic currents cause additional harmonic losses in the utility system and may excite electrical resonances, leading to large overvoltages. Another problem caused by harmonics in the line current is to overload the circuit wiring, for example, a 1.2-kW load supplied by a 120 V, 15 A service may cause the circuit breaker to trip.

Standards for harmonics and the remedies for a poor line-current waveform and the input power factor are important concerns of power electronic systems. These are discussed in detail in Chapter 17.

3-9   SUMMARY

1. Line-frequency diode rectifier circuits are used to convert 60-Hz ac input into a dc voltage in an uncontrolled manner. A large filter capacitor is connected across the rectifier output since in most power electronic applications, a low ripple in the output dc voltage V_d is desirable.

2. Equivalent circuits are derived for the commonly used full-bridge circuit topologies with single-phase and three-phase inputs.

3. A graphical method is presented to obtain V_d–I_d rectifier characteristic.

4. In diode rectifier circuits with small L_s or L_d, the currents i_d and i_s are highly discontinuous, and as a consequence, the rms value of the input current I_s becomes large, and the power is drawn from the utility source at a very poor power factor.

5. In case of a single-phase ac input, voltage-doubler rectifiers can be used to approximately double the output dc voltage magnitude compared to a full-bridge rectifier. These are sometimes used in low-power equipment, which may be required to operate from dual voltages of 115 V and 230 V.

6. Comparison of single-phase and three-phase diode rectifiers shows that the three-phase rectifiers are preferable in most respects.

7. Both single-phase and three-phase diode rectifiers inject large amounts of harmonic currents into the utility system. As the power electronic systems proliferate, remedies for the poor input current waveform would have to be implemented. These topics are discussed in Chapter 17.

PROBLEMS

3-1. In the single-phase rectifier of Fig. 3-2a, $V_s = 120$ V at 60 Hz, and $L_s = 1$ mH. Calculate the waveform for i_d shown in Fig. 3-3a and indicate the values of θ_b, θ_f, and $I_{d,\,peak}$ on the waveform, if $V_d = 150$ V.

3-2. In the rectifier of Problem 3-1, $I_{\text{load}} = 8.2$ A. Use Figs. 3-6 and 3-7 to find V_d, power factor, displacement power factor, crest factor, and form factor. Using I_d, crest factor, and form factor, obtain $I_{d,\text{peak}}$. Also calculate I_s, I_{s1}, I_{dis} and %THD in the input current.

3-3. In the three-phase rectifier of Fig. 3-11a, derive the minimum value of L_d in terms of V_{LL}, ω, and I_d, which will result in a continuous i_d.

3-4. Show that in the three-phase rectifier of Fig. 3-11a, as i_d becomes a constant dc current for a large L_d, the input power factor approaches 0.955.

3-5. Consider a balanced three-phase, 60-Hz system shown in Fig. P3-5. Calculate the rms value of the neutral wire current I_N, if the three single-phase rectifiers are identical to the rectifier in Problem 3-2. Draw the i_N waveform.

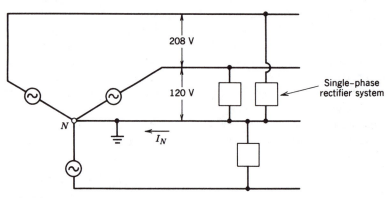

FIGURE P3-5:

3-10 REFERENCES

1. PHIL M. CAMP, "Input Current Analysis of Motor Drives with Rectifier Converters," *IEEE-IAS Conference Record*, 1985, pp. 672–675.
2. BOGDAN BRAKUS, "100 Amp Switched Mode Charging Rectifier for Three-Phase Mains," *IEEE-Intelec* 1984, pp. 72–78.
3. T. M. UNDELAND and N. MOHAN, "Overmodulation and Loss Considerations in High Frequency Modulated Transistorized Induction Motor Drives," IEEE *Transactions on Power Electronics*, Vol. 3, No. 4, pp. 447–452, October 1988.

4

Line-frequency
Phase-controlled
Rectifiers and Inverters:
60-Hz ac ↔ Controlled dc

4-1 INTRODUCTION

In Chapter 3 we discussed the line-frequency diode rectifiers that are increasingly being used at the front end of the switch-mode power electronic systems to convert 60-Hz ac input to an uncontrolled dc output voltage.

However, in some applications such as a class of dc-motor and ac-motor drives, it is necessary that the dc output voltage be controllable. The ac to controlled dc conversion is accomplished in line-frequency phase-controlled converters by means of thyristors. In the past these converters were used in a large number of applications for controlling the flow of electrical power. Owing to the availability of controllable switches in high voltage and current ratings, the use of these thyristor converters nowadays is primarily in three-phase, high-power applications. This is particularly true in applications, most of them at high power levels, where it is necessary or desirable to be able to control the power flow in both directions between the ac and the dc sides. Examples of such applications are converters in high-voltage dc power transmission (Chapter 16) and some dc-motor and ac-motor drives with regenerative capabilities (Chapters 11 through 13).

As the name of these converters implies, the line-frequency voltages are present on their ac side. In these converters, the instant at which a thyristor begins or ceases to conduct depends on the line-frequency ac voltage waveforms, and the control inputs. Moreover, the transfer of current from one device to the next occurs naturally because of the presence of these ac voltages.

It should be noted that the uncontrollable, line-frequency diode rectifiers of Chapter 3 are a subset of the controlled converters discussed in this chapter. The

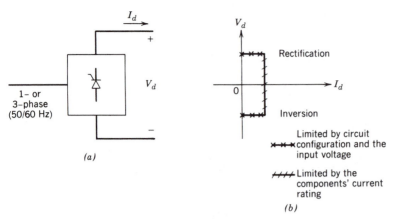

FIGURE 4-1: Line-frequency controlled converter.

reasons for discussing the diode rectifiers separately in Chapter 3 have to do with their increasing importance, and the way in which the dc current i_d flows through them. In Chapter 3 it was pointed out that in many applications of line-frequency diode rectifiers, the current i_d through them does not flow continuously and therefore the emphasis was on the discontinuous-current-conduction mode. However, in the controlled converters of this chapter, in most of their applications, the current i_d through them flows continuously and hence the emphasis is on the continuous-current-conduction mode, although the discontinuous-current-conduction mode is also briefly discussed.

A fully controlled converter is shown in Fig. 4-1a in block diagram form. For given ac line voltages, the average dc-side voltage can be controlled from a positive maximum to a negative maximum value in a continuous manner. The converter dc current I_d (or i_d on an instantaneous basis) cannot change direction, as will be explained later. Therefore, a converter of this type can operate in only two quadrants (of the V_d–I_d plane) as shown in Fig. 4-1b. Here, the positive values of V_d and I_d imply *rectification* where the power flow is from the ac to the dc side. In an *inverter* mode, V_d becomes negative (but I_d stays positive) and the power is transferred from the dc to the ac side. The inverter mode of operation on a sustained basis is possible only if a source of power, such as batteries, is present on the dc side.

In some applications, such as in reversible-speed dc-motor drives with regenerative braking, the converter must be capable of operating in all four quadrants. This is accomplished by connecting two two-quadrant converters described above in antiparallel (or back-to-back).

Because of the trend to use these line-frequency controlled converters only at high power levels, very often the ac side consists of three phases. Therefore, only the three-phase full-bridge converters are discussed.

In analyzing the converters in this chapter, the thyristors are assumed to be ideal, except for the consideration of the thyristor turn-off time t_q, which was described in Chapter 2.

4-2   CONTROL OF LINE-FREQUENCY CONTROLLED RECTIFIERS AND INVERTERS

For given ac input voltages, the magnitude of the average output voltage in thyristor converters can be controlled by delaying the instants at which the thyristors are allowed to start conduction. This is illustrated by the simple circuits of Fig. 4-2.

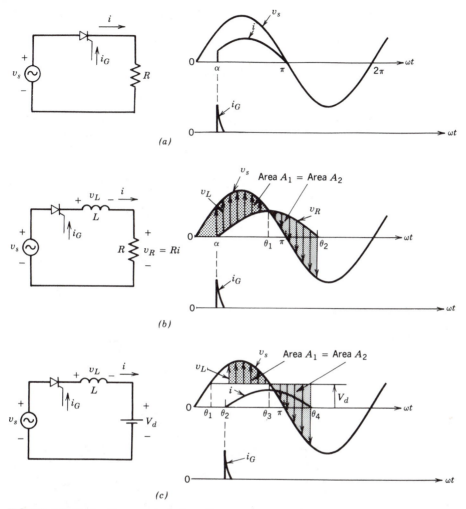

FIGURE 4-2: Basic phase-controlled rectifier concepts. By varying α from 0 to π in (a) and (b), and from θ_1 to θ_3 in (c), the average load current is varied from a maximum to zero.

In Fig. 4-2a, a thyristor connects the line-frequency source v_s to a load resistance. In the positive half-cycle of v_s, the current is zero until $\omega t = \alpha$, at which time the thyristor is supplied a positive gate pulse of a short duration. For the rest of the positive half-cycle, the current waveform follows the ac voltage waveform and becomes zero at $\omega t = \pi$. Then the thyristor blocks the current from flowing during the negative half-cycle of v_s. The current stays zero until $\omega t = 2\pi + \alpha$, at which time another short-duration gate pulse is applied and the next cycle of the waveform begins (not shown for simplicity).

In Fig. 4-2b, the load consists of both R and L. Initially, the current is zero. The thyristor conduction is delayed until $\omega t = \alpha$. Once the thyristor is fired or gated on at $\omega t = \alpha$ during the positive half-cycle of v_s, the current begins to flow and the voltage across the inductor can be written as

$$v_L(t) = L\frac{di}{dt} = v_s - v_R, \tag{4-1}$$

where $v_R = Ri$. In Fig. 4-2b, v_R (which is proportional to the current) is plotted and v_L

is shown as the difference between v_s and v_R. During α to θ_1, v_L is positive and the current increases since

$$i(t) = \frac{1}{L} \int_{\alpha/\omega}^{t} v_L(\lambda) \cdot d\lambda \qquad \frac{\alpha}{\omega} < t \qquad (4\text{-}2)$$

where λ is a dummy variable of integration. Beyond $\omega t = \theta_1$, v_L becomes negative and the current (as well as v_R) begins to decline. The instant at which the current becomes zero, and stays zero due to the thyristor, is dictated by Eq. 4-2. Graphically in Fig. 4-2b, the current becomes zero at $\omega t = \theta_2$ corresponding to area $A_1 = $ area A_2. These areas represent the time integral of v_L, which must be zero over one cycle of repetition in steady state. It should be noted that the current will continue to flow for a while, even after v_s has become negative. The reason for this has to do with the stored energy in the inductor, a part of which is supplied to R, and the other part is absorbed by v_s when v_s becomes negative.

In Fig. 4-2c, the load consists of an inductor and a dc voltage V_d. Here, with the current initially being zero, the thyristor is reverse biased until $\omega t = \theta_1$ and therefore cannot conduct. The thyristor conduction is further delayed until θ_2, when a positive gate pulse is applied. With the current flowing

$$v_L(t) = L\frac{di}{dt} = v_s - V_d \qquad (4\text{-}3)$$

and

$$i(t) = \frac{1}{L} \int_{\theta_2/\omega}^{t} [v_s(\lambda) - V_d] \cdot d\lambda, \qquad (4\text{-}4)$$

where λ is a dummy variable of integration.

The current peaks at θ_3. The current goes to zero at $\omega t = \theta_4$. The instant $\omega t = \theta_4$ corresponds to area A_1 equal to area A_2, due to the fact that the time integral of inductor voltage over one time period of repetition must be zero.

By controlling the instant at which the thyristor is gated on, the average current in the circuits of Fig. 4-2 can be controlled in a continuous manner from 0 to a maximum value. The same is true for the power supplied by the ac source.

Versatile integrated circuits, such as the TCA780, are available to provide delayed gate-trigger signals to the thyristors. A simplified block diagram of a gate trigger control circuit is shown in Fig. 4-3. Here, a sawtooth waveform (synchronized to the ac input) is compared with the control signal $v_{control}$, and the delay angle α with respect to the positive zero crossing of the ac line voltage is obtained in terms of $v_{control}$ and the peak of the sawtooth waveform $\hat{V}_{st}$

$$\alpha° = 180° \cdot \frac{v_{control}}{\hat{V}_{st}} \qquad (4\text{-}5)$$

Another gate trigger signal can easily be obtained, delayed with respect to the negative zero crossing of the ac line voltage.

A three-phase, six-pulse, full-bridge converter shown in Fig. 4-4 is a very commonly used circuit arrangement. In the following analysis, it is assumed that i_d flows continuously; the effect of discontinuous i_d is described separately in Section 4-5.

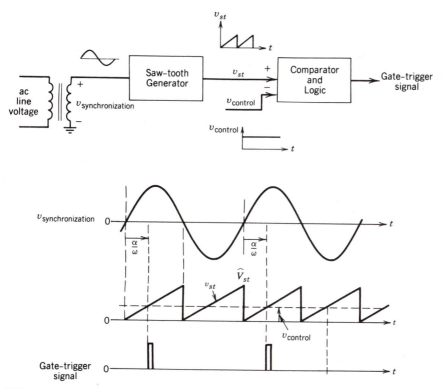

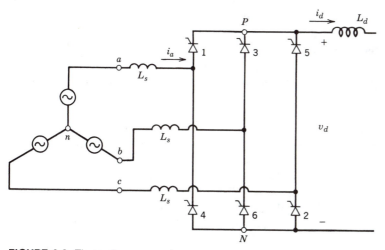

FIGURE 4-3: Gate Trigger Control Circuit.

FIGURE 4-4: Three-phase converter.

4-3   THREE-PHASE CONVERTER ANALYSIS WITH $L_s = 0$

The commutating inductance L_s is neglected in the initial analysis for simplicity. Its effect is considered in Section 4-4. In the circuit of Fig. 4-4, by controlling the instant at which the thyristor gate pulses are applied with respect to the ac voltage waveforms, the dc-side voltage V_d can be controlled in magnitude. The ac voltage waveforms are shown in Fig. 4-5a, where the thyristors T_5 and T_6 are assumed to be

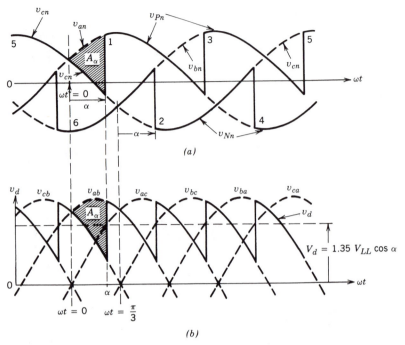

FIGURE 4-5: Converter waveforms ($L_s = 0$).

conducting initially. At the crossing of v_{an} and v_{cn} waveforms, which is defined as ωt equal to zero, T_1 becomes forward biased. If a gate current was continuously applied to T_1, the instant $\omega t = 0$ is the earliest at which T_1 would have started to conduct. This instant for T_1 we refer to as its instant of earliest possible conduction. The beginning of conduction for T_1 can be delayed by an angle α as shown in Fig. 4-5a, if the pulse current to its gate is applied at this delay angle α (also called the firing angle). Similar instants of earliest possible conduction can be defined for the rest of the thyristors. In practice, in steady state, all thyristors are gated with the same delay angle α. This results in a voltage waveform for v_{Pn} and v_{Nn}, as shown in Fig. 4-5a. To explain further, due to continuous flow of i_d during the delay interval (between $\omega t = 0$ and $\omega t = \alpha$), v_{Pn} follows the phase voltage v_{cn} waveform as shown. By recognizing that $v_d = v_{Pn} - v_{Nn}$, the v_d waveform is as shown in Fig. 4-5b.

The average voltage V_d can be calculated as follows: if α is equal to zero, v_d during the interval between 0 and $\pi/3$ will be equal to v_{ab} in Fig. 4-5b. Then, the waveform during this interval would be identical to v_d waveform for a diode bridge with a continuous i_d, which was shown in Fig. 3-11d. Therefore, the average dc voltage with $\alpha = 0$ is as calculated by Eqs. 3-22 and 3-23:

$$V_{do} = \frac{3\sqrt{2}}{\pi} V_{LL} = 1.35 \, V_{LL} \tag{4-6}$$

where V_{LL} is the rms line-to-line voltage.

Due to the delay angle α, the integral of v_d during the interval between 0 to $\pi/3$ is less than the integral of v_d for $\alpha = 0$ by an area A_α (in volt-radians) as shown in Fig. 4-5. Since v_d waveform repeats every $\pi/3$ radians, A_α results in a drop in the average dc voltage by $A_\alpha/(\pi/3)$:

$$V_{d\alpha} = V_{do} - \frac{A_\alpha}{\pi/3} \tag{4-7}$$

The area A_α is the time integral of $v_{an} - v_{cn}(= v_{ac})$. With the time origin chosen as in Figs. 4-5a and 4-5b to coincide with the instant of earliest possible conduction for T_1, the line-line voltage between phases a and c (also called the commutation voltage) can be written as

$$v_{ac} = \sqrt{2}\, V_{LL} \sin \omega t \tag{4-8}$$

Therefore

$$A_\alpha = \int_0^\alpha \sqrt{2}\, V_{LL} \sin \omega t \cdot d(\omega t) = \sqrt{2}\, V_{LL} \left(1 - \cos \alpha\right) \tag{4-9}$$

Substitution of A_α in Eq. 4-7 results in

$$V_{d\alpha} = \frac{3\sqrt{2}}{\pi} V_{LL} \cos \alpha = 1.35\, V_{LL} \cos \alpha \tag{4-10}$$

Equation 4-10 shows that V_d is independent of I_d and the load, as long as i_d flows continuously. Use of thyristors allows V_d to be controlled continuously by means of α, where the delay angle in a $0°$ to $90°$ range results in a positive V_d representing a rectifier mode of operation. The inverter mode of operation with α in a range of $90°$ to $180°$ is described in Section 4-6.

4-4   EFFECT OF AC-SIDE INDUCTANCE L_s

The effect of L_s on the current commutation interval is analyzed in the circuit of Fig. 4-4 for a continuous-current-conduction mode. Moreover, to simplify the discussion, the dc current is assumed to be constant, that is $i_d(t) \simeq I_d$. For generality, an arbitrary value of delay angle (firing angle) α is chosen.

Consider the situation where thyristors T_5 and T_6 have been conducting previously, and at $\omega t = \alpha$ the current begins to commutate (or shift) from T_5 to T_1. Only the thyristors involved in current conduction are drawn in Fig. 4-6a. The instant when v_{an} becomes more positive than v_{cn} (the instant of earliest possible conduction for T_1) is chosen as the time origin $\omega t = 0$ in Fig. 4-6b.

During the current-commutation interval u, T_1 and T_5 conduct simultaneously and the phase voltages v_{an} and v_{cn} are shorted together through L_s in each phase. The current i_a builds up from 0 to I_d, whereas i_c decreases from I_d to 0, at which instant the current commutation from T_5 to T_1 is completed. Currents i_5 and i_1 through thyristors 5 and 1 are drawn in Fig. 4-6c.

In Fig. 4-6a during the commutation interval $\alpha < \omega t < \alpha + u$

$$v_{Pn} = v_{an} - v_{L_s} \tag{4-11}$$

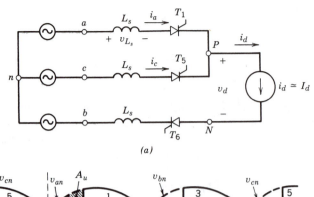

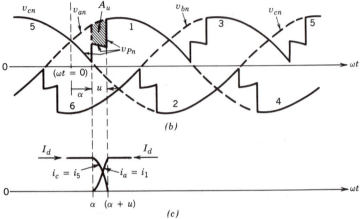

(c)

FIGURE 4-6: Commutation in the presence of L_s.

· where

$$v_{L_s} = L_s \frac{di_a}{dt} \tag{4-12}$$

The reduction in volt-radians area due to the commutation interval (in Fig. 4-6b) is

$$A_u = \int_\alpha^{\alpha+u} v_{L_s} \cdot d(\omega t) \tag{4-13}$$

Using Eq. 4-12 in Eq. 4-13 and recognizing that i_a changes from 0 to I_d in the interval $\omega t = \alpha$ to $\omega t = \alpha + u$

$$A_u = \omega L_s \int_0^{I_d} di_a = \omega L_s I_d \tag{4-14}$$

Therefore, the average dc output voltage is reduced from $V_{d\alpha}$ (given by Eq. 4-10) by $A_u/(\pi/3)$:

$$V_d = \frac{3\sqrt{2}}{\pi} V_{LL} \cos \alpha - \frac{3\omega L_s}{\pi} I_d \tag{4-15}$$

In the foregoing derivation, the following should be noted: During the current commutation, phases a and c are shorted together. Therefore, during commutation

$$v_{Pn} = v_{an} - L_s \frac{di_a}{dt} \qquad (4\text{-}16)$$

Also

$$v_{Pn} = v_{cn} - L_s \frac{di_c}{dt} \qquad (4\text{-}17)$$

Therefore, from Eqs. 4-16 and 4-17

$$v_{Pn} \ (\text{during commutation}) = \frac{v_{an} + v_{cn}}{2} - \frac{L_s}{2} \left(\frac{di_a}{dt} + \frac{di_c}{dt} \right) \qquad (4\text{-}18)$$

Since $i_d \ (= i_a + i_c)$ is assumed to be constant during the commutation interval

$$\frac{di_a}{dt} = -\frac{di_c}{dt} \qquad (4\text{-}19)$$

Therefore, Eq. 4-18 reduces to

$$v_{Pn} = \frac{v_{an} + v_{cn}}{2} \qquad (4\text{-}20)$$

The v_{Pn} waveform during the commutation interval is shown in Fig. 4-6b. The importance of L_s in the inverter operation is discussed in the next section.

Even though an explicit expression for the commutation interval u is not needed to calculate V_d (see Eq. 4-15), it is required to ensure reliable operation in the inverter mode. Thus, this is the appropriate place to calculate this interval u.

Combining Eqs. 4-16 and 4-20 in the circuit of Fig. 4-6a

$$L_s \frac{di_a}{dt} = \frac{v_{an}}{2} - \frac{v_{cn}}{2} = \frac{v_{ac}}{2} \qquad (4\text{-}21)$$

With the time origin chosen in Fig. 4-6b, $v_{ac} = \sqrt{2} \, V_{LL} \sin \omega t$. Therefore

$$\frac{di_a}{d(\omega t)} = \sqrt{2} \, \frac{V_{LL} \sin \omega t}{2\omega L_s}$$

Its integration between $\omega t = \alpha$ and $\omega t = \alpha + u$, recognizing that during this interval i_a changes from 0 to I_d, results in

$$\int_0^{I_d} di_a = \sqrt{2} \, \frac{V_{LL}}{2\omega L_s} \int_\alpha^{\alpha+u} \sin \omega t \cdot d(\omega t)$$

or

$$\cos (\alpha + u) = \cos \alpha - \frac{2\omega L_s}{\sqrt{2} \, V_{LL}} I_d \qquad (4\text{-}22)$$

Thus, knowing α and I_d, the commutation interval u can be calculated.

EXAMPLE 4-1

Consider a three-phase, 60-Hz converter system whose block diagram is shown in Fig. 4-7a. At the point where the converter is connected to the 460-V line-to-line voltage, the three-phase short-circuit capacity of the supply system is 12 MVA.

Plot V_d as a function of I_d for α equal to 0°, 30°, and 60°. Assume that the dc-side converter inductance is such that $i_d \approx I_d$. The maximum value of I_d is 1000 A.

SOLUTION

The short-circuit current at a point refers to the per-phase current that would flow if all three phases are shorted together. At a given voltage level, it is a measure of the system impedance seen from that point. The three-phase short-circuit capacity is the product of the short-circuit current and $\sqrt{3}\, V_{LL}$.

The internal impedance Z_s of the ac system (assumed to be purely inductive) can be obtained in the following manner:

$$\text{Short-circuit current} = \frac{12 \times 10^6}{\sqrt{3} \times 460}$$

$$Z_s = \frac{V_{\text{phase}}}{\text{Short-circuit current}} = \frac{460/\sqrt{3}}{12 \times 10^6/(\sqrt{3} \times 460)} = \frac{460^2}{12 \times 10^6} = 0.0176\Omega$$

and

$$L_s = \frac{Z_s}{\omega} = 47\,\mu H$$

Therefore, from Eq. 4-15

$$V_d = \frac{3\sqrt{2}}{\pi}(460)\cos\alpha - \frac{3 \times 377 \times 47 \times 10^{-6}}{\pi}I_d$$

$$= 621\cos\alpha - 16.9 \times 10^{-3}I_d$$

The results are plotted in Fig. 4-7b. In this example, at $\alpha = 0$ and $I_d = 1000$ A, the rectifier power is 604 kW, which is a large load on the system with a short-circuit capacity of 12 MVA. For $\alpha = 0$, the dc voltage regulation is 2.7% from 0 to full load,

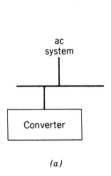

(a)

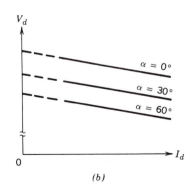

(b)

FIGURE 4-7: Example 4-1.

which is approximately as large as encountered in practice. The characteristics for low values of I_d are shown as dotted because of the discontinuous current conduction described in the next section. ∎

4-5 EFFECT OF DISCONTINUOUS CURRENT

The current i_d becomes discontinuous under light-load conditions corresponding to low values of I_d. As an example, Fig. 4-8a shows a circuit where the load on the converter is a dc motor, which is represented by its dc back-emf E_a and its inductance (its resistance is neglected). L_a in Fig. 4-8a is the sum of the motor inductance and some additional inductance that may be purposely introduced. V_d is plotted as a function of the average current I_d in Fig. 4-8b for $\alpha < 30°$. It shows that for $\alpha < 30°$, V_d can increase up to ($\sqrt{2}\, V_{LL}$), depending on how small I_d becomes. The converter and its control must be designed to handle the effects of any discontinuity in i_d.

4-6 INVERTER OPERATION

In an inverter mode of operation, V_d reverses polarity, whereas i_d must flow in the same direction as in the rectifier mode discussed in the previous sections. This

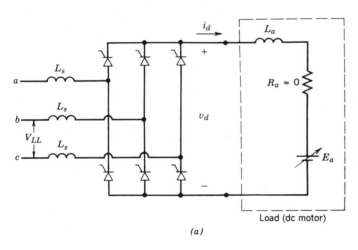

(a)

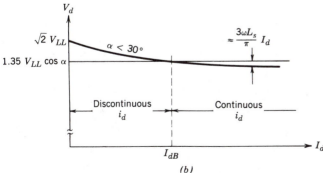

(b)

FIGURE 4-8: Effect of discontinuous current: *(a)* circuit, *(b)* V_d versus I_d.

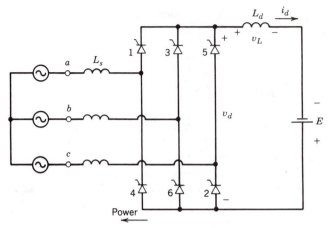

FIGURE 4-9: Inverter circuit.

results in an average power transfer from the dc to the ac side of the converter. This mode of operation on a long-term steady-state basis is possible only if a source of energy exists on the dc side, as shown in Fig. 4-9. This mode of operation may be used in interconnecting a photovoltaic or a wind electric system to the utility grid. These systems are discussed in Chapter 16.

The voltage waveforms are shown in Figs. 4-10a and 4-10b for comparison purposes for two values of delay angle $\alpha(= 45°$ and $135°)$, assuming that L_s is zero and i_d is continuous. It can be observed that for any value of α in a range of $0°$ to $180°$, the thyristor to be gated on is forward biased and the commutation voltage

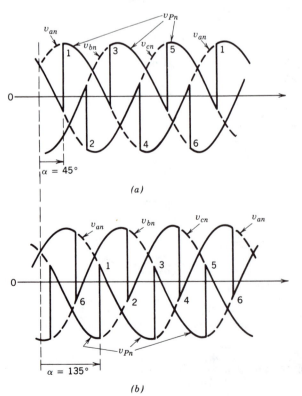

FIGURE 4-10: Inverter waveforms ($L_s = 0$):
(a) $\alpha = 45°$, (b) $\alpha = 135°$.

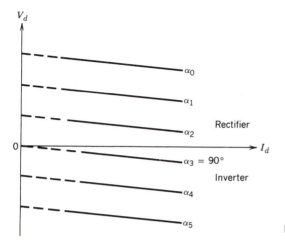

FIGURE 4-11: V_d versus I_d (L_s = finite).

exists with a correct polarity to cause current commutation (instantaneously in this case due to $L_s = 0$). In Fig. 4-10b, $\alpha = 135°$ corresponds to the inverter mode of operation since it results in the $v_d(= v_{Pn} - v_{Nn})$ waveform which is mostly negative and, hence, has a negative value of V_d. The delay angle $\alpha = 90°$ is on the border of rectification and inversion, resulting in $V_d = 0$. The voltage expression given by Eq. 4-10 is valid:

$$0 < \alpha < 90° \qquad V_d = \text{positive} \quad (\text{rectifier mode})$$

$$\tag{4-23}$$

$$90° < \alpha < 180° \qquad V_d = \text{negative} \quad (\text{inverter mode})$$

In practice, the upper limit on α is approximately $150°$, as will be discussed later.

In the presence of a finite L_s, the derivation of the voltage expression leading to Eq. 4-15 is valid in the inverter mode as well.

V_d is plotted as a function of I_d for various values of delay angle in Fig. 4-11, with a finite L_s. Both the rectifier and the inverter modes are identified. The characteristics are shown dotted for smaller value of I_d, which may cause i_d to become discontinuous and hence V_d to increase.

In the inverter mode, the cathode to anode voltage across one of the thyristors (in this case, v_5 across T_5) is drawn in Fig. 4-12 along with other converter waveforms. Figure 4-12b shows that subsequent to the commutation of current from T_5 to T_1 (i.e., after $\omega t = \alpha + u$), the voltage across T_5 is negative only for a short time γ/ω, where γ is called the extinction angle which is measured with respect to $\omega t = \pi$. Therefore, $\gamma = 180° - (\alpha + u)$. Beyond $\omega t = 180°$, T_5 must block a positive voltage across it.

As discussed briefly in Chapter 2, this time interval γ/ω should be greater than the turn-off time t_q (also called the circuit commutated recovery time) of the thyristors; otherwise thyristor 5 would restart conduction beyond $\omega t = \pi$, resulting in a commutation failure. During commutation failure, the voltage across the converter would reverse in polarity, resulting in a large i_d. To avoid such commutation failures, a sufficiently large value of the minimum extinction angle $\gamma_{\min}$ (for example, $\gamma_{\min} \approx 15°-30°$, or $\gamma_{\min}/\omega$ in the neighborhood of 1 ms) is used. This fairly large safety margin makes the commutation process less susceptible to sudden glitches in the ac voltage waveforms (which in practice can be caused by a number of factors) in order to make the inverter operation reliable. It is easy to see that the extinction angle γ is not a concern in the rectifier mode of operation, since with $\alpha \leq 90°$, there is always a

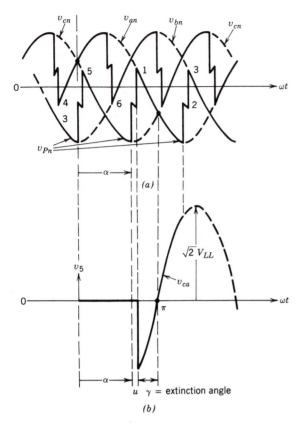

FIGURE 4-12: Extinction angle γ in inverters.

sufficient turn-off time available to the thyristors. To summarize,

$$(\alpha + u) < 180° - \gamma_{min} \tag{4-24}$$

where $\gamma_{min} = 15°$ to $30°$.

EXAMPLE 4-2

In a three-phase inverter of Fig. 4-9, $V_{LL} = 460$ V at 60 Hz, $E = 550$ V, and $L_s = 0.5$ mH. L_d is assumed to be very large to yield $i_d(t) = I_d$. Calculate α and γ, if the power flow is 55 kW.

SOLUTION

$$I_d = \frac{P}{E} = \frac{55 \times 10^3}{550} = 100 \ A$$

Using Eq. 4-15

$$V_d = 1.35 \times 460 \times \cos\alpha - \frac{3 \times 377 \times 0.5 \times 10^{-3}}{\pi} \times 100 = -E = -550$$

Therefore

$$\text{delay angle } \alpha = 149°$$

From Eq. 4-22

$$\cos\left(\alpha + u\right) = \cos 149° - \frac{2 \times 377 \times 0.5 \times 10^{-3}}{\sqrt{2} \times 460} \times 100$$

Therefore

$$\alpha + u \simeq 156°$$

and

$$\gamma = 180° - (\alpha + u) = 24° \qquad \blacksquare$$

4-7   AC-SIDE WAVEFORMS

Three-phase converters result in ac-side voltages and line currents that are nonsinusoidal. Alternatives to the three-phase line-frequency converters to minimize the problems of current harmonics, voltage distortion, line-voltage notching, and poor power factor are presented in Chapter 17.

In this section both the ac current and voltage waveforms of the line-frequency converter are discussed and their effect on the power factor of operation is evaluated.

4-7-1   AC-Side Current Waveforms

In the converter of Fig. 4-4, only one of the phases, for example phase a, is considered; the other two phase quantities will be identical to those of phase a, except for the $\pm 120°$ phase displacement under a balanced steady-state operation.

Initially, L_s is assumed to be zero and L_d is assumed to be very large so that $i_d \simeq I_d$. The v_{an} and i_a waveforms are shown in Fig. 4-13a for α equal to zero. The ac input voltage is assumed to be a pure sinusoid and the fundamental frequency component I_{a1} of the ac current is shown by means of the dotted curve. It shows that i_{a1} is in phase with v_{an}, and the phase angle ϕ_1 by which i_{a1} lags v_{an} is zero. Figure 4-13b shows the waveforms for an arbitrary delay angle α. Now, the i_a waveform and, hence, the i_{a1} waveform lags the v_{an} waveform by the delay angle α. Therefore, $\phi_1 = \alpha$.

Fourier components of the current waveforms in Figs. 4-13a and 4-13b have the following rms values:

$$I_{a1} = \frac{2\sqrt{3}}{\sqrt{2}\,\pi} I_d = 0.78 I_d \qquad (4\text{-}25)$$

and the rms values of the harmonic components are inversely proportional to their harmonic order:

$$I_{ah} = \frac{I_{a1}}{h} \qquad \text{where} \qquad h = 6n \pm 1 \qquad (n = 1, 2, 3, \dots) \qquad (4\text{-}26)$$

From the i_a waveform of Fig. 4-13a or 4-13b, the total rms value of the phase current

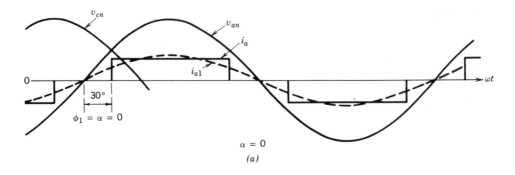

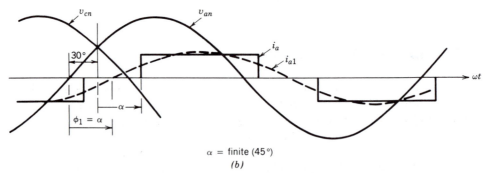

FIGURE 4-13: AC-side current waveforms ($L_s = 0$): (a) $\alpha = 0°$, (b) α = finite (45°).

can be calculated as

$$I_a = \sqrt{\frac{2}{3}}\, I_d = 0.816 I_d \tag{4-27}$$

Therefore, for $i_d \simeq I_d$ with $L_s = 0$, from Eqs. 4-25 and 4-27

$$\frac{I_{a1}}{I_a} = \frac{3}{\pi} = 0.955 \tag{4-28}$$

In the absence of L_s, ϕ_1 equals α. Therefore,

$$\text{Displacement power factor} = \cos \phi_1 = \cos \alpha \tag{4-29}$$

Using Eq. 3-14 of the previous chapter for the power factor, Eqs. 4-28 and 4-29 yield

$$\text{Power factor} = \frac{3}{\pi} \cos \alpha \tag{4-30}$$

In all practical circuits, L_s is finite, which results in a finite commutation interval u. Still assuming $i_d \simeq I_d$, the ac-side current waveform is shown in Fig. 4-14a. The commutation interval introduces a slightly additional phase lag in i_{a1} waveform of Fig. 4-14b compared to Fig. 4-13b (with $L_s = 0$). If the current waveform during the commutation interval can be approximated to be linear, then i_a has a trapezoidal waveform as shown in Fig. 4-14b. By inspection of Fig. 4-14b, where i_a waveform has

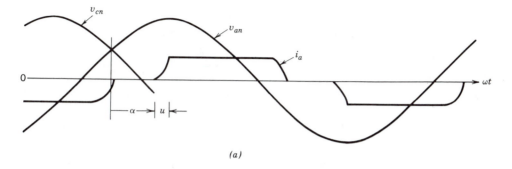

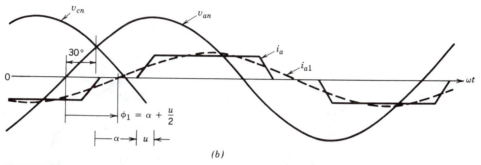

FIGURE 4-14: AC-side current waveforms, finite L_s: (a) actual, (b) trapezoidal approximation.

been assumed to be trapezoidal

$$\text{Displacement power factor} \simeq \cos\left(\alpha + \frac{u}{2}\right) \qquad (4\text{-}31)$$

A more accurate (but not exact) value of the displacement power factor, by equating ac-side and dc-side powers, is given in Eq. 4-32; its derivation is left as an exercise (see Problem 4-1):

$$\text{Displacement power factor} \simeq \frac{\cos\alpha + \cos(\alpha + u)}{2} \qquad (4\text{-}32)$$

The presence of L_s, and therefore u (compared to the case with $L_s = 0$) results in a smaller displacement power factor for a constant α.

EXAMPLE 4-3

In typical applications, L_s is finite. Moreover, i_d is not a pure dc current, as was assumed in the foregoing analysis. Table 4-1 lists typical and idealized values of ac-side current harmonics in a six-pulse full-bridge controlled converter, as functions of its fundamental current component.

Calculate the ratio I_1/I and the total harmonic distortion (THD) in the current for typical, as well as idealized harmonics.

TABLE 4-1

Typical and Idealized Harmonics

h	5	7	11	13	17	19	23	25
Typical	I_h/I_1	0.17	0.10	0.04	0.03	0.02	0.01	0.01	0.01
Idealized	I_h/I_1	0.20	0.14	0.09	0.07	0.06	0.05	0.04	0.04

SOLUTION

In Table 4-1, let $I_1 = 1.0$ per unit. Then, in the typical case, the rms current I is

$$I \simeq \sqrt{\sum_{h=1}^{25} I_h^2} = 1.023 \text{ per unit}$$

Therefore, for typical harmonics, $I_1/I = 1/1.023 = 0.977$. Using Eqs. 3-17 and 3-18, the total harmonic distortion in the current is

$$\text{Current } \%\text{THD} = \frac{\sqrt{\sum_{h \neq 1} I_h^2}}{I_1} \times 100 = 21.6\%$$

Comparing these with the idealized cases where $L_s = 0$ and $i_d = I_d$, $I_1/I = 0.955$ as in Eq. 4-28, and the current THD = 31%.   ∎

4-7-2   AC Voltage Waveform (Line Notching and Distortion)

Figure 4-15a shows a practical arrangement with $L_s = L_{s1} + L_{s2}$, where L_{s1} is the per-phase internal inductance of the ac source and L_{s2} is the inductance associated with the converter. The junction of L_{s1} and L_{s2} is also the point of common coupling where other loads may be connected, as shown in Fig. 4-15a. A thyristor converter results in line noise. Two main reasons for line noise are line notching and waveform distortion, both of which are considered in the following sections.

4-7-2-1   LINE NOTCHING

In the converter of Fig. 4-15a, there are six commutations per line-frequency cycle. The converter voltage waveforms are shown in Fig. 4-15b. During each commutation, two out of three phase voltages are shorted together by the converter thyristors through L_s in each phase. Considering a line-to-line voltage, for example V_{AB} at the converter internal terminals as shown in Fig. 4-15c. It is short-circuited twice per cycle, resulting in deep notches. There are four other notches where either phase A or phase B (but not both) is involved in commutation. Ringing due to stray capacitances C_s and snubbers in Fig. 4-15a has been omitted for clarity.

The area for each of the deep notches in Fig. 4-15c, where both phases A and B are shorted, would be twice that of A_u in Fig. 4-6b. Therefore, from Eq. 4-14

$$\text{Deep notch area } A_n = 2\omega L_s I_d \quad [\text{volt-radians}] \tag{4-33}$$

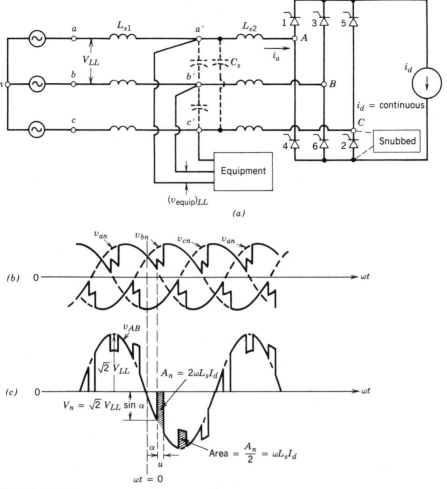

FIGURE 4-15: Line notching in other equipment voltage. (*a*) circuit, (*b*) phase voltages, (*c*) line-line voltage v_{AB}.

The notch-width u can be calculated from Eq. 4-22. Assuming that u is small, as an approximation we may write

$$\text{Deep notch depth} \approx \sqrt{2}\, V_{LL} \sin \alpha \quad (\alpha = \text{delay angle}) \qquad (4\text{-}34)$$

Therefore, from Eq. 4-33 and 4-34, considering one of the deep notches

$$\text{Notch-width } u = \frac{\text{Notch area}}{\text{Notch depth}} \approx \frac{2\omega L_s I_d}{\sqrt{2}\, V_{LL} \sin \alpha} \quad [\text{radians}] \qquad (4\text{-}35)$$

The shallow notches in Fig. 4-15*c* have the same width u as is calculated for the deep notches in Eq. 4-35. However, the depth and the area for each of the shallow notches is one-half compared with those of the deep notches.

In practice, the line notching at the point of common coupling is of concern. The notches in the line-to-line voltage $v_{a'b'}$ have the same width u as shown in Fig. 4-15*c* and given by Eq. 4-35. However, the notch depths and the notch areas are a

TABLE 4-2

Line Notching and Distortion Limits for 460 V Systems

Class	Line-notch Depth $\rho(\%)$	Line-notch Area (V · μs)	Voltage Total Harmonic Distortion (%)
Special applications	10	16,400	3
General system	20	22,800	5
Dedicated system	50	36,500	10

factor ρ times the corresponding depth and areas, respectively in Fig. 4-15c, where

$$\rho = \frac{L_{s1}}{L_{s1} + L_{s2}} \tag{4-36}$$

Therefore, for a given ac system (i.e., given L_{s1}), a higher value of L_{s2} will result in smaller notches at the point of common coupling. The German VDE standards recommend that ωL_{s2} must be a minimum of 5%, that is in Fig. 4-15a

$$\omega L_{s2} I_{a1} \geq 0.05 \frac{V_{LL}}{\sqrt{3}} \tag{4-37}$$

In Table 4-2, guidelines suggested by the IEEE standard 519–1981 for the line noise at the point of common coupling are given.

This IEEE standard 519–1981 also suggests that the converter equipment should be capable of performing satisfactorily on supply systems containing line notches of 250 μs width (5.4 electrical degrees), and a notch depth of 70% of the rated maximum line voltage.

Due to the stray capacitances (or if any filter capacitor is used at the input) and snubbers across thyristors, there would be ringing at the end of each commutation interval. These transient voltages can overload the transient suppressors within the equipment.

4-7-2-2 VOLTAGE DISTORTION

The voltage distortion at the point of common coupling can be calculated by means of phase quantities by knowing the harmonic components I_h of the converter input current and the ac source inductance (L_{s1}):

$$\text{Voltage \% THD} = \frac{\left[\sum_{h \neq 1} (I_h \times h\omega L_{s1})^2 \right]^{1/2}}{V_{\text{phase}} \text{(fundamental)}} \times 100 \tag{4-38}$$

Its recommended limits are given in Table 4-2.

It should be noted that the total harmonic distortion THD in the voltage at the point of common coupling can also be calculated by means of the notches in its line-to-line voltage waveform as follows: The total rms value of the harmonic components (other than the fundamental) can be approximately obtained by root-mean-squaring the six notches per cycle in the line-to-line voltage waveform; moreover the fundamental frequency component of the line-to-line voltage at the point of common coupling can be approximated as V_{LL} (see Problem 4-6).

4-8   OTHER THREE-PHASE CONVERTERS

Only the three-phase, full-bridge, six-pulse converters have been analyzed in detail in the preceding sections. There are several other types of converters as follows: 12-pulse and higher-pulse-number bridge converters, six-pulse bridge converter with a star-connected transformer, six-pulse bridge converter with an interphase transformer, half-controlled bridge rectifiers, and so on. The choice of converter topology depends on application, for example, 12-pulse bridge converters are used for high-voltage dc transmission, as discussed in Chapter 16. A detailed description of these converters is presented in Reference 1.

4-9   SUMMARY

1. Line-frequency controlled rectifiers and inverters are used for controlled transfer of power between the line-frequency ac and the adjustable-magnitude dc. By controlling the delay angle of thyristors in these converters, a smooth transition can be made from the rectification mode to the inversion mode, or vice versa. The dc-side voltage can reverse polarity but the dc-side current remains unidirectional.

2. Because of the increasing importance of diode rectifiers, discussed in Chapter 3, the phase-controlled converters are mostly used at high power levels. Therefore, only three-phase full-bridge converters are analyzed, and it is assumed that i_d flows continuously. The effect of discontinuous current conduction on the waveforms and V_d is briefly discussed.

3. Phase-controlled converters inject large harmonics into the utility system. At small values of V_d (compared to its maximum possible value), these operate at a very poor power factor as well as a poor displacement power factor. Additionally, these converters produce notches in the line-voltage waveform.

PROBLEMS

4-1. In the three-phase converter of Fig. 4-4, derive the expression for the displacement power factor given by Eq. 4-32.

4-2. In the three-phase rectifier circuit of Fig. 4-4, $V_{LL} = 460$ V at 60 Hz, and $L_s = 25$ μH. Calculate the commutation angle u, if $V_d = 525$ V and $P_d = 500$ kW.

4-3. In Fig. 4-8b, derive the expression for the minimum dc current I_{dB} that results in a continuous-current conduction for given V_{LL}, ω, L_d, and $\alpha = 30°$. Assume that L_s is negligible.

4-4. In the circuit of Fig. 4-15a, L_{s1} corresponds to the leakage inductance of a 60 Hz transformer with the following ratings: three-phase kVA rating of 500 kVA, line-to-line voltage of 480 V, and an impedance of 6%. L_{s2} is due to a 200-foot-long cable, with a per-phase inductance of 0.1 μH/ft. The ac input voltage is 460 V(L-L) and the dc-side of the rectifier is delivering 25 kW at a voltage of 525 V.

Calculate the notch-widths in μs and the line-notch depth ρ in percentage at the point of common coupling. Also, calculate the area for a deep line-notch at the point of common coupling in V $\cdot$ μs and compare the answers with the recommended limits in Table 4-2.

4-5. Repeat Problem 4-4, if a 40-kVA, 480-V 1:1 transformer is also used at the input to the rectifier, which has a leakage impedance of 3%. The 3-phase rating of the transformer equals 40-kVA.

4-6. Calculate the total harmonic distortion in the voltage at the point of common coupling in Problems 4-4 and 4-5.

4-7. Using the typical harmonics in the input current given in Table 4-1, obtain the total harmonic distortion in the voltage at the point of common coupling in Problem 4-4.

4-10 REFERENCES

1. E. W. KIMBARK, *Direct Current Transmission*, Vol. 1, Wiley-Interscience, New York, 1971.
2. B. M. BIRD and K. G. KING, *An Introduction to Power Electronics*, John Wiley & Sons, New York, 1983.
3. GENERAL ELECTRIC Co., "AF-400 Adjustable Speed Drive Application Guide," GET-6655A, Speed Variators Products Operation, Erie, PA 16531 (no date).
4. GENERAL ELECTRIC Co., "SCR Drives—AC Line Disturbances, Isolation Transformers, Short Circuit Protection, Power Factor, Grounding," GET-6468A, Speed Variator Products Operation, Erie, PA 16531 (no date).
5. IEEE, "IEEE Guide for Harmonic Control and Reactive Compensation of Static Power Converters," ANSI/IEEE Std 519–1981.
6. D. A. JARC and R. G. SCHIEMAN, "Power Line Considerations for Variable Frequency Drives," *IEEE Transactions on Industry Applications*, Vol. IAS, No. 5, Sept./Oct. 1985, pp. 1099–1105.
7. IEEE, "IEEE Standard Practice and Requirements for General Purpose Thyristor DC Drives," IEEE Standard 597–1983.

5

DC-to-DC Switch-Mode Converters

5-1 INTRODUCTION

Dc–dc converters are widely used in regulated switch-mode dc power supplies and in dc-motor drive applications. As shown in Fig. 5-1, often the input to these converters is an unregulated dc voltage, which is obtained by rectifying the line voltage and therefore it will fluctuate due to changes in the line-voltage magnitude. Switch-mode dc-to-dc converters are used to convert the unregulated dc input into a controlled dc output at a desired voltage level.

Looking ahead to the application of these converters, we find that these converters are very often used with an electrical isolation transformer in the switch-mode dc power supplies, and almost always without an isolation transformer in case of dc-motor drives. Therefore, to discuss these circuits in a generic manner, only the nonisolated converters are considered in this chapter, since the electrical isolation is an added modification.

The following dc–dc converters are discussed in this chapter:

1. Step-down (buck) converter
2. Step-up (boost) converter
3. Step-down/up (buck–boost) converter
4. Cúk converter
5. Full-bridge converter

Of these five converters, only the step-down and the step-up are the basic converter topologies. Both the buck–boost and the Cúk converters are combinations of the two basic topologies. The full-bridge converter is derived from the step-down converter.

The converters listed are discussed in detail in this chapter. Their variations, as they apply to specific applications, are described in the chapters dealing with switch-mode dc power supplies and dc-motor drives.

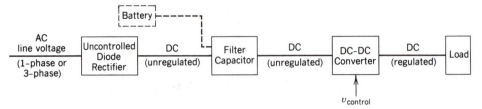

FIGURE 5-1: DC – DC converter system.

In this chapter, the converters are analyzed in steady state. The switches are treated as being ideal, and the losses in the inductive and the capacitive elements are neglected. Such losses can limit the operational capacity of some of these converters and are discussed separately.

The dc input voltage to the converters is assumed to have zero internal impedance. It could be a battery source; however, in most cases, the input is a diode rectified ac line voltage (as is discussed in Chapter 3) with a large filter capacitance, as shown in Fig. 5-1 to provide a low internal impedance and a low-ripple dc voltage source.

In the output stage of the converter, a small filter is treated as an integral part of the dc-to-dc converter. The output is assumed to supply a load that can be represented by an equivalent resistance, as is usually the case in switch-mode dc power supplies. A dc-motor load (the other application of these converters) can be represented by a dc voltage in series with the motor winding resistance and inductance.

5-2   CONTROL OF dc–dc CONVERTERS

In dc–dc converters, the average dc output voltage must be controlled to equal a desired level, though the input voltage and the output load may fluctuate. Switch-mode dc–dc converters utilize one or more switches to transform dc from one level to another. In a dc–dc converter with a given input voltage, the average output voltage is controlled by controlling the switch on and off durations (t_{on} and t_{off}). To illustrate the switch-mode conversion concept, consider a basic dc–dc converter shown in Fig. 5-2a. The average value V_o of the output voltage v_o in Fig. 5-2b depends on t_{on} and t_{off}. One of the methods for controlling the output voltage employs switching at a constant frequency (hence, a constant switching time period $T_s = t_{on} + t_{off}$), and adjusting the on-duration of the switch to control the average output voltage. In this

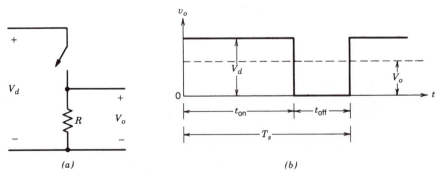

FIGURE 5-2: Switch-mode dc – dc conversion.

method, called Pulse-Width Modulation (PWM) switching, the switch duty ratio D, which is defined as the ratio of the on-duration to the switching time period, is varied.

The other control method is more general, where both the switching frequency (and hence the time period) and the on-duration of the switch are varied. This method is used only in dc–dc converters utilizing force-commutated thyristors, and therefore, will not be discussed in this book.

In the PWM switching at a constant switching frequency, the switch control signal, which controls the state (on or off) of the switch, is generated by comparing a signal level control voltage $v_{control}$ with a repetitive waveform as shown in Figs. 5-3a and 5-3b. The control voltage signal generally is obtained by amplifying the error, or the difference between the actual output voltage and its desired value. The frequency of the repetitive waveform with a constant peak, which is shown to be a sawtooth, establishes the switching frequency. This frequency is kept constant in a PWM control and is chosen to be in a few kilohertz to a few hundred kilohertz range. When the amplified error signal, which varies very slowly with time relative to the switching frequency, is greater than the sawtooth waveform, the switch control signal becomes high, causing the switch to turn on. Otherwise, the switch is off. In terms of $v_{control}$ and the peak of the sawtooth waveform $\hat{V}_{st}$ in Fig. 5-3, the switch duty ratio can be expressed as

$$D = \frac{t_{on}}{T_s} = \frac{v_{control}}{\hat{V}_{st}} \qquad (5\text{-}1)$$

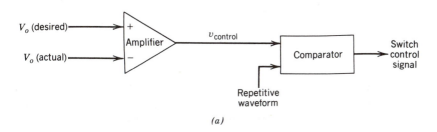

(a)

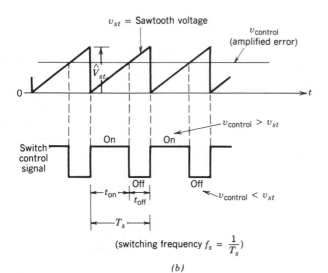

(b)

FIGURE 5-3: Pulse-width modulator (PWM): (*a*) block diagram, (*b*) comparator signals.

The dc–dc converters can have two distinct modes of operation: (1) continuous-current conduction and (2) discontinuous-current conduction. In practice, a converter may operate in both modes, which have significantly different characteristics. Therefore, a converter and its control should be designed based on both modes of operation.

5-3   STEP-DOWN (BUCK) CONVERTER

As the name implies, a step-down converter produces a lower average output voltage than the dc input voltage V_d. Its main application is in regulated dc power supplies and dc-motor speed control.

Conceptually, the basic circuit of Fig. 5-2a constitutes a step-down converter for a purely resistive load. Assuming an ideal switch and a purely resistive load, the instantaneous output voltage depends on the switch position. From Fig. 5-2b, the average output voltage can be calculated in terms of the switch-duty ratio:

$$V_o = \frac{1}{T_s} \int_0^{T_s} v_o(t)\, dt = \frac{1}{T_s} \left(\int_0^{t_{on}} V_d\, dt + \int_{t_{on}}^{T_s} 0 \cdot dt \right) = \frac{t_{on}}{T_s} V_d = D V_d \qquad (5\text{-}2)$$

Substituting for D in Eq. 5-2 from Eq. 5-1,

$$V_o = \frac{V_d}{\hat{V}_{st}} \cdot v_{control} = k \cdot v_{control}$$

where

$$k = \frac{V_d}{\hat{V}_{st}} = \text{constant}$$

By varying the duty ratio t_{on}/T_s of the switch, V_o can be controlled. Another important observation is that the average output voltage V_o varies linearly with the control voltage, as is the case in linear amplifiers. In an actual application, the foregoing circuit has two drawbacks: (1) In practice the load would be inductive. Even with a resistive load, there would always be certain associated stray inductance. This means that the switch would have to absorb (or dissipate) the inductive energy and therefore it may be destroyed. (2) The output voltage fluctuates between 0 and V_d, which is not acceptable in most applications. The problem of stored inductive energy is overcome by using a diode as shown in Fig. 5-4a. The output voltage fluctuations are very much diminished by using a low-pass filter, consisting of an inductor and a capacitor. Figure 5-4b shows the waveform of the input v_{oi} to the low-pass filter (same as the output voltage in Fig. 5-2b without a low-pass filter) which consists of a dc component V_o, and the harmonics at the switching frequency f_s and its multiples, as shown in Fig. 5-4b. The low-pass filter characteristic with the damping provided by the load-resistor R is shown in Fig. 5-4c. The corner frequency f_c of this low-pass filter is selected to be much lower than the switching frequency, thus essentially eliminating the switching frequency ripple in the output voltage.

During the interval when the switch is on, the diode in Fig. 5-4a becomes reverse biased and the input provides energy to the load as well as to the inductor. During the interval when the switch is off, the inductor current flows through the diode, transferring some of its stored energy to the load.

In the steady-state analysis presented here, the filter capacitor at the output is assumed to be very large, as is normally the case in applications requiring a nearly

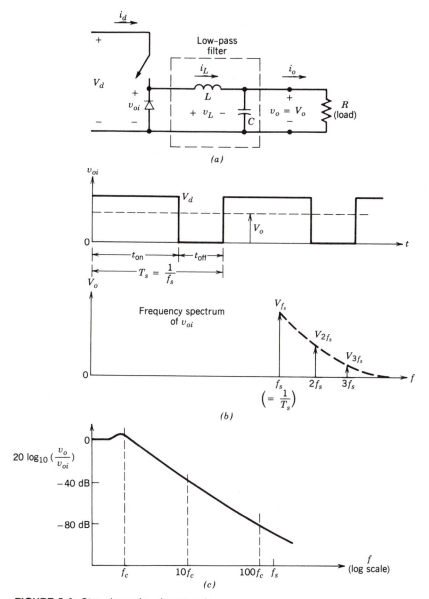

FIGURE 5-4: Step-down dc–dc converter.

constant instantaneous output voltage $v_o(t) \simeq V_o$. The ripple in the capacitor voltage (output voltage) is calculated later.

From Fig. 5-4a we observe that in a step-down converter, the average inductor current is equal to the average output current I_o, since the average capacitor current in steady state is zero.

5-3-1 Continuous-Conduction Mode

Figure 5-5 shows the waveforms for the continuous-conduction mode of operation where the inductor current flows continuously ($i_L(t) > 0$). When the switch is on for a time duration t_{on}, the switch conducts the inductor current and the diode

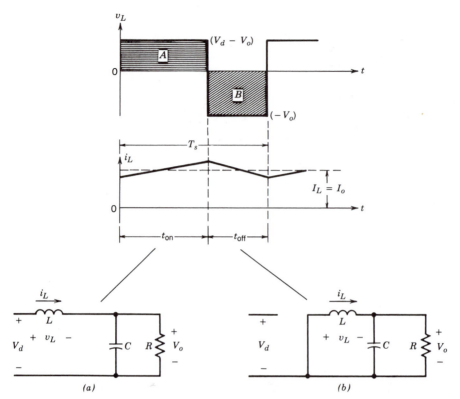

FIGURE 5-5: Step-down converter circuit states (assuming i_L flows continuously): (a) switch on, (b) switch off.

becomes reverse biased. This results in a positive voltage $v_L = V_d - V_o$ across the inductor in Fig. 5-5a. This voltage causes a linear increase in the inductor current i_L. When the switch is turned off, because of the inductive energy storage, i_L continues to flow. This current now flows through the diode, and $v_L = -V_o$ in Fig. 5-5b.

Since in steady-state operation the waveform must repeat from one time period to the next, the integral of the inductor voltage v_L over one time period must be zero, as discussed in Chapter 3 (Eq. 3-3), where $T_s = t_{on} + t_{off}$:

$$\int_0^{T_s} v_L \, dt = \int_0^{t_{on}} v_L \, dt + \int_{t_{on}}^{T_s} v_L \, dt = 0$$

In Fig. 5-5, the foregoing equation implies that the areas A and B must be equal. Therefore,

$$\left(V_d - V_o\right)t_{on} = V_o\left(T_s - t_{on}\right)$$

or

$$\frac{V_o}{V_d} = \frac{t_{on}}{T_s} = D \quad \text{duty ratio} \tag{5-3}$$

Therefore, in this mode, the voltage output varies linearly with the duty ratio of the switch for a given input voltage. It does not depend on any other circuit parameter. The foregoing equation can also be derived by simply averaging the voltage v_{oi} in Fig.

5-4b and recognizing that the average voltage across the inductor in steady-state operation is zero:

$$\frac{V_d t_{\text{on}} + 0 \cdot t_{\text{off}}}{T_s} = V_o$$

or

$$\frac{V_o}{V_d} = \frac{t_{\text{on}}}{T_s} = D$$

Neglecting power losses associated with all the circuit elements, the input power P_d equals the output power P_o:

$$P_d = P_o$$

Therefore,

$$V_d I_d = V_o I_o$$

and

$$\frac{I_o}{I_d} = \frac{V_d}{V_o} = \frac{1}{D} \tag{5-4}$$

Therefore, in the continuous-conduction mode, the step-down converter is equivalent to a dc transformer where the turns ratio of this equivalent transformer can be continuously controlled electronically in a range of 0 to 1 by controlling the duty ratio of the switch.

We observe that even though the average input current I_d follows the transformer relationship, the instantaneous input current waveform jumps from a peak value to zero every time the switch is turned off. An appropriate filter at the input may be required to eliminate the undesirable effects of the current harmonics.

5-3-2 Boundary Between Continuous and Discontinuous Conduction

In this section, we will develop equations that show the influence of various circuit parameters on the conduction mode of the inductor current (continuous or discontinuous). At the edge of the continuous-current-conduction mode, Fig. 5-6a shows the waveforms for v_L and i_L. Being at the boundary between the continuous and the discontinuous mode, by definition, the inductor current i_L goes to zero at the end of the off period.

At this boundary, the average inductor current, where the subscript B refers to the boundary is,

$$I_{LB} = \frac{1}{2} i_{L,\text{peak}} = \frac{t_{\text{on}}}{2L}(V_d - V_o) = \frac{DT_s}{2L}(V_d - V_o) = I_{oB} \tag{5-5}$$

Therefore, during an operating condition (with a given set of values for T_s, V_d, V_o, L, and D), if the average output current (and, hence, the average inductor current) becomes less than I_{LB} given by Eq. 5-5, then i_L will become discontinuous.

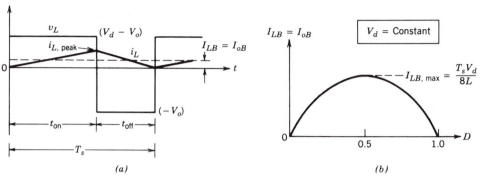

FIGURE 5-6: Current at the boundary of continuous / discontinuous conduction: (a) current waveform, (b) I_{LB} versus D, keeping V_d constant.

5-3-3 Discontinuous-Conduction Mode

Depending on the application of these converters, either the input voltage V_d, or the output voltage V_o remains constant during the converter operation. Both of these types of operation are discussed below.

5-3-3-1 DISCONTINUOUS-CONDUCTION MODE WITH CONSTANT V_d

In an application such as a dc-motor speed control, V_d remains essentially constant and V_o is controlled by adjusting the converter duty ratio D.

Since $V_o = DV_d$, the average inductor current at the edge of the continuous conduction mode from Eq. 5-5 is

$$I_{LB} = \frac{T_s V_d}{2L} D(1 - D) \tag{5-6}$$

Using this equation, we find that Fig. 5-6b shows the plot of I_{LB} as a function of the duty ratio D, keeping V_d and all other parameters constant. It shows that the output current required for a continuous conduction mode is maximum at $D = 0.5$:

$$I_{LB,\,\mathrm{max}} = \frac{T_s V_d}{8L} \tag{5-7}$$

From Eqs. 5-6 and 5-7

$$I_{LB} = 4I_{LB,\,\mathrm{max}} D(1 - D) \tag{5-8}$$

Next the voltage ratio V_o/V_d will be calculated in the discontinuous mode. Let us assume that initially the converter is operating at the edge of continuous conduction as in Fig. 5-6a, for given values of T, L, V_d, and D. If these parameters are kept constant and the output load power is decreased (i.e., the load resistance goes up), then the average inductor current will decrease. As is shown in Fig. 5-7, this dictates a higher value of V_o than before and results in a discontinuous inductor current.

During the interval $\Delta_2 T_s$ where the inductor current is zero, the power to the load resistance is supplied by the filter capacitor alone. The inductor voltage v_L during this interval is zero. Again, equating the integral of the inductor voltage over one time

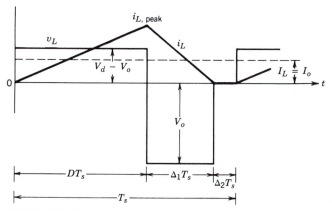

FIGURE 5-7: Discontinuous conduction in step-down converter.

period to zero

$$\left(V_d - V_o\right) DT_s + \left(-V_o\right)\Delta_1 T_s = 0 \tag{5-9}$$

$$\therefore \frac{V_o}{V_d} = \frac{D}{D + \Delta_1} \tag{5-10}$$

where $D + \Delta_1 < 1.0$.
From Fig. 5-7,

$$i_{L,\text{peak}} = \frac{V_o}{L} \Delta_1 T_s \tag{5-11}$$

Therefore,

$$I_o = i_{L,\text{peak}} \frac{D + \Delta_1}{2} \tag{5-12}$$

$$= \frac{V_o T_s}{2L}(D + \Delta_1)\Delta_1 \quad \left(\text{using Eq. 5-11}\right) \tag{5-13}$$

$$= \frac{V_d T_s}{2L} D\Delta_1 \quad \left(\text{using Eq. 5-10}\right) \tag{5-14}$$

$$= 4I_{LB,\text{max}} D\Delta_1 \quad \left(\text{using Eq. 5-7}\right) \tag{5-15}$$

$$\therefore \Delta_1 = \frac{I_o}{4I_{LB,\text{max}} D} \tag{5-16}$$

From Eqs. 5-10 and 5-16

$$\frac{V_o}{V_d} = \frac{D^2}{D^2 + \dfrac{1}{4}\left(\dfrac{I_o}{I_{LB,\text{max}}}\right)} \tag{5-17}$$

Figure 5-8 shows the step-down converter characteristic in both modes of operation, for a constant V_d. (V_o/V_d) is plotted as a function of $I_o/I_{LB,\text{max}}$ for various

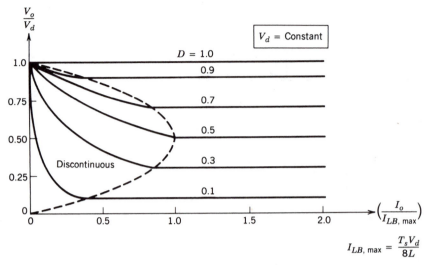

FIGURE 5-8: Step-down converter characteristics, keeping V_d constant.

values of duty ratio using Eqs. 5-3 and 5-17. The boundary between the continuous and the discontinuous mode, shown by the dotted curve, is established by Eqs. 5-3 and 5-8.

5-3-3-2 DISCONTINUOUS-CONDUCTION MODE WITH CONSTANT V_o

In applications such as regulated dc power supplies, V_d may fluctuate but V_o is kept constant by adjusting the duty ratio D.

Since $V_d = V_o/D$, the average inductor current at the edge of continuous-conduction mode from Eq. 5-5 is

$$I_{LB} = \frac{T_s V_o}{2L}(1 - D) \tag{5-18}$$

Equation 5-18 shows that if V_o is kept constant, the maximum value of I_{LB} occurs at $D = 0$:

$$I_{LB, \max} = \frac{T_s V_o}{2L} \tag{5-19}$$

It should be noted that the operation corresponding to $D = 0$ and a finite V_o is, of course, hypothetical because it would require V_d to be infinite.

From Eqs. 5-18 and 5-19

$$I_{LB} = (1 - D)I_{LB, \max} \tag{5-20}$$

For the converter operation where V_o is kept constant, it will be useful to obtain the required duty ratio D as a function of $I_o/I_{LB, \max}$. Using Eqs. 5-10 and 5-13 (which are valid in the discontinuous-conduction mode whether V_o or V_d is kept

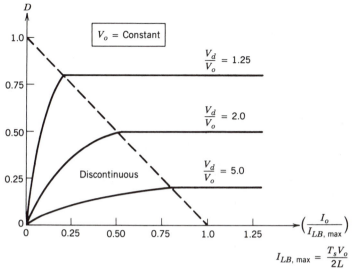

$$I_{LB,\,max} = \frac{T_s V_o}{2L}$$

FIGURE 5-9: Step-down converter characteristics, keeping V_o constant.

constant), along with Eq. 5-19 for the case where V_o is kept constant,

$$D = \frac{V_o}{V_d} \left(\frac{I_o/I_{LB,\,max}}{1 - \dfrac{V_o}{V_d}} \right)^{1/2} \tag{5-21}$$

D as a function of $I_o/I_{LB,\,max}$ is plotted in Fig. 5-9 for various values of V_d/V_o, keeping V_o constant. The boundary between the continuous and the discontinuous mode of operation is obtained by using Eq. 5-20.

5-3-4 Output Voltage Ripple

In the previous analysis, the output capacitor is assumed to be so large as to yield $v_o(t) = V_o$. However, the ripple in the output voltage with a practical value of capacitance can be calculated by considering the waveforms shown in Fig. 5-10 for a continuous-conduction mode of operation. Assuming that all of the ripple component in i_L flows through the capacitor and its average component flows through the load resistor, the shaded area in Fig. 5-10 represents an additional charge ΔQ. Therefore, the peak-to-peak voltage ripple ΔV_o can be written as

$$\Delta V_o = \frac{\Delta Q}{C} = \frac{1}{C} \cdot \frac{1}{2} \cdot \frac{\Delta I_L}{2} \cdot \frac{T_s}{2}$$

From Fig. 5-5 during t_{off}

$$\Delta I_L = \frac{V_o}{L}(1 - D)T_s \tag{5-22}$$

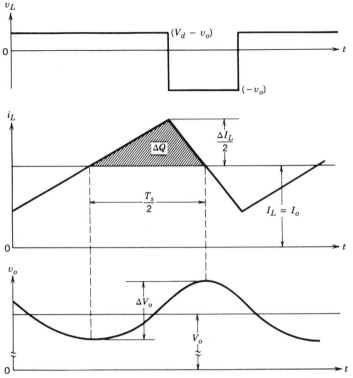

FIGURE 5-10: Output voltage ripple in a step-down converter.

Therefore, substituting ΔI_L from Eq. 5-22 into the previous equation

$$\Delta V_o = \frac{T_s}{8C}\frac{V_o}{L}(1 - D)T_s \tag{5-23}$$

$$\therefore \frac{\Delta V_o}{V_o} = \frac{1}{8}\frac{T_s^2(1 - D)}{LC} = \frac{\pi^2}{2}(1 - D)\left(\frac{f_c}{f_s}\right)^2 \tag{5-24}$$

where, switching frequency $f_s = 1/T_s$ and,

$$f_c = \frac{1}{2\pi\sqrt{LC}}. \tag{5-25}$$

Equation 5-24 shows that the voltage ripple can be minimized by selecting a corner frequency f_c of the low-pass filter at the output such that $f_c \ll f_s$. Also, the ripple is independent of the output load power, so long as the converter operates in the continuous-conduction mode. A similar analysis can be performed for the discontinuous-conduction mode.

We should note that in switch-mode dc power supplies, the percentage ripple in the output voltage is usually specified to be less than, for instance 1%. Therefore, the analysis in the previous sections assuming $v_o(t) = V_o$ is valid. It should be noted that the output ripple in Eq. 5-24 is consistent with the discussion of the low-pass filter characteristic in Fig. 5-4c.

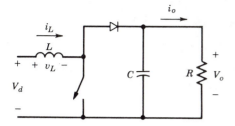

FIGURE 5-11: Step-up dc – dc converter.

5-4   STEP-UP (BOOST) CONVERTER

Figure 5-11 shows a step-up converter. Its main application is in regulated dc power supplies and the regenerative braking of dc motors. As the name implies, the output voltage is always greater than the input voltage. When the switch is on, the diode is reversed biased, thus isolating the output stage. The input supplies energy to the inductor. When the switch is off, the output stage receives energy from the inductor as well as from the input. In the steady-state analysis presented here, the output filter capacitor is assumed to be very large to ensure a constant output voltage $v_o(t) \simeq V_o$.

5-4-1   Continuous-Conduction Mode

Figure 5-12 shows the steady-state waveforms for this mode of conduction where the inductor current flows continuously $(i_L(t) > 0)$.

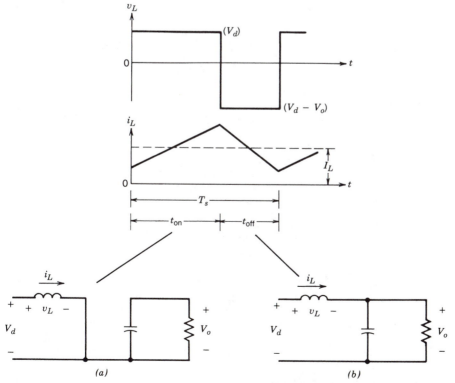

FIGURE 5-12: Continuous conduction mode: *(a)* switch on, *(b)* switch off.

Since in steady-state the time integral of the inductor voltage over one time period must be zero

$$V_d t_{on} + (V_d - V_o)t_{off} = 0$$

Dividing both sides by T_s and rearranging terms

$$\frac{V_o}{V_d} = \frac{T_s}{t_{off}} = \frac{1}{1 - D} \qquad (5\text{-}26)$$

Assuming a lossless circuit, $P_d = P_o$

$$\therefore V_d I_d = V_o I_o$$

and

$$\frac{I_o}{I_d} = (1 - D) \qquad (5\text{-}27)$$

5-4-2   Boundary Between Continuous and Discontinuous Conduction

Figure 5-13a shows the waveforms at the edge of continuous conduction. By definition, in this mode i_L goes to zero at the end of the off interval. The average value of the inductor current at this boundary is

$$I_{LB} = \frac{1}{2} i_{L,peak} \quad (\text{Fig. 5-13}a)$$

$$= \frac{1}{2} \frac{V_d}{L} t_{on}$$

$$= \frac{T_s V_o}{2L} D(1 - D) \quad (\text{using Eq. 5-26}) \qquad (5\text{-}28)$$

Recognizing that in a step-up converter the inductor current and the input current are the same ($i_d = i_L$) and using Eqs. 5-27 and 5-28, we find that the average

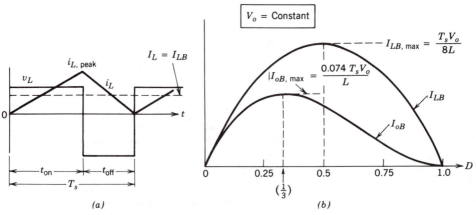

FIGURE 5-13: Step-up dc – dc converter, at the boundary of continuous / discontinuous conduction.

output current at the edge of continuous conduction is

$$I_{oB} = \frac{T_s V_o}{2L} D(1 - D)^2 \tag{5-29}$$

Most applications in which a step-up converter is used require that V_o be kept constant. Therefore, with V_o constant, I_{LB} and I_{oB} are plotted in Fig. 5-13b as a function of duty ratio D. Keeping V_o constant and varying the duty ratio implies that the input voltage is varying.

Figure 5-13b shows that I_{LB} reaches a maximum value at $D = 0.5$:

$$I_{LB,\,\text{max}} = \frac{T_s V_o}{8L} \tag{5-30}$$

Also, I_{oB} has its maximum at $D = \frac{1}{3} = 0.333$:

$$I_{oB,\,\text{max}} = \frac{2}{27} \frac{T_s V_o}{L} = 0.074 \frac{T_s V_o}{L} \tag{5-31}$$

In terms of their maximum values, I_{LB} and I_{oB} can be expressed as

$$I_{LB} = 4D(1 - D)I_{LB,\,\text{max}} \tag{5-32}$$

and

$$I_{oB} = \frac{27}{4} D(1 - D)^2 I_{oB,\,\text{max}} \tag{5-33}$$

Figure 5-13b shows that for a given D, with constant V_o, if the average load current drops below I_{oB} (and, hence, the average inductor current below I_{LB}), the current conduction will become discontinuous.

5-4-3 Discontinuous-Conduction Mode

To understand the discontinuous current conduction mode, we would assume that as the output load power decreases, V_d and D remain constant (even though in practice, D would vary in order to keep V_o constant). Figure 5-14 compares the waveforms at the boundary of continuous-conduction and discontinuous-conduction, assuming that V_d and D are constant.

In Fig. 5-14b, the discontinuous current conduction occurs due to decreased $P_o\ (= P_d)$ and, hence, a lower $I_L\ (= I_d)$, since V_d is constant. Since $i_{L,\,\text{peak}}$ is the same in both modes in Fig. 5-14, a lower value of I_L (and, hence, a discontinuous i_L) is possible only if V_o goes up in Fig. 5-14b.

If we equate the integral of the inductor voltage over one time period to zero,

$$V_d D T_s + (V_d - V_o)\Delta_1 T_s = 0$$

$$\therefore \frac{V_o}{V_d} = \frac{\Delta_1 + D}{\Delta_1} \tag{5-34}$$

and

$$\frac{I_o}{I_d} = \frac{\Delta_1}{\Delta_1 + D} \quad (\text{since } P_d = P_o) \tag{5-35}$$

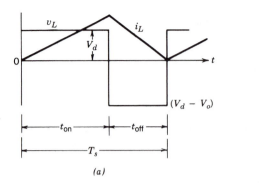

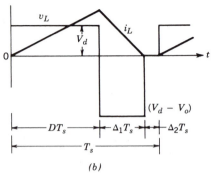

FIGURE 5-14: Step-up converter waveforms: (a) boundary of continuous / discontinuous conduction, (b) discontinuous conduction.

From Fig. 5-14b, the average input current, which is also equal to the inductor current, is

$$I_d = \frac{V_d}{2L} DT_s \cdot (D + \Delta_1) \tag{5-36}$$

Using Eq. 5-35 in the foregoing equation

$$I_o = \left(\frac{T_s V_d}{2L} \right) D\Delta_1 \tag{5-37}$$

In practice, since V_o is held constant and D varies in response to the variation in V_d, it is more useful to obtain the required duty ratio D as a function of load current for various values of V_o/V_d. By using Eqs. 5-34, 5-37, and 5-31, we determine that

$$D = \left[\frac{4}{27} \left(\frac{V_o}{V_d} \right) \left(\frac{V_o}{V_d} - 1 \right) \frac{I_o}{I_{oB,\max}} \right]^{1/2} \tag{5-38}$$

In Fig. 5-15, D is plotted as a function of $I_o/I_{oB,\max}$ for various values of V_d/V_o. The boundary between continuous and discontinuous-conduction is shown by the dotted curve.

In the discontinuous mode if V_o is not controlled during each switching time period, at least

$$\frac{L}{2} i_{L,\text{peak}}^2 = \frac{(V_d DT_s)^2}{2L} \text{ W-s}$$

are transferred from the input to the output capacitor and to the load. If the load is not able to absorb this energy, the capacitor voltage V_o would increase until an energy balance is established. If the load becomes very light, the increase in V_o may cause a capacitor breakdown or a dangerously high voltage to occur.

EXAMPLE 5-1

In a step-up converter, the duty ratio is adjusted to regulate the output voltage V_o at 48 V. The input voltage varies in a wide range from 12 V to 36 V. The maximum power output is 120 W. For stability reasons, it is required that the converter always operate in a discontinuous current conduction mode. The switching frequency is 50 kHz.

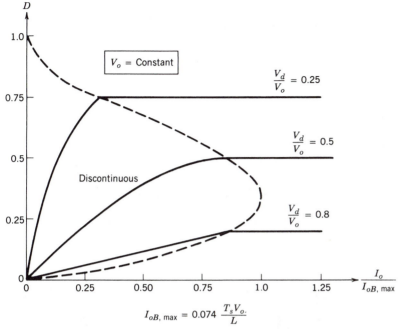

$$I_{oB, \, max} = 0.074 \, \frac{T_s V_{o.}}{L}$$

FIGURE 5-15: Step-up converter characteristics, keeping V_o = constant.

Assuming ideal components and C as very large, calculate the maximum value of L that can be used.

SOLUTION

In this converter, $V_o = 48$ V, $T_s = 20$ μs, and $I_{o, \, max} = 120$ W/48 V = 2.5 A. To find the maximum value of L that keeps the current conduction discontinuous, we will assume that at the extreme operating condition, the inductor current is at the edge of continuous conduction.

For the given range of V_d (12 V to 36 V), D is in a range of 0.75 to 0.25 (corresponding to the current conduction bordering on being continuous). For this range of D, from Fig. 5-13b, I_{oB} has the smallest value at $D = 0.75$.

Therefore, by substituting $D = 0.75$ in Eq. 5-29 for I_{oB} and equating it to $I_{o, \, max}$ of 2.5 A, we can calculate

$$L = \frac{20 \times 10^{-6} \times 48}{2 \times 2.5} 0.75 (1 - 0.75)^2$$

$$= 9 \, \mu\text{H}$$

Therefore, if $L = 9$ μH is used, the converter operation will be at the edge of continuous conduction with $V_d = 12$ V and $P_o = 120$ W. Otherwise, the conduction will be discontinuous. To further ensure a discontinuous-conduction mode, a smaller than 9 μH inductance may be used. ∎

5-4-4 Effect of Parasitic Elements

The parasitic elements in a step-up converter are due to the losses associated with the inductor, the capacitor, the switch, and the diode. Figure 5-16 qualitatively

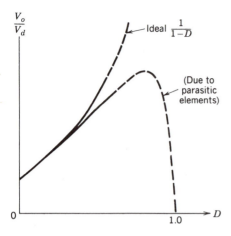

FIGURE 5-16: Effect of parasitic elements on voltage conversion ratio (step-up converter).

shows the effect of these parasitics on the voltage transfer ratio. Unlike the ideal characteristic, in practice V_o/V_d declines as the duty ratio approaches unity. Because of very poor switch utilization at high values of duty ratio (as discussed in Section 5-8), the curves in this range are shown as dotted. These parasitic elements have been ignored in the simplified analysis presented here; however, these can be easily incorporated into circuit simulation programs on computers for designing such converters.

5-4-5 Output Voltage Ripple

The peak-to-peak ripple in the output voltage can be calculated by considering the waveforms shown in Fig. 5-17 for a continuous mode of operation. Assuming that all the ripple current component of the diode current i_D flows through the capacitor and its average value flows through the load resistor, the shaded area in Fig. 5-17

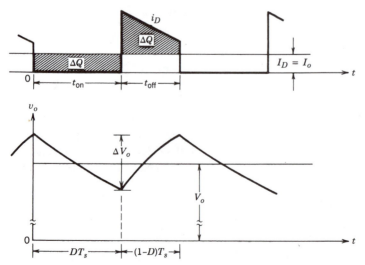

FIGURE 5-17: Step-up converter output voltage ripple.

represents charge ΔQ. Therefore, the peak–peak voltage ripple is given by

$$\Delta V_o = \frac{\Delta Q}{C} = \frac{I_o DT_s}{C} \quad \text{(assuming a constant output current)}$$

$$= \frac{V_o}{R} \frac{DT_s}{C} \tag{5-39}$$

$$\therefore \frac{\Delta V_o}{V_o} = \frac{DT_s}{RC}$$

$$= D\frac{T_s}{\tau} \quad \text{(where } \tau = RC \text{ time constant)} \tag{5-40}$$

A similar analysis can be performed for the discontinuous mode of conduction.

5-5 BUCK–BOOST CONVERTER

The main application of a step-down/up or buck–boost converter is in regulated dc power supplies, where a negative polarity output may be desired with respect to the common terminal of the input voltage, and the output voltage can be either higher or lower than the input voltage.

A buck–boost converter can be obtained by the cascade connection of the two basic converters: the step-down converter and the step-up converter. In steady state, the output-to-input voltage conversion ratio is the product of the conversion ratios of the two converters in cascade (assuming that switches in both converters have the same duty ratio):

$$\frac{V_o}{V_d} = D \cdot \frac{1}{1 - D} \quad \text{(from Eqs. 5-3 and 5-26)} \tag{5-41}$$

This allows the output voltage to be higher or lower than the input voltage, based on the duty ratio D.

The cascade connection of the step-down and the step-up converters can be combined into a single buck-boost converter shown in Fig. 5-18. When the switch is closed, the input provides energy to the inductor and the diode is reverse biased. When the switch is open, the energy stored in the inductor is transferred to the output. No

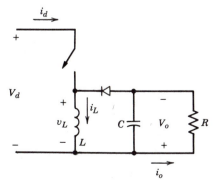

FIGURE 5-18: Buck–boost converter.

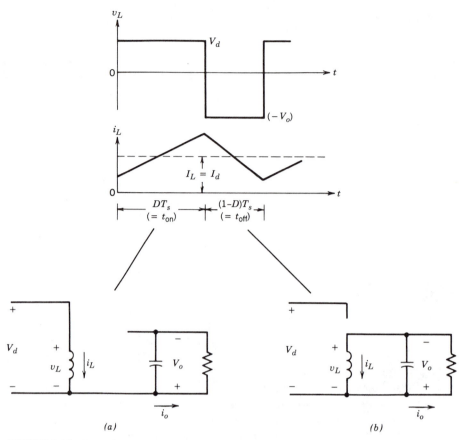

FIGURE 5-19: Buck–boost converter ($i_L > 0$): (*a*) switch on, (*b*) switch off.

energy is supplied by the input during this interval. In the steady-state analysis presented here, the output capacitor is assumed to be very large, which results in a constant output voltage $v_o(t) \simeq V_o$.

5-5-1   Continuous-Conduction Mode

Figure 5-19 shows the waveforms for the continuous conduction mode where the inductor current flows continuously.

Equating the integral of the inductor voltage over one time period to zero

$$V_d D T_s + (-V_o)(1 - D)T_s = 0$$

$$\therefore \frac{V_o}{V_d} = \frac{D}{1 - D} \tag{5-42}$$

and

$$\frac{I_o}{I_d} = \frac{1 - D}{D} \quad \left(\text{assuming } P_d = P_o\right) \tag{5-43}$$

Equation 5-42 implies that depending on the duty ratio, the output voltage can be either higher or lower than the input.

5-5-2　Boundary Between Continuous and Discontinuous Conduction

Figure 5-20a shows the waveforms at the edge of continuous conduction. By definition, in this mode i_L goes to zero at the end of the off interval.

From Fig. 5-20a

$$I_{LB} = \frac{1}{2} i_{L,\text{peak}}$$

$$= \frac{T_s V_d}{2L} D \tag{5-44}$$

From Fig. 5-18,

$$I_o = I_L - I_d \tag{5-45}$$

(since the average capacitor current is zero).

By using Eqs. 5-42 through 5-45, we can obtain the average inductor current and the output current at the border of continuous conduction in terms of V_o

$$I_{LB} = \frac{T_s V_o}{2L}(1 - D) \tag{5-46}$$

and

$$I_{oB} = \frac{T_s V_o}{2L}(1 - D)^2 \tag{5-47}$$

Most applications in which a buck–boost converter may be used require that V_o be kept constant, though V_d (and, hence, D) may vary. Our inspection of Eqs. 5-46

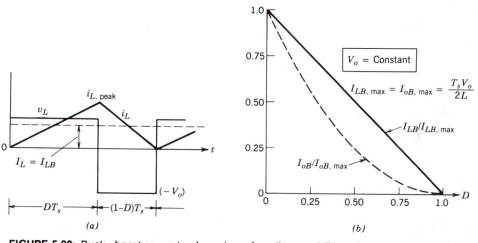

FIGURE 5-20: Buck–boost converter: boundary of continuous / discontinuous conduction.

and 5-47 shows that both I_{LB} and I_{oB} result in their maximum values at $D = 0$:

$$I_{LB,\,\text{max}} = \frac{T_s V_o}{2L} \tag{5-48}$$

and

$$I_{oB,\,\text{max}} = \frac{T_s V_o}{2L} \tag{5-49}$$

Using Eqs. 5-46 through 5-49,

$$I_{LB} = I_{LB,\,\text{max}}(1 - D) \tag{5-50}$$

and

$$I_{oB} = I_{oB,\,\text{max}}(1 - D)^2 \tag{5-51}$$

Figure 5-20b shows I_{LB} and I_{oB} as a function of D, keeping $V_o = $ a constant.

5-5-3   Discontinuous-Conduction Mode

Figure 5-21 shows the waveforms with a discontinuous i_L. If we equate the integral of the inductor voltage over one time period to zero,

$$V_d DT_s + (-V_o)\Delta_1 T_s = 0$$

$$\therefore \frac{V_o}{V_d} = \frac{D}{\Delta_1} \tag{5-52}$$

and

$$\frac{I_o}{I_d} = \frac{\Delta_1}{D} \quad (\text{since } P_d = P_o) \tag{5-53}$$

From Fig. 5-21

$$I_L = \frac{V_d}{2L} DT_s(D + \Delta_1) \tag{5-54}$$

Since V_o is kept constant, it is useful to obtain D as a function of the output load current I_o for various values of V_o/V_d. Using the equations derived earlier, we find

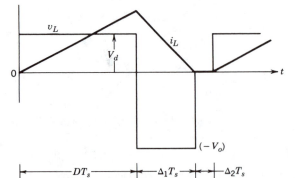

FIGURE 5-21: Buck – boost converter waveforms in a discontinuous conduction mode.

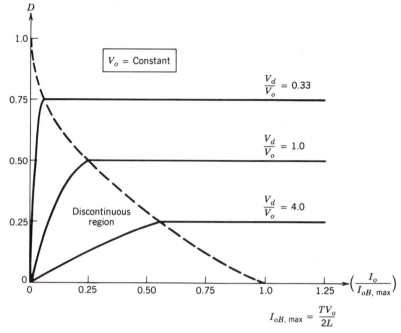

$$I_{oB,\,max} = \frac{TV_o}{2L}$$

FIGURE 5-22: Buck – boost converter characteristics, keeping V_o constant.

that

$$D = \frac{V_o}{V_d} \sqrt{\frac{I_o}{I_{oB,\,max}}} \tag{5-55}$$

Figure 5-22 shows the plot of D as a function of $I_o/I_{oB,\,max}$ for various values of V_d/V_o. The boundary between the continuous and the discontinuous-mode is shown by the dotted curve.

EXAMPLE 5-2

In a buck–boost converter operating at 20 kHz, L is equal to 0.05 mH. The output capacitor C is sufficiently large and $V_d = 15$ V. The output is to be regulated at 10 V and the converter is supplying a load of 10 W.

Calculate the duty ratio D.

SOLUTION

In this example, $I_o = 10/10 = 1$ A. Initially, the mode of conduction is not known. If the current is assumed to be at the edge of continuous conduction, from Eq. 5-42

$$\frac{D}{1 - D} = \frac{10}{15}$$

$$D = 0.4 \quad \text{(initial estimate)}$$

From Eq. 5-49

$$I_{oB,\max} = \frac{0.05 \times 10}{2 \times 0.05} = 5\text{ A}$$

Using $D = 0.4$ and $I_{oB,\max} = 5$ A in Eq. 5-51, we find that

$$I_{oB} = 5(1 - 0.4)^2$$

$$= 1.8\text{ A}$$

Since the output current $I_o = 1$ A is less than I_{oB}, the current conduction is discontinuous.

Therefore, using Eq. 5-55

$$\text{Duty ratio } D = \frac{10}{15}\sqrt{\frac{1.0}{5.0}}$$

$$= 0.3 \quad (\text{discontinuous conduction}) \qquad \blacksquare$$

5-5-4   Effect of Parasitic Elements

Analogous to the step-up converter, the parasitic elements have significant impact on the voltage conversion ratio and the stability of the feedback regulated buck–boost converter. As an example, Fig. 5-23 qualitatively shows the effect of these parasitic elements. The curves are shown as dotted because of the very poor switch utilization, making very high duty ratios impractical. The effect of these parasitic elements can be modelled in the circuit simulation programs for designing such converters.

5-5-5   Output Voltage Ripple

The ripple in the output voltage can be calculated by considering the waveform shown in Fig. 5-24 for a continuous mode of operation. Assuming that all the ripple

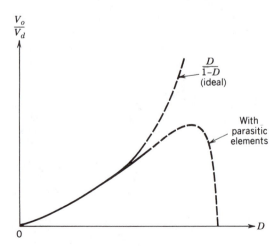

FIGURE 5-23: Effect of parasitic elements on the voltage conversion ratio in a buck – boost converter.

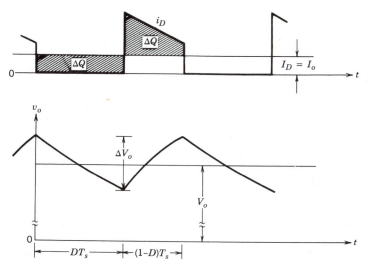

FIGURE 5-24: Output voltage ripple in a buck–boost converter.

current component of i_D flows through the capacitor and its average value flows through the load resistor, the shaded area in Fig. 5-24 represents charge ΔQ. Therefore, the peak-to-peak voltage ripple

$$\Delta V_o = \frac{\Delta Q}{C} = \frac{I_o D T_s}{C} \quad \text{(assuming a constant output current)}$$

$$= \frac{V_o}{R}\frac{D T_s}{C} \tag{5-56}$$

$$\frac{\Delta V_o}{V_o} = \frac{D T_s}{RC}$$

$$= D\frac{T_s}{\tau} \tag{5-57}$$

where $\tau = RC$ time constant.

A similar analysis can be performed for the discontinuous mode of operation.

5-6 CÚK dc–dc CONVERTER

Named after its inventor, the Cúk converter is shown in Fig. 5-25. This converter is obtained by using the duality principle on the circuit of buck–boost converter, discussed in the previous section. Similar to the buck–boost converter, the Cúk converter provides a negative polarity regulated output voltage with respect to the common terminal of the input voltage. Here, the capacitor C_1 acts as the primary means of storing and transferring energy from the input to the output.

In steady state, the average inductor voltages V_{L1} and V_{L2} are zero. Therefore, by inspection of Fig. 5-25

$$V_{C1} = V_d + V_o \tag{5-58}$$

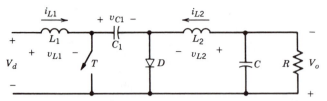

FIGURE 5-25: Cúk converter.

Therefore, V_{C1} is larger than both V_d and V_o. Assuming C_1 to be sufficiently large, in steady state the variation in v_{C1} from its average value V_{C1} can be assumed to be negligibly small (i.e., $v_{C1} \simeq V_{C1}$), even though it stores and transfers energy from the input to the output.

When the switch is off, the inductor currents i_{L1} and i_{L2} flow through the diode. The circuit is shown in Fig. 5-26a. C_1 is charged through the diode by energy from both the input and L_1. i_{L1} decreases, because V_{C1} is larger than V_d. Energy stored in L_2 feeds the output. Therefore, i_{L2} also decreases.

When the switch is on, V_{C1} reverse biases the diode. The inductor currents i_{L1} and i_{L2} flow through the switch as shown in Fig. 5-26b. Since $V_{C1} > V_o$, C_1 discharges through the switch, transferring energy to the output and L_2. Therefore, i_{L2} increases. The input feeds energy to L_1 causing i_{L1} to increase.

The inductor currents i_{L1} and i_{L2} are assumed to be continuous. The voltage and current expressions in steady state can be obtained in two different ways.

If we assume the capacitor voltage V_{C1} to be constant, then equating the integral of the voltages across L_1 and L_2 over one time period to zero:

$$L_1: \qquad V_d D T_s + (V_d - V_{C1})(1 - D)T_s = 0$$

$$\therefore V_{C1} = \frac{1}{1 - D} V_d \tag{5-59}$$

$$L_2: \qquad (V_{C1} - V_o)D T_s + (-V_o)(1 - D)T_s = 0$$

$$\therefore V_{C1} = \frac{1}{D} V_o \tag{5-60}$$

From Eqs. 5-59 and 5-60

$$\frac{V_o}{V_d} = \frac{D}{1 - D} \tag{5-61}$$

Assuming $P_d = P_o$

$$\frac{I_o}{I_d} = \frac{1 - D}{D} \tag{5-62}$$

where $I_{L1} = I_d$ and $I_{L2} = I_o$.

There is another way to obtain these expressions. Assume that the inductor currents i_{L1} and i_{L2} are essentially ripple-free (i.e., $i_{L1} = I_{L1}$ and $i_{L2} = I_{L2}$). When the switch is off, the charge delivered to C_1 equals $I_{L1}(1 - D)T_s$. When the switch is on, the capacitor discharges by an amount $I_{L2}D T_s$. Since, in steady state the net

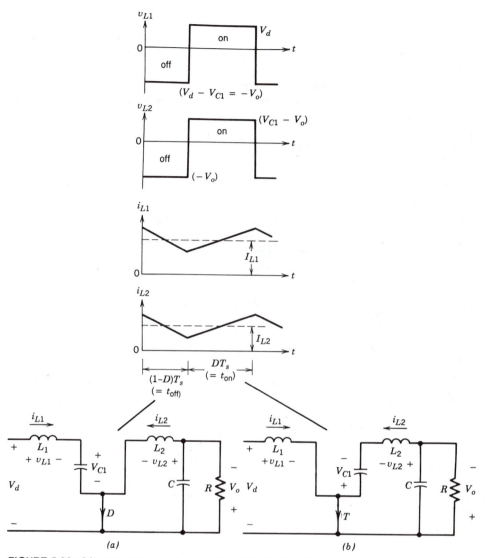

FIGURE 5-26: Cúk converter waveforms: (*a*) switch off, (*b*) switch on.

change of charge associated with C_1 over one time period must be zero

$$I_{L1}(1 - D)T_s = I_{L2}DT_s \tag{5-63}$$

$$\therefore \frac{I_{L2}}{I_{L1}} = \frac{I_o}{I_d} = \frac{1 - D}{D} \tag{5-64}$$

and

$$\frac{V_o}{V_d} = \frac{D}{1 - D} \quad (\text{since } P_o = P_d) \tag{5-65}$$

Both methods of analysis yield identical results. The average input and output relations are similar to that of a buck–boost converter.

In practical circuits, the assumption of a nearly constant V_{C1} is reasonably valid. An advantage of this circuit is that both the input current and the current feeding the output stage are reasonably ripple-free (unlike the buck–boost converter where both these currents are highly discontinuous). It is possible to simultaneously eliminate the ripples in i_{L1} and i_{L2} completely, leading to lower external filtering requirements. A significant disadvantage is the requirement of a capacitor C_1 with a large ripple-current carrying capability.

Detailed analysis of this converter has been adequately reported in the technical literature and will not be undertaken here.

EXAMPLE 5-3

In a Cúk converter operating at 50 kHz, $L_1 = L_2 = 1$ mH and $C_1 = 5$ μF. The output capacitor is sufficiently large to yield an essentially constant output voltage. $V_d = 10$ V and the output V_o is regulated to be constant at 5 V. It is supplying 5 W to a load. Assume ideal components.

Calculate the percentage errors in assuming a constant voltage across C_1 or in assuming constant currents i_{L1} and i_{L2}.

SOLUTION

(a) If the voltage across C_1 is assumed to be constant, from Eq. 5-58

$$v_{C1} = V_{C1} = 10 + 5 = 15 \text{ V}$$

Initially, we will assume the current conduction to be continuous. Therefore, from Eq. 5-61

$$\frac{D}{1 - D} = \frac{5}{10}$$

$$\therefore D = 0.333$$

Therefore, from Fig. 5-26 during the off interval

$$\Delta i_{L1} = \frac{V_{C1} - V_d}{L_1}(1 - D)T_s$$

$$= \frac{(15 - 10)}{10^{-3}}(1 - 0.333) \times 20 \times 10^{-6}$$

$$= 0.067 \text{ A}$$

and

$$\Delta i_{L2} = \frac{V_o}{L_2}(1 - D)T_s$$

$$= \frac{5}{10^{-3}}(1 - 0.333) \times 20 \times 10^{-6}$$

$$= 0.067 \text{ A}$$

Note that Δi_{L1} and Δi_{L2} would be equal (since $V_{C1} - V_d = V_o$), provided that $L_1 = L_2$.

At an output load of 5 W, using Eq. 5-62

$$I_o = 1 \text{ A} \quad \text{and} \quad I_d = 0.5 \text{ A}$$

Since $\Delta i_{L1} < I_d \, (= I_{L1})$ and $\Delta i_{L2} < I_o \, (= I_{L2})$, the mode of operation is continuous, as assumed earlier.

Therefore, the percentage errors in assuming constant I_{L1} and I_{L2} would be

$$\frac{\Delta i_{L1}}{I_{L1}} = \frac{0.067 \times 100}{0.5} = 13.4\%$$

and

$$\frac{\Delta i_{L2}}{I_{L2}} = \frac{0.067 \times 100}{1.0} = 6.7\%$$

(b) If i_{L1} and i_{L2} are assumed to be constant, from Fig. 5-26 during the off interval

$$\Delta V_{C1} = \frac{1}{C} \int_0^{(1-D)T_s} i_{L1} \, dt$$

Assuming

$$i_{L1} = I_{L1} = 0.5 \text{ A}$$

$$\Delta V_{C1} = \frac{1}{5 \times 10^{-6}} \times 0.5 \times (1 - 0.333)20 \times 10^{-6}$$

$$= 1.33 \text{ V}$$

Therefore, the percentage error in assuming a constant voltage across C_1 is

$$\frac{\Delta V_{C1}}{V_{C1}} = \frac{1.33 \times 100}{15} = 8.87\%$$

This example shows that as a first approximation to illustrate the principle of operation it is reasonable to assume either a constant v_{C1} or constant i_{L1} and i_{L2} in this problem. ∎

5-7 FULL-BRIDGE dc–dc CONVERTER

There are three distinct applications of the full-bridge switch-mode converters shown in Fig 5-27:

- dc-motor drives
- dc-to-ac (sine-wave) conversion in single-phase uninterruptible ac power supplies
- dc-to-ac (high intermediate frequency) conversion in switch-mode transformer-isolated dc power supplies

Even though the full-bridge topology remains the same in each of these three applications, the type of control depends on the application. However, the full-bridge

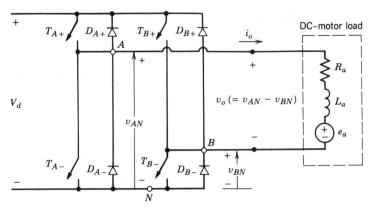

FIGURE 5-27: Full-bridge dc – dc converter.

converter, as used in dc-motor drives, is covered in this "generic" chapter because it provides a good basis for understanding the switch-mode dc-to-ac (sine-wave) converters of Chapter 6.

In the full-bridge converter shown in Fig. 5-27, the input is a fixed magnitude dc voltage V_d. The output of the converter in this chapter is a dc voltage V_o, which can be controlled in magnitude as well as polarity.

In a converter topology such as that of a full-bridge converter shown in Fig. 5-27 where diodes are connected in antiparallel with the switches, a distinction must be made between the on-state versus the conducting-state of a switch. Because of the diodes in antiparallel with the switches, when a switch is turned on, it may or may not conduct a current, depending on the direction of the output current i_o. If the switch conducts a current, then it is in a conducting state. No such distinction is required when the switch is turned off.

The full-bridge converter consists of two legs, A and B. Each leg consists of two switches and their antiparallel diodes. The two switches in each leg are switched in such a way that when one of them is in its off state, the other switch is on. Therefore, the two switches are never off simultaneously. In practice, they are both off for a short time interval, known as the blanking time, to avoid short-circuiting of the dc input. This blanking time is neglected in this chapter since we are assuming the switches to be ideal, capable of turning off instantaneously.

We should note that if the converter switches in each leg are switched in such a way that both the switches in a leg are not off simultaneously, then the output current i_o in Fig. 5-27 will flow continuously. Therefore, the output voltage is solely dictated by the status of the switches. For example, consider leg A in Fig. 5-27. The output voltage v_{AN}, with respect to the negative dc bus N, is dictated by the switch states as follows: when T_{A+} is on, the output current will flow through T_{A+} if i_o is positive, or it will flow through D_{A+} if i_o is negative. In either case, T_{A+} being on ensures that point A in Fig. 5-27 is at the same potential as the positive terminal of the dc input and therefore

$$v_{AN} = V_d \quad (\text{if } T_{A+} \text{ is on and } T_{A-} \text{ is off}) \tag{5-66a}$$

Similarly, when T_{A-} is on, a negative i_o will flow through T_{A-} (since D_{A+} is reverse biased) and a positive i_o will flow through D_{A-}. Therefore

$$v_{AN} = 0 \quad (\text{if } T_{A-} \text{ is on and } T_{A+} \text{ is off}) \tag{5-66b}$$

Equations 5-66a and 5-66b show that v_{AN} depends only on the switch status and is independent of the direction of i_o. Therefore, the output voltage of the converter leg A, averaged over one switching frequency time period T_s, depends only on the input voltage V_d and the duty ratio of T_{A+}:

$$V_{AN} = \frac{V_d \cdot t_{\text{on}} + 0 \cdot t_{\text{off}}}{T_s} = V_d \cdot \text{duty ratio of } T_{A+} \qquad (5\text{-}67)$$

where t_{on} and t_{off} are the on and off intervals of T_{A+}, respectively.

Similar arguments apply to the converter leg B and V_{BN} depends on V_d and the duty ratio of the switch T_{B+}:

$$V_{BN} = V_d \cdot \text{duty ratio of } T_{B+} \qquad (5\text{-}68)$$

independent of the direction of i_o. Therefore, the converter output V_o $(= V_{AN} - V_{BN})$ can be controlled by controlling the switch duty ratios and is independent of the magnitude and the direction of i_o.

(It should be noted that it is possible to control the output voltage of a converter leg by turning both switches off simultaneously for some time interval. However, this scheme would make the output voltage dependent on the direction of i_o in Fig. 5-27. This is obviously undesirable, since it would introduce nonlinearity in the relationship between the control voltage and the average output voltage. Therefore, such a scheme will not be considered with a load, such as a dc-motor load, shown in Fig. 5-27.)

In the single-switch converters discussed previously, the polarity of the output voltage is unidirectional and, therefore, the converter switch is pulse-width modulated by comparing a switching-frequency sawtooth waveform with the control voltage v_{control}. In contrast, the output voltage of the full-bridge converter is reversible in polarity and therefore, a switching-frequency triangular waveform is used for pulse-width modulation (PWM) of the converter switches. Two such PWM switching strategies are described below:

1. *PWM with bipolar voltage switching*, where (T_{A+}, T_{B-}) and (T_{A-}, T_{B+}) are treated as two switch pairs; switches in each pair are turned on and off simultaneously.

2. *PWM with unipolar voltage switching* is also referred to as the Double PWM Switching. Here the switches in each inverter leg are controlled independently of the other leg.

As we mentioned earlier, the output current through these PWM full-bridge dc–dc converters, while supplying dc loads of the type shown in Fig. 5-27, does not become discontinuous at low values of I_o, unlike the single-switch converters discussed in the previous sections.

5-7-1 PWM With Bipolar Voltage Switching

In this type of voltage switching, switches (T_{A+}, T_{B-}) and (T_{B+}, T_{A-}), are treated as two switch pairs (two switches in a pair are simultaneously turned on and off). One of the two switch pairs is always on.

The switching signals are generated by comparing a switching-frequency triangular waveform (v_{tri}) with the control voltage v_{control}. When $v_{\text{control}} > v_{\text{tri}}$, T_{A+} and T_{B-} are turned on. Otherwise, T_{A-} and T_{B+} are turned on. The switch duty ratios can be obtained from the waveforms in Fig. 5-28a as follows by arbitrarily choosing a

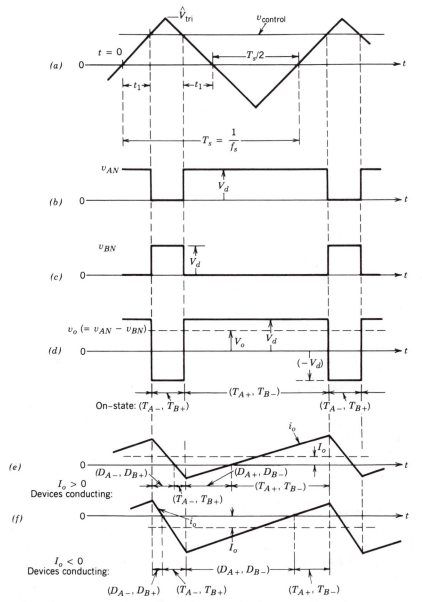

FIGURE 5-28: PWM with bipolar voltage switching.

time origin as shown in the figure:

$$v_{\text{tri}} = \hat{V}_{\text{tri}} \frac{t}{T_s/4} \qquad 0 < t < \frac{T_s}{4} \tag{5-69}$$

At $t = t_1$ in Fig. 5-28a, v_{tri} equals v_{control}. Therefore, from Eq. 5-69

$$t_1 = \frac{v_{\text{control}}}{\hat{V}_{\text{tri}}} \cdot \frac{T_s}{4} \tag{5-70}$$

By studying Fig. 5-28, we find that the on duration t_{on} of switch pair 1 (T_{A+}, T_{B-}) is

$$t_{on} = 2t_1 + T_s/2 \qquad (5\text{-}71)$$

Therefore, their duty ratio from Eq. 5-71 is

$$D_1 = \frac{t_{on}}{T_s} = \frac{1}{2}\left(1 + \frac{v_{control}}{\hat{V}_{tri}}\right) \qquad (T_{A+}, T_{B-}) \qquad (5\text{-}72)$$

Therefore, the duty ratio D_2 of the switch pair 2 (T_{B+}, T_{A-}) is

$$D_2 = 1 - D_1 \qquad (T_{B+}, T_{A-}) \qquad (5\text{-}73)$$

By using the foregoing duty ratios, we can obtain V_{AN} and V_{BN} in Fig. 5-28 from Eqs. 5-67 and 5-68, respectively. Therefore,

$$V_o = V_{AN} - V_{BN} = D_1 V_d - D_2 V_d = (2D_1 - 1)V_d \qquad (5\text{-}74)$$

Substituting D_1 from Eq. 5-72 into Eq. 5-74

$$V_o = \frac{V_d}{\hat{V}_{tri}} v_{control} = k v_{control} \qquad (5\text{-}75)$$

where $k = V_d/\hat{V}_{tri}$ = constant. This equation shows that in this switch-mode converter, similar to the single-switch converters discussed previously, the average output voltage varies linearly with the input control signal, similar to a linear amplifier. We will learn in Chapter 6 that a finite blanking time has to be used between turning off one switch pair and turning on the other switch pair. This blanking time introduces a slight nonlinearity in the relationship between $v_{control}$ and V_o.

The waveform for the output voltage v_o in Fig. 5-28d shows that the voltage jumps between $+V_d$ and $-V_d$. This is the reason why this switching strategy is referred to as the "bipolar voltage switching PWM."

We should also note that the duty ratio D_1 in Eq. 5-72 can vary between 0 and 1, depending on the magnitude and the polarity of $v_{control}$. Therefore, V_o can be continuously varied in a range from $-V_d$ to V_d. Here, the output voltage of the converter is independent of the output current i_o, since the blanking time has been neglected.

The average output current I_o can be either positive or negative. For small values of I_o, i_o during a cycle can be both positive and negative; this is shown in Fig. 5-28e for $I_o > 0$ where the average power flow is from V_d to V_o, and in Fig. 5-28f for $I_o < 0$ where the average power flow is from V_o to V_d.

5-7-2 PWM With Unipolar Voltage Switching

An inspection of Fig. 5-27 shows that regardless of the direction of i_o, $v_o = 0$ if T_{A+} and T_{B+} are both on. Similarly, $v_o = 0$, if T_{A-} and T_{B-} are both on. This property can be exploited to improve the output voltage waveform.

In Fig. 5-29, a triangular waveform is compared with the control voltage $v_{control}$ and $-v_{control}$ for determining the switching signals for leg A and leg B, respectively. A comparison of $v_{control}$ with v_{tri} controls leg A switches, whereas leg B switches are

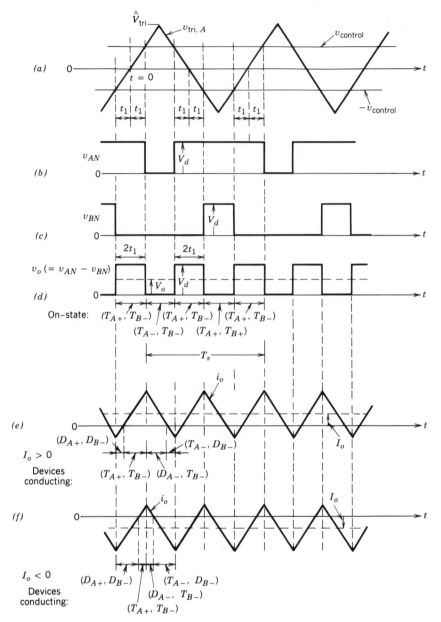

FIGURE 5-29: PWM with unipolar voltage switching.

controlled by comparing $-v_{\text{control}}$ with v_{tri} in the following manner:

$$T_{A+}: \text{on}; \quad \text{if } v_{\text{control}} > v_{\text{tri}} \tag{5-76}$$

and

$$T_{B+}: \text{on}; \quad \text{if } -v_{\text{control}} > v_{\text{tri}} \tag{5-77}$$

Output voltages of each leg and v_o are shown in Fig. 5-29. By examining Fig. 5-29 and comparing it to Fig. 5-28, it can be seen that the duty ratio D_1 of the switch T_{A+} is given by Eq. 5-72 of the previous switching strategy. Similarly, the duty ratio D_2 of the switch T_{B+} is given by Eq. 5-73. That is

$$D_1 = \frac{1}{2}\left(\frac{v_{control}}{\hat{V}_{tri}} + 1\right) \qquad T_{A+} \tag{5-78}$$

and

$$D_2 = 1 - D_1 \qquad T_{B+} \tag{5-79}$$

Therefore, from Eq. 5-74, which is also valid in this case

$$V_o = (2D_1 - 1)V_d = \frac{V_d}{\hat{V}_{tri}}v_{control} \tag{5-80}$$

Therefore, the average output voltage V_o in this switching scheme is the same as in the bipolar-voltage switching scheme and varies linearly with $v_{control}$.

Figures 5-29e and 5-29f show the current waveforms and the devices conducting for $I_o > 0$ and $I_o < 0$, respectively, where V_o is positive in both cases.

If the switching frequencies of the switches are the same in these two PWM strategies, then the unipolar voltage switching results in a better output voltage waveform and in a better frequency response, since the "effective" switching frequency of the output voltage waveform is doubled and the ripple is reduced. This is illustrated by means of the following example.

EXAMPLE 5-4

In a full-bridge dc–dc converter, the input V_d is constant and the output voltage is controlled by varying the duty ratio. Calculate the rms value of the ripple V_r in the output voltage as a function of the average V_o for

(a) PWM with bipolar voltage switching and
(b) PWM with unipolar voltage switching.

SOLUTION

(a) Using PWM with bipolar voltage switching, the waveform of the output voltage v_o is shown in Fig. 5-28d. For such a waveform, independent of the value of $v_{control}/\hat{V}_{tri}$, the rms value of the output voltage is

$$V_{o,rms} = V_d \tag{5-81}$$

and the average value V_o is given by Eq. 5-74.

By using the definition of an rms value and Eqs. 5-74 and 5-81, we can calculate the ripple component V_r in the output voltage as

$$V_{r,rms} = \sqrt{V_{o,rms}^2 - V_o^2} = V_d\sqrt{1 - (2D_1 - 1)^2} = 2V_d\sqrt{D_1 - D_1^2} \tag{5-82}$$

As D_1 varies from 0 to 1, V_o varies from $-V_d$ to V_d. Plot of $V_{r,rms}$ as a function of V_o is shown in Fig. 5-30 by the solid curve.

(b) Using PWM with unipolar voltage switching, the waveform of the output voltage v_o is shown in Fig. 5-29d. To find the rms value of the output voltage, the interval t_1 in Fig. 5-29 can be written as

$$t_1 = \frac{v_{control}}{\hat{V}_{tri}} \cdot \frac{T_s}{4} \qquad \text{for} \quad v_{control} > 0 \tag{5-83}$$

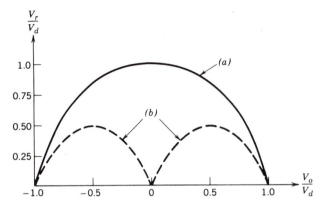

FIGURE 5-30: V_r in a full-bridge converter using PWM: (a) with bipolar voltage switching, (b) with unipolar voltage switching.

By examining the v_o waveform in Fig. 5-29d for $v_{control} > 0$, the rms voltage can be obtained as

$$V_{o,\,rms} = \sqrt{\frac{4t_1 V_d^2}{T_s}}$$

$$= \sqrt{\frac{v_{control}}{\hat{V}_{tri}}}\, V_d$$

$$= \sqrt{(2D_1 - 1)}\, V_d \quad \text{(using Eq. 5-80)} \tag{5-84}$$

Therefore, using Eqs. 5-80 and 5-84

$$V_{r,\,rms} = \sqrt{V_{o,\,rms}^2 - V_o^2} = \sqrt{6D_1 - 4D_1^2 - 2}\, V_d \tag{5-85}$$

where $v_{control} > 0$ and, $0.5 < D_1 < 1$. As $v_{control}/\hat{V}_{tri}$ varies from 0 to 1, D_1 varies from 0.5 to 1.0 and the plot of $V_{r,\,rms}$ as a function of V_o is shown by the dotted curve in Fig. 5-30. Similarly, the curve corresponding to $v_{control}/\hat{V}_{tri}$ in the range from -1.0 to 0 can be obtained.

Figure 5-30 shows that with the same switching frequency, PWM with unipolar voltage switching results in a lower rms ripple component in the output voltage. ∎

5-8   dc-dc CONVERTER COMPARISON

The step-down, step-up, buck–boost, and Cúk converters in their basic forms are capable of transferring energy only in one direction. This is a consequence of their capability to produce only unidirectional voltage and unidirectional current. A full-bridge converter is capable of a bidirectional power flow, where both V_o and I_o can be reversed independent of one another. This capability to operate in four-quadrants of V_o–I_o plane allows a full-bridge converter to be used as a dc-to-ac

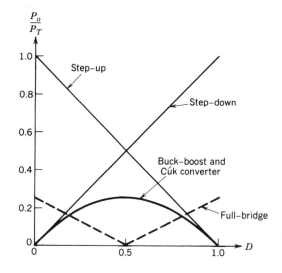

FIGURE 5-31: Switch utilization in dc-dc converters.

inverter. In dc-to-ac inverters, which are described in Chapter 6, the full-bridge converter operates in all four quadrants during each cycle, of the ac output.

To evaluate how well the switch is utilized in the previously discussed converter circuits, we make the following assumptions:

1. The average current is at its rated (designed maximum) value I_o. The ripple in the inductor current is negligible; therefore $i_L(t) = I_L$. This condition implies a continuous-conduction mode for all converters.

2. The output voltage v_o is at its rated (designed maximum) value V_o. The ripple in v_o is assumed to be negligible; therefore $v_o(t) = \hat{V}_o$.

3. The input voltage V_d is allowed to vary. Therefore, the switch duty ratio must be controlled to hold V_o constant.

With the foregoing steady-state operating conditions, the switch peak voltage rating V_T and the peak current rating I_T are calculated. The switch power rating is calculated as $P_T = V_T \cdot I_T$. The switch utilization is expressed as P_o/P_T where $P_o = V_o I_o$ is the rated output power.

In Fig. 5-31, the switch utilization factor P_o/P_T is plotted for the previously considered converters. This shows that in the step-down and the step-up converters, if the input and the output voltages are of the same order of magnitude, then the switch utilization is very good. In the buck–boost and the Cúk converter, the switch is poorly utilized. The maximum switch utilization of 0.25 is realized at $D = 0.5$, which corresponds to $V_o = V_d$.

In the nonisolated full-bridge converter, the switch-utilization factor is plotted as a function of the duty ratio of one of the switches (for example, switch T_{A+} in Fig. 5-27). Here, the overall switch utilization is also poor. It is maximum at $V_o = -V_d$ and $V_o = V_d$, respectively.

In conclusion, for these *nonisolated* dc–dc converters, if at all possible, it is preferable to use either the step-down or the step-up converter from the switch utilization consideration. If both higher as well as lower output voltages compared to the input are necessary, or a negative polarity output compared to the input is desired, then the buck–boost or the Cúk converter should be used. Similarly, the nonisolated full-bridge converter should only be used if four-quadrant operation is required.

5-9   SUMMARY

In this chapter various nonisolated converter topologies for dc-to-dc conversion are discussed. Except for the full-bridge converter, all others operate in a single quadrant of the output voltage-current plane, thereby allowing power to flow only in one direction.

PROBLEMS

Step-Down Converters

5-1. In a step-down converter, consider all components to be ideal. $v_o \simeq V_o$ is held constant at 5 V by controlling the switch duty ratio D. Calculate the minimum inductance L required to keep the converter operation in a continuous conduction mode under all conditions if $V_d = 10$ V to 40 V, $P_o \geq 5$ W, and $f_s = 50$ kHz.

5-2. Consider all components to be ideal. $V_o = 5$ V, $f_s = 20$ kHz, $L = 1$ mH, and $C = 470$ μF. Calculate ΔV_o (peak-peak), if $V_d = 12.6$ V and $I_o = 200$ mA.

5-3. In Problem 5-2, calculate the rms value of the ripple current through L and, hence, through C.

5-4. Derive an expression for ΔV_o (peak-peak) in a discontinuous-conduction mode in terms of the circuit parameters.

5-5. In Problem 5-2, calculate ΔV_o (peak-peak) if I_o (instead of being 200 mA) is equal to $I_{oB}/2$.

Step-Up Converters

5-6. In a step-up converter, consider all components to be ideal. $V_d = 8$ V to 16 V, $V_o = 24$ V (regulated), $f_s = 20$ kHz, and $C = 470$ μF. Calculate L_{min} that will keep the converter operating in a continuous conduction mode if $P_o \geq 5$ W.

5-7. In a step-up converter, $V_d = 12$ V, $V_o = 24$ V, $I_o = 0.5$ A, $L = 150$ μH, $C = 470$ μF, and $f_s = 20$ kHz. Calculate ΔV_o (peak-peak).

5-8. In Problem 5-7, calculate the rms value of the ripple in the diode current (which also flows through the capacitor).

5-9. Derive an expression for ΔV_o (peak-peak) in a discontinuous-conduction mode in terms of the circuit parameters.

5-10. In Problem 5-7, calculate ΔV_o peak-peak, if I_o (instead of being 0.5 A) is equal to $I_{oB}/2$.

Buck – Boost Converters

5-11. In a buck–boost converter, consider all components to be ideal. $V_d = 8$ V to 40 V, $V_o = 15$ V (regulated), $f_s = 20$ kHz, and $C = 470$ μF. Calculate L_{min} that will keep the converter operating in a continuous-conduction mode if $P_o \geq 2$ W.

5-12. In a buck–boost converter, $V_d = 12$ V, $V_o = 15$ V, $I_o = 250$ mA, $L = 150$ μH, $C = 470$ μF, and $f_s = 20$ kHz. Calculate ΔV_o (peak-peak).

5-13. Calculate the rms value of the ripple current in Problem 5-12 through the diode and, hence, through the capacitor.

5-14. Derive an expression for ΔV_o (peak-peak) in a discontinuous-conduction mode in terms of the circuit parameters.

5-15. In Problem 5-12, calculate ΔV_o (peak-peak), if I_o (instead of being 250 mA) is equal to $I_{oB}/2$.

Cúk Converter

5-16. In the circuit of Example 5-3, calculate the rms current flowing through the capacitor C_1.

Full-Bridge dc–dc Converters

5-17. In a full-bridge dc–dc converter using PWM bipolar voltage switching, $v_{control} = 0.5\,\hat{V}_{tri}$. Obtain V_o and by Fourier analysis, calculate the amplitudes of the switching frequency harmonics in v_o.

5-18. Repeat Problem 5-17 for a PWM unipolar voltage switching scheme.

5-19. Plot instantaneous power output $p_o(t)$ and the average power P_o, corresponding to i_o in Figs. 5-28e and 5-28f.

5-20. Repeat Problem 5-19 for Figs. 5-29e and 5-29f.

5-10 REFERENCES

1. R. P. Severns and Ed Bloom, *Modern DC-to-DC Switchmode Power Converter Circuits*, Van Nostrand Reinhold Company, New York, 1985.
2. G. Chryssis, *High Frequency Switching Power Supplies: Theory and Design*, McGraw-Hill, New York, 1984.
3. R. E. Tarter, *Principles of Solid-State Power Conversion*, Sams and Co., Indianapolis, IN, 1985.
4. K. Kit Sum, *SWITCH-MODE Power Conversion: Basic Theory and Design*, Marcel Dekker, Inc., New York, 1984.
5. R. D. Middlebrook and S. Cúk, *Advances in Switched-Mode Power Conversion*, Vol. I and II, TESLAco, 490 S. Rosemead Blvd., Suite 6, Pasadena, CA 91107, 1981.

6

Switch-Mode DC-to-AC Inverters: DC ↔ Sinusoidal AC

6-1 INTRODUCTION

Switch-mode dc-to-ac inverters are used in ac-motor drives and uninterruptible ac power supplies where the objective is to produce a sinusoidal ac output whose magnitude and frequency can both be controlled. As an example, consider an ac-motor drive, shown in Fig. 6-1 in a block diagram form. The dc voltage is obtained by rectifying and filtering the line voltage, most often by the diode-rectifier circuits discussed in Chapter 3. In an ac-motor load, as will be discussed in Chapters 12 and 13, the voltage at its terminals is desired to be sinusoidal and adjustable in its magnitude and frequency. This is accomplished by means of the switch-mode dc-to-ac inverter of Fig. 6-1, which accepts a dc voltage as the input and produces the desired ac voltage output.

To be precise, the switch-mode inverter in Fig. 6-1 is a converter through which the power flow is reversible. However, most of the time the power flow is from the dc side to the motor on the ac side, requiring an inverter mode of operation. Therefore, these switch-mode converters are often referred to as switch-mode inverters.

To slow down the ac motor in Fig. 6-1, the kinetic energy associated with the inertia of the motor and its load is recovered and the ac motor acts as a generator. During the so-called braking of the motor, the power flows from the ac side to the dc side of the switch-mode converter and it operates in a rectifier mode. The energy recovered during the braking of the ac motor can be dissipated in a resistor, which can be switched in parallel with the dc-bus capacitor for this purpose in Fig. 6-1. However, in applications where this braking is performed frequently, a better alternative is regenerative braking where the energy recovered from the motor-load inertia is fed back to the utility grid, as shown in the system of Fig. 6-2. This requires that the converter connecting the drive to the utility grid be a two-quadrant converter with a reversible dc current, which can operate as a rectifier during the motoring mode of the

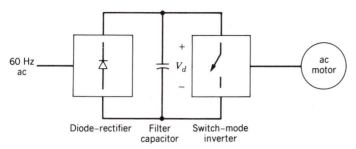

FIGURE 6-1: Switch-mode inverter in ac-motor drive.

ac motor, and as an inverter during the braking of the motor. Such a reversible-current two-quadrant converter can be realized by two back-to-back connected line-frequency thyristor converters of the type discussed in Chapter 4 or by means of a switch-mode converter as shown in Fig. 6-2. There are other reasons for using such a switch-mode rectifier (called a rectifier because most of the time, the power flows from the ac line input to the dc bus) to interface the drive with the utility system. A detailed discussion of switch-mode rectifiers is deferred to Chapter 17, which deals with issues regarding the interfacing of power electronics equipment with the utility grid.

In this chapter, we will discuss inverters with single-phase and three-phase ac outputs. The input to switch-mode inverters will be assumed to be a dc-voltage source as was assumed in the block diagrams of Figs. 6-1 and 6-2. Such inverters are referred to as voltage-source inverters (VSI). The other type of inverters, currently used only for very-high-power ac-motor drives, are the current-source inverters (CSI) where the dc input to the inverter is a dc-current source. Because of their limited applications, the current-source inverters are not discussed in this chapter, and their discussion is deferred to ac-motor drives Chapters 12 and 13.

The voltage-source inverters can be further divided into the following three general categories:

1. *Pulse-width Modulated (PWM) Inverters.* In these inverters, the input dc voltage is essentially constant in magnitude, such as in the circuit of Fig. 6-1, where a diode-rectifier is used to rectify the line voltage. Therefore, the inverter must control the magnitude and the frequency of the ac output voltages. This is achieved by pulse-width modulation (PWM) of the inverter switches and hence such inverters are called PWM inverters. There are various schemes to pulse-width modulate the inverter switches in order to shape the output ac voltages to be as close to a sine wave as possible. Out of these various PWM schemes, a

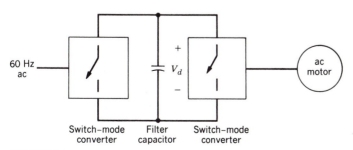

FIGURE 6-2: Switch-mode converters for motoring and regenerative braking in ac-motor drive.

scheme called the sinusoidal-PWM will be discussed in detail and some of the other PWM techniques will be described in a separate section at the end of this chapter.

2. *Square-wave Inverters.* In these inverters, the input dc voltage is controlled in order to control the magnitude of the output ac voltage, and therefore the inverter has to control only the frequency of the output voltage. The output ac voltage has a waveform similar to a square wave and hence these inverters are called square-wave inverters.

3. *Single-phase Inverters with Voltage Cancellation.* In case of inverters with single-phase output, it is possible to control the magnitude and the frequency of the inverter output voltage, even though the input to the inverter is a constant dc voltage and the inverter switches are not pulse-width modulated (and hence the output voltage waveshape is like a square wave). Therefore, these inverters combine the characteristics of the previous two inverters. It should be noted the voltage-cancellation technique works only with single-phase inverters and not with three-phase inverters.

6-2   BASIC CONCEPTS OF SWITCH-MODE INVERTERS

In this section, we will consider the requirements on the switch-mode inverters. For simplicity, let us consider a single-phase inverter, which is shown in block-diagram form in Fig. 6-3a, where the output voltage of the inverter is filtered so that v_o can be assumed to be sinusoidal. Since the inverter supplies an inductive load such as an ac motor, i_o will lag v_o, as shown in Fig. 6-3b. The output waveforms of Fig. 6-3b show that during interval 1, v_o and i_o are both positive, whereas during interval 3, v_o and i_o are both negative. Therefore, during intervals 1 and 3, the instantaneous power flow $p_o(= v_o \cdot i_o)$ is from the dc side to the ac side, corresponding to an inverter-mode of operation. In contrast, v_o and i_o are of opposite signs during intervals 2 and 4, and therefore p_o flows from the ac side to the dc side of the inverter, corresponding to a rectifier mode of operation. Therefore, the switch-mode inverter of Fig. 6-3a must be capable of operating in all four quadrants of the i_o-v_o plane as shown in Fig. 6-3c during each cycle of the ac output. Such a four-quadrant inverter was first introduced in Chapter 5 where it was shown that in a full-bridge converter of Fig. 5-27, i_o is reversible and v_o can be of either polarity independent of the direction of i_o. Therefore, the full-bridge converter of Fig. 5-27 meets the switch-mode inverter requirements. Only one of the two legs of full-bridge converter, for example leg A, is shown in Fig. 6-4. All the dc-to-ac inverter topologies described in this chapter are derived from the one-leg converter of Fig. 6-4. For the ease of explanation, it will be assumed that in the inverter of Fig. 6-4, the midpoint "o" of the dc input voltage is available, although in most inverters it is not needed and also not available.

To understand the dc-to-ac inverter characteristics of the one-leg inverter of Fig. 6-4, we will first assume that the input dc voltage V_d is constant and that the inverter switches are pulse-width modulated to shape and control the output voltage. Later on, it will be shown that the square-wave switching is a special case of the PWM-switching scheme.

6-2-1   Pulse-Width Modulated (PWM) Switching Scheme

We discussed the pulse-width modulation of full-bridge dc–dc converters in Chapter 5. There, a control signal $v_{control}$ (constant or slowly varying in time) was compared with a repetitive switching frequency triangular waveform, in order to

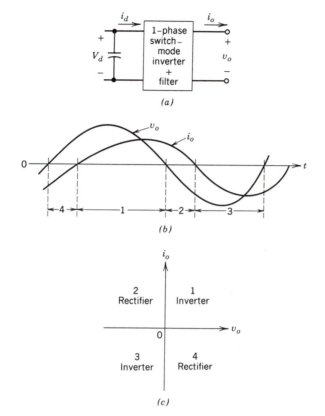

FIGURE 6-3: Single-phase switch-mode inverter.

generate the switching signals. Controlling the switch duty ratios in this way allowed the average dc voltage output to be controlled.

In inverter circuits, the PWM is a bit more complex, since as mentioned earlier, we would like the inverter output to be sinusoidal whose magnitude and frequency should both be controllable. In order to produce a sinusoidal output voltage waveform at a desired frequency, a sinusoidal control signal at the desired frequency is compared with a triangular waveform, as shown in Fig. 6-5a. The frequency of the triangular waveform establishes the inverter switching frequency and is generally kept constant along with its amplitude $\hat{V}_{\text{tri}}$.

Before discussing the PWM behavior, it is necessary to define a few terms. The triangular waveform v_{tri} in Fig. 6-5a is at a switching frequency f_s, which establishes the frequency with which the inverter switches are switched (f_s is also called the carrier frequency). The control signal v_{control} is used to modulate the switch duty ratio and has a frequency f_1, which is the desired fundamental frequency of the inverter

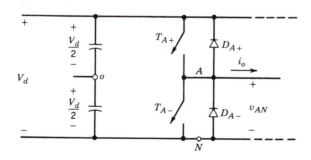

FIGURE 6-4: One-leg switch-mode inverter.

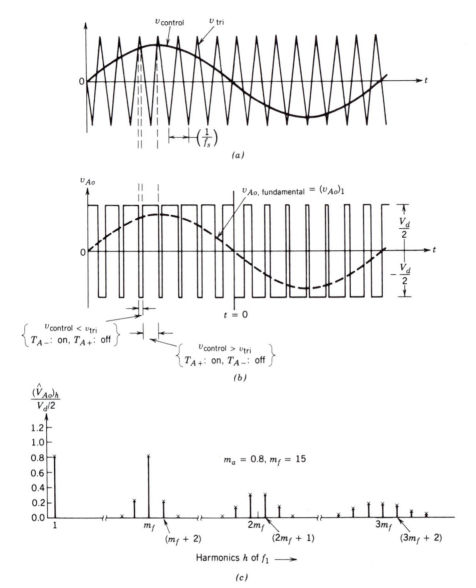

FIGURE 6-5: Pulse-width modulation (PWM).

voltage output (f_1 is also called the modulating frequency), recognizing that the inverter output voltage will not be a perfect sine wave and will contain voltage components at harmonic frequencies of f_1. The amplitude modulation ratio m_a is defined as

$$m_a = \frac{\hat{V}_{control}}{\hat{V}_{tri}} \tag{6-1}$$

where $\hat{V}_{control}$ is the peak amplitude of the control signal. The amplitude $\hat{V}_{tri}$ of the triangular signal is generally kept constant.

The frequency-modulation ratio m_f is defined as

$$m_f = \frac{f_s}{f_1} \qquad (6\text{-}2)$$

In the inverter of Fig. 6-4b, the switches T_{A+} and T_{A-} are controlled based on the comparison of $v_{control}$ and v_{tri}, and the following output voltage results, independent of the direction of i_o:

$$v_{control} > v_{tri}, \qquad T_{A+} \text{ is on}, \qquad v_{Ao} = \frac{V_d}{2}$$

or   (6-3)

$$v_{control} < v_{tri}, \qquad T_{A-} \text{ is on}, \qquad v_{Ao} = -\frac{V_d}{2}$$

Since the two switches are never off simultaneously, the output voltage v_{Ao} fluctuates between two values ($V_d/2$ and $-V_d/2$). v_{Ao} and its fundamental frequency component (dotted curve) are shown in Fig. 6-5b, which are drawn for $m_f = 15$ and $m_a = 0.8$.

The harmonic spectrum of v_{Ao} under the conditions indicated in Figs. 6-5a and 6-5b is shown in Fig. 6-5c, where the normalized harmonic voltages $(\hat{V}_{Ao})_h/(V_d/2)$ having significant amplitudes are plotted. This plot (for $m_a \le 1.0$) shows three items of importance:

1. The peak amplitude of the fundamental frequency component $(\hat{V}_{Ao})_1$ is m_a times $(V_d/2)$. This can be explained by first considering a constant $v_{control}$ as shown in Fig. 6-6a. This results in an output waveform v_{Ao}. From the discussion of Chapter 5 regarding the pulse-width modulation in a full-bridge dc–dc converter, it can be noted that the average output voltage (or more specifically, the output voltage, averaged over one switching time period $T_s = 1/f_s$) V_{Ao} depends

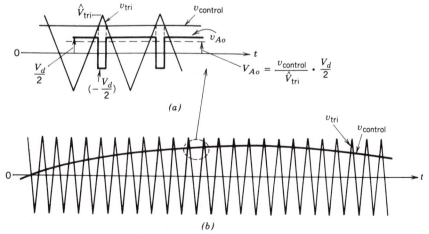

(a)

(b)

FIGURE 6-6: Sinusoidal pulse-width modulation.

on the ratio of $v_{control}$ to $\hat{V}_{tri}$ for a given V_d:

$$V_{Ao} = \frac{v_{control}}{\hat{V}_{tri}} \cdot \frac{V_d}{2} \qquad v_{control} \leq \hat{V}_{tri} \qquad (6\text{-}4)$$

Let us assume (though this assumption is not necessary) that $v_{control}$ varies very little during a switching time period, that is, m_f is large, as shown in Fig. 6-6b. Therefore, assuming $v_{control}$ to be constant over a switching time period, Eq. 6-4 indicates how the "instantaneous-average" value of v_{Ao} (averaged over one switching time period T_s) varies from one switching time period to the next. This "instantaneous-average" is the same as the fundamental frequency component of v_{Ao}.

The foregoing argument shows why $v_{control}$ is chosen to be sinusoidal to provide a sinusoidal output voltage with fewer harmonics. Let the control voltage vary sinusoidally at the frequency $f_1 = \omega_1/2\pi$, which is the desired (or the fundamental) frequency of the inverter output:

$$v_{control} = \hat{V}_{control} \sin \omega_1 t$$

where   (6-5)

$$\hat{V}_{control} \leq \hat{V}_{tri}$$

Using Eqs. 6-4 and 6-5, and the foregoing arguments which show that the fundamental-frequency component $(v_{Ao})_1$ varies sinusoidally and in phase with $v_{control}$ as a function of time:

$$\left(v_{Ao}\right)_1 = \frac{\hat{V}_{control}}{\hat{V}_{tri}} \cdot \sin \omega_1 t \cdot \frac{V_d}{2}$$

$$= m_a \cdot \sin \omega_1 t \cdot \frac{V_d}{2} \quad \left(\text{for } m_a \leq 1.0\right) \qquad (6\text{-}6)$$

Therefore

$$\left(\hat{V}_{Ao}\right)_1 = m_a \cdot \frac{V_d}{2} \quad \left(m_a \leq 1.0\right) \qquad (6\text{-}7)$$

which shows that in a sinusoidal PWM, the amplitude of the fundamental-frequency component of the output voltage varies linearly with m_a (provided $m_a \leq 1.0$). Therefore, the range of m_a from 0 to 1 is referred to as the linear range.

2. The harmonics in the inverter output voltage waveform appear as sidebands, centered around the switching frequency and its multiples, that is, around harmonics m_f, $2m_f$, $3m_f$, and so on. This general pattern holds true for all values of m_a in a range 0 to 1.

For a frequency modulation ratio $m_f \geq 9$ (which is always the case, except in very high power ratings), the harmonic amplitudes are almost independent of m_f, though m_f defines the frequencies at which they occur. Theoretically, the frequencies at which voltage harmonics occur can be indicated as

$$f_h = \left(jm_f \pm k \right) f_1$$

that is, the harmonic order h corresponds to the kth sideband of the j times the frequency-modulation ratio m_f:

$$h = j(m_f) \pm k \tag{6-8}$$

where the fundamental frequency corresponds to $h = 1$. For odd values of j, the harmonics exist only for even values of k. For even values of j, the harmonics exist only for odd values of k.

In Table 6-1, the normalized harmonics $(\hat{V}_{Ao})_h/(V_d/2)$ are tabulated as a function of the amplitude modulation ratio m_a, assuming $m_f \geq 9$. Only those with significant amplitudes up to $j = 4$ in Eq. 6-8 are shown. A detailed discussion is provided in Reference 6.

It will be useful later on to recognize that in the inverter circuit of Fig. 6-4

$$v_{AN} = v_{Ao} + \frac{V_d}{2} \tag{6-9}$$

Therefore, the harmonic voltage components in v_{AN} and v_{Ao} are the same:

$$\left(\hat{V}_{AN} \right)_h = \left(\hat{V}_{Ao} \right)_h \tag{6-10}$$

TABLE 6-1

Generalized Harmonics of v_{Ao} for a Large m_f. $(\hat{V}_{Ao})_h / (V_d / 2) [= (\hat{V}_{AN})_h / (V_d / 2)]$ Is Tabulated as a Function of m_a.

h \ m_a	0.2	0.4	0.6	0.8	1.0
1 fundamental	0.2	0.4	0.6	0.8	1.0
m_f	1.242	1.15	1.006	0.818	0.601
$m_f \pm 2$	0.016	0.061	0.131	0.220	0.318
$m_f \pm 4$					0.018
$2m_f \pm 1$	0.190	0.326	0.370	0.314	0.181
$2m_f \pm 3$		0.024	0.071	0.139	0.212
$2m_f \pm 5$				0.013	0.033
$3m_f$	0.335	0.123	0.083	0.171	0.113
$3m_f \pm 2$	0.044	0.139	0.203	0.176	0.062
$3m_f \pm 4$		0.012	0.047	0.104	0.157
$3m_f \pm 6$				0.016	0.044
$4m_f \pm 1$	0.163	0.157	0.008	0.105	0.068
$4m_f \pm 3$	0.012	0.070	0.132	0.115	0.009
$4m_f \pm 5$			0.034	0.084	0.119
$4m_f \pm 7$				0.017	0.050

Table 6-1 shows that Eq. 6-7 is followed almost exactly and the amplitude of the fundamental component in the output voltage varies linearly with m_a.

3. m_f *should be an odd integer.* Choosing m_f as an odd integer results in an odd symmetry $[f(-t) = -f(t)]$ as well as a half-wave symmetry $[f(t) = -f(t + T_s/2)]$ with the time origin shown in Fig. 6-5*b*, which is plotted for $m_f = 15$. Therefore, only odd harmonics are present and the even harmonics disappear from the waveform of v_{Ao}. Moreover, only the coefficients of the sine series in the Fourier analysis are finite; those for the cosine series are zero. The harmonic spectrum is plotted in Fig. 6-5*c*.

EXAMPLE 6-1

In the circuit of Fig. 6-4, $V_d = 300$ V, $m_a = 0.8$, $m_f = 39$, and the fundamental frequency is 47 Hz. Calculate the rms values of the fundamental frequency voltage and some of the dominant harmonics in v_{Ao}, using Table 6-1.

SOLUTION

From Table 6-1, the rms voltage at any value of h is given as

$$(V_{Ao})_h = \frac{1}{\sqrt{2}} \cdot \frac{V_d}{2} \cdot \frac{(\hat{V}_{Ao})_h}{(V_d/2)}$$

$$= 106.07 \frac{(\hat{V}_{Ao})_h}{(V_d/2)} \qquad (6\text{-}11)$$

Therefore, from Table 6-1 the rms voltages are

fundamental: $(V_{Ao})_1 = 106.07 \times 0.8 = 84.86\ V$ at   47 Hz

$(V_{Ao})_{37} = 106.07 \times 0.22 = 23.33\ V$ at   1739 Hz

$(V_{Ao})_{39} = 106.07 \times 0.818 = 86.76\ V$ at   1833 Hz

$(V_{Ao})_{41} = 106.07 \times 0.22 = 23.33\ V$ at   1927 Hz

$(V_{Ao})_{77} = 106.07 \times 0.314 = 33.31\ V$ at   3619 Hz

$(V_{Ao})_{79} = 106.07 \times 0.314 = 33.31\ V$ at   3713 Hz

etc.   ∎

Now we discuss the selection of the switching frequency and the frequency-modulation ratio m_f. Because of the relative ease in filtering harmonic voltages at high frequencies, it is desirable to use as high a switching frequency as possible, except for one significant drawback: switching losses in the inverter switches increase proportionally with the switching frequency f_s. Therefore, in most applications, the switching frequency is selected to be either less than 6 kHz or greater than 20 kHz to be above the audible range. If the optimum switching frequency (based on the overall system performance) turns out to be somewhere in the 6 to 20 kHz range, then the disadvantages of increasing it to 20 kHz are often outweighed by the advantage of no audible noise with f_s of 20 kHz or greater. Therefore, in 50- or 60-Hz type applications, such as ac-motor drives (where the fundamental frequency of the inverter output may be required to be as high as 200 Hz), the frequency-modulation ratio m_f may be 9 or even less for the switching frequencies of less than 2 kHz. On the other hand, m_f will be larger than 100 for switching frequencies higher than 20 kHz. The desirable relationships between the triangular-waveform signal and the control voltage signal are dictated by how large m_f is. In the discussion here, $m_f = 21$ is treated as the borderline between large and small, though its selection is somewhat arbitrary. Here, it is assumed that the amplitude-modulation ratio m_a is less than 1.

6-2-1-1 SMALL m_f ($m_f \leq 21$)

1. *Synchronous PWM.* For small values of m_f, the triangular-waveform signal and the control signal should be synchronized to each other (synchronous PWM) as shown in Fig. 6-5a. This synchronous PWM requires that m_f be an integer. The reason for using the synchronous PWM is that the asynchronous PWM (where m_f is not an integer) results in subharmonics (of the fundamental frequency) that are very undesirable in most applications. This implies that the triangular waveform frequency varies with the desired inverter frequency (for example, if the inverter output frequency and hence the frequency of $v_{control}$ is 65.42 Hz and $m_f = 15$, the triangular-wave frequency should be exactly $15 \times 65.42 = 981.3$ Hz).

2. *m_f should be an odd integer.* As discussed previously, m_f should be an odd integer except in single-phase inverters with PWM unipolar voltage switching to be discussed in Section 6-3-2-2.

3. *Slopes of $v_{control}$ and v_{tri} should be of the opposite polarity* at the coincident zero crossings as shown in Fig. 6-5a. This is particularly important at very low values of m_f.

6-2-1-2 LARGE m_f ($m_f > 21$)

The amplitudes of subharmonics due to asynchronous PWM are small at large values of m_f. Therefore, at large values of m_f, the asynchronous PWM can be used where the frequency of the triangular-waveform is kept constant, whereas the frequency of $v_{control}$ varies, resulting in noninteger values of m_f (so long as they are large). However, if the inverter is supplying a load such as an ac motor, the subharmonics at zero or close to zero frequency, even though small in amplitude, will result in large currents which will be highly undesirable. Therefore, the asynchronous PWM should be avoided.

6-2-1-3 OVERMODULATION ($m_a > 1.0$)

In the previous discussion, it was assumed that $m_a \leq 1.0$, corresponding to a sinusoidal PWM in the linear range. Therefore, the amplitude of the fundamental-frequency voltage varies linearly with m_a, as derived in Eq. 6-7. In this range of $m_a \leq 1.0$, PWM pushes the harmonics into a high frequency range around the switching frequency and its multiples. In spite of this desirable feature of a sinusoidal PWM in the linear range, one of the drawbacks is that the maximum available amplitude of the fundamental frequency component is not as high as we wish. This is a natural consequence of the notches in the output voltage waveform of Fig. 6-5b.

To increase further the amplitude of the fundamental-frequency component in the output voltage, m_a is increased beyond 1.0, resulting in what is called *overmodulation*. Overmodulation causes the output voltage to contain many more harmonics in the sidebands as compared with the linear range (with $m_a \leq 1.0$), as shown in Fig. 6-7.

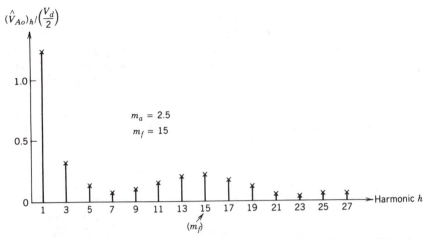

FIGURE 6-7: Harmonics due to overmodulation; figure is drawn for $m_a = 2.5$ and $m_f = 15$.

The harmonics with dominant amplitudes in the linear range may not be dominant during overmodulation. More significantly, with overmodulation, the amplitude of the fundamental frequency component does not vary linearly with the amplitude modulation ratio m_a. Figure 6-8 shows the normalized peak amplitude of the fundamental frequency component $(\hat{V}_{Ao})_1/(V_d/2)$ as a function of the amplitude modulation ratio m_a. Even at reasonably large values of m_f, $(\hat{V}_{Ao})_1/(V_d/2)$ depends on m_f in the overmodulation region. This is contrary to the linear range ($m_a \leq 1.0$) where $(\hat{V}_{Ao})_1/(V_d/2)$ varies linearly with m_a, almost independent of m_f (provided $m_f > 9$).

With overmodulation regardless of the value of m_f, it is recommended that a synchronous PWM operation be used, thus meeting the requirements indicated previously for a small value of m_f.

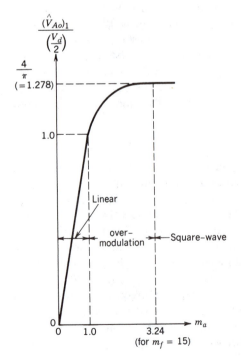

FIGURE 6-8: Voltage control by varying m_a.

As will be described in Chapter 9, the overmodulation region is avoided in uninterruptible power supplies because of a stringent requirement on minimizing the distortion in the output voltage. In induction motor drives described in Chapter 12, overmodulation is normally used.

For sufficiently large values of m_a, the inverter voltage waveform degenerates from a pulse-width modulated waveform into a square wave, which is discussed in detail in the next section. From Fig. 6-8 and the discussion of square-wave switching to be presented in the next section, it can be concluded that in the overmodulation region with $m_a > 1$

$$\frac{V_d}{2} < \left(\hat{V}_{Ao}\right)_1 < \frac{4}{\pi}\frac{V_d}{2} \tag{6-12}$$

6-2-2 Square-Wave Switching Scheme

In the square-wave switching scheme, each switch of the inverter-leg of Fig. 6-4 is on for one-half cycle (180°) of the desired output frequency. This results in an output-voltage waveform as shown in Fig. 6-9a. From Fourier analysis, the peak values of the fundamental frequency and the harmonic components in the inverter output waveform can be obtained for a given input V_d as

$$\left(\hat{V}_{Ao}\right)_1 = \frac{4}{\pi}\frac{V_d}{2} = 1.273\left(\frac{V_d}{2}\right) \tag{6-13}$$

and

$$\left(\hat{V}_{Ao}\right)_h = \frac{\left(\hat{V}_{Ao}\right)_1}{h} \tag{6-14}$$

where the harmonic order h takes on only odd values, as shown in Fig. 6-9b. It should be noted that the square-wave switching is also a special case of the sinusoidal PWM switching when m_a becomes so large that the control-voltage waveform intersects with the triangular waveform in Fig. 6-5a only at the zero-crossing of $v_{control}$. Therefore, the output voltage is independent of m_a in the square-wave region, as shown in Fig. 6-8.

One of the advantages of the square-wave operation is that each inverter switch changes its state only twice per cycle, which is important at very high power levels

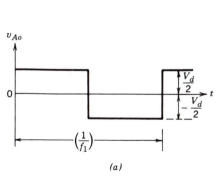
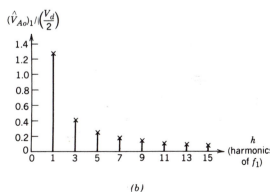

FIGURE 6-9: Square-wave switching.

where the solid-state switches generally have slower turn-on and turn-off speeds. One of the serious disadvantages of the square-wave switching is that the inverter is not capable of regulating the output voltage magnitude. Therefore, the dc input voltage V_d to the inverter must be adjusted in order to control the magnitude of the inverter-output voltage.

6-3   SINGLE-PHASE INVERTERS

6-3-1   Half-Bridge Inverters (Single-Phase)

Figure 6-10 shows the half-bridge inverter. Here, two equal capacitors are connected in series across the dc input and their junction is at a midpotential, with a voltage ($V_d/2$) across each capacitor. Sufficiently large capacitances should be used such that it is reasonable to assume that the potential at point "o" remains essentially constant with respect to the negative dc bus N. Therefore, this circuit configuration is identical to the basic one-leg inverter discussed in detail earlier, and $v_o = v_{Ao}$.

Assuming a PWM switching, we find that the output-voltage waveform will be exactly as in Fig. 6-5b. It should be noted that regardless of the switch states, the current between the two capacitors C_+ and C_- (which have equal and very large values) divides equally. When T_+ is on, either T_+ or D_+ conducts depending on the direction of the output current, and i_o splits equally between the two capacitors. Similarly, when the switch T_- is in its on-state, either T_- or D_- conducts depending on the direction of i_o, and i_o splits equally between the two capacitors. Therefore, the capacitors C_+ and C_- are "effectively" connected in parallel in the path of i_o. This also explains why the junction "o" in Fig. 6-10 stays at midpotential.

Since i_o must flow through the parallel combination of C_+ and C_-, i_o in steady state cannot have a dc component. Therefore, these capacitors act as dc blocking capacitors, thus eliminating the problem of transformer saturation from the primary side, if a transformer is used at the output to provide electrical isolation. Since the current in the primary winding of such a transformer would not be forced to zero with each switching, the transformer leakage inductance energy does not present a problem to the switches.

In a half-bridge inverter, the peak voltage and current ratings of the switches are as follows:

$$V_T = V_d \tag{6-15}$$

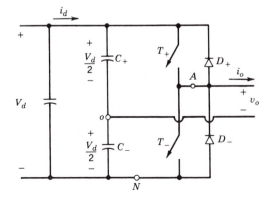

FIGURE 6-10: Half-bridge inverter.

and

$$I_T = i_{o,\text{peak}} \tag{6-16}$$

6-3-2 Full-Bridge Inverters (Single-Phase)

A full-bridge inverter is shown in Fig. 6-11. This inverter consists of two one-leg inverters of the type discussed in Section 6-2 and is preferred over other arrangements in higher power ratings. With the same dc input voltage, the maximum output voltage of the full-bridge inverter is twice that of the half-bridge inverter. This implies that for the same power, the output current and the switch currents are one-half of those for a half-bridge inverter. At high power levels, this is a distinct advantage, since it requires less paralleling of devices.

6-3-2-1 PWM WITH BIPOLAR VOLTAGE SWITCHING

This PWM scheme was first discussed in connection with the full-bridge dc–dc converters in Chapter 5. Here, the diagonally opposite switches (T_{A+}, T_{B-}) and (T_{A-}, T_{B+}) from the two legs in Fig. 6-11 are switched as switch-pairs 1 and 2, respectively. With this type of PWM switching, the output voltage waveform of leg A is identical to the output of the basic one-leg inverter in Section 6-2, which is determined in the same manner by comparison of v_{control} and v_{tri} in Fig. 6-12a. The output of inverter leg B is negative of the leg A output, for example, when T_{A+} is on and v_{Ao} is equal to $+V_d/2$, T_{B-} is also on and $v_{Bo} = -V_d/2$. Therefore

$$v_{Bo}(t) = -v_{Ao}(t) \tag{6-17}$$

and

$$v_o(t) = v_{Ao}(t) - v_{Bo}(t) = 2v_{Ao}(t) \tag{6-18}$$

The v_o waveform is shown in Fig. 6-12b. The analysis carried out in Section 6-2 for the basic one-leg inverter completely applies to this type of PWM switching. Therefore, the peak of the fundamental frequency component in the output voltage $(\hat{V}_{o1})$ can be obtained from Eqs. 6-7, 6-12, and 6-18 as

$$\hat{V}_{o1} = m_a V_d \quad (m_a \leq 1.0) \tag{6-19}$$

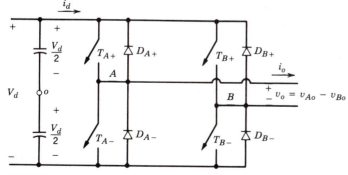

FIGURE 6-11: Single-phase full-bridge inverter.

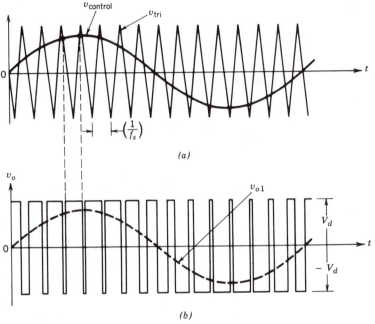

FIGURE 6-12: PWM with bipolar voltage switching.

and

$$V_d < \hat{V}_{o1} < \frac{4}{\pi} V_d \quad (m_a > 1.0) \tag{6-20}$$

In Fig. 6-12*b*, we observe that the output voltage v_o switches between $-V_d$ and $+V_d$ voltage levels. That is the reason why this type of switching is called a **PWM with bipolar voltage switching**. The amplitudes of harmonics in the output voltage can be obtained by using Table 6-1, as illustrated by the following example.

EXAMPLE 6-2

In the full-bridge converter circuit of Fig. 6-11, $V_d = 300$ V, $m_a = 0.8$, $m_f = 39$, and the fundamental frequency is 47 Hz. Calculate the rms values of the fundamental frequency voltage and some of the dominant harmonics in the output voltage v_o, if a PWM bipolar-voltage switching scheme is used.

SOLUTION

From Eq. 6-18, the harmonics in v_o can be obtained by multiplying the harmonics in Table 6-1 and Example 6-1 by a factor of 2. Therefore from Eq. 6-11, the rms voltage at any harmonic h is given as

$$(V_o)_h = \frac{1}{\sqrt{2}} \cdot 2 \cdot \frac{V_d}{2} \cdot \frac{\left(\hat{V}_{Ao}\right)_h}{\left(V_d/2\right)} = \frac{V_d}{\sqrt{2}} \frac{\left(\hat{V}_{Ao}\right)_h}{\left(V_d/2\right)}$$

$$= 212.13 \frac{\left(\hat{V}_{Ao}\right)_h}{\left(V_d/2\right)} \tag{6-21}$$

Therefore, the rms voltages are as follows:

fundamental: $V_{o1} = 212.13 \times 0.8 = 169.7\ V$ at 47 Hz

$\qquad (V_o)_{37} = 212.13 \times 0.22 = 46.67\ V$ at 1739 Hz

$\qquad (V_o)_{39} = 212.13 \times 0.818 = 173.52\ V$ at 1833 Hz

$\qquad (V_o)_{41} = 212.13 \times 0.22 = 46.67\ V$ at 1927 Hz

$\qquad (V_o)_{77} = 212.13 \times 0.314 = 66.60\ V$ at 3619 Hz

$\qquad (V_o)_{79} = 212.13 \times 0.314 = 66.60\ V$ at 3713 Hz

etc.   ■

DC-Side Current i_d. It is informative to look at the dc-side current i_d in the PWM biopolar-voltage-switching scheme.

For simplicity, fictitious *L-C* high-frequency filters will be used at the dc side as well as at the ac side, as shown in Fig. 6-13. The switching frequency is assumed to be very high, approaching infinity. Therefore, to filter out the high switching frequency components in v_o and i_d, the filter components *L* and *C* required in both ac- and dc-side filters approach zero. This implies that the energy stored in the filters is negligible. Since the converter itself has no energy-storage elements, the instantaneous power input must equal the instantaneous power output.

Having made these assumptions, v_o in Fig. 6-13 is a pure sine wave, at the fundamental output frequency ω_1

$$v_{o1} = v_o = \sqrt{2}\ V_o \sin \omega_1 t \tag{6-22}$$

If the load is as shown in Fig. 6-13 where e_o is a sine wave at frequency ω_1, then the output current would also be sinusoidal and would lag v_o for an inductive load such as an ac motor:

$$i_o = \sqrt{2}\ I_o \sin(\omega_1 t - \phi) \tag{6-23}$$

where ϕ is the angle by which i_o lags v_o.

On the dc side, the *L-C* filter will filter the high switching-frequency components in i_d, and i_d^* would only consist of the low frequency and dc components.

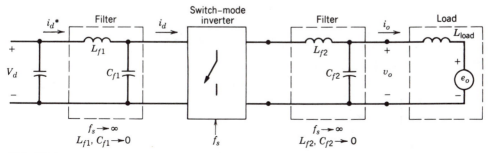

FIGURE 6-13: Inverter with "fictitious" filters.

Assuming that no energy is stored in the filters

$$V_d \cdot i_d^*(t) = v_o(t) \cdot i_o(t) = \sqrt{2}\, V_o \sin \omega_1 t \cdot \sqrt{2}\, I_o \sin(\omega_1 t - \phi) \qquad (6\text{-}24)$$

Therefore

$$i_d^*(t) = \frac{V_o I_o}{V_d}\cos\phi - \frac{V_o I_o}{V_d}\cos(2\omega_1 t - \phi) = I_d + i_{d2} \qquad (6\text{-}25)$$

$$= I_d - \sqrt{2}\, I_{d2}\cos(2\omega_1 t - \phi) \qquad (6\text{-}26)$$

where

$$I_d = \frac{V_o I_o}{V_d}\cos\phi \qquad (6\text{-}27)$$

and

$$I_{d2} = \frac{1}{\sqrt{2}}\cdot\frac{V_o I_o}{V_d} \qquad (6\text{-}28)$$

Equation 6-26 for i_d^* shows that it consists of a dc component I_d, which is responsible for the power transfer from V_d on the dc side of the inverter to the ac side. i_d^* also contains a sinusoidal component at twice the fundamental frequency. The inverter input current i_d consists of i_d^* and the high frequency components due to inverter switchings, as shown in Fig. 6-14.

In practical systems, the previous assumption of a constant dc voltage as the input to the inverter is not entirely valid. Normally, this dc voltage is obtained by rectifying the ac utility line voltage. A large capacitor is used across the rectifier

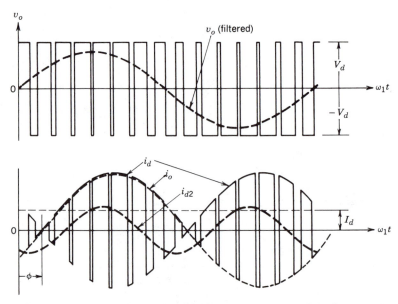

FIGURE 6-14: DC-side current in a single-phase inverter with PWM bipolar voltage switching.

output terminals to filter the dc voltage. The ripple in the capacitor voltage, which is also the dc input voltage to the inverter is due to two reasons: (1) The rectification of the line voltage to produce dc does not result in a pure dc as discussed in Chapters 3 and 4, dealing with the line-frequency rectifiers, and (2) as shown earlier by Eq. 6-26, the current drawn by a single-phase inverter from the dc-side is not a constant dc but has a second harmonic component (of the fundamental frequency at the inverter output) in addition to the high switching-frequency components. The second harmonic current component results in a ripple in the capacitor voltage, although the voltage ripple due to the high switching frequencies is essentially negligible.

6-3-2-2 PWM WITH UNIPOLAR VOLTAGE SWITCHING

In PWM with unipolar voltage switching, the switches in the two legs of the full-bridge inverter of Fig. 6-11 are not switched simultaneously, as in the previous PWM scheme. Here, the legs A and B of the full-bridge inverter are controlled separately by comparing v_{tri} with $v_{control}$ and $-v_{control}$, respectively. As shown in Fig. 6-15a, the comparison of $v_{control}$ with the triangular waveform results in the following logic signals to control the switches in leg A:

$$v_{control} > v_{tri} : T_{A+} \text{ on } \text{ and } v_{AN} = V_d$$

and

$$v_{control} < v_{tri} : T_{A-} \text{ on } \text{ and } v_{AN} = 0$$

(6-29)

The output voltage of inverter leg A with respect to the negative dc bus N is shown in Fig. 6-15b. For controlling the leg B switches, $(-v_{control})$ is compared with the same triangular waveform, which yields the following:

$$(-v_{control}) > v_{tri} : T_{B+} \text{ on } \text{ and } v_{BN} = V_d$$

and

$$(-v_{control}) < v_{tri} : T_{B-} \text{ on } \text{ and } v_{BN} = 0$$

(6-30)

Because of the feedback diodes in antiparallel with the switches, the foregoing voltages given by Eqs. 6-29 and 6-30 are independent of the direction of the output current i_o.

The waveforms of Fig. 6-15 show that there are four combinations of switch on-states and the corresponding voltage levels:

(1) T_{A+}, T_{B-} on: $v_{AN} = V_d$, $v_{BN} = 0$; $v_o = V_d$

(2) T_{A-}, T_{B+} on: $v_{AN} = 0$, $v_{BN} = V_d$; $v_o = -V_d$

(3) T_{A+}, T_{B+} on: $v_{AN} = V_d$, $v_{BN} = V_d$; $v_o = 0$

(4) T_{A-}, T_{B-} on: $v_{AN} = 0$, $v_{BN} = 0$; $v_o = 0$

(6-31)

We notice that when both the upper switches are on, the output voltage is zero. The output current circulates in a loop through $(T_{A+}$ and $D_{B+})$ or $(D_{A+}$ and $T_{B+})$ depending on the direction of i_o. During this interval, the input current i_d is zero. A similar condition occurs when both bottom switches T_{A-} and T_{B-} are on.

In this type of PWM scheme, when a switching occurs, the output voltage changes between 0 and $+V_d$ or between 0 and $-V_d$ voltage levels. For this reason, this type of PWM scheme is called the pulse-width modulation with a unipolar

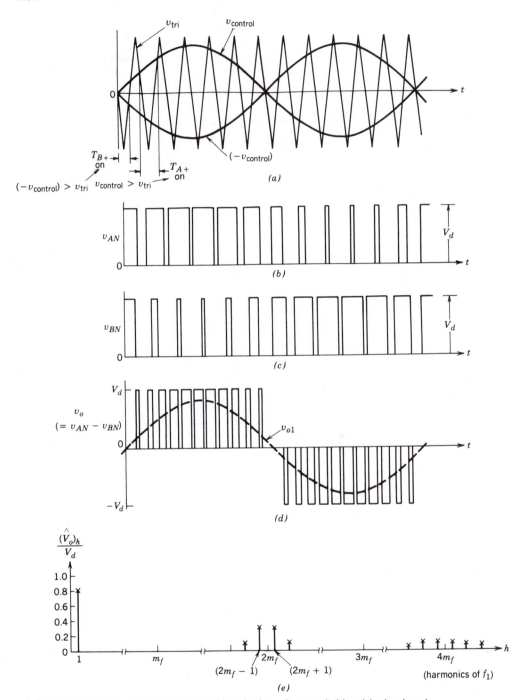

FIGURE 6-15: Pulse-width modulation with unipolar voltage switching (single-phase).

voltage switching, as opposed to the PWM with bipolar (between $+V_d$ and $-V_d$) voltage switching scheme described earlier. This scheme has the advantage of "effectively" doubling the switching frequency as far as the output harmonics are concerned, compared to the bipolar-voltage-switching scheme. Also, the voltage jumps in the output voltage at each switching are reduced to V_d, as compared to $2V_d$ in the previous scheme.

The advantage of "effectively" doubling the switching frequency appears in the harmonic spectrum of the output voltage waveform, where the lowest harmonics (in the idealized circuit) appear as sidebands of twice the switching frequency. It is easy to understand this if we choose the frequency-modulation ratio m_f to be even (m_f should be odd for PWM with bipolar voltage switching) in a single-phase inverter. The voltage waveforms v_{AN} and v_{BN} are displaced by 180° of the fundamental frequency f_1, with respect to each other. Therefore, the harmonic components at the switching frequency in v_{AN} and v_{BN} have the same phase ($\phi_{AN} - \phi_{BN} = 180° \cdot m_f = 0°$, since the waveforms are 180° displaced and m_f is assumed to be even). This results in the cancellation of the harmonic component at the switching frequency in the output voltage $v_o = v_{AN} - v_{BN}$. In addition, the sidebands of the switching frequency harmonics disappear. In a similar manner, the other dominant harmonic at twice the switching frequency cancels out, while its sidebands do not. Here also

$$\hat{V}_{o1} = m_a V_d \qquad (m_a \leq 1.0) \tag{6-32}$$

and

$$V_d < \hat{V}_{o1} < \frac{4}{\pi} V_d \qquad (m_a > 1.0) \tag{6-33}$$

EXAMPLE 6-3

In Example 6-2, suppose that a PWM with unipolar-voltage-switching scheme is used, with $m_f = 38$. Calculate the rms values of the fundamental frequency voltage and some of the dominant harmonics in the output voltage.

SOLUTION

Based on the discussion of unipolar voltage switching, the harmonic order h can be written as

$$h = j(2m_f) \pm k \tag{6-34}$$

where the harmonics exist as sidebands around $2m_f$ and the multiples of $2m_f$. Since h is odd, k in Eq. 6-34 attains only odd values. From Example 6-2

$$(V_o)_h = 212.13 \frac{(\hat{V}_{Ao})_h}{(V_d/2)} \tag{6-35}$$

Using Eq. 6-35 and Table 6-1, we find that the rms voltages are as follows:

at fundamental or 47 Hz, $V_{o1} = 0.8 \times 212.13 = 169.7 \, V$

at $h = 2m_f - 1 = 75$ or 3525 Hz, $(V_o)_{75} = 0.314 \times 212.13 = 66.60 \, V$

at $h = 2m_f + 1 = 77$ or 3619 Hz, $(V_o)_{77} = 0.314 \times 212.13 = 66.60 \, V$

etc.

Comparison of the unipolar voltage switching with the bipolar-voltage-switching scheme of Example 6-2 shows that in both cases, the fundamental frequency voltages are the same for equal m_a. However, with unipolar voltage switching, the dominant harmonic voltages centered around m_f disappear, thus resulting in a significantly lower harmonic content. ∎

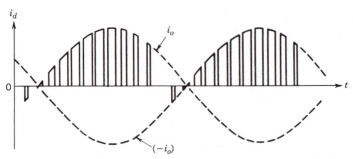

FIGURE 6-16: DC-side current in a single-phase inverter with PWM unipolar voltage switching.

DC-Side Current i_d. Under conditions similar to those in the circuit of Fig. 6-13 for the PWM with bipolar voltage switching, Fig. 6-16 shows the dc-side current i_d for the PWM unipolar-voltage-switching scheme, where $m_f = 14$ (instead of $m_f = 15$ for the bipolar voltage switching).

By comparing Figs. 6-14 and 6-16, it is clear that using PWM with unipolar voltage switching results in a smaller ripple in the current on the dc side of the inverter.

6-3-2-3 SQUARE-WAVE OPERATION

The full-bridge inverter can also be operated in a square wave mode. Both types of PWM discussed earlier degenerate into the same square-wave mode of operation, where the switches (T_{A+}, T_{B-}) and (T_{B+}, T_{A-}) are operated as two pairs with a duty ratio of 0.5.

As is the case in square-wave mode of operation, the output voltage magnitude given below is regulated by controlling the input dc voltage:

$$\hat{V}_{o1} = \frac{4}{\pi} V_d \qquad (6\text{-}36)$$

6-3-2-4 OUTPUT CONTROL BY VOLTAGE CANCELLATION

This type of control is feasible only in a single-phase, full-bridge inverter circuit. It is based on the combination of square-wave switching and PWM with a unipolar voltage switching. In the circuit of Fig. 6-17a, the switches in the two inverter legs are controlled separately (similar to PWM unipolar voltage switching). But all switches have a duty ratio of 0.5 similar to a square-wave control. This results in waveforms for v_{AN} and v_{BN} shown in Fig. 6-17b, where the waveform overlap angle α can be controlled. During this overlap interval, the output voltage is zero as a consequence of either both top switches or both bottom switches being on. With $\alpha = 0$, the output waveform is similar to a square-wave inverter with the maximum possible fundamental output magnitude.

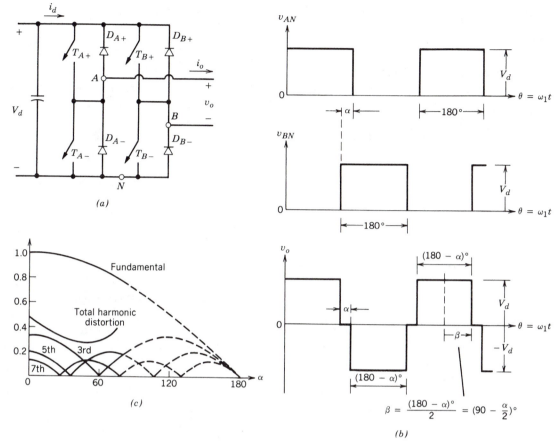

FIGURE 6-17: Full-bridge, single-phase inverter control by voltage cancellation: (a) power circuit, (b) waveforms, (c) normalized fundamental and harmonic voltage output and total harmonic distortion as a function of α.

It is easier to derive the fundamental and the harmonic frequency components of the output voltage in terms of $\beta = 90° - \alpha/2$, as is shown in Fig. 6-17b:

$$\left(\hat{V}_o\right)_h = \frac{2}{\pi} \int_{-\pi/2}^{\pi/2} v_o \cdot \cos\left(h\theta\right) d\theta$$

$$= \frac{2}{\pi} \int_{-\beta}^{\beta} V_d \cdot \cos\left(h\theta\right) d\theta$$

$$\therefore \left(\hat{V}_o\right)_h = \frac{4}{\pi h} V_d \sin\left(h\beta\right) \tag{6-37}$$

where $\beta = 90° - (\alpha/2)$, and $h =$ an odd integer.

Figure 6-17c shows the variation in the fundamental frequency component as well as the harmonic voltages as a function of α. These are normalized with respect to the fundamental frequency component for the square-wave ($\alpha = 0$) operation. The total harmonic distortion, which is the ratio of the rms value of the harmonic distortion to the rms value of the fundamental frequency component, is also plotted as a function of α. Because of a large distortion, the curves are shown as dotted for large values of α.

6-3-2-5 SWITCH UTILIZATION IN FULL-BRIDGE INVERTERS

Similar to a half-bridge inverter, if a transformer is utilized at the output of a full-bridge inverter, the transformer leakage inductance does not present a problem to the switches.

Independent of the type of control and the switching scheme used, the peak switch voltage and current ratings required in a full-bridge inverter are as follows:

$$V_T = V_d \tag{6-38}$$

and

$$I_T = i_{o,\text{peak}} \tag{6-39}$$

6-3-2-6 RIPPLE IN THE SINGLE-PHASE INVERTER OUTPUT

The *ripple* in a repetitive waveform refers to the difference between the instantaneous values of the waveform and its fundamental frequency component.

Figure 6-18a shows a single-phase switch-mode inverter. It is assumed to be supplying an induction-motor load, which is shown by means of a simplified equivalent circuit with a counter-emf e_o. Since $e_o(t)$ is sinusoidal, only the sinusoidal (fundamental frequency) components of the inverter output voltage and current are responsible for the real power transfer to the load.

Therefore, in a phasor form (with the fundamental frequency components designated by subscript "1") as shown in Fig. 6-18b:

$$\mathbf{V}_{o1} = \mathbf{E}_o + \mathbf{V}_{L1} = \mathbf{E}_o + j\omega_1 L\mathbf{I}_{o1} \tag{6-40}$$

Since the superposition principle is valid here, all the ripple in v_o appears across L, where

$$v_{\text{ripple}}(t) = v_o - v_{o1} \tag{6-41}$$

The output current ripple can be calculated as

$$i_{\text{ripple}}(t) = \frac{1}{L} \int_0^t v_{\text{ripple}}(t) \cdot dt + k \tag{6-42}$$

where k is a constant.

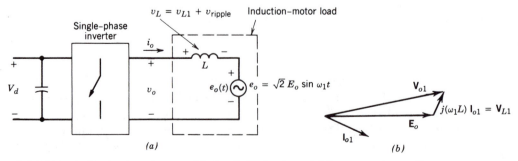

FIGURE 6-18: Single-phase inverter: (*a*) circuit, (*b*) fundamental frequency phasor diagram.

With a properly selected time origin $t = 0$, the constant k in Eq. 6-42 will be zero. Therefore, Eqs. 6-41 and 6-42 show that the current ripple is independent of the power being transferred to the load.

As an example, Fig. 6-19a shows the ripple current for a square-wave inverter output. Figure 6-19b shows the ripple current in a PWM bipolar voltage switching. In drawing Figs. 6-19a and 6-19b, the fundamental frequency components in the inverter output voltages are kept equal in magnitude (this requires a higher value of V_d in the PWM inverter). The PWM inverter results in a substantially smaller peak ripple current compared to the square-wave inverter. This shows the advantage of pushing the harmonics in the inverter output voltage to as high frequencies as feasible, thereby reducing the losses in the load by reducing the output-current harmonics. This is achieved by using higher inverter switching frequencies, which would result in more frequent switchings, hence higher switching losses in the inverter. Therefore, from the viewpoint of the overall system energy efficiency, a compromise must be made in selecting the inverter switching frequency.

6-3-3 Push – Pull Inverters

Figure 6-20 shows a push–pull inverter circuit. It requires a transformer with a center-tapped primary. We will initially assume that the output current i_o flows continuously. With this assumption, when the switch T_1 is on (and T_2 is off), T_1 would conduct for a positive value of i_o, and D_1 would conduct for a negative value of i_o. Therefore, regardless of the direction of i_o, $v_o = V_d/n$, where n is the transformer turns ratio between the primary-half and the secondary winding, as shown in Fig. 6-20. Similarly, when T_2 is on (and T_1 is off), $v_o = -V_d/n$. A push–pull inverter can be operated in a PWM or a square-wave mode and the waveforms are identical to those in Figs. 6-5 and 6-12 for half-bridge and full-bridge inverters. The output voltage in Fig. 6-20 equals

$$\hat{V}_{o1} = m_a \frac{V_d}{n} \quad (m_a \leq 1.0) \tag{6-43}$$

and

$$\frac{V_d}{n} < \hat{V}_{o1} < \frac{4}{\pi} \frac{V_d}{n} \quad (m_a > 1.0) \tag{6-44}$$

In a push–pull inverter, the peak switch voltage and current ratings are

$$V_T = 2V_d$$

and

$$\tag{6-45}$$

$$I_T = i_{o,\text{peak}}/n$$

The main advantage of the push–pull circuit is that no more than one switch in series conducts at any instant of time. This can be important if the dc input to the converter is from a low-voltage source, such as a battery, where the voltage drops across more than one switch in series would result in a significant reduction in energy efficiency. Also, the control drives for the two switches have a common ground. It is, however, difficult to avoid the dc saturation of the transformer in a push–pull inverter.

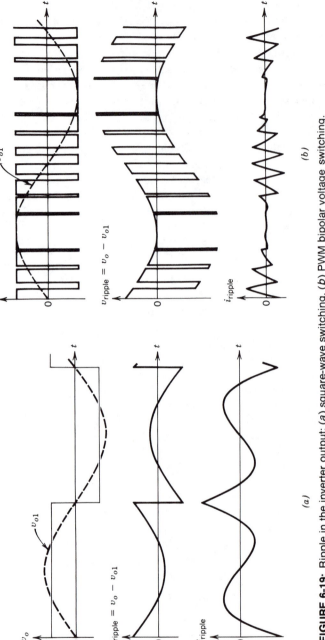

FIGURE 6-19: Ripple in the inverter output: (*a*) square-wave switching, (*b*) PWM bipolar voltage switching.

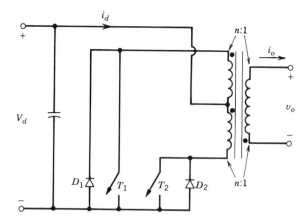

FIGURE 6-20: Push – pull inverter (single-phase).

The output current, which is the secondary current of the transformer, is a slowly varying current at the fundamental output frequency. It can be assumed to be a constant during a switching interval. When a switching occurs, the current shifts from one half to the other half of the primary winding. This requires very good magnetic coupling between these two half-windings in order to reduce the energy associated with the leakage inductance of the two primary windings. This energy will be dissipated in the switches or in snubber circuits used to protect the switches. This is a general phenomenon associated with all converters (or inverters) with isolation where the current in one of the windings is forced to go to zero with every switching. This phenomenon is very important in the design of such converters.

In a pulse-width-modulated push–pull inverter for producing sinusoidal output (unlike those used in switch-mode dc power supplies), the transformer must be designed for the fundamental output frequency. The number of turns will therefore be high compared to a transformer designed to operate at the switching frequency in a switch-mode dc power supply. This will result in a high transformer leakage inductance, which is proportional to the square of the number of turns, provided all other dimensions are kept constant. This makes it difficult to operate a sine-wave-modulated PWM push–pull inverter at switching frequencies higher than approximately 1 kHz.

6-3-4 Switch Utilization in Single-Phase Inverters

Since the intent in this section is to compare the utilization of switches in various single-phase inverters, the circuit conditions are idealized. We will assume that $V_{d,\max}$ is the highest value of the input voltage, which establishes the switch voltage ratings. In the PWM mode, the input remains constant at $V_{d,\max}$. In the square-wave mode, the input voltage is decreased below $V_{d,\max}$ to decrease the output voltage from its maximum value. Regardless of the PWM or the square-wave mode of operation, we assume that there is enough inductance associated with the output load to yield a purely sinusoidal current (an idealized condition indeed for a square-wave output) with an rms value of $I_{o,\max}$ at the maximum load.

If the output current is assumed to be purely sinusoidal, the inverter rms volt-ampere output at the fundamental frequency equals $V_{o1}I_{o,\max}$ at the maximum rated output, where the subscript "1" designates the fundamental frequency component of the inverter output. With V_T and I_T as the peak voltage and current ratings of a switch, the combined utilization of all the switches in an inverter can be defined as

$$\text{Switch-utilization ratio} = \frac{V_{o1}I_{o,\max}}{qV_T I_T} \tag{6-46}$$

where q = number of switches in an inverter.

To compare the utilization of switches in various single-phase inverters, we will initially compare them for a square-wave mode of operation at the maximum rated output. (The maximum switch utilization occurs at $V_d = V_{d,\max}$.)

Push – Pull Inverter

$$V_T = 2V_{d,\max}; \quad I_T = \sqrt{2}\,\frac{I_{o,\max}}{n}; \quad V_{o1,\max} = \frac{4}{\pi\sqrt{2}}\frac{V_{d,\max}}{n}; \quad q = 2 \tag{6-47}$$

$$(n = \text{turns ratio, Fig. 6-20})$$

$$\therefore \text{Maximum switch-utilization ratio} = \frac{1}{2\pi} \simeq 0.16 \tag{6-48}$$

Half-Bridge Inverter

$$V_T = V_{d,\max}; \quad I_T = \sqrt{2}\,I_{o,\max}; \quad V_{o1,\max} = \frac{4}{\pi\sqrt{2}}\left(\frac{V_{d,\max}}{2}\right); \quad q = 2 \tag{6-49}$$

$$\therefore \text{Maximum switch-utilization ratio} = \frac{1}{2\pi} \simeq 0.16 \tag{6-50}$$

Full-Bridge Inverter

$$V_T = V_{d,\max}; \quad I_T = \sqrt{2}\,I_{o,\max}; \quad V_{o1,\max} = \frac{4}{\pi\sqrt{2}}V_{d,\max}; \quad q = 4 \tag{6-51}$$

$$\therefore \text{Maximum switch-utilization ratio} = \frac{1}{2\pi} \simeq 0.16 \tag{6-52}$$

This shows that in each inverter, the switch utilization is the same with

$$\text{Maximum switch-utilization ratio} = \frac{1}{2\pi} \simeq 0.16 \tag{6-53}$$

In practice, the switch-utilization ratio would be much smaller than 0.16 for the following reasons: (1) switch ratings are chosen conservatively to provide safety margins; (2) in determining the switch-current rating in a PWM inverter, one would have to take into account the variations in the input dc voltage available; and (3) the ripple in the output current would influence the switch current rating. Moreover, the inverter may be required to supply a short-term overload. Thus, the switch-utilization ratio, in practice, would be substantially less than the 0.16 calculated.

At the lower output volt-amperes compared to the maximum rated output, the switch utilization decreases linearly. It should be noted that using a PWM switching with $m_a \leq 1.0$, this ratio would be smaller by a factor of $[(\pi/4)m_a]$ as compared to the square-wave switching:

$$\text{Maximum switch-utilization ratio} = \frac{1}{2\pi} \cdot \frac{\pi}{4} \cdot m_a = \frac{1}{8}m_a \qquad (6\text{-}54)$$

$$(\text{PWM, } m_a \leq 1.0)$$

Therefore, the theoretical maximum switch-utilization ratio in a PWM switching is only 0.125 at m_a equal to 1, as compared with 0.16 in a square-wave inverter.

EXAMPLE 6-4

In a single-phase full-bridge PWM inverter, V_d varies in a range of 295 to 325 V. The output voltage is required to be constant at 200 V (rms), and the maximum load current (assumed to be sinusoidal) is 10 A (rms). Calculate the combined switch-utilization ratio (under these idealized conditions, not accounting for any overcurrent capabilities).

SOLUTION

In this inverter

$$V_T = V_{d,\max} = 325 \, V$$

$$I_T = \sqrt{2}\,I_o = \sqrt{2} \times 10 = 14.14$$

$$q = \text{no. of switches} = 4$$

The maximum output volt-ampere (fundamental frequency)

$$V_{o1}I_{o,\max} = 200 \times 10 = 2000 \, VA \qquad (6\text{-}55)$$

Therefore, from Eq. 6-46

$$\text{Switch-utilization ratio} = \frac{V_{o1}I_{o,\max}}{qV_TI_T} = \frac{2000}{4 \times 325 \times 14.14} \approx 0.11 \qquad \blacksquare$$

6-4 THREE-PHASE INVERTERS

In applications such as uninterruptible ac power supplies and ac-motor drives, three-phase inverters are commonly used to supply three-phase loads. It is possible to supply a three-phase load by means of three separate single-phase inverters, where each inverter produces an output displaced by 120° (of the fundamental frequency) with respect to each other. Though this arrangement may be preferable under certain conditions, it requires either a three-phase output transformer or separate access to each of the three phases of the load. In practice, such access is generally not available. Moreover, it requires 12 switches.

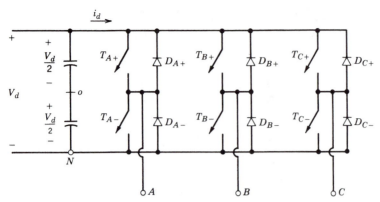

FIGURE 6-21: Three-phase inverter.

The most frequently used three-phase inverter circuit consists of three legs, one for each phase, as shown in Fig. 6-21. Each inverter leg is similar to the one used for describing the basic one-leg inverter in Section 6-2. Therefore, the output of each leg, for example v_{AN} (with respect to the negative dc bus) depends only on V_d, and the switch status; the output voltage is independent of the output load current since one of the two switches in a leg is always on at any instant. Here, we again ignore the blanking time required in practical circuits by assuming the switches to be ideal. Therefore, the inverter output voltage is independent of the direction of the load current.

6-4-1 PWM in Three-Phase Voltage-Source Inverters

Similar to the single-phase inverters, the objective in pulse-width-modulated three-phase inverters is to shape and control the three-phase output voltages in magnitude and frequency with an essentially constant input voltage V_d. To obtain balanced three-phase output voltages in a three-phase pulse-width-modulated (PWM) inverter, the same triangular voltage waveform is compared with three sinusoidal control voltages that are 120° out of phase, as shown in Fig. 6-22a (which is drawn for $m_f = 15$).

It should also be noted from Fig. 6-22b that an identical amount of average dc component is present in the output voltages v_{AN} and v_{BN}, which are measured with respect to the negative dc bus. These dc components are cancelled out in the line-line voltages, for example in v_{AB} shown in Fig. 6-22b. This is similar to what happens in a single-phase full-bridge inverter utilizing a PWM switching.

In the three-phase inverters, only the harmonics in the line-to-line voltages are of concern. The harmonics in the output of any one of the legs, for example v_{AN} in Fig. 6-22b, are identical to the harmonics in v_{Ao} in Fig. 6-5, where only the odd harmonics exist as sidebands, centered around m_f and its multiples, provided m_f is odd. Only considering the harmonic at m_f (the same applies to its odd multiples), the phase difference between the m_f harmonic in v_{AN} and v_{BN} is $(120 \, m_f)°$. This phase difference will be equivalent to zero (a multiple of 360°) if m_f is odd and a multiple of 3. As a consequence, the harmonic at m_f is suppressed in the line-to-line voltage v_{AB}. The same argument applies in the suppression of harmonics at the odd multiples of m_f, if m_f is chosen to be an odd multiple of 3 (where the reason for choosing m_f to be an odd multiple of 3 is to keep m_f odd and, hence, eliminate even harmonics).

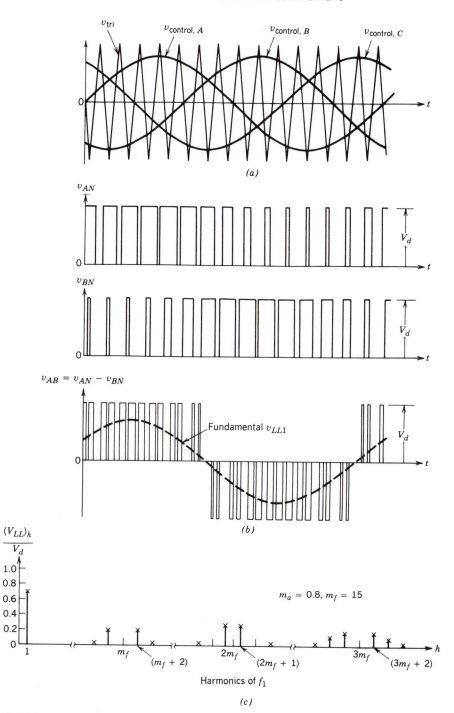

FIGURE 6-22: Three-phase PWM waveforms and harmonic spectrum.

Thus, some of the dominant harmonics in the one-leg inverter can be eliminated from the line-to-line voltage of a three-phase inverter.

PWM considerations are summarized as follows:

1. *For low values of m_f*, to eliminate the even harmonics, a synchronized PWM should be used and m_f should be an odd integer. Moreover, m_f should be a multiple of 3 to cancel out the most dominant harmonics in the line-line voltage. Moreover, the slopes of $v_{control}$ and v_{tri} should be of the opposite polarity at the coincident zero crossings.

2. *For large values of m_f*, the comments in Section 6-2-1-2 for a single-phase PWM apply.

3. *During overmodulation ($m_a > 1.0$)*, regardless of the value of m_f, the conditions pertinent to a small m_f should be observed.

6-4-1-1   LINEAR MODULATION ($m_a \leq 1.0$)

In the linear region ($m_a \leq 1.0$), the fundamental frequency component in the output voltage varies linearly with the amplitude-modulation ratio m_a. From Figs. 6-5b and 6-22b, the peak value of the fundamental frequency component in one of the inverter legs is

$$\left(\hat{V}_{AN}\right)_1 = m_a \cdot \frac{V_d}{2} \tag{6-56}$$

Therefore, the line-line rms voltage at the fundamental frequency, due to 120° phase displacement between phase voltages, can be written as

$$\begin{aligned} V_{LL_1} \atop \text{(line-line, rms)} \quad &= \frac{\sqrt{3}}{\sqrt{2}} \cdot \left(\hat{V}_{AN}\right)_1 \\[2mm] &= \frac{\sqrt{3}}{2\sqrt{2}} m_a \cdot V_d \quad (m_a \leq 1.0) \\[2mm] &\simeq 0.612 m_a \cdot V_d \end{aligned} \tag{6-57}$$

The harmonic components of the line-line output voltages can be calculated in a similar manner from Table 6-1, recognizing that some of the harmonics are cancelled out in the line-line voltages. These rms harmonic voltages are listed in Table 6-2.

6-4-1-2   OVERMODULATION ($m_a > 1.0$)

In PWM overmodulation, the peak of the control voltages are allowed to exceed the peak of the triangular waveform. Unlike the linear region, in this mode of operation the fundamental frequency voltage magnitude does not increase proportionally with m_a. This is shown in Fig. 6-23, where the rms value of the fundamental frequency line-to-line voltage V_{LL1} is plotted as a function of m_a. Similar to a single-phase PWM, for sufficiently large values of m_a, the PWM degenerates into a square-wave inverter waveform. This results in the maximum value of V_{LL_1} equal to 0.78 V_d as explained in the next section.

In the overmodulation region compared to the region with $m_a \leq 1.0$, more sideband harmonics appear centered around the frequencies of harmonics m_f and its

TABLE 6-2

Generalized Harmonics of v_{LL} for a Large and odd m_f
that Is a Multiple of 3. $(V_{LL})_h / V_d$ are tabulated as a function of m_a,
where $(V_{LL})_h$ are the rms values of the harmonic voltages.

h	m_a 0.2	0.4	0.6	0.8	1.0
1	0.122	0.245	0.367	0.490	0.612
$m_f \pm 2$	0.010	0.037	0.080	0.135	0.195
$m_f \pm 4$				0.005	0.011
$2m_f \pm 1$	0.116	0.200	0.227	0.192	0.111
$2m_f \pm 5$				0.008	0.020
$3m_f \pm 2$	0.027	0.085	0.124	0.108	0.038
$3m_f \pm 4$		0.007	0.029	0.064	0.096
$4m_f \pm 1$	0.100	0.096	0.005	0.064	0.042
$4m_f \pm 5$			0.021	0.051	0.073
$4m_f \pm 7$				0.010	0.030

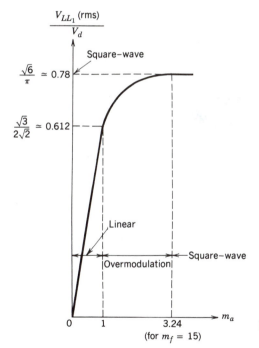

FIGURE 6-23: Three-phase inverter; V_{LL_1}(rms) $/ V_d$ as a function of m_a.

multiples. However, the dominant harmonics may not have as large an amplitude as with $m_a \leq 1.0$. Therefore, the power loss in the load due to the harmonic frequencies may not be as high in the overmodulation region as the presence of additional sideband harmonics would suggest. Depending on the nature of the load and on the switching frequency, the losses due to these harmonics in overmodulation may be even less than those in the linear region of the PWM.

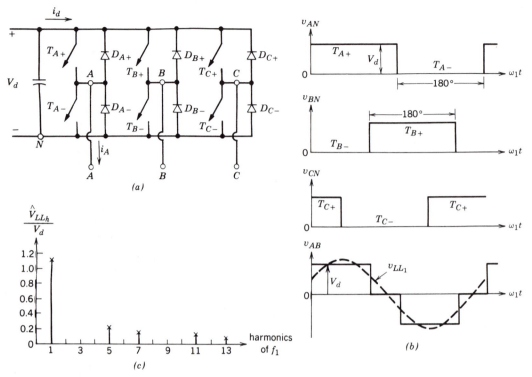

FIGURE 6-24: Square-wave inverter (three-phase).

6-4-2   Square-Wave Operation in Three-Phase Inverters

If the input dc voltage V_d is controllable, the inverter in Fig. 6-24a can be operated in a square wave mode. Also, for sufficiently large values of m_a, PWM degenerates into square wave operation and the voltage waveforms are shown in Fig. 6-24b. Here, each switch is on for 180° (i.e., its duty ratio is 50%). Therefore, at any instant of time, three switches are on.

In the square-wave mode of operation, the inverter itself cannot control the magnitude of the output ac voltages. Therefore, the dc input voltage must be controlled in order to control the output in magnitude. Here, the fundamental frequency line-to-line rms voltage component in the output can be obtained from Eq. 6-13 for the basic one-leg inverter operating in a square-wave mode:

$$V_{LL_1} = \frac{\sqrt{3}}{\sqrt{2}} \frac{4}{\pi} \frac{V_d}{2}$$
(rms)

$$= \frac{\sqrt{6}}{\pi} V_d$$

$$\approx 0.78 V_d \tag{6-58}$$

The line-to-line output voltage waveform does not depend on the load and contains harmonics ($6n \pm 1$; $n = 1, 2, \ldots$), whose amplitudes decrease inversely proportional to their harmonic order, as shown in Fig. 6-24c:

$$V_{LL_h} = \frac{0.78}{h} V_d$$

(6-59)

where

$$h = 6n \pm 1 \qquad (n = 1, 2, 3, \ldots)$$

It should be noted that it is *not* possible to control the output magnitude in a three-phase, square-wave inverter by means of voltage cancellation as described in Section 6-3-2-4.

6-4-3 Switch Utilization in Three-Phase Inverters

We will assume that $V_{d,\max}$ is the maximum input voltage that remains constant during PWM, and is decreased below this level to control the output voltage magnitude in a square-wave mode. We will also assume that there is sufficient inductance associated with the load to yield a pure sinusoidal output current with an rms value of $I_{o,\max}$ (both in the PWM and the square-wave mode) at maximum loading. Therefore, each switch would have the following peak ratings:

$$V_T = V_{d,\max}$$

(6-60)

and

$$I_T = \sqrt{2} I_{o,\max}$$

(6-61)

If V_{LL_1} is the rms value of the fundamental frequency line-to-line voltage component, the three-phase output volt-amperes (rms) at the fundamental frequency at the rated output is

$$(VA)_{3\text{-phase}} = \sqrt{3} V_{LL_1} I_{o,\max}$$

(6-62)

Therefore, the total switch utilization ratio of all six switches combined is

$$\text{Switch-utilization ratio} = \frac{(VA)_{3\text{-phase}}}{6V_T I_T}$$

$$= \frac{\sqrt{3} V_{LL_1} I_{o,\max}}{6 \cdot V_{d,\max} \cdot \sqrt{2} I_{o,\max}}$$

$$= \frac{1}{2\sqrt{6}} \cdot \frac{V_{LL_1}}{V_{d,\max}}$$

(6-63)

In the PWM-linear region ($m_a \leq 1.0$) using Eq. 6-57 and noting that the maximum switch utilization occurs at $V_d = V_{d,\max}$

$$\begin{array}{c}\text{Maximum switch-utilization ratio} \\ \text{(PWM)}\end{array} = \frac{1}{2\sqrt{6}} \cdot \frac{\sqrt{3}}{2\sqrt{2}} m_a$$

$$= \frac{1}{8} \cdot m_a \quad (m_a \leq 1.0)$$

(6-64)

In the square-wave mode, this ratio is $1/2\pi \simeq 0.16$ compared to a maximum of 0.125 for a PWM-linear region with $m_a = 1.0$.

In practice, the same derating in the switch-utilization ratio applies as discussed in Section 6-3-4 for single-phase inverters.

Comparing Eqs. 6-54 and 6-64, we observe that the maximum switch-utilization ratio is the same in a three-phase, three-leg inverter as in a single-phase inverter. In other words, using the switches with identical ratings, a three-phase inverter with 50% increase in the number of switches results in a 50% increase in the output volt-ampere, compared to a single-phase inverter.

6-4-4   Ripple in the Inverter Output

Figure 6-25a shows a three-phase, three-leg, voltage-source, switch-mode inverter in a block-diagram form. It is assumed to be supplying a three-phase ac-motor load. Each phase of the load is shown by means of its simplified equivalent circuit with respect to the load neutral "n." The induced back-emfs $e_A(t)$, $e_B(t)$, and $e_C(t)$ are assumed to be sinusoidal.

Under balanced operating conditions, it is possible to express the inverter phase output voltages v_{AN}, and so on (with respect to the load neutral n), in terms of the inverter output voltages with respect to the negative dc bus N:

$$v_{kn} = v_{kN} - v_{nN} \quad (k = A, B, C) \tag{6-65}$$

Each phase voltage can be written as

$$v_{kn} = L\frac{di_k}{dt} + e_{kn} \quad (k = A, B, C) \tag{6-66}$$

In a three-phase, three-wire load

$$i_A + i_B + i_C = 0 \tag{6-67a}$$

and

$$\frac{d}{dt}(i_A + i_B + i_C) = 0 \tag{6-67b}$$

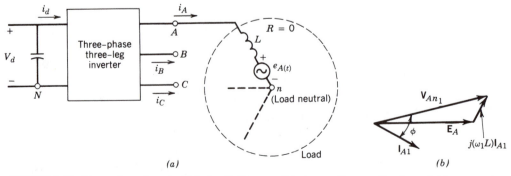

(a)   *(b)*

FIGURE 6-25: Three-phase inverter: (*a*) circuit diagram, (*b*) phasor diagram (fundamental frequency).

Similarly, under balanced operating conditions, the three back-emfs are a balanced three-phase set of voltages, and therefore

$$e_{An} + e_{Bn} + e_{Cn} = 0 \qquad (6\text{-}68)$$

From the foregoing equations, the following condition for the inverter voltages can be written:

$$v_{An} + v_{Bn} + v_{Cn} = 0 \qquad (6\text{-}69)$$

Using Eqs. 6-65 through 6-69,

$$v_{nN} = \frac{1}{3}(v_{AN} + v_{BN} + v_{CN}) \qquad (6\text{-}70)$$

Substituting v_{nN} from Eq. 6-70 into Eq. 6-65, we can write the phase-to-neutral voltage for phase A as

$$v_{An} = \frac{2}{3}v_{AN} - \frac{1}{3}(v_{BN} + v_{CN}) \qquad (6\text{-}71)$$

Similar equations can be written for phase B and C voltages.

Similar to the discussion in Section 6-3-2-6 for the ripple in the single-phase inverter output, only the fundamental frequency components of the phase voltage v_{An_1} and the output current i_{A1} are responsible for the real power transformer since the back-emf $e_A(t)$ is assumed to be sinusoidal and the load resistance is neglected. Therefore, in a phasor form as shown in Fig. 6-25b

$$\mathbf{V}_{An_1} = \mathbf{E}_A + j\omega_1 L \mathbf{I}_{A1} \qquad (6\text{-}72)$$

By using the principle of superposition, all the ripple in v_{An} appears across the load inductance L. Using Eq. 6-71, the waveform for the phase to load-neutral voltage V_{An} is shown in Figs. 6-26a and 6-26b for square-wave and PWM operations, respectively. Both inverters have identical magnitudes of the fundamental frequency voltage component V_{An_1}, which requires a higher V_d in the PWM operation. The voltage ripple v_{ripple} ($= v_{An} - v_{An_1}$) is the ripple in the phase to neutral voltage. Assuming identical loads in these two cases, the output current ripple is obtained by using Eq. 6-42 and plotted in Fig. 6-26. This current ripple is independent of the power being transferred, that is, the current ripple would be the same so long as for a given load inductance L, the ripple in the inverter output voltage remains constant in magnitude and frequency. This comparison indicates that for large values of m_f, the current ripple in the PWM inverter will be significantly lower compared to a square-wave inverter.

6-4-5 DC-Side Current i_d

Similar to the treatment of a single-phase inverter, we now look at the voltage and current waveforms associated with the dc side of a pulse-width-modulated, three-phase inverter. The input voltage V_d is assumed to be dc without any ripple. If the switching frequency in Fig. 6-25a is assumed to approach infinity, then similar to Fig. 6-13, a fictitious filter with negligible energy storage can be inserted on the ac side and the current at the inverter output will be sinusoidal with no ripple. Because of the

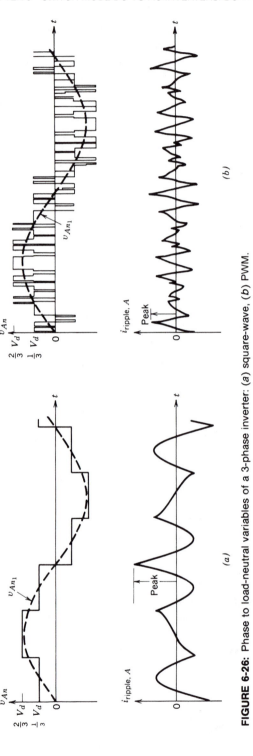

FIGURE 6-26: Phase to load-neutral variables of a 3-phase inverter: (*a*) square-wave, (*b*) PWM.

assumption of no energy storage in the fictitious ac-side filter, the instantaneous ac power output can be expressed in terms fundamental frequency output voltages and currents. Similarly, on the dc side, a fictitious filter with no energy storage, as shown in Fig. 6-13, can be assumed. Then, the high switching frequency components in i_d are filtered. Now equating the instantaneous power input to the instantaneous power output, we get

$$V_d i_d^* = v_{An_1}(t) i_A(t) + v_{Bn_1}(t) i_B(t) + v_{Cn_1}(t) i_C(t) \tag{6-73}$$

In a balanced steady-state operation, the three phase quantities are displaced by 120° with respect to each other. Assuming that ϕ is the phase angle by which a phase current lags the inverter phase voltage, and $\sqrt{2}\, V_o$ and $\sqrt{2}\, I_o$ are the amplitudes of the phase voltages and currents, respectively:

$$i_d^* = \frac{2V_o I_o}{V_d} \big[\cos \omega_1 t \cdot \cos(\omega_1 t - \phi) + \cos(\omega_1 t - 120°) \cdot \cos(\omega_1 t - 120° - \phi)$$

$$+ \cos(\omega_1 t + 120°) \cdot \cos(\omega_1 t + 120° - \phi)\big]$$

$$= \frac{3V_o I_o}{V_d} \cdot \cos \phi = I_d \quad (\text{a dc quantity}) \tag{6-74}$$

The foregoing analysis shows that i_d^* is a dc quantity, unlike in the single-phase inverter, where i_d^* contained a component at twice the output frequency. However, i_d consists of high-frequency switching components as shown in fig. 6-27, in addition to i_d^*. These high-frequency components, due to their high frequencies, would have a negligible effect on the capacitor voltage V_d.

6-4-6   Conduction of Switches in Three-Phase Inverters

We discussed earlier that the output voltage does not depend on the load. However, the duration of each switch conduction is dependent on the power factor of the load.

6-4-6-1   SQUARE-WAVE OPERATION

Here, each switch is in its on-state for 180°. To determine the switch conduction interval, a load with a fundamental frequency displacement angle of 30° (lagging) is assumed (as an example). The waveforms are shown in Fig. 6-28 for one of the three phases. The phase to neutral voltages V_{An} and V_{An_1} are shown in Fig. 6-28a. In Fig.

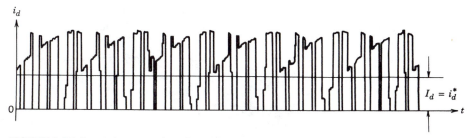

FIGURE 6-27: Input dc current in a three-phase inverter.

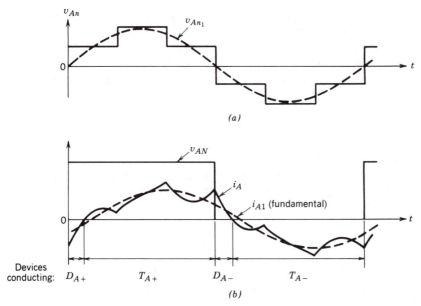

Devices
conducting: D_{A+} T_{A+} D_{A-} T_{A-}

(b)

FIGURE 6-28: Square-wave inverter: phase A waveforms.

6-28*b*, V_{AN} (with respect to the negative dc bus), i_A, and its fundamental component i_{A_1} are plotted. Even though the switches T_{A+} and T_{A-} are in their on state for 180°, due to the lagging power factor of the load, their actual conduction intervals are smaller than 180°. It is easy to interpret that as the power factor (lagging) of the load decreases, the diode-conduction intervals will increase, and the switch conduction intervals will decrease. On the other hand, with a purely resistive load, theoretically the feedback diodes would not conduct at all.

6-4-6-2 PWM OPERATION

The voltage and current waveforms associated with a PWM inverter are shown in Fig. 6-29. Here, as an example, the load displacement power factor angle is assumed to be 30° (lagging). Also, the output current is assumed to be a perfect sinusoid. In Figs. 6-29*a* through 6-29*c*, the phase to the negative dc-bus voltages and the phase current (v_{AN}, i_A, etc.) are plotted for approximately one-fourth of the fundamental frequency cycle.

By looking at the devices conducting in Figs. 6-29*a* through 6-29*c*, we notice that there are intervals during which the phase currents i_A, i_B, i_C flow through only the devices connected to the positive dc bus (i.e., three out of T_{A+}, D_{A+}, T_{B+}, D_{B+}, T_{C+}, and D_{C+}). This implies that during these intervals, all three phases of the load are short-circuited and there is no power input from the dc bus (i.e., $i_d = 0$) as shown in Fig. 6-30*a*. Similarly, there are intervals during which all conducting devices are connected to the negative dc bus resulting in the circuit of Fig. 6-30*b*.

The output voltage magnitude is controlled by controlling the duration of these short-circuit intervals. Such intervals of three-phase short circuit do not exist in a square-wave mode of operation. Therefore, the output voltage magnitude in an inverter operating in a square-wave mode must be controlled by controlling the input voltage V_d.

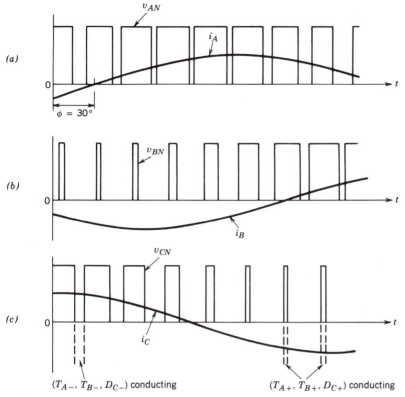

(T_{A-}, T_{B-}, D_{C-}) conducting (T_{A+}, T_{B+}, D_{C+}) conducting

FIGURE 6-29: PWM inverter waveforms: load power factor angle = 30° (lag).

6-5   EFFECT OF BLANKING TIME ON OUTPUT VOLTAGE IN PWM INVERTERS

The effect of blanking time on the output voltage is described by means of one leg of a single-phase or a three-phase full-bridge inverter, as shown in Fig. 6-31a. In the previous discussion, the switches were assumed to be ideal, which allowed the status of the two switches in an inverter leg to change simultaneously from on to off and vice versa. Concentrating on one switching time period, v_{control} is a constant dc voltage, as explained in Fig. 6-6; its comparison with a triangular waveform v_{tri}

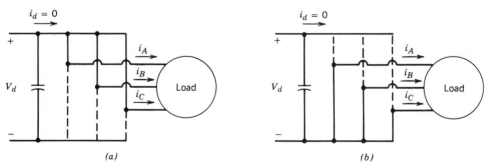

FIGURE 6-30: Short-circuit states in a three-phase PWM inverter.

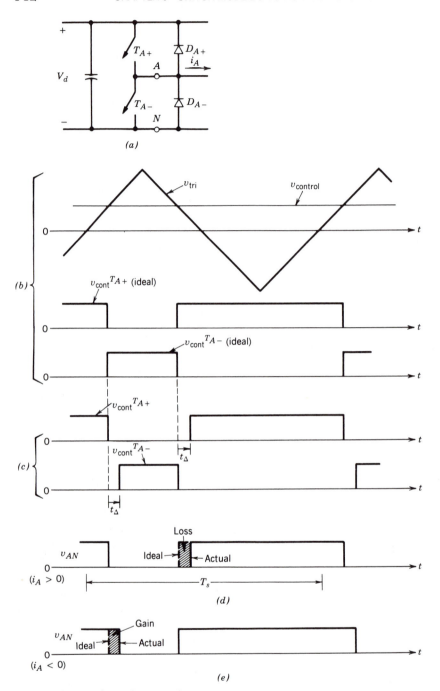

FIGURE 6-31: Effect of blanking time t_Δ.

determines the switching instants and the switch-control signals v_{cont} (ideal) as shown in Fig. 6-31b, assuming ideal switches.

In practice, because of the finite turn-off and turn-on times associated with any type of switch, a switch is turned off at the switching instant determined in Fig. 6-31b. However, the turn-on of the other switch in that inverter leg is delayed by a blanking time t_Δ, which is conservatively chosen to avoid a "shoot through" or cross-conduction current through the leg. This blanking time is chosen to be just a few microseconds for fast switching devices like MOSFETs, and larger for slower switching devices. The switch-control signals for the two switches in the presence of a blanking time are shown in Fig. 6-31c.

Since both the switches are off during the blanking time, v_{AN} during that interval depends on the direction of i_A, as shown in Fig. 6-31d for $i_A > 0$, and in Fig. 6-31e for $i_A < 0$. The ideal waveforms (without the blanking time) are shown as dotted. Comparing the ideal waveform of v_{AN} without the blanking time to the actual waveform with the blanking time, the difference between the ideal and the actual output voltage is

$$v_\varepsilon = (v_{AN})_{\text{ideal}} - (v_{AN})_{\text{actual}}$$

By averaging v_ε over one time period of the switching frequency, we can obtain the change (defined as a drop if positive) in the output voltage due to t_Δ

$$\Delta V_{AN} = + \frac{t_\Delta}{T_s} V_d \qquad i_A > 0$$

$$= - \frac{t_\Delta}{T_s} V_d \qquad i_A < 0$$

(6-75)

Equation 6-75 shows that ΔV_{AN} does not depend on the magnitude of current but its polarity depends on the current direction. Moreover, ΔV_{AN} is proportional to the blanking time t_Δ and the switching frequency $f_s(= 1/T_s)$, which suggests that at higher switching frequencies, faster switching devices that allow t_Δ to be small should be used.

Applying the same analysis to the leg B of the single-phase inverter of Fig. 6-32a, and recognizing that $i_A = -i_B$, we determine that

$$\Delta V_{BN} = - \frac{t_\Delta}{T_s} V_d \qquad i_A > 0$$

$$= + \frac{t_\Delta}{T_s} V_d \qquad i_A < 0$$

(6-76)

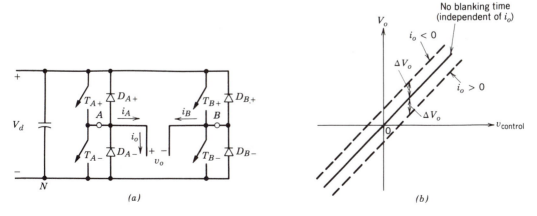

FIGURE 6-32: Effect of t_Δ on V_o where ΔV_o is defined as a voltage drop if positive.

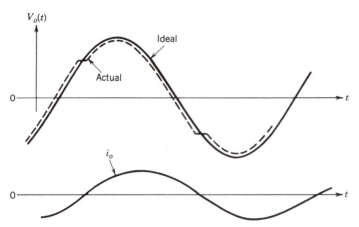

FIGURE 6-33: Effect of t_Δ on the sinusoidal output.

Since $v_o = v_{AN} - v_{BN}$ and $i_o = i_A$, the "instantaneous-average" value of the voltage difference, that is, the average value during T_s of the idealized waveform minus the actual waveform, is

$$\Delta V_o = \Delta V_{AN} - \Delta V_{BN} = +\frac{2t_\Delta}{T_s}V_d \qquad i_o > 0$$

$$= -\frac{2t_\Delta}{T_s}V_d \qquad\qquad i_o < 0 \tag{6-77}$$

A plot of "instantaneous-average" value V_o as a function of $v_{control}$ is shown in Fig. 6-32b, with and without the blanking time.

If the full-bridge converter in Fig. 6-32a is pulse-width modulated for a dc-to-dc conversion as discussed in Chapter 5, then $v_{control}$ is a constant dc voltage in steady state. The plot of Fig. 6-32b is useful in determining the effect of blanking time in applications of such converters for dc-motor drives, as will be discussed in Chapter 11.

For a sinusoidal $v_{control}$ in a single-phase full-bridge PWM inverter, the "instantaneous-average" output $V_o(t)$ is shown in Fig. 6-33 for a load current i_o which is assumed to be sinusoidal and lagging behind $V_o(t)$. The distortion in $V_o(t)$ at the current zero-crossings results in low order harmonics such as third, fifth, seventh, and so on of the fundamental frequency in the inverter output. Similar distortions occur in the line-to-line voltages at the output of a three-phase PWM inverter, where the low-order harmonics are of the order $6m \pm 1$ ($m = 1, 2, 3, \ldots$) of the fundamental frequency. The effect of these distortions due to the blanking time is further discussed in Chapter 9, dealing with uninterruptible power supplies, Chapter 11, dealing with dc motor drives, and Chapter 13, dealing with synchronous motor drives.

6-6   OTHER INVERTER SWITCHING SCHEMES

In the previous sections, two commonly used inverter switching schemes, sinusoidal PWM and square-wave, have been analyzed in detail. In this section, some other PWM schemes are briefly discussed. A detailed discussion of these techniques is presented in the references cited at the end of the chapter.

6-6-1 Square-Wave Pulse Switching

Here, each phase voltage output is essentially square-wave except for a few notches (or pulses) to control the fundamental amplitude. These notches are introduced without any regard to the harmonic content in the output, and therefore this type of scheme is no longer employed except in some thyristor inverters. A serious drawback of the foregoing techniques is that no attention is paid to the output harmonic content, which can become unacceptable. The advantage is in their simplicity and a small number of switchings required (which is significant in high-power thyristor inverters).

6-6-2 Programmed Harmonic Elimination Switching

This technique combines the square-wave switching and the pulse-width modulation to control the fundamental output voltage as well as to eliminate the designated harmonics from the output.

The voltage v_{Ao}, of an inverter leg, normalized by $V_d/2$, is plotted in Fig. 6-34a, where six notches are introduced in the otherwise square-wave output, to control the magnitude of the fundamental voltage and to eliminate fifth and seventh harmonics. On a half-cycle basis, each notch provides one degree of freedom, that is, having three notches per half-cycle provides control of fundamental and elimination of two harmonics (in this case fifth and seventh).

Figure 6-34a shows that the output waveform has odd half-wave symmetry (sometimes it is referred to as odd quarter-wave symmetry). Therefore, only odd harmonics (coefficients of sine series) will be present. Since in a three-phase inverter (consisting of three such inverter legs), the third harmonic and its multiples are cancelled out in the output, these harmonics need not be eliminated from the output of the inverter leg by means of waveform notching.

A careful examination shows that the switching frequency of a switch in Fig. 6-34a is seven times the switching frequency associated with a square-wave operation.

In a square-wave operation, the fundamental frequency voltage component is

$$\frac{(\hat{V}_{Ao})_1}{(V_d/2)} = \frac{4}{\pi} = 1.273 \qquad \text{(Eq. 6-13, repeated)}$$

Because of the notches to eliminate fifth and seventh harmonics, the maximum available fundamental amplitude is reduced. It can be shown that

$$\frac{(\hat{V}_{Ao})_{1,\max}}{(V_d/2)} = 1.188 \qquad (6\text{-}78)$$

The required values of α_1, α_2, and α_3 are plotted in Fig. 6-34b, as a function of the normalized fundamental in the output voltage (see References 8 and 9 for details).

To allow control over the fundamental output and to eliminate fifth, seventh, eleventh, and the thirteenth order harmonics, five notches per half-cycle would be needed. In that case, each switch would have 11 times the switching frequency compared with a square wave operation.

With the help of VLSI circuits and microcontrollers, this programmed harmonic elimination scheme can be implemented. Without making the switching frequency (and therefore the switching losses) very high, it allows the undesirable lower order

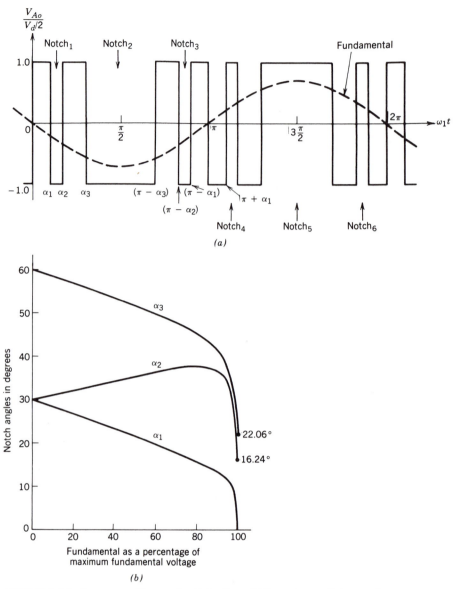

FIGURE 6-34: Programmed harmonic elimination of fifth and seventh harmonics.

harmonics to be eliminated. The higher order harmonics can be filtered by a small filter, if necessary. However, before deciding on this technique, it should be compared with a sinusoidal PWM technique with a low m_f, to evaluate which one is better. It should be noted that the distortions due to the blanking time, as discussed in Section 6-5, occur here as well.

6-6-3 Current-Regulated (Current-Mode) Modulation

In applications such as dc- and ac-motor servo drives, it is the motor current (supplied by the switch-mode converter or inverter) that needs to be controlled, even though a voltage-source inverter (VSI) is often used. In this regard, it is similar to the switching dc power supplies of Chapter 8, where the output-stage current can be controlled in order to regulate the output voltage.

There are various ways to obtain the switching signals for the inverter switches in order to control the inverter output current. Two such methods are described.

6-6-3-1 TOLERANCE BAND CONTROL

This is illustrated by Fig. 6-35 for a sinusoidal reference current $i_A{}^*$, where the actual phase current i_A is compared with the tolerance band around the reference current associated with that phase. If the actual current in Fig. 6-35a tries to go beyond the upper tolerance band, T_{A-} is turned on (i.e., T_{A+} is turned off). The opposite switching occurs if the actual current tries to go below the lower tolerance band. Similar actions take place in the other two phases. This control is shown in a block-diagram form in Fig. 6-35b.

The switching frequency depends on how fast the current changes from the upper limit to the lower limit and vice versa. This, in turn, depends on V_d, the load back-emf and the load inductance. Moreover, the switching frequency does not remain constant but varies along the current waveform.

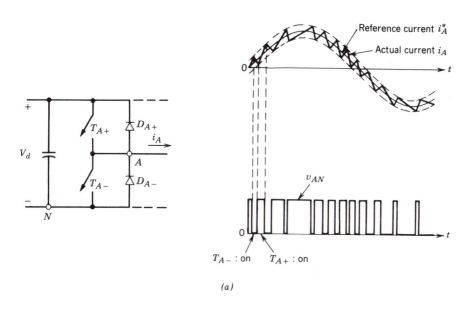

(a)

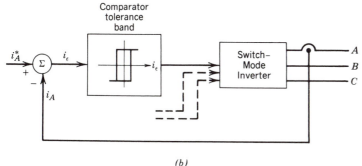

(b)

FIGURE 6-35: Tolerance-band current control.

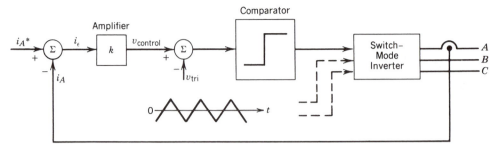

FIGURE 6-36: Fixed-frequency current control.

6-6-3-2 FIXED-FREQUENCY CONTROL

The fixed-frequency current control is shown in a block-diagram form in Fig. 6-36. The error between the reference and the actual current is amplified or fed through a proportional-integral (PI) regulator. The output $v_{control}$ of the amplifier is compared with a fixed-frequency (switching frequency f_s) triangular waveform v_{tri}. A positive error ($i_A^* - i_A$) and, hence, a positive $v_{control}$ results in a larger inverter output voltage, thus bringing i_A to its reference value. Similar action takes place in the other two phases.

It should be noted that many sophisticated switching techniques that minimize the total number of combined switchings in all three phases are discussed in the literature.

6-6-4   Switching Scheme Incorporating Harmonic Neutralization by Modulation and Transformer Connections

In some applications such as three-phase uninterruptible ac power supplies, it is usually required to have isolation transformers at the output. In such applications, the presence of output transformers is utilized in eliminating certain harmonics. In addition, the programmed harmonic elimination technique can be used to control the fundamental output and to eliminate (or reduce) a few more harmonics.

This arrangement is discussed in connection with the uninterruptible ac power supplies in Chapter 9.

6-7   RECTIFIER MODE OF OPERATION

As we discussed in the introduction in Section 6-1, these switch-mode converters can make a smooth transition from the inverter mode to the rectifier mode. The rectifier mode of operation results, for example, during braking (slowing down) of induction motors supplied through a switch-mode converter. This mode of operation is briefly discussed in this section. The switch-mode rectifiers, used for interfacing power electronics equipment with the utility grid, operate on the same basic principle and are discussed in detail in Chapter 17.

The rectifier mode of operation is discussed only for the three-phase converters; the same principle applies to single-phase converters. Assuming a balanced steady-state operating condition, a three-phase converter is discussed on a per-phase basis.

As an example, consider the three-phase system shown in Fig. 6-25a, which is redrawn in Fig. 6-37a. Consider only the fundamental frequency (where the subscript

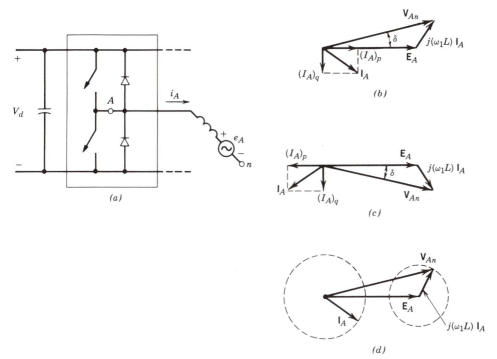

FIGURE 6-37: Operation modes: (*a*) circuit, (*b*) inverter mode, (*c*) rectifier mode, (*d*) constant I_A.

1 is omitted), neglecting the switching frequency harmonics. In Fig. 6-37*b*, a motoring mode of operation is shown where the converter voltage $\mathbf{V}_{An}$ applied to the motor leads $\mathbf{E}_A$ by an angle δ. The active (real) component $(I_A)_p$ of $\mathbf{I}_A$ is in phase with $\mathbf{E}_A$ and therefore the converter is operating in an inverter mode.

The phase angle (as well as the magnitude) of the ac voltage produced by the converter can be controlled. If the converter voltage $\mathbf{V}_{An}$ is now made to lag $\mathbf{E}_A$ by the same angle δ as before (keeping V_{An} constant), the phasor diagram in Fig. 6-37*c* shows that the active component $(I_A)_p$ of $\mathbf{I}_A$ is now 180° out of phase with $\mathbf{E}_A$, resulting in a rectifier mode of operation where the power flows from the motor to the dc side of the converter.

In fact, $\mathbf{V}_{An}$ can be controlled both in magnitude (within limits) and phase, thus allowing a control over the current magnitude and the power level, for example during the ac-motor braking. Assuming that $\mathbf{E}_A$ cannot change instantaneously, Fig. 6-37*d* shows the locus of the $\mathbf{V}_{An}$ phasor, which would keep the magnitude of the current constant.

The waveforms of Fig. 6-22 can be used for explaining how to control $\mathbf{V}_{An}$ in magnitude, as well as in phase, with a given (fixed) dc voltage V_d. It is obvious that by controlling the amplitude of the sinusoidal reference waveform $v_{\text{control, }A}$, V_{An} can be varied. Similarly, by shifting the phase of $v_{\text{control, }A}$ with respect to $\mathbf{E}_A$, the phase angle of $\mathbf{V}_{An}$ can be varied. For a balanced operation, the control voltages for phases B and C are equal in magnitude, but $\pm 120°$ displaced with respect to the control voltage of phase A.

Switch-mode rectifiers, where the rectifier is the primary mode of operation, are further discussed in Chapter 17, which deals with circuits for interfacing power electronics equipment with the utility grid.

6-8   SUMMARY

1. Switch-mode, voltage-source dc-to-ac inverters are described, which accept dc voltage source as input and produce either single-phase or three-phase sinusoidal output voltages at a low frequency relative to the switching frequency (current-source inverters are described in Chapter 12).

2. These dc-to-ac inverters can make a smooth transition into the rectification mode, where the flow of power reverses to be from the ac side to the dc side. This occurs, for example, during braking of an induction motor supplied through such an inverter.

3. The sinusoidal-PWM switching scheme allows control of the magnitude and the frequency of the output voltage. Therefore, the input to the PWM inverters is an uncontrolled, essentially constant dc voltage source. This switching scheme results in harmonic voltages in the range of the switching frequency and higher, which can be easily filtered out.

4. The square-wave switching scheme controls only the frequency of the inverter output. Therefore, the output magnitude must be controlled by controlling the magnitude of the input dc voltage source. The square-wave output voltage contains low-order harmonics. A variation of the square-wave switching scheme, called the voltage cancellation technique, can be used to control both the frequency and the magnitude of the single-phase (but not three-phase) inverter output.

5. As a consequence of the harmonics in the inverter output voltage, the ripple in the output current does not depend on the level of power transfer at the fundamental frequency; instead the ripple depends inversely on the load inductance, which is more effective at higher frequencies.

6. In practice, if a switch turns off in an inverter leg, the turn-on of the other switch is delayed by a blanking time, which introduces low-order harmonics in the inverter output.

7. There are many other switching schemes in addition to the sinusoidal-PWM. For example, the programmed harmonic elimination switching technique can be easily

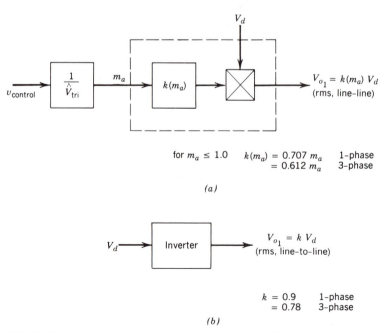

FIGURE 6-38: Summary of inverter output voltages: (a) PWM operation $(m_a \leq 1)$, (b) square-wave operation.

implemented with the help of VLSI circuits to eliminate specific harmonics from the inverter output.

8. The current-regulated (current-mode) modulation allows the inverter output current(s) to be controlled directly by comparing the measured actual current with the reference current, and using the error to control the inverter switches. As will be discussed in Chapters 11 to 13, this technique is extensively used for dc and ac servodrives. The current-mode control is also used in dc-to-dc converters, as discussed in Chapter 8, dealing with switching dc power supplies.

9. The relationship between the control input and the full-bridge inverter output magnitude can be summarized as shown in Fig. 6-38a, assuming a sinusoidal-PWM in the linear range of $m_a \leq 1.0$. For a square-wave switching, the inverter does not control the magnitude of the inverter output and the relationship between the dc input voltage and the output magnitude are summarized in Fig. 6-38b.

10. These converters can be used for interfacing power electronics equipment with the utility source. As discussed in Chapter 17, because of rectification being the primary mode of operation, these are called switch-mode ac-to-dc rectifiers.

PROBLEMS

6-1. In a single-phase full-bridge PWM inverter, the input dc voltage varies in a range of 295 V to 325 V. Because of the low distortion required in the output v_o, $m_a \leq 1.0$.
(a) What is the highest V_{o1} that can be obtained and stamped on its nameplate as its voltage rating?
(b) Its nameplate volt-ampere rating is specified as 2000 VA, that is, $V_{o1,\max} I_{o,\max} = 2000$ VA, where i_o is assumed to be sinusoidal.
    Calculate the combined switch-utilization ratio when the inverter is supplying its rated volt-amperes.

6-2. Consider the problem of ripple in the output current of a single-phase full-bridge inverter. $V_{o1} = 220$ V at a frequency of 47 Hz and the type of load is as shown in Fig. 6-18a with $L = 100$ mH.
    If the inverter is operating in a square-wave mode, calculate the peak value of the ripple current.

6-3. Repeat Problem 6-2 with the inverter operating in a sinusoidal-PWM mode, with $m_f = 21$ and $m_a = 0.8$.

6-4. Repeat Problem 6-2 but assume that the output voltage is controlled by voltage cancellation and V_d has the same value as required in the PWM inverter of Problem 6-3.

6-5. Calculate and compare the peak values of the ripple currents in Problems 6-2 through 6-4.

6-6. Consider the problem of ripple in the output current of a three-phase square-wave inverter. $(V_{LL})_1 = 200$ V at a frequency of 52 Hz and the type of load is as shown in Fig. 6-25a with $L = 100$ mH. Calculate the peak ripple current defined in Fig. 6-26a.

6-7. Repeat Problem 6-6 if the inverter of Problem 6-6 is operating in a synchronous-PWM mode with $m_f = 39$ and $m_a = 0.8$. Calculate the peak ripple current defined in Fig. 6-26b.

6-8. Obtain an expression for the Fourier components in the waveform of Fig. 6-34a for programmed harmonic elimination of the fifth and the seventh order harmonics. Show that for $\alpha_1 = 0$, $\alpha_2 = 16.24°$, and $\alpha_3 = 22.06°$, the fifth and the seventh harmonics are eliminated and the fundamental frequency output of the inverter has a maximum amplitude given by Eq. 6-78.

6-9. In the three-phase, square-wave inverter of Fig. 6-24a, consider the load to be balanced

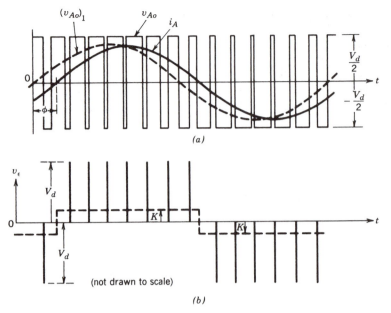

FIGURE P6-11:

and purely resistive with a load-neutral n. Draw the steady-state v_{An}, i_A, $i_{D_{A+}}$, and i_d waveforms where $i_{D_{A+}}$ is the current through D_{A+}.

6-10. Repeat Problem 6-9 by assuming that the load is purely inductive, where the load resistance, though finite, can be neglected.

6-11. Consider only one inverter leg as shown in Fig 6-4, where the output current lags $(v_{Ao})_1$ by an angle ϕ, as shown in Fig. P6-11a, and "o" is the fictitious midpoint of the dc input. Because of the blanking time t_Δ, the instantaneous error voltage v_ϵ is plotted in Fig. P6-11b, where

$$v_\epsilon = (v_{Ao})_{\text{ideal}} - (V_{Ao})_{\text{actual}}$$

Each v_ϵ pulse, either positive or negative, has an amplitude of V_d and a duration of t_Δ. In order to calculate the low-order harmonics of the fundamental frequency in the output voltage due to blanking time, these pulses can be replaced by an equivalent rectangular pulse (shown dotted in Fig. P6-11b) of amplitude K whose volt-second area per half-cycle equals that of v_ϵ pulses.

Derive the following expression for the harmonics of the fundamental frequency in v_{Ao} introduced by the blanking time:

$$\left(\hat{V}_{Ao}\right)_h = \frac{4}{\pi h} V_d \cdot t_\Delta \cdot f_s \qquad (h = 1, 3, 5, \dots)$$

where f_s is the switching frequency.

6-9 REFERENCES

1. K. THORBORG, *Power Electronics*, Prentice Hall International (UK) Ltd, 1988.
2. A. B. PLUNKETT, "A Current-Controlled PWM Transistor Inverter Drive," IEEE/IAS 1979 Annual Meeting, pp. 785–792.

3. T. KENJO and S. NAGAMORI, *Permanent Magnet and Brushless DC Motors,* Clarendon Press, Oxford, 1985.

4. H. AKAGI, A. NABAE and S. ATOH, "Control Strategy of Active Filters Using Multiple Voltage-Source PWM Converters," *IEEE Transactions on Industry Applications*, Vol. IA-22, No. 3, May/June 1986, pp. 460–465.

5. T. KATO, "Precise PWM Waveform Analysis of Inverter for Selected Harmonic Elimination," 1986 IEEE/IAS Annual Meeting, pp. 611–616.

6. J. W. A. WILSON and J. A. YEAMANS, "Intrinsic Harmonics of Idealized Inverter PWM Systems," 1976 IEEE/IAS Annual Meeting, pp. 967–973.

7. Y. MURAI, T. WATANABE, and H. IWASAKI, "Waveform Distortion and Correction Circuit for PWM Inverters with Switching Lag-Times," 1985 IEEE/IAS Annual Meeting, pp. 436–441.

8. H. PATEL and R. G. HOFT, "Generalized Techniques of Harmonic Elimination and Voltage Control in Thyristor Inverters: Part I—Harmonic Elimination," *IEEE Transactions on Industry Applications*, Vol. IA-9, No. 3, May/June 1973.

9. H. PATEL and R. G. HOFT, "Generalized Techniques of Harmonic Elimination and Voltage Control in Thyristor Inverters: Part II—Voltage Control Techniques," *IEEE Transactions on Industry Applications,* Vol. IA-10, No. 5, September/October 1974.

10. I. J. PITEL, S. N. TALUKDAR, and P. WOOD, "Characterization of Programmed-Waveform Pulsewidth Modulation," *IEEE Transactions on Industry Applications,* Vol. IA-16, No. 5, September/October 1980.

11. M. BOOST and P. D. ZIOGAS, "State-of-the-Art PWM Techniques: A Critical Evaluation," IEEE Power Electronics Specialists Conference 1986, pp. 425–433.

12. J. ROSA, "The Harmonic Spectrum of D.C. Link Currents in Inverters," *Fourth International PCI Conference on Power Conversion,* March 1982, San Francisco, pp. 38–52, Published by Intertec Communications, Inc. 2909 Ocean Drive, Oxnard, CA 93030.

13. EPRI REPORT, "AC/DC Power Converter for Batteries and Fuel Cells," Project 841-1, Final Report, September 1981, EPRI, Palo Alto, CA.

7

Resonant Converters: Zero-Voltage and / or Zero-Current Switchings

_____
_____
_____

7-1   INTRODUCTION

In all the pulse-width-modulated dc-to-dc and dc-to-ac converter topologies discussed in Chapters 5 and 6, the controllable switches are operated in a switch mode where they are required to turn on and turn off the entire load current during each switching. In this switch-mode operation, as explained further in Section 7-1-1, the switches are subjected to high switching stresses and high switching power loss that increases linearly with the switching frequency of the pulse-width modulation. Another significant drawback of the switch-mode operation is the electromagnetic interference (EMI) produced due to large di/dt and dv/dt caused by a switch-mode operation.

These shortcomings of switch-mode converters are exacerbated if the switching frequency is increased in order to reduce the converter size and weight, and hence to increase the power density. Therefore, to realize high switching frequencies in converters, the aforementioned shortcomings are minimized if each switch in a converter changes its status (from on to off or vice versa) when the voltage across it and/or the current through it is zero at the switching instant. The converter topologies and the switching strategies, which result in zero-voltage and/or zero-current switchings, are discussed in this chapter. Since most of these topologies (but not all) require some form of L-C resonance, these are broadly classified as "resonant converters."

7-1-1   Switch-Mode Inductive Current Switching

This topic was briefly reviewed in Chapter 2. To illustrate further the problems associated with switch-mode operation, consider one of the legs of a full-bridge dc–dc converter or a dc-to-ac inverter (single-phase or three-phase), as shown in Fig. 7-1. The output current can be in either direction and can be assumed to have a constant magnitude I_o due to the load inductance, during the very brief switching interval. The

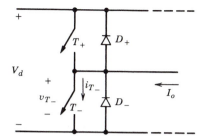

FIGURE 7-1: One inverter leg.

linearized voltage and current waveforms, for example, for the lower switch T_- are shown in Fig. 7-2a.

Initially, I_o is assumed to be flowing through T_-. If a control signal is applied to turn T_- off, the switch voltage v_{T_-} increases to V_d (it overshoots V_d due to stray inductances) and then, the switch current i_{T_-} decays to zero. After the turn-off of T_-, I_o flows through D_+. The power loss $P_{T_-}(= v_{T_-} \cdot i_{T_-})$ in the switch during turn-off is shown in Fig. 7-2a.

Now consider the turn-on of T_-. Prior to the turn-on of T_-, I_o is flowing through D_+. When the switch control signal is applied to turn T_- on, i_{T_-} increases to I_o plus the peak-reverse-recovery current of the diode D_+, as shown in Fig. 7-2a. Subsequently, the diode recovers and the switch voltage v_{T_-} decreases to essentially zero. Similar to the turn-off interval, the simultaneous presence of v_{T_-} and i_{T_-} results in a switching power loss in T_- during turn-on.

The average value the switching loss P_{T_-}, being proportional to the switching frequency, limits how high the switching frequency can be pushed, without significantly degrading the system efficiency. With the availability of fast switches (with the switching times as low as a few tens of nanoseconds), the present limit seems to be up to approximately 500 kHz with a reasonable energy efficiency.

Another significant disadvantage of the switch-mode operation is that it results in large di/dt and dv/dt due to fast switching transitions required to keep the

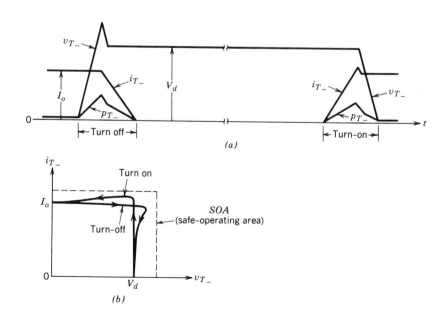

FIGURE 7-2: Switch-mode inductive current switchings.

switching losses in the switch as low as possible. Diodes with poor reverse recovery characteristics significantly add to this phenomenon, which produces electromagnetic interference (EMI).

Switch-mode inductive current switching results in switching loci in the v_T-i_T plane, as shown in Fig. 7-2b. Because a large switch-voltage and a large switch-current occur simultaneously, the switch must be capable of withstanding high switching stresses, with a safe operating area (SOA) as shown by the dotted lines. This requirement to be able to withstand such large stresses results in undesirable design compromises in other characteristics of the power semiconductor devices.

7-1-2   Zero-Voltage and Zero-Current Switchings

Switching frequencies in the megahertz range, even tens of megahertz, are being contemplated to reduce the size and the weight of transformers and filter components and, hence, to reduce the cost as well as the size and the weight of power electronics converters. Realistically, the switching frequencies can be increased to such high values only if the problems of switch stresses, switching losses, and the EMI associated with the switch-mode converters can be overcome.

The switch stresses, as discussed in later chapters in this book, can be reduced by connecting simple dissipative snubber circuits (consisting of diodes and passive components) in series and parallel with the switches in the switch-mode converters. Such snubber circuits are shown in Fig. 7-3a and the switching loci that result in reduced switch stresses are shown in Fig. 7-3b. However, these dissipative snubbers shift the switching power loss from the switch to the snubber circuit, and therefore do not provide a reduction in the overall switching power loss.

In contrast to dissipative snubbers in switch-mode converters, the combination of proper converter topologies and switching strategies can overcome the problems of switching stresses, switching power losses, and the EMI by turning on and turning off each of the converter switches when either the switch voltage or the switch current is zero. Ideally, both the switch voltage and current should be zero when the switching transition occurs.

As a brief introduction, once again consider a one-leg inverter of Fig. 7-1. If both the turn-on and turn-off switchings occur under a zero-voltage and/or a zero-current condition, then the switching loci are shown in Fig. 7-4, where the switching loci in the switch-mode are shown (by dotted curves) for comparison purposes. Such switching loci, without dissipative snubbers, reduce switch stresses, switching power losses, and the EMI.

7-2   CLASSIFICATION OF RESONANT CONVERTERS

The resonant converters are defined here as the combination of converter topologies and switching strategies that result in zero-voltage and/or zero-current switchings. One way to categorize these converters is as follows:

1. Load-resonant converters
2. Resonant-switch converters
3. Resonant-dc-link converters
4. High-frequency-link integral-half-cycle converters

These classifications are explained further.

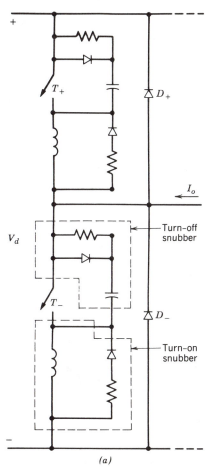

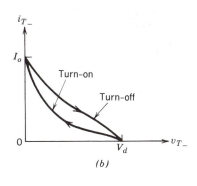

FIGURE 7-3: Dissipative snubbers: (a) snubber circuits, (b) switching loci with snubbers.

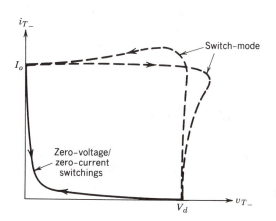

FIGURE 7-4: Zero-voltage / zero-current switching loci.

7-2-1   Load-Resonant Converters

These converters consist of an L-C resonant tank circuit. Oscillating voltage and current, due to L-C resonance in the tank, are applied to the load, and the converter switches can be switched at zero voltage and/or zero current. Either a series L-C or a parallel L-C circuit can be used. In these converter circuits, the power flow to the load is controlled by the resonant tank impedance, which in turn is controlled by the switching frequency f_s in comparison to the resonant frequency f_0 of the tank. These dc-to-dc and dc-to-ac converters can be subclassified as follows:

1. Voltage-source series-resonant converters
   a. Series-loaded resonant (SLR) converters
   b. Parallel-loaded resonant (PLR) converters
   c. Hybrid-resonant converters
2. Current-source parallel-resonant converters
3. Class-E and subclass-E resonant converters

7-2-2 Resonant-Switch Converters

In certain switch-mode converter topologies, an L-C resonance can be utilized primarily to shape the switch voltage and current to provide zero-voltage and/or zero-current switchings. In such resonant-switch converters, during one switching-frequency time period, there are resonant as well as nonresonant operating intervals. Therefore, these converters in the literature have also been termed quasi-resonant converters. They can be subclassified as follows:

1. Resonant-switch dc–dc converters
   a. Zero-current-switching (ZCS) converters
   b. Zero-voltage-switching (ZVS) converters
2. Zero-voltage-switching, clamped-voltage (ZVS-CV) converters which are also referred to as pseudo-resonant converters and resonant-transition converters, respectively in References 34 and 31.

7-2-3 Resonant-dc-Link Converters

In the conventional switch-mode PWM dc-to-ac inverters, the input V_d to the inverter is a fixed magnitude dc, and the sinusoidal output (single-phase or three-phase) is obtained by switch-mode PWM switchings. However, in the resonant-dc-link converters, the input voltage is made to oscillate around V_d by means of an L-C resonance so that the input voltage remains zero for a finite duration during which the status of the inverter switches can be changed, thus resulting in zero-voltage switch-ings.

7-2-4 High-Frequency-Link Integral-Half-Cycle Converters

If the input to a single-phase or three-phase inverter is a high-frequency sinusoidal ac, then by using bidirectional switches it is possible to synthesize a low-frequency ac of adjustable magnitude and frequency or an adjustable magnitude dc, where the switches are turned on and turned off at the zero-crossings of the input voltage.

7-3   BASIC RESONANT CIRCUIT CONCEPTS

Some basic configurations encountered in the resonant converters discussed in this chapter are analyzed in a generic fashion. Appropriate assumptions are made to keep the analysis simple.

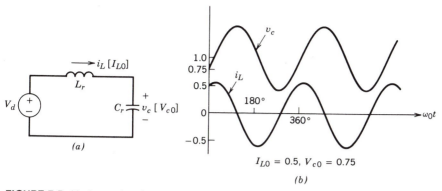

FIGURE 7-5: Undamped series-resonant circuit; i_L and v_c are normalized: (a) circuit, (b) waveforms with $I_{L0} = 0.5$, $V_{c0} = 0.75$.

The initial conditions are indicated by the uppercase letters, subscript "0" and the square brackets, for example, $[V_{c0}]$ and $[I_{c0}]$.

7-3-1 Series-Resonant Circuits

7-3-1-1 UNDAMPED SERIES-RESONANT CIRCUIT

Figure 7-5a shows an undamped series resonant circuit where the input voltage is V_d at time t_0. The initial conditions are I_{L0} and V_{c0}. With the inductor current i_L and the capacitor voltage v_c as the state variables, the circuit equations are

$$L_r \frac{di_L}{dt} + v_c = V_d \tag{7-1}$$

and

$$C_r \frac{dv_c}{dt} = i_L \tag{7-2}$$

The solution of this set of equations for $t \geq t_0$ is as follows:

$$i_L(t) = I_{L0} \cos \omega_0(t - t_0) + \frac{V_d - V_{c0}}{Z_0} \sin \omega_0(t - t_0) \tag{7-3}$$

and

$$v_c(t) = V_d - (V_d - V_{c0}) \cos \omega_0(t - t_0) + Z_0 I_{L0} \sin \omega_0(t - t_0) \tag{7-4}$$

where

$$\text{Angular resonance frequency} \quad \omega_0 = 2\pi f_0 = \frac{1}{\sqrt{L_r C_r}} \tag{7-5}$$

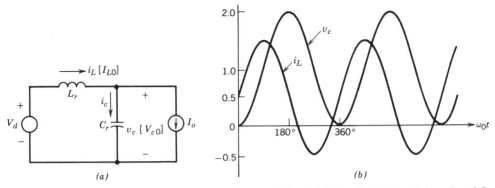

FIGURE 7-6: Series-resonant circuit with capacitor-parallel load: (a) circuit, (b) $V_{c0} = 0$, $I_{L0} = I_o = 0.5$; i_L and v_c are normalized.

and

$$\text{Characteristic impedance} \quad Z_0 = \sqrt{\frac{L_r}{C_r}} \text{ ohms} \qquad (7\text{-}6)$$

To plot normalized v_c and i_L, the following base quantities are chosen:

$$V_{\text{base}} = V_d \qquad (7\text{-}7)$$

and

$$I_{\text{base}} = \frac{V_d}{Z_0} \qquad (7\text{-}8)$$

As an example, the normalized i_L and v_c are plotted in Fig. 7-5b for $I_{L0} = 0.5$ and $V_{c0} = 0.75$.

7-3-1-2 SERIES-RESONANT CIRCUIT WITH A CAPACITOR-PARALLEL LOAD

Figure 7-6a shows a circuit where in a series resonant circuit, the capacitor is in parallel with a current I_o, which represents the load. In this circuit, V_d and I_o are dc quantities. The initial conditions are I_{L0} and V_{c0} at the initial time t_0. Therefore

$$v_c = V_d - L_r \frac{di_L}{dt} \qquad (7\text{-}9)$$

and

$$i_L - i_c = I_o \qquad (7\text{-}10)$$

By differentiating Eq. 7-9

$$i_c = C_r \frac{dv_c}{dt} = -L_r C_r \frac{d^2 i_L}{dt^2} \qquad (7\text{-}11)$$

Substituting i_c from Eq. 7-11 into Eq. 7-10

$$\frac{d^2 i_L}{dt^2} + \omega_0^2 i_L = \omega_0^2 I_o \qquad (7\text{-}12)$$

where ω_0 is the same as in Eq. 7-5. Solution of these equations for $t \geq t_0$ is as follows:

$$i_L(t) = I_o + (I_{L0} - I_o)\cos\omega_0(t - t_0) + \frac{V_d - V_{c0}}{Z_0}\sin\omega_0(t - t_0) \qquad (7\text{-}13)$$

and

$$v_c(t) = V_d - (V_d - V_{c0})\cos\omega_0(t - t_0) + Z_0(I_{L0} - I_o)\sin\omega_0(t - t_0) \qquad (7\text{-}14)$$

where ω_0 is the angular resonant frequency as defined in Eq. 7-5 and Z_0 is the characteristic impedance defined in Eq. 7-6.

In a special case with $V_{c0} = 0$ and $I_{L0} = I_o$,

$$i_L(t) = I_o + \frac{V_d}{Z_0}\sin\omega_0(t - t_0) \qquad (7\text{-}15)$$

and

$$v_c(t) = V_d\big[1 - \cos\omega_0(t - t_0)\big] \qquad (7\text{-}16)$$

For this special case, Fig. 7-6b shows the plot of i_L and v_c, which are normalized by using Eqs. 7-7 and 7-8, respectively, and $I_{L0} = I_o = 0.5$ per unit.

7-3-1-3 FREQUENCY CHARACTERISTICS OF A SERIES RESONANT CIRCUIT

It is informative to obtain the frequency characteristics of the series-resonant circuit of Fig. 7-7a. The resonance frequency ω_0 and the characteristic impedance Z_0 are defined by Eqs. 7-5 and 7-6, respectively. In the presence of a load resistance R,

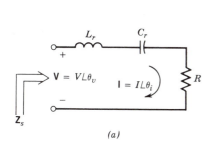

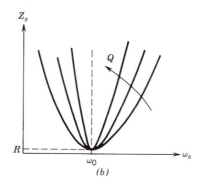

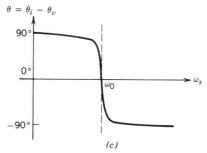

FIGURE 7-7: Frequency characteristics of a series-resonant circuit.

another quantity called the quality factor Q is defined as

$$Q = \frac{\omega_0 L_r}{R} = \frac{1}{\omega_0 C_r R} = \frac{Z_0}{R} \tag{7-17}$$

Figure 7-7b shows the magnitude Z_s of the circuit impedance as a function of frequency with Q as a parameter, keeping R constant. It shows that Z_s is a pure resistance equal to R at $\omega_s = \omega_0$ and is very sensitive to frequency deviation from ω_0 at higher values of Q.

Figure 7-7c shows the current phase angle $\theta(= \theta_i - \theta_v)$ as a function of frequency. The current leads at frequencies below ω_0 ($\omega_s < \omega_0$), where the capacitor impedance dominates over inductor impedance. At frequencies above ω_0 ($\omega_s > \omega_0$), the inductor impedance dominates over the capacitor impedance and the current lags the voltage, with the current phase angle θ approaching $-90°$.

7-3-2 Parallel-Resonant Circuits

7-3-2-1 UNDAMPED PARALLEL-RESONANT CIRCUIT

Figure 7-8a shows an undamped parallel-resonant circuit supplied by a dc current I_d. The initial conditions at time $t = t_0$ are I_{L0} and V_{c0}. With the inductor current i_L and the capacitor voltage v_c as the state variables, the circuit equations are

$$i_L + C_r \frac{dv_c}{dt} = I_d \tag{7-18}$$

and

$$v_c = L_r \frac{di_L}{dt} \tag{7-19}$$

The solution of the foregoing sets of equations for $t \geq t_0$ is as follows:

$$i_L(t) = I_d + (I_{L0} - I_d) \cos \omega_0(t - t_0) + \frac{V_{c0}}{Z_0} \sin \omega_0(t - t_0) \tag{7-20}$$

and

$$v_c(t) = Z_0(I_d - I_{L0}) \sin \omega_0(t - t_0) + V_{c0} \cos \omega_0(t - t_0) \tag{7-21}$$

where

$$\omega_0 = \frac{1}{\sqrt{L_r C_r}} \tag{7-22}$$

and

$$Z_0 = \sqrt{\frac{L_r}{C_r}} \tag{7-23}$$

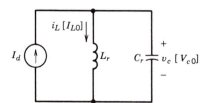

FIGURE 7-8: Undamped parallel resonant circuit.

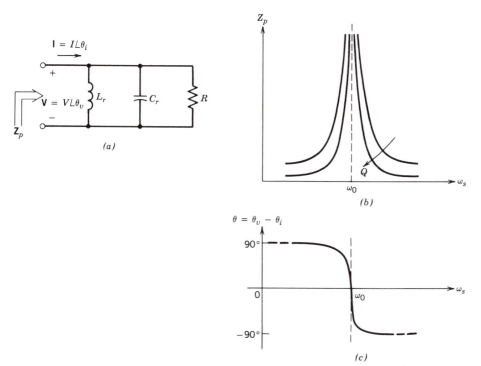

FIGURE 7-9: Frequency characteristics of a parallel-resonant circuit.

7-3-2-2 FREQUENCY CHARACTERISTICS OF PARALLEL-RESONANT CIRCUIT

It is informative to obtain the frequency characteristics of the parallel-resonance circuit of Fig. 7-9a. The resonance frequency ω_0 and Z_0 are as defined by Eqs. 7-22 and 7-23, respectively. In the presence of a load resistor R, another quantity called the quality factor Q is defined, where

$$Q = \omega_0 R C_r = \frac{R}{\omega_0 L_r} = \frac{R}{Z_0} \qquad (7\text{-}24)$$

Figure 7-9b shows the magnitude Z_p of the circuit impedance as a function of frequency with Q as a parameter, keeping R constant.

Figure 7-9c shows the voltage phase angle $\theta(= \theta_v - \theta_i)$ as a function of frequency. The voltage leads the current at frequencies below ω_0 ($\omega_s < \omega_0$), where the inductor impedance is lower than the capacitor impedance, and hence the inductor current dominates. At frequencies above ω_0 ($\omega_s > \omega_0$), the capacitor impedance is lower and the voltage lags the current, with the voltage phase angle θ approaching $-90°$.

7-4 LOAD-RESONANT CONVERTERS

In these resonant converters, an L-C tank is used that results in oscillating load voltage and current, and thus provides zero-voltage and/or zero-current switchings. Each circuit in this category is analyzed with a load that is most practical for the converter topology being considered. Only the steady-state operation is considered.

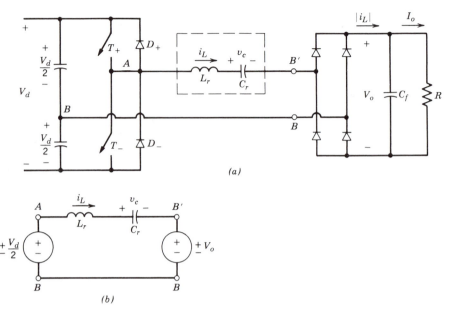

FIGURE 7-10: SLR dc–dc converter: (*a*) half-bridge, (*b*) equivalent circuit.

7-4-1   Series-Loaded Resonant (SLR) DC–DC Converters

A half-bridge configuration of the SLR converter is shown in Fig. 7-10*a*. The waveforms and the operating principles are the same for the full-bridge configurations. A transformer can be included to provide the output voltage of a desired magnitude, as well as the electrical isolation between the input and the output.

L_r and C_r form the series-resonant tank, and the current through the resonant-tank circuit is full-wave rectified at the output, and $|i_L|$ feeds the output stage. Therefore, as the name suggests, the output load appears in series with the resonant tank.

The filter capacitor C_f at the output is usually very large, and therefore the output voltage across the capacitor can be assumed to be a dc voltage without any ripple. The resistive power loss in the resonant circuit is assumed to be negligible, which greatly simplifies the analysis. The output voltage V_o is reflected across the rectifier input as $v_{B'B}$ where $v_{B'B} = V_o$ if i_L is positive and $v_{B'B} = -V_o$ if i_L is negative.

When i_L is positive, it flows through T_+ if it is on, otherwise through the diode D_-. Similarly, when i_L is negative, it flows through T_- if it is on, otherwise through the diode D_+. Therefore, in the circuit of Fig. 7-10*a*:

for $i_L > 0$

$$T_+ \text{ conducting:} \quad v_{AB} = +\frac{V_d}{2}, \quad v_{AB'} = +\frac{V_d}{2} - V_o \qquad (7\text{-}25)$$

$$D_- \text{ conducting:} \quad v_{AB} = -\frac{V_d}{2}, \quad v_{AB'} = -\frac{V_d}{2} - V_o \qquad (7\text{-}26)$$

for $i_L < 0$

$$T_- \text{ conducting:} \quad v_{AB} = -\frac{V_d}{2}, \quad v_{AB'} = -\frac{V_d}{2} + V_o \qquad (7\text{-}27)$$

$$D_+ \text{ conducting:} \quad v_{AB} = +\frac{V_d}{2}, \quad v_{AB'} = +\frac{V_d}{2} + V_o \qquad (7\text{-}28)$$

The foregoing equations show that the voltage applied across the tank ($v_{AB'}$) depends on which device is conducting and on the direction of i_L. The conditions described by Eqs. 7-25 through 7-28 can be represented by an equivalent circuit of Fig. 7-10b. The solution for the circuit of Fig. 7-5a is applied to the equivalent circuit of Fig. 7-10b for each interval, based on the initial conditions and the voltages V_{AB} and $V_{B'B}$ which appear as dc voltages for a given interval.

In the steady-state symmetrical operation, both the switches are operated identically. Similarly, the two diodes operate identically. Therefore, it is sufficient to analyze only one-half cycle of operation, since the other half is symmetrical. It can be shown that in an SLR converter of Fig. 7-10a, the output voltage V_o cannot exceed the input voltage $V_d/2$, that is, $V_o \le V_d/2$.

The switching frequency $f_s(= \omega_s/2\pi)$, with which the circuit waveforms repeat, can be controlled to be less than or greater than the resonance frequency $f_0(= \omega_0/2\pi)$, if the converter consists of self-controlled switches. There are three possible modes of operation based on the ratio of switching frequency ω_s to the resonance frequency ω_0, which determines if i_L flows continuously or discontinuously.

7-4-1-1 DISCONTINUOUS-CONDUCTION MODE WITH $\omega_s < \omega_0/2$

By using Eqs. 7-3 and 7-4, Fig. 7-11 shows the circuit waveforms in steady state where at $\omega_0 t_0$, switch T_+ is turned on and the inductor current builds up from its zero value. The capacitor voltage builds up from its initial negative value $V_{c0} = -2V_o$. Figure 7-11 also shows the circuits during various intervals with corresponding v_{AB} and $v_{B'B}$.

At $\omega_0 t_1$, 180° subsequent to $\omega_0 t_0$, the inductor current reverses and now must flow through D_+ since the other switch T_- is not yet turned on. After another 180° subsequent to $\omega_0 t_1$ with a smaller peak current in this half-cycle, the current goes to zero and remains zero as no switches are on. A symmetrical operation requires that v_c during the discontinuous-interval $\omega_0(t_3 - t_2)$ be negative of V_{c0}, that is, equal to $2V_o$. During this interval, the capacitor voltage equal to $2V_o$ is less than $(V_d/2) + V_o$ (since $V_o \le V_d/2$); therefore the current becomes discontinuous. At $\omega_0 t_3$, the next switch T_- is turned on and the next half cycle ensues.

Because of the discontinuous interval in Fig. 7-11, one-half cycle of the operating frequency exceeds 360° of the resonance frequency f_0, and therefore in this mode of operation, $\omega_s < \omega_0/2$. The average of the rectified inductor current $|i_L|$ equals the output dc current I_o, which is supplied to the load at a voltage of V_o.

Note that in this mode of operation, the switches turn off naturally at zero current and at zero voltage, since the inductor current goes through zero. The switches turn on at zero current but not at zero voltage. Also the diodes turn on at zero current and turn off naturally at zero current. Since the switches turn off naturally in this mode of operation, it is possible to use thyristors in low switching-frequency applications.

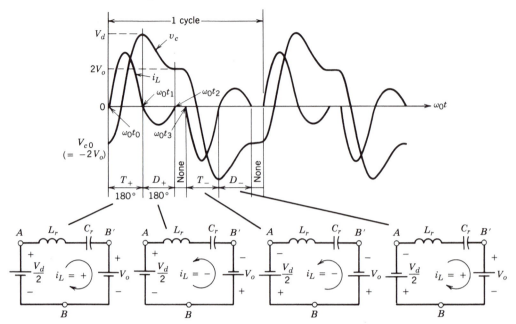

FIGURE 7-11: SLR dc–dc converter; discontinuous-conduction mode with $\omega_s < (\omega_0 / 2)$.

The disadvantage of this mode is the relatively large peak current in the circuit and, therefore, higher conduction losses, compared with the continuous-conduction mode.

7-4-1-2 CONTINUOUS-CONDUCTION MODE WITH $\omega_0/2 < \omega_s < \omega_0$

The waveforms are shown in Fig. 7-12 where T_+ turns on at $\omega_0 t_0$, with a finite value of the inductor current and at a preconduction switch voltage of V_d. T_+ conducts for less than 180°. At $\omega_0 t_1$, i_L reverses and flows through D_+, thus naturally turning off T_+. At $\omega_0 t_2$, T_- is turned on and i_L transfers from D_+ to T_-. In this mode, D_+ conducts for less than 180° because T_- is switched on early, compared with the discontinuous-conduction mode.

In this mode of operation, the switches turn on at a finite current and at a finite voltage, thus resulting in a turn-on switching loss. Moreover, the freewheeling diodes must have good reverse-recovery characteristics to avoid large reverse current spikes flowing through the switches, for example at $\omega_0 t_2$ through D_+ and T_-, and to minimize the diode turn-off losses. However, the turn-off of switches occurs naturally at zero current and at zero voltage as the inductor current through them goes to zero and reverses through the freewheeling diodes. Therefore, it is possible to use thyristors as switches in low switching-frequency applications.

7-4-1-3 CONTINUOUS CONDUCTION MODE WITH $\omega_s > \omega_0$

Compared with the previous continuous-conduction mode, where the switches turn off naturally but turn on at a finite current, the switches in this mode with $\omega_s > \omega_0$ are forced to turn off a finite current, but they are turned on at zero current and zero voltage.

Figure 7-13 shows the circuit waveforms where T_+ starts conduction at $\omega_0 t_0$ at zero current, when the inductor current reverses in direction. At $\omega_0 t_1$, before the half-cycle of the current oscillation ends, T_+ is forced to turn off, thus forcing the

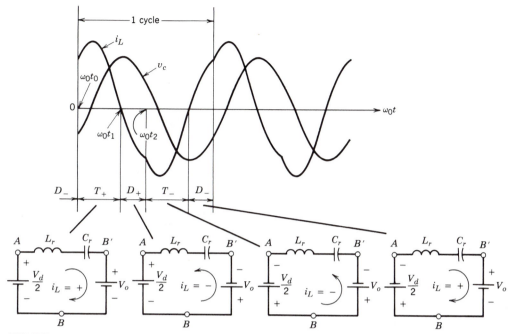

FIGURE 7-12: SLR dc–dc converter; continuous-conduction mode with $(\omega_0 / 2) < \omega_s < \omega_0$.

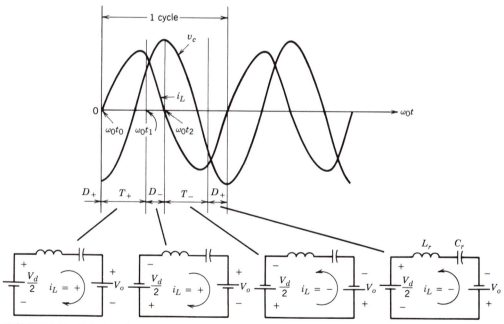

FIGURE 7-13: SLR dc–dc converter; continuous-conduction mode with $\omega_s > \omega_0$.

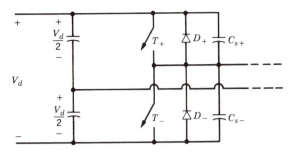

FIGURE 7-14: Lossless snubbers in an SLR converter at $\omega_s > \omega_0$.

positive i_L to flow through D_-. Because of the large negative dc voltage applied across the L-C tank ($v_{AB'} = -(V_d/2) - V_o$), the current through the diode goes to zero quickly (note that its frequency of oscillation ω_0 does not change) at $\omega_0 t_2$. T_- is gated on as soon as D_- begins to conduct so it can conduct when i_L reverses. The combined conduction interval for T_+ and D_- is equal to one-half cycle of operation at the switching frequency of ω_s. This half-cycle is less than 180° of the resonance frequency ω_0, thus resulting in $\omega_s > \omega_0$.

There are several advantages in operating at $\omega_s > \omega_0$. Unlike the continuous-conduction mode with ω_s less than ω_0, the switches turn on at a zero current and zero voltage; thus, the freewheeling diodes do not need to have very fast reverse-recovery characteristics. A significant disadvantage would appear to be that the switches need to force turn-off near the peak of i_L, thus causing a large turn-off switching loss. However, since the switches turn on not only at zero current but also at zero voltage (note that prior to turn-on of T_-, the freewheeling diode D_- across it is conducting), it is possible to use lossless snubber capacitors C_s in parallel with the switches as shown in Fig. 7-14, which act as lossless turn-off snubbers for the switches.

Operation above the resonance frequency requires that the controllable switches be used.

7-4-1-4   STEADY-STATE OPERATING CHARACTERISTICS

It is useful to know the relationship of the peak and the average values of the circuit voltages and currents to the operating conditions (V_d, V_o, I_o, ω_0, etc.). The voltages, currents, and switching angular frequency ω_s are normalized by the following base quantities:

$$V_{\text{base}} = \frac{V_d}{2} \qquad (7\text{-}29)$$

$$I_{\text{base}} = \frac{V_d/2}{Z_0} \qquad (7\text{-}30)$$

and

$$\omega_{\text{base}} = \omega_0 \qquad (7\text{-}31)$$

Figure 7-15 shows normalized I_o versus ω_0 for two values of V_o. This figure shows that a SLR dc–dc converter in the discontinuous-conduction mode (corresponding to $\omega_s < 0.5$) operates as a current source, that is, I_o stays constant even though the load resistance and hence V_o may change. Because of this property, this converter exhibits an inherent overload protection capability in the discontinuous-conduction mode.

It should be noted that in Fig. 7-10a, I_o is the average value of the full-wave rectified inductor current $|i_L|$, where the ripple in $|i_L|$ is assumed to flow through the

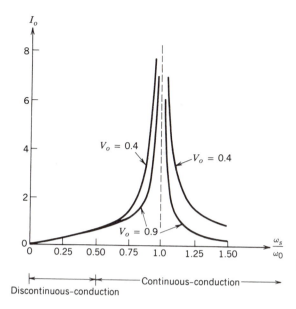

FIGURE 7-15: Steady-state characteristics of an SLR dc–dc converter; all parameters are normalized.

output filter-capacitor and its average value I_o flows through the output load resistance. In this converter, the peak value of the inductor current (which also is the peak value of the current through the switches) and the peak voltage across the capacitor C_r can be several times higher than I_o and V_d, respectively (see Problems). This aspect must be considered in comparing this converter with other converter topologies.

7-4-1-5 CONTROL OF SLR DC–DC CONVERTERS

As shown in Section 7-3-1-3 dealing with the frequency characteristics of series-resonant circuits, the resonant-tank impedance depends on the frequency of operation. Therefore, for a given applied input voltage V_d and a load resistance, V_o can be regulated by controlling the switching frequency f_s. This is shown in block-diagram form in Fig. 7-16, where the error between the sensed output voltage and the reference voltage determines the output frequency f_s of the voltage-controlled oscillator, which in turn controls the two switches.

The variable frequency control described before is not optimum because of the complexity of its analysis and the design of EMI filters. As discussed in Reference 7, a constant frequency control can be implemented in a full-bridge version of the SLR converter, where the switches in each leg of the converter operate at the 50% duty ratio at a constant frequency of $\omega_s > \omega_0$, but the phase delay between the output of the two converter legs is controlled. Such a control restricts the load to be in a limited range, beyond which the zero-voltage/zero-current switching characteristics of the converter do not hold.

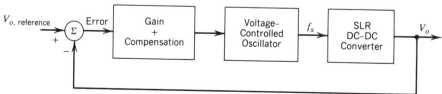

FIGURE 7-16: Control of SLR dc–dc converter.

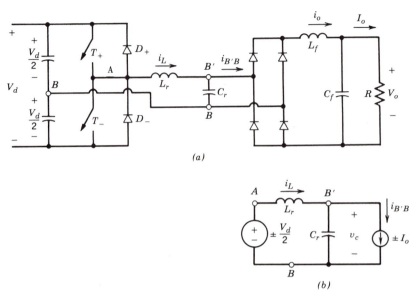

FIGURE 7-17: PLR dc–dc converter: (*a*) half-bridge, (*b*) equivalent circuit.

It should be noted that the SLR converter can be used where the output is not a rectified dc; for example, SLR inverters are used for induction heating applications, where the load appears as a resistance rather than a dc voltage V_o.

7-4-2   Parallel-Loaded Resonant (PLR) DC–DC Converters

These converters are similar to the series-loaded resonant (SLR) converters in terms of operating with a series resonant L-C tank circuit. However, unlike the SLR converters, where the output stage or the load appears in series with the resonant tank, here the output stage is connected in parallel with the resonant-tank capacitor C_r, as shown in Fig. 7-17a. The isolation transformer is omitted for simplicity.

PLR converters differ from the SLR converters in many important respects, for example (1) PLR converters appear as a voltage source, and, hence, are better suited for multiple outputs; (2) unlike the SLR converters, the PLR converters do not possess inherent short-circuit protection capability, which obviously is a drawback; and (3) PLR converters can step up as well as step down the voltage, unlike the SLR converters, which can operate only as a step-down converter (not counting the transformer turns-ratio).

In the following subsections, only the modes in which a PLR converter is likely to operate are discussed. The discussion on the other modes can be found in the literature.

The voltage across the resonant-tank capacitor C_r is rectified, filtered, and then supplied to the load. To develop an equivalent circuit, the current through the output filter-inductor in Fig. 7-17a can be assumed to be a ripple-free dc current I_o during a switching frequency time period. This is a reasonable assumption, based on a high switching frequency and a sufficiently large value of the filter inductor. The voltage across the resonant tank depends on the devices conducting as follows:

$$T_+ \text{ or } D_+: \quad v_{AB} = +V_d/2 \tag{7-32}$$

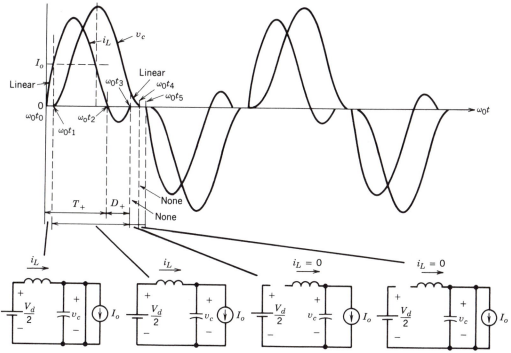

FIGURE 7-18: PLR dc–dc converter in a discontinuous mode.

and

$$T_- \text{ or } D_-: \quad v_{AB} = -V_d/2 \tag{7-33}$$

Based on the previous discussion, an equivalent circuit of Fig. 7-17b can be obtained where the input voltage to the tank (v_{AB}) is equal in magnitude to $V_d/2$ but its polarity depends on which switch is turned on (T_+ or T_-). The current $i_{B'B}$, defined in Fig. 7-17a, equals I_o in magnitude, but its direction depends on the polarity of the voltage v_c across C_r at the input to the bridge rectifier.

The equivalent circuit of Fig. 7-17b is identical to that discussed in Section 7-3-1-2. Therefore, Eqs. 7-13 and 7-14 can be applied with the appropriate v_{AB}, $i_{B'B}$ and the initial conditions.

Unlike SLR converters, a PLR dc–dc converter can operate in a large number of combinations consisting of the states of i_L and v_c. However, only three modes are considered in the following subsections.

7-4-2-1 DISCONTINUOUS MODE OF OPERATION

In this mode of operation, both i_L and v_c remain zero simultaneously for some length of time. The steady-state waveforms for this mode of operation are plotted in Fig. 7-18, based on Eqs. 7-13 and 7-14. During steady-state operation, initially both i_L and v_c are zero and T_+ is turned on at $\omega_0 t_0$. So long as $|i_L| < I_o$, the output current circulates through the rectifier bridge, which appears as a short circuit across C_r and keeps its voltage at zero, as shown in Fig. 7-18. At $\omega_0 t_1$, i_L exceeds I_o and the difference ($i_L - I_o$) flows through C_r, and v_c increases. Due to L-C resonance, i_L reverses at $\omega_0 t_2$ and flows through D_+, since T_- is not turned on until some time later. During the interval $\omega_0(t_3 - t_1)$, i_L and v_c can be calculated from Eqs. 7-13 and

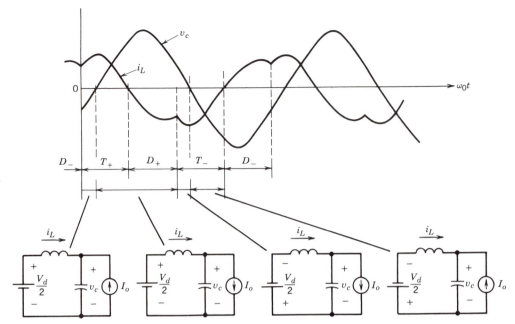

FIGURE 7-19: PLR dc–dc converter in a continuous mode with $\omega_s < \omega_0$.

7-14 using $i_{L0} = I_o$ and $v_{c0} = 0$ as the initial conditions at time $\omega_0 t_1$. If the gate/base drive of T_+ is removed prior to $\omega_0 t_3$, i_L can no longer flow after $\omega_0 t_3$ and stays at zero. With $i_L = 0$, I_o flows through C_r, and v_c decays linearly to zero during the interval $\omega_0 t_3$ to $\omega_0 t_4$.

In this discontinuous mode of operation, both v_c and i_L stay at zero for an interval that can be varied in order to control the output voltage. Beyond this discontinuous interval, T_- is gated on at $\omega_0 t_5$ and the next half-cycle ensues with identical initial conditions of zero i_L and v_c as for the first half-cycle.

Clearly, the foregoing operation corresponds to ω_s in a range from zero to approximately $\omega_0/2$. Also, there are no turn-on or turn-off stresses on the switches or the diodes.

7-4-2-2 CONTINUOUS MODE OF OPERATION BELOW ω_0

At switching frequencies higher than those in the discontinuous mode but less than ω_0, both v_c and i_L become continuous. The waveforms are shown in Fig. 7-19, where a switch turns on at a finite i_L and the current commutates from the diode connected in antiparallel with the other switch. This results in turn-on losses in the switches, and the diodes must have good reverse-recovery characteristics. However, there are no turn-off losses in the switches since the current through them commutates naturally when i_L reverses in direction.

7-4-2-3 CONTINUOUS MODE OF OPERATION ABOVE ω_0

This mode with continuous v_c and i_L occurs at $\omega_s > \omega_0$. The circuit waveforms are shown in Fig. 7-20. Here, the turn-on losses in the switches are eliminated since the switches turn on naturally when i_L, initially flowing through the diodes, reverses. However, this operating mode results in the turn-off losses in the switches, since a switch is forced to turn off, thus transferring its current to the diode connected in antiparallel with the other switch.

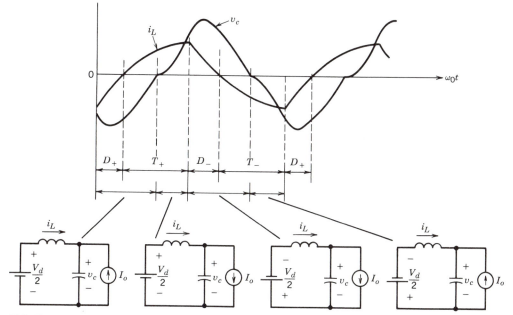

FIGURE 7-20: PLR dc–dc converter in a continuous mode with $\omega_s > \omega_0$.

Similar to the SLR converter operating in a continuous-conduction mode with $\omega_s > \omega_0$, the switches here turn on at zero voltage, thus at the switching instant the snubber capacitor in parallel has no stored energy. Therefore, it is possible to eliminate the turn-off losses by connecting a lossless snubber consisting of a capacitor (with no series resistor) in parallel with each switch, as in an SLR converter in Fig. 7-14.

7-4-2-4 STEADY-STATE OPERATING CHARACTERISTICS

The steady-state operating characteristics of the PLR dc–dc converters are shown in Fig. 7-21 for two values of I_o, where the variables are normalized by using the base quantities defined in Eqs. 7-29 through 7-31. Figure 7-21 shows the following important properties of the PLR converter:

- In the discontinuous mode of operation with $\omega_s < \omega_0/2$, this converter exhibits a good voltage-source characteristic and V_o remains independent of I_o. This property is useful in designing a converter with multiple outputs.

- Also in the frequency range $\omega_s < \omega_0/2$, the output varies linearly with ω_s, thus simplifying the output regulation.

- It is also possible to operate in the high frequency range ($\omega_s > \omega_0$) and the maximum change required in the operating frequency is less than 50% to compensate for the output loading for a normalized output voltage of 1.0.

- It is possible to step up or step down the output voltage, that is, V_o can be less than or greater than 1.0.

In this converter, the peak inductor current (which is also the peak current through the switches) and the peak capacitor voltage can be several times higher than I_o and V_d, respectively (see Problems).

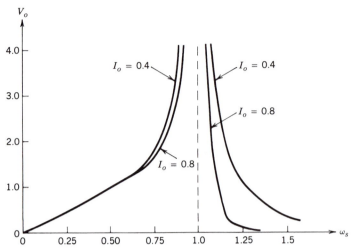

FIGURE 7-21: Steady-state characteristics of a PLR dc–dc converter. All quantities are normalized.

The converter characteristics shown in Fig. 7-21 suggest than an effective way to regulate the output is by controlling the frequency of operation ω_s.

7-4-3 Hybrid-Resonant DC–DC Converter

This topology consists of a series-resonant circuit as shown in Fig. 7-22 but the load is connected in parallel with only part of the capacitance, for example one-third of the total capacitance, and the other two-thirds of the capacitance appears in series. The purpose of this topology is to benefit from the advantageous properties of both the SLR and the PLR converters, namely that a SLR converter offers an inherent current limiting under short-circuit conditions and a PLR converter acts as a voltage source, and thus regulating its voltage at no-load with a high Q resonant tank is not a problem. These converters can be analyzed based on the discussion presented in the previous two sections. These are analyzed in detail in Reference 15.

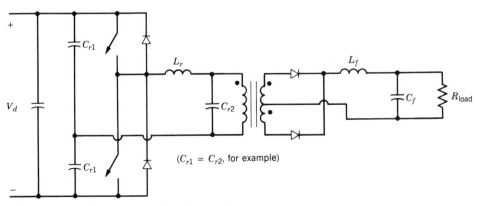

FIGURE 7-22: Hybrid resonant dc–dc converter.

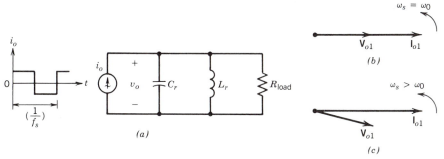

FIGURE 7-23: Basic circuit for current-source, parallel-resonant converter for induction heating: (*a*) basic circuit, (*b*) phasor diagram (at $\omega_s = \omega_0$), (*c*) phasor diagram (at $\omega_s > \omega_0$).

7-4-4 Current-Source, Parallel-Resonant DC-to-AC Inverters for Induction Heating

The basic principle of such an inverter is illustrated by means of the circuit of Fig. 7-23a, where a square-wave current source is applied to a parallel-resonant load. The induction coil and the load (R-L combination) are modeled by means of a parallel combination of equivalent R_{load} and L_r, rather than a series R-L. The capacitor C_r is added to resonate with L_r in parallel. It is assumed that the harmonic impedance of the parallel-resonant load at the harmonic frequencies of the input current source is negligibly small, thus, resulting in an essentially sinusoidal voltage v_o. Therefore, the analysis of Section 7-3-2-2 applies.

When the fundamental frequency ω_s of the source-current i_o equals the natural resonance frequency $\omega_0 = (1/\sqrt{L_rC_r})$, the circuit phasor diagram is shown in Fig. 7-23b where the fundamental frequency component $\mathbf{V}_{o1}$ of the resulting voltage is in phase with the fundamental frequency component $\mathbf{I}_{o1}$ of the input current.

Since the square-wave input current in practice is supplied by a thyristor inverter, the resonant load must supply the capacitive vars (volt-amperes reactive) to the inverter. This implies that the load voltage $\mathbf{V}_{o1}$ should lag the input current $\mathbf{I}_{o1}$, which is possible only at a frequency $\omega_s > \omega_0$, as shown in Fig. 7-23c.

A current-source inverter consisting of thyristors is shown in Fig. 7-24a. To avoid a large di/dt (during current commutation) through the inverter thyristors, a small inductance L_c in series with the resonant load is purposely introduced. The inverter output current i_o therefore deviates from its idealized square-wave shape and becomes trapezoidal, as shown in Fig. 7-24b.

The voltage across one of the thyristors T_1 shows that after it stops conducting, a reverse voltage appears across it for a time interval equal to (γ/ω_s); subsequently it is required to block a forward polarity voltage. Therefore, (γ/ω_s) should be sufficiently larger than the specified turn-off time t_q of the thyristor that is being used.

One of the techniques to control the power output of this inverter is by controlling its switching frequency. As the switching frequency f_s is increased further above the natural resonance frequency f_0, the power output decreases if I_d is held constant by means of a controlled dc supply. Another obvious technique to control the power output is to control I_d, keeping the switching frequency of the inverter constant.

7-4-4-1 START-UP

In case of a current-fed parallel-resonant inverter, the load must be in resonance with C_r prior to the inverter operation in Fig. 7-24a. This is accomplished by means of

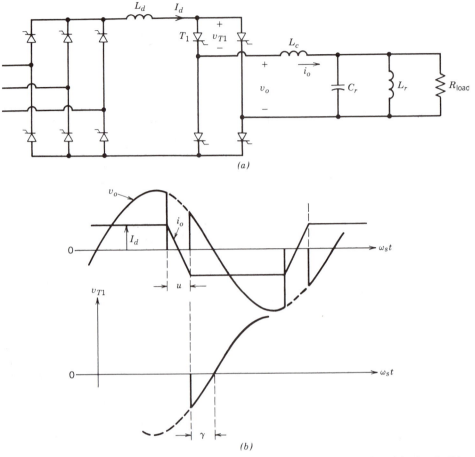

FIGURE 7-24: Current-source, parallel-resonant inverter for induction heating: (a) circuit, (b) waveforms.

a precharged capacitor dumping its charge onto the parallel-resonant load circuit, thus establishing oscillating load voltages and currents. Shortly after that, the inverter operation is initiated.

7-4-5 Class-E Converters

In class-E converters, the load is supplied through a sharply tuned series resonant circuit shown in Fig. 7-25a. This results in an essentially sinusoidal current i_o. The input to the converter is through a sufficiently large inductor to allow the assumption that in steady state, the input to the converter is a dc current source I_d as shown in Fig. 7-25a, where the current magnitude depends on the power output. The waveforms are shown in Fig. 7-25b for an optimum mode, which is discussed later on. When the switch is on, $I_d + i_o$ flows through the switch, as shown in Fig. 7-25c. When the switch is turned off, because of the capacitor C_1, the voltage across the switch builds up slowly, thus allowing a zero-voltage turn-off of the switch. With the switch off, the oscillating circuit is as shown in Fig. 7-25d, where the voltage across capacitor C_1 builds up, reaches its peak, and eventually comes back to zero, at which instant the switch is turned back on.

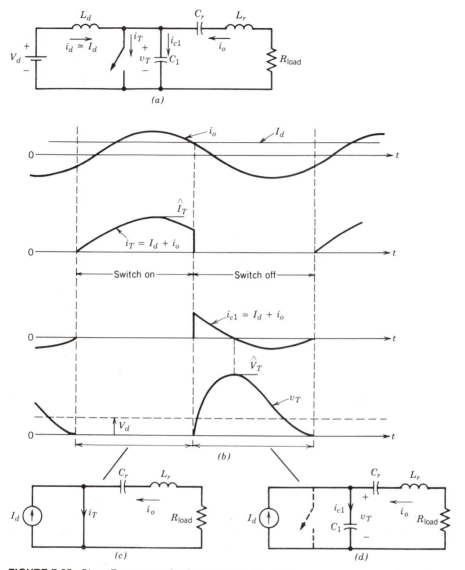

FIGURE 7-25: Class-E converter (optimum mode, $D = 0.5$).

A class-E converter operates at a switching f_s, which is slightly higher than the resonant frequency $f_0 = 1/(2\pi\sqrt{L_rC_r})$. During the interval when the switch is off, the input supplies power to the circuit since v_T is positive, as shown in Fig. 7-25b. For a high quality factor of the series L_r-C_r-R circuit ($Q \geq 7$), which results in an essentially sinusoidal load current i_o, only a slight variation in f_s is needed to vary the output voltage. As f_s increases (where $f_s > f_0$), i_o and therefore v_R decrease.

Another observation that can be made is as follows: the average value of v_T equals V_d. If i_o is assumed to be purely sinusoidal, the average voltage across the load resistance R is zero. The average voltage across L_r is also zero in steady state. Therefore, C_r blocks the dc voltage V_d in addition to providing a resonant circuit.

The operation of a class-E converter can be categorized in optimum and suboptimum modes. The circuit and the waveforms shown in Fig. 7-25 belong to the optimum mode of operation where the switch voltage returns to zero with a zero slope

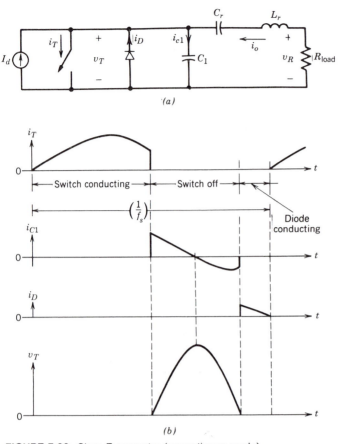

FIGURE 7-26: Class-E converter (nonoptimum mode).

($i_{c1} = 0$), and there is no need for a diode in antiparallel with the switch. This mode of operation requires that the load resistance R be equal to an optimum value R_{opt}. The switch duty ratio $D = 0.5$ results in a maximum power capability or, in other words, the maximum switch-utilization ratio, where the switch-utilization ratio is defined as the ratio of the output power P_o to the product of the peak switch voltage and the peak switch current. It is shown in the literature that the peak switch current is approximately $3I_d$ and the peak switch-voltage is approximately $3.5V_d$.

The nonoptimum mode of operation occurs if $R < R_{opt}$. Here, the switch voltage reaches zero with a negative slope ($dv_T/dt < 0$ and hence, $i_{c1} = C_1[dv_T/dt] < 0$). A diode is connected in antiparallel with the switch as shown in Fig. 7-26a to allow this current to flow while keeping the switch voltage at zero (at one diode drop). The waveforms are shown in Fig. 7-26b, where the switch is turned on as soon as the diode starts to conduct. In a circuit with a high input voltage, it is important to reduce the peak switch voltage $\hat{V}_T$. It can be shown that for a smaller switch duty ratio, $\hat{V}_T$ decreases but the peak switch current $\hat{I}_T$ goes up.

The advantage of a class-E converter is the elimination of switching losses and the reduction in EMI. Also, it is a single-switch topology and produces a sinusoidal output current. Significant disadvantages are high peak voltage and current associated with the switch, and large voltages and currents through the resonant L-C elements. For a resistive load as shown in Fig. 7-26a (the optimum mode without an antiparallel diode with the switch is very restrictive), class-E converters have been considered for high-frequency electronic lamp ballasts.

It is possible to obtain a dc–dc voltage conversion by rectifying the output current. Since the output load may vary over a large range, an impedance matching network is required between the output of the class-E converter and the output rectification stage to ensure a lossless switching operation of the class-E converter. Various suboptimum class-E topologies are described in Reference 22.

7-5 RESONANT-SWITCH CONVERTERS

Historically, prior to the availability of controllable switches with appreciable voltage- and current-handling capability, the switch-mode converters consisted of thyristors (currently, thyristors in switch-mode converters are used only at very high power levels). Such converters had topologies and control schemes similar to those described in Chapters 5 and 6 for switch-mode dc–dc converters and dc-to-ac inverters. Each thyristor in such a converter required a current-commutation circuit, which consisted of an L-C resonant circuit plus other auxillary thyristors and diodes, which turned the main thyristor off by forcing the current through it to go to zero. Because of the complexity and substantial losses in the commutation circuits, thyristors were replaced by controllable switches, as their power-handling capability improved.

A need to increase switching frequencies and to reduce EMI led to augmenting the controllable switches in certain basic switch-mode converter topologies of Chapters 5 and 6 by a simple L-C resonant circuit, and thereby shaping the switch voltage and current in order to yield zero-voltage and/or zero-current switchings. Such converters are termed resonant-switch converters. Often, the diode needed for the resonant-switch circuit operation is the same as that in the original switch-mode converter topology. Similarly, inductors (such as the transformer leakage inductance) and the capacitors (such as the output capacitance of the semiconductor switch), which appear as undesirable parasitics in switch-mode topologies, can be utilized to provide the resonant inductor and the capacitor needed for the resonant-switch circuit.

The output in some of these circuits is controlled by controlling the operating frequency; in others a constant frequency square-wave or PWM control can be used with some additional constraints to provide zero voltage and/or zero current switchings.

A majority of such converters can be divided into three switching categories:

1. Zero-current-switching (ZCS) topology where the switch turns on and turns off at zero current. The peak resonant current flows through the switch but the peak switch voltage remains the same as in its switch-mode counterpart. Such a topology is shown in Fig. 7-27a for a step-down dc–dc converter.

2. Zero-voltage-switching (ZVS) topology where the switch turns on and turns off at zero voltage. The peak resonant voltage appears across the switch, but the peak switch current remains the same as in its switch-mode counterpart. Such a topology is shown in Fig. 7-27b for a step-down dc–dc converter.

3. Zero-voltage-switching, clamped-voltage (ZVS-CV) topology where the switch turns on and turns off at zero voltage as in category (2) above. However, a converter of this topology consists of at least one converter leg made up of two such switches. The peak switch voltage remains the same as in its switch-mode counterpart, but the peak switch current is generally higher. Such a converter topology is shown in Fig. 7-27c for a step-down dc–dc converter.

In the following sections, the operating principles of the converters belonging to all three switching categories are discussed.

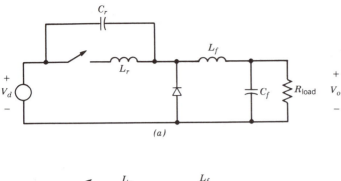

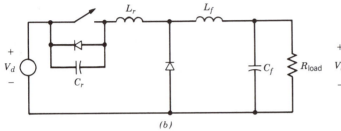

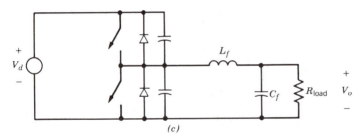

FIGURE 7-27: Resonant-switch converters: (a) ZCS dc–dc converter
(step-down), (b) ZVS dc–dc converter (step-down), (c)
ZVS-CV dc–dc converter (step-down).

7-5-1 ZCS Resonant-Switch Converters

In such converters, the current produced by L-C resonance flows through the switch, thus causing it to turn on and off at zero current. This can be easily explained in a step-down dc–dc converter of Fig. 7-28a, which has been modified as shown in Fig. 7-28b by the addition of L_r and C_r. The filter inductor L_f is sufficiently large such that the current i_o can be assumed to be a current of constant magnitude I_o in Fig. 7-28b. The circuit waveforms in steady state are shown in Fig. 7-28c and the subcircuits are shown in Fig. 7-28d.

Prior to turning the switch on, the output current I_o freewheels through the diode D, and the voltage v_c across C_r equals V_d. At t_0, the switch is turned on at zero current. So long as i_T is less than I_o, D keeps on conducting and v_c stays at V_d. Therefore, i_T rises linearly and at t_1, i_T equals I_o, which causes D to stop conducting. Now, L_r and C_r form a parallel resonant circuit and the analysis of Section 7-3-2-1 applies. Use of Eq. 7-20 shows that at t_1', i_T peaks at $V_d/Z_0 + I_o$ and v_c reaches zero. The negative peak of v_c occurs at t_1'' when $i_T = I_o$. At t_2, i_T reaches zero and cannot reverse its direction. Thus the switch T is naturally turned off.

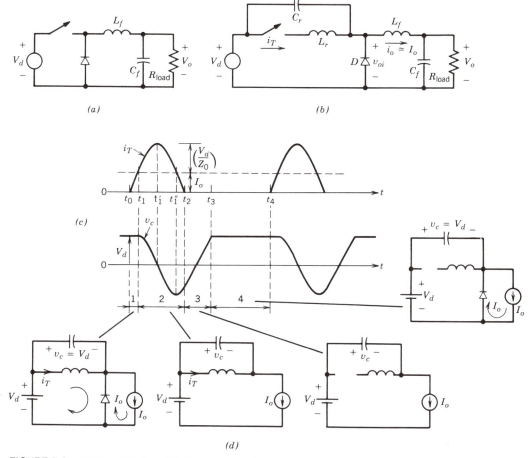

FIGURE 7-28: ZCS resonant-switch dc–dc converter.

Beyond t_2, the gate pulse from T is removed. Now I_o flows through C_r and v_c rises linearly to V_d at t_3, at which point the diode D turns on and v_c stays at V_d. After an interval during which i_T is zero and $v_c = V_d$, the gate pulse to T is again applied at t_4 to turn it on, and the next cycle ensues.

It is clear from the waveforms of Fig. 7-28c that the forward switch voltage is limited to V_d. The instantaneous voltage $v_{oi} = V_d - v_c$ across the output diode, as defined in Fig. 7-28b, is plotted in Fig. 7-29. By controlling the switch-off time interval $(t_4 - t_3)$, or in other words the switching frequency of operation, the average value of v_{oi} and, hence, the average power supplied to the output stage can be controlled. This in turn regulates the output voltage V_o for a given load current I_o.

From the waveforms of Fig. 7-28c, it can be seen that if $I_o > V_d/Z_0$, i_T will not come back to zero naturally and the switch will have to be forced off, thus resulting in turn-off losses.

Zero-current switching can also be obtained by connecting C_r in parallel with D as shown in Fig. 7-30a. As discussed previously, i_o can be assumed to be a current of constant magnitude I_o in Fig. 7-30a during a high-frequency resonant cycle.

Initially both the capacitor voltage (across C_r) and the inductor current (through L_r) are assumed to be zero and the load current I_o freewheels through the diode D.

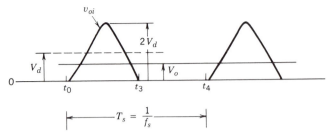

FIGURE 7-29: v_{oi} waveform in a ZCS resonant-switch dc – dc converter.

The converter operation can be divided into the following intervals for which the converter waveforms as well as the corresponding circuit states are shown in Figs. 7-30*b* and 7-30*c*:

1. *Time interval 1 (between t_0 and t_1).* At time t_0, the switch is turned on. Because of I_o flowing through the diode, it appears as a short circuit and the entire input voltage V_d appears across L_r. Therefore, the switch current builds up linearly until it becomes equal to I_o at time t_1. Beyond this time, the diode turns off and the voltage clamp across C_r is removed.

2. *Time interval 2 (between t_1 and t_2).* Beyond t_1, $i_T > I_o$ and their difference $(i_T - I_o)$ flows through C_r. At t_1', i_T peaks and $v_c = V_d$. At time t_1'', the switch current drops from its peak value to I_o and the capacitor voltage reaches $2V_d$. The switch current eventually drops to zero at t_2 and cannot reverse through the switch (if a BJT or a MOSFET is used as a switch, then a diode in series with it must be used to block a negative voltage and to prevent the flow of reverse current through the switch). Thus, the switch current is commutated off naturally and the gate/base drive from the switch should be removed at this point.

3. *Time interval 3 (between t_2 and t_3).* Beyond the time t_2 with the switch off, the capacitor C_r discharges into the output load and the capacitor voltage linearly drops to zero at t_3.

4. *Time interval 4 (between t_3 and t_4).* Beyond t_3, the load current just freewheels through the diode until a time t_4, when the switch is turned on and the next switching cycle begins. This time interval is controlled to adjust the output voltage.

Under a steady-state operating condition, the average voltage across the filter inductor is zero, therefore the voltage across C_r averaged over one switching cycle equals the output voltage V_o. By controlling the freewheeling time interval 4 (i.e., by controlling the switching frequency), the output voltage V_o can be regulated.

From the waveforms of Fig. 7-30*b*, the following circuit properties can be observed:

- L_r and C_r together determine the natural resonance frequency $\omega_0 = 1/(2\pi\sqrt{L_r C_r})$, which can be made to be large (in megahertz range) by proper selection of L_r and C_r. Both the switch turn-on and turn-off occur at zero current, thus reducing the switching losses. It should be noted that at turn-on, the voltage across the switch equals V_d. This results in losses, as discussed in Section 7-5-3.

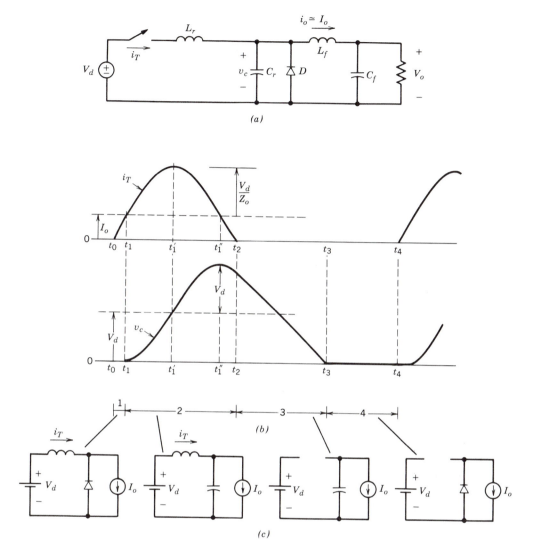

FIGURE 7-30: ZCS resonant-switch dc–dc converter; alternate configuration.

- The load current I_o must be less than a maximum value of V_d/Z_0, which depends on the circuit parameters. Otherwise, the switch would have to turn off a finite amount of current.

- At a given switching frequency of operation, V_o declines with increasing load. Therefore, the switching frequency ω_s must be increased to regulate V_o. The opposite is true if the load decreases.

- By placing a diode in antiparallel across the switch in Fig. 7-30a, the inductor current is allowed to reverse. This permits excess energy in the resonant circuit at light loads to be transferred back to the voltage source V_d. This significantly reduces the dependence of V_o on the output load.

Since the switching losses are minimized and the EMI is reduced, very high switching frequencies can be attained. One drawback of such a converter is that the switch peak current rating required is significantly higher than the load current. This

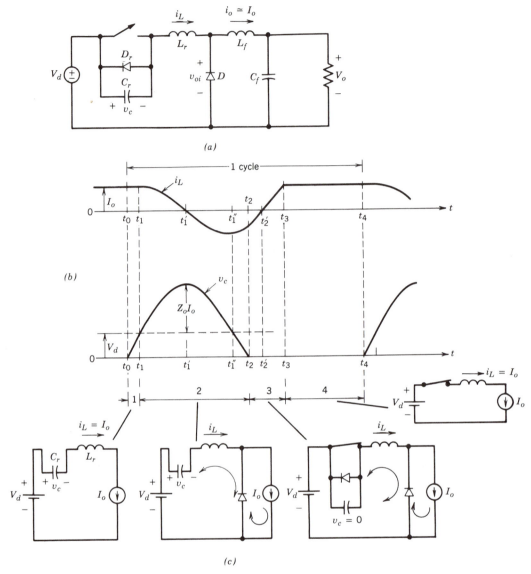

FIGURE 7-31: ZVS resonant-switch dc–dc converter.

also implies that the conduction losses in the switch would be higher compared with its switch-mode counterpart. In References 25 and 26, it has been shown that this principle can be applied to various other single-switch dc–dc converter topologies.

7-5-2   ZVS Resonant-Switch Converters

In these converters, the resonant capacitor produces a zero voltage across the switch at which instant the switch can be turned on or off. Such a step-down dc–dc converter circuit is shown in Fig. 7-31a, where a diode D_r is connected in antiparallel with the switch. As discussed previously, the output current i_o can be assumed to be a current of constant magnitude I_o in Fig. 7-31a during a high-frequency resonant cycle.

Initially, the switch is conducting I_o and therefore, $I_{L0} = I_o$ and $V_{c0} = 0$. The converter operation can be divided into the following intervals for which the converter waveforms as well as the corresponding circuit states are shown in Fig. 7-31b and 7-31c, respectively:

1. *Time interval 1 (between t_0 and t_1).* At time t_0, the switch is turned off. Because of C_r, the voltage across the switch builds up slowly but linearly from 0 to V_d at t_1. This results in a zero voltage turn-off of the switch.

2. *Time interval 2 (between t_1 and t_2).* Beyond t_1, since $v_c > V_d$, the diode D becomes forward biased, C_r and L_r resonate, and the analysis of Section 7-3-1-1 applies. At t_1', i_L goes through zero and v_c reaches its peak of $V_d + Z_0 I_o$. At t_1'', $v_c = V_d$ and $i_L = -I_o$. At t_2, the capacitor voltage reaches zero and cannot reverse its polarity because the diode D_r begins to conduct.

   Note that the load current I_o should be sufficiently large so that $Z_0 I_o > V_d$. Otherwise, the switch voltage will not come back to zero naturally and the switch will have to be turned on at a nonzero voltage, resulting in turn-on losses (the energy stored in C_r will dissipate in the switch).

3. *Time interval 3 (between t_2 and t_3).* Beyond t_2, the capacitor voltage is clamped to zero by D_r, which conducts the negative i_L. The gate drive to the switch is applied once its antiparallel diode begins to conduct. i_L now increases linearly and goes through zero at time t_2', at which instant i_L begins to flow through the switch. Therefore, the switch turns on at a zero voltage and zero current. i_L increases linearly to I_o at t_3.

4. *Time interval 4 (between t_3 and t_4).* Once i_L reaches I_o at t_3, the freewheeling diode D turns off. Because a small negative slope is associated with di/dt through the diode at turn off, there are no diode reverse-recovery problems like the ones encountered in the switch mode. The switch conducts I_o as long as it is kept on until t_4. The interval $(t_4 - t_3)$ can be controlled. At t_4, the switch is turned off and the next cycle ensues.

It is clear from the waveforms of Fig. 7-31b that the switch current is limited to I_o. The voltage v_{oi} across the output diode, as defined in Fig. 7-31a, is plotted in Fig. 7-32. By controlling the on interval $(t_4 - t_3)$ of the switch, the average value of v_{oi} and, hence, the average power supplied to the output stage can be controlled. This in turn regulates the output voltage V_o for a given load current I_o.

This zero-voltage switching approach can also be applied to various other single-switch dc–dc converter topologies, as described in Reference 27.

7-5-3 Comparison of ZCS and ZVS Topologies

Both of these techniques require a variable frequency control to regulate the output voltage.

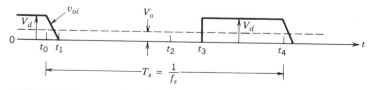

FIGURE 7-32: v_{oi} waveform in a ZVS resonant-switch dc–dc converter.

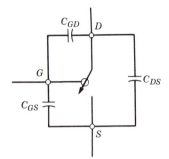

FIGURE 7-33: Switch internal capacitances.

In the ZCS, the switch is required to conduct a peak current that is higher than the load current I_o by an amount V_d/Z_0. For natural turn-off of the switch at zero current, the load current I_o must not exceed V_d/Z_0. Therefore, there is a limit on how low the load resistance can become. By placing a diode in antiparallel with the switch, the output voltage can be made insensitive to the load variations.

In the ZVS topology discussed here, the switch is required to withstand a forward voltage that is higher than V_d by an amount $Z_0 I_o$. For zero voltage (lossless) turn-on of the switch, the load current I_o must be greater than V_d/Z_0. Therefore, if the output load current I_o varies in a wide range, then the foregoing two conditions result in a very large voltage rating of the switch (see Problem 7-13). Therefore, this technique is limited to an essentially constant load application. To overcome this limitation, a zero-voltage switching multi-resonant technique is described in Reference 29.

In general, ZVS is preferable over ZCS at high switching frequencies. The reason has to do with the internal capacitances of the switch, as shown in Fig. 7-33. When the switch turns on at zero current but at a finite voltage, the charge on the internal capacitances is dissipated in the switch. As discussed in Reference 30, this loss becomes significant at very high switching frequencies. However, no such loss occurs if the switch turns on at a zero voltage.

7-6   ZERO-VOLTAGE-SWITCHING, CLAMPED-VOLTAGE (ZVS-CV) TOPOLOGIES

In the literature, these topologies have been referred to as the pseudo-resonant and the resonant-transition topologies. In this topology, the switches turn on and turn off at zero voltage. But unlike the ZVS topology discussed in Section 7-5-2, the peak voltage of a switch is clamped at the input dc voltage. Such converters consist of at least one converter leg having two switches.

7-6-1   ZVS-CV DC–DC Converters

The basic principle is shown by means of the dc–dc step-down converter shown in Fig. 7-34a, consisting of two switches. The filter inductor L_f is very small compared with the normal switch-mode topology so that i_L becomes positive as well as negative, during each cycle of operation. Assuming C_f to be large, the filter capacitor and the load can be replaced by a dc voltage V_o in steady-state as shown in the equivalent circuit of Fig. 7-34b. The waveforms are shown in Fig. 7-34c.

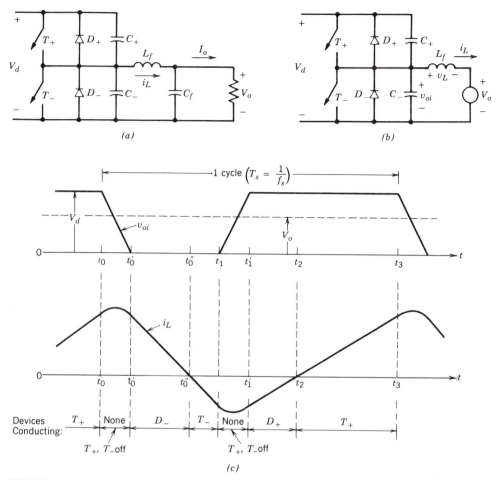

FIGURE 7-34: ZVS-CV dc–dc converter.

Initially, T_+ is conducting a positive i_L and $v_L(= V_d - V_o)$ is positive. At time t_0, T_+ is turned off at zero voltage because of C_+ in Fig. 7-34a; the voltage across T_+ builds up slowly compared with its switching time. With T_- off and T_+ just off, the subcircuit is as shown in Fig. 7-35a. It can be redrawn as in Fig. 7-35b, where the initial voltage V_d on C_- is shown explicitly by means of a voltage source. Since $C_+ = C_- = C/2$, the Thévenin equivalent of the circuit on the left results in the circuit of Fig. 7-35c. Since C is very small, the resonant frequency $f_0 = 1/(2\pi\sqrt{L_f C})$ is much larger than the switching frequency of the converter. Moreover, $Z_0 = \sqrt{L_f/C}$ in this circuit is very large, resulting in a small variation in i_L during the time interval shown in Fig. 7-35d. At t_0', v_{oi}, the voltage across C_- reaches zero, beyond which time this subcircuit has to be modified since v_{oi} cannot become negative because of the presence of D_- in the original circuit. During the time interval $(t_0' - t_0)$, the magnitude of dv/dt across both the capacitors is the same; therefore, $i_L/2$ flows through each of the capacitors during this interval since $C_+ = C_-$.

The preceding discussion can be simplified if i_L during this interval is assumed to be essentially constant. This allows the assumption that v_{oi} changes linearly, as shown in Fig. 7-34c during the blanking-time interval where both the switches are off.

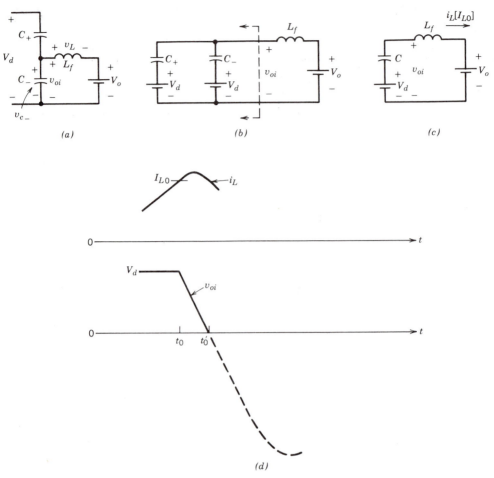

FIGURE 7-35: ZVS-CV dc–dc converter; T_+, T_- off.

Beyond t_0', i_L decreases linearly, since it flows through D_- and, therefore, $v_L = -V_o$. Once D_- begins to conduct, T_- is gated on. At t_0'', i_L reverses direction and flows through T_-.

At t_1, T_- is turned off at zero voltage and after a capacitor charging/discharging interval $(t_1' - t_1)$, similar to $(t_0' - t_0)$, the negative i_L flows through D_+. Since $v_L(= V_d - V_o)$ is positive beyond t_1', i_L increases. T_+ is gated on at zero voltage, as soon as D_+ begins to conduct. i_L becomes positive at t_2 and flows through T_+.

At t_3, T_+ is turned off at zero voltage, thus completing a cycle with a time period $T_s = (t_3 - t_0)$.

As shown by the waveforms in Fig. 7-34c, the switch voltage peak in this topology is clamped at V_d.

An important observation is that for the zero voltage turn-off of a switch, a capacitor is connected directly across the switch. Therefore, the switch must be turned on only at zero voltage; otherwise the energy stored in the capacitor will be dissipated in the switch. Therefore, the diode in antiparallel with the switch must conduct prior to the closing of the switch. This requires that in the circuit of Fig. 7-34a, i_L must flow in both directions during each cycle to satisfy this requirement for both the switches.

In such a circuit, the output voltage can be regulated by means of a constant switching frequency, PWM control. Assuming the blanking-time intervals $(t'_0 - t_0)$ and $(t'_1 - t_1)$, during which the resonant transition occurs, to be much smaller than the time period T_s of the switching frequency, the voltage v_{oi} in Fig. 7-34c is of a rectangular waveform. Since the average voltage across L_f is zero, the average value of v_{oi} equals V_o. Therefore, $V_o = D V_d$, where D is the duty ratio of the switch T_+ and DT_s is the time interval during which either T_+ or D_+ is conducting. The average value of i_L equals the output current I_o.

If a constant frequency PWM control is used to regulate V_o, then L_f needs to be chosen such that even under the minimum value of V_d and the highest load (that is, the minimum load resistance), i_L reaches a value less than zero.

The advantage of this ZVS-CV is that the switch voltages are clamped to V_d. The disadvantage is that because of the higher ripple in i_L, the switches need to carry higher peak currents as compared to the switch-mode of operation.

7-6-2 ZVS-CV DC-to-AC Inverters

It should be noted that the dc–dc converter discussed in Section 7-6-1 is capable of a two-quadrant operation where i_o can reverse. Therefore, such a converter can be modified as shown in Fig. 7-36a, which results in a half-bridge square-wave dc-to-ac inverter to supply an inductive load. The resulting waveforms with equal switch duty ratios are shown in Fig. 7-36b, and the switching losses are eliminated, since the switches turn on and turn off at zero voltage. The load current must lag the voltage (that is, the load must be inductive like a motor load) for the switchings to occur at a zero voltage.

It is possible to operate the inverter of Fig. 7-36a in a current-regulated mode, similar to that discussed in Chapter 6. However, to achieve zero voltage switchings, both switches must conduct every switching cycle, and therefore i_o must flow in each direction during every switching cycle. The waveforms for square-wave and PWM modes are shown in Figs. 7-36b and 7-36c, respectively. This concept can be extended to a three-phase inverter as shown in Fig. 7-37.

7-6-3 ZVS-CV DC–DC Converter with Voltage Cancellation

ZVS-CV technique can be extended to the single-phase dc-to-ac inverter with voltage cancellation. The switch-mode circuit, which was discussed in detail in Chapter 6, is shown in Fig. 7-38a and the resulting waveforms are shown in Fig. 7-38b, where both switches in each leg operate at a 50% duty ratio but the phase delay ϕ between the outputs of the two legs is controlled in order to control the output v_{AB} of the full-bridge. The output voltage of the full-bridge is stepped-down through an isolation transformer and then rectified to yield an overall dc-dc converter.

The switch-mode circuit of Fig. 7-38a can be modified to provide clamped-voltage, zero-voltage switching by adding L_A, C_{A+}, C_{A-} to leg A and L_B, C_{B+}, C_{B-} to leg B, as shown in Fig. 7-39a. For simplicity, the transformer is replaced by its magnetizing inductance L_m and its leakage inductance is neglected. The output stage is represented by the output current I_o. The resulting waveforms are shown in Fig. 7-39c, where the idealized switch-mode waveforms of Fig. 7-38 are repeated in Fig. 7-39b for comparison. Proper selection of inductance and capacitance values and a proper switching strategy result in a ZVS-CV switching, as discussed in Reference 34.

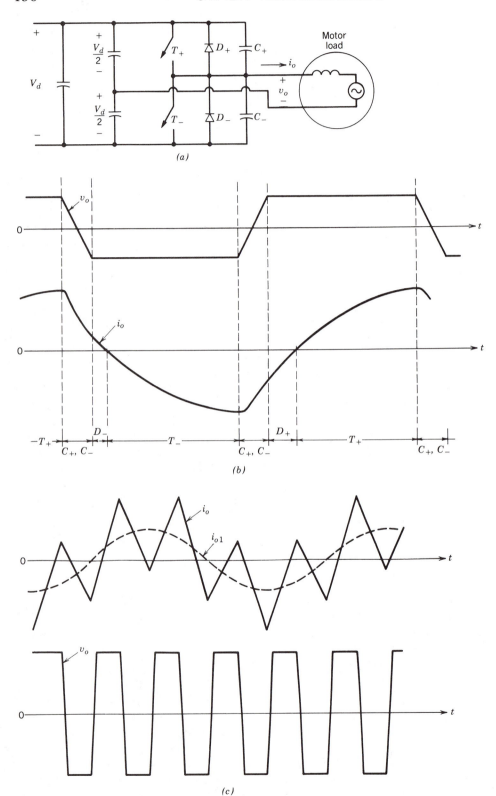

FIGURE 7-36: ZVS-CV dc-to-ac inverter: (*a*) half-bridge, (*b*) square-wave mode, (*c*) current-regulated mode.

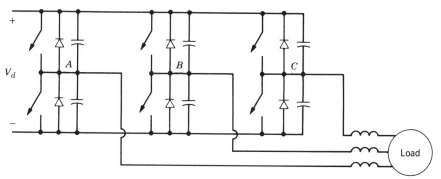

FIGURE 7-37: Three-phase, ZVS-CV dc-to-ac inverter.

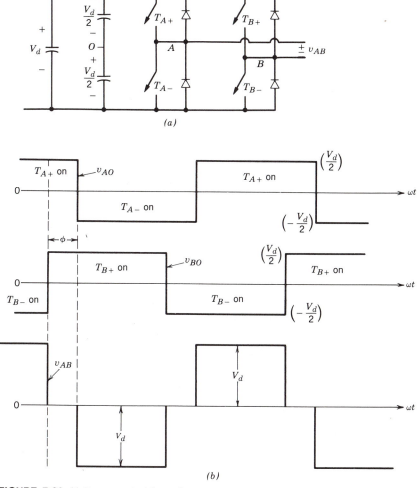

FIGURE 7-38: Voltage control by voltage cancellation: conventional switch-mode converter.

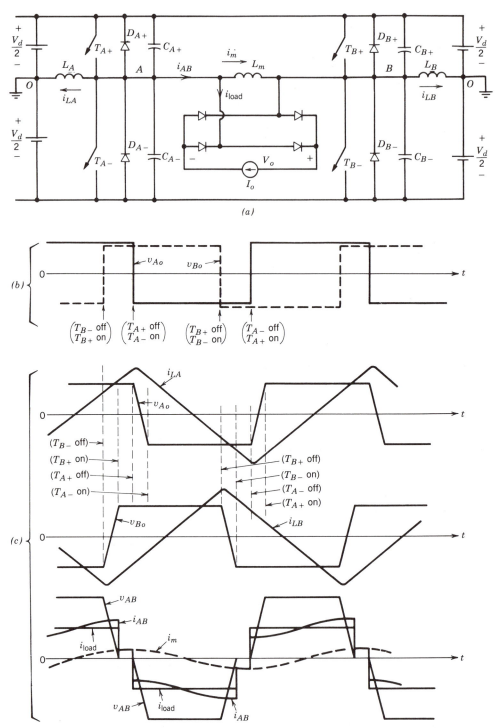

FIGURE 7-39: ZVS-CV full-bridge dc–dc converter: (a) circuit, (b) idealized switch-mode waveforms, (c) ZVS-CV waveforms.

7-7 RESONANT-DC-LINK INVERTERS WITH ZERO-VOLTAGE SWITCHINGS

In the conventional switch-mode PWM inverters of the type discussed in Chapter 6, the input is a dc voltage. To avoid the switching losses in the inverter, a new topology has been recently proposed in Reference 35, where a resonant circuit is introduced in between the dc input voltage and the PWM inverter. As a result, the input voltage to the inverter in the basic configuration oscillates between 0 and slightly greater than twice the dc input voltage. The inverter switches are turned on and turned off at zero voltage.

The basic concept is illustrated by means of the circuit of Fig. 7-40a. The resonant circuit consists of L_r, C_r, and a switch with an antiparallel diode. The load

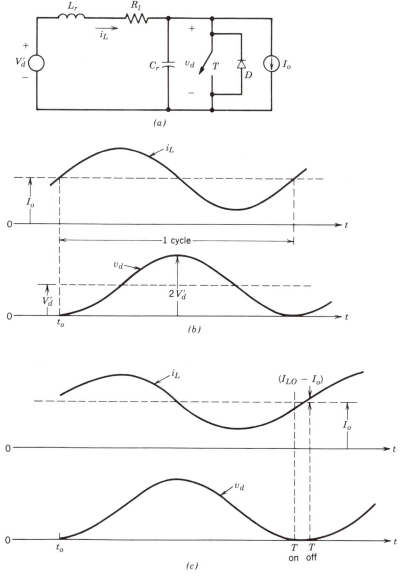

FIGURE 7-40: Resonant-dc-link inverter — basic concept: (a) basic circuit, (b) lossless $R_l = 0$, (c) losses are present.

on the circuit is represented by a current I_o, which represents, for example, the current being supplied by the inverter to a motor load. Because of the internal inductance of the load, it is reasonable to assume I_o to be constant in magnitude during a resonant frequency cycle.

As a first step, R_l is assumed to be zero. Initially, the switch is closed and the difference of i_L and I_o flows through the diode-switch combination. i_L builds up linearly. At time t_0, with $i_L = I_{L0}$, the switch is turned off at zero voltage. The equations for the resonant circuit are as follows for $t > t_0$:

$$t > t_0 \qquad i_L(t) = I_o + \left[\frac{V_d'}{\omega_0 L_r} \sin \omega_0(t - t_0) + (I_{L0} - I_o) \cos \omega_0(t - t_0) \right] \quad (7\text{-}34)$$

and

$$v_d(t) = V_d' + \left[\omega_0 L_r (I_{L0} - I_o) \sin \omega_o t - V_d' \cos \omega_0 t \right] \quad (7\text{-}35)$$

where

$$\omega_0 = \frac{1}{\sqrt{L_r C_r}} \quad (7\text{-}36)$$

The waveforms in Fig. 7-40b for $I_{L0} = I_o$ show that v_d returns to zero and i_L returns to I_o after one resonant cycle from the switch opening. Therefore, in this idealized circuit without any losses, the switch T and diode D can be removed once the oscillations start.

In the basic circuit of Fig. 7-40a, R_l represents the losses. In order for the zero-voltage turn on and zero-voltage turn off of the switch to occur, v_d must return to zero. In the presence of losses in R_l, I_{L0} must be greater than I_o at the instant the switch is turned off. The waveforms are shown in Fig. 7-40c. If the switch is kept on too long and I_{L0} is much larger than I_o, then v_d will peak at a value significantly larger than $2V_d'$. Therefore, $(I_{L0} - I_o)$ must be controlled by controlling the time interval during which the switch remains closed.

The foregoing concept can be applied to the 3-phase PWM inverter of Fig. 7-41. The resonant switch T and D of Fig. 7-40a are not needed since their function can be fulfilled by any of two switches comprising an inverter leg. The switches in any of the three inverter legs can be turned on and turned off at zero voltage when v_d reaches zero.

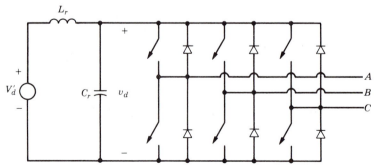

FIGURE 7-41: Three-phase resonant-dc-link inverter.

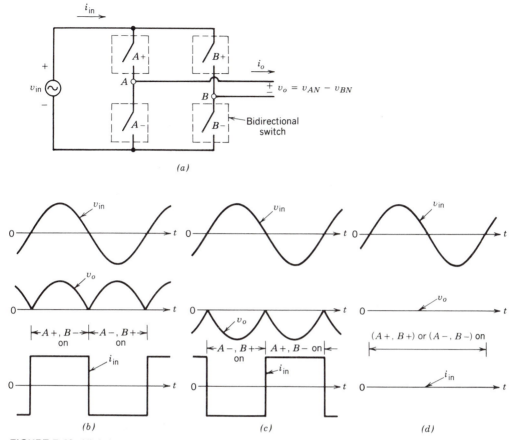

FIGURE 7-42: High-frequency-link integral-half-cycle inverter.

Further modifications to clamp the peak voltage across the switches to less than twice the input dc voltage have been discussed in the literature.

7-8   HIGH-FREQUENCY-LINK INTEGRAL-HALF-CYCLE CONVERTERS

Unlike the resonant-dc-link converters where the input to the single-phase or three-phase converter oscillates between zero and a value higher than the average input dc voltage, in the high-frequency-link converters the input to the single-phase or three-phase converter is a single-phase, high-frequency sinusoidal ac, as shown in Fig. 7-42a. As discussed in Reference 38, by turning the inverter switches on or off when the input voltage passes through zero, the switching losses can be minimized.

Figure 7-42a shows a single-phase converter of this type with a high-frequency sinusoidal input voltage v_{in}. The output is synthesized to be a low-frequency ac, for example, to supply a motor load. This requires that all four switches be bidirectional. Each bidirectional switch in Fig. 7-42a can be obtained by connecting two unidirectional switches with reverse blocking capability in antiparallel.

To describe the operating principle, the output load current is assumed to be a constant I_o during a cycle of high-frequency ac input. I_o can be positive or negative. For either direction of I_o, v_{AB} can consist of two positive half-cycles, two negative

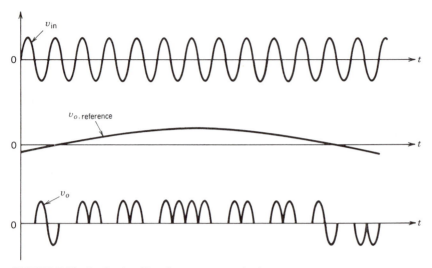

FIGURE 7-43: Synthesis of low-frequency ac output.

half-cycles, or zero (or any combination of these three options). These three options and the corresponding i_{in} are shown in Fig. 7-42b through 7-42d for a positive I_o, for example. This control over v_o to be positive, negative, or zero during each high-frequency half-cycle allows a low-frequency output to be synthesized to be of the desirable frequency and magnitude, as shown in Fig. 7-43. This control is discussed in detail in Reference 38. Since the low-frequency output consists of an integral number of half-cycles of the high-frequency input, these converters are labeled high-frequency-link integral-half-cycle converters.

   This concept can also be extended to deliver three-phase ac output by the circuit of Fig. 7-44. It should be noted that in both the single-phase and three-phase converters, a parallel resonant filter of the type shown in Fig. 7-44 must be used. It is tuned to be parallel-resonant at the input voltage frequency f_{in}. Therefore, it does not draw any current from the high-frequency ac input. However, the capacitor C_f provides a low impedance path to all other frequency components in i_{in} so that they do not have to be supplied by v_{in} through the stray inductance L_{stray}.

   The low-frequency output may in fact be dc in the circuit of Fig. 7-42a. Also, the power can flow in either direction in these converters.

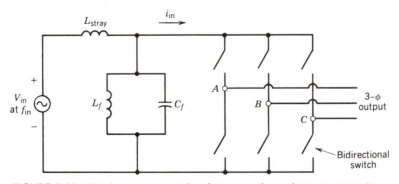

FIGURE 7-44: High-frequency ac to low-frequency three-phase ac converter.

These high-frequency-link converters are a form of cycloconverters, whereby the power is transferred between two ac systems operating at two different frequencies without an intermediate dc link. Unlike the phase-controlled, line-frequency cyclocon-verters using thyristors, which are described in Chapter 13, the bidirectional control-lable switches in these high-frequency-link converters are turned on or off when the input high-frequency ac passes through zero.

7-9 SUMMARY

In this chapter, various techniques are discussed that can either eliminate or diminish the stresses and the switching losses in the semiconductor devices. The following converters are described:

1. Load-resonant converters
   a. Series-loaded resonant (SLR) dc–dc converters
   b. Parallel-loaded resonant (PLR) dc–dc converters
   c. Hybrid resonant dc–dc converters
   d. Current-source, parallel-resonant dc-to-ac inverters for induction heating
   e. Class-E converters

2. Resonant-switch converters
   a. Zero-current switching (ZCS) resonant-switch converters
   b. Zero-voltage-switching (ZVS) resonant-switch converters
   c. Zero-voltage-switching clamped-voltage (ZVS-CV) converters
      c1. ZVS-CV dc–dc converters
      c2. ZVS-CV dc-to-ac inverters
      c3. ZVS-CV dc–dc converters with zero voltage cancellation

3. Resonant-dc-link inverters with zero-voltage switchings

4. High-frequency-link integral-half-cycle converters

An overview of these converters is provided in Reference 41.

PROBLEMS

SLR dc – dc CONVERTERS

7-1. An SLR dc–dc converter of Fig. 7-10a is operating in a discontinuous-conduction mode with $\omega_s < 0.5\omega_0$. In the waveforms of Fig. 7-11 (with $t_0 = 0$), the initial conditions in terms of normalized quantities are always as follows: $V_{c0} = -2V_o$ and $I_{L0} = 0$. Show that in terms of normalized quantities, $V_{c,\,\mathrm{peak}} = 2$ and $I_{L,\,\mathrm{peak}} = 1 + V_o$.

7-2. Design an SLR dc–dc converter of Fig. 7-10a with an isolation transformer of turns-ratio $n:1$, where $V_d = 155$ V, and the operating frequency $f_s = 100$ kHz. The output is at 5 V and 20 A.
   (a) The foregoing converter is to operate in a discontinuous-conduction mode with $\omega_s < 0.5\omega_0$. The normalized output voltage V_o is chosen to be 0.9 and the normalized frequency to be 0.45. Using the curves of Fig. 7-15, obtain turns-ratio n, L_r, and C_r.
   (b) Obtain the numerical value for the sum of peak energies stored in L_r and C_r:

$$S = \tfrac{1}{2}L_r I_{L,\,\mathrm{peak}}^2 + \tfrac{1}{2}C_r V_{c,\,\mathrm{peak}}^2$$

7-3. Repeat Problem 7-2 if the SLR converter is designed to operate in a continuous-conduction mode at below the resonant frequency.

 (a) Choose normalized output voltage as 0.9 and the normalized output current as 1.4. Use the design curves of Fig. 7-15. Obtain n, L_r, and C_r.

 (b) Obtain S as defined in Problem 7-2(b), by means of the design curves of Fig. P7-3.

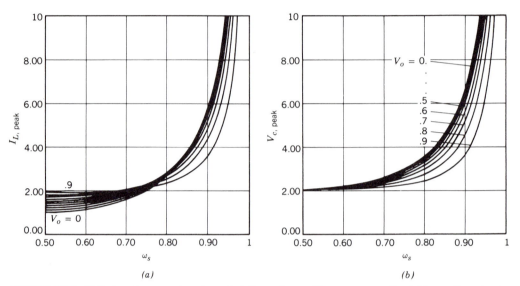

FIGURE P7-3: SLR dc – dc converter characteristics; all quantities are normalized (*Source*: Ramesh Oruganti, Ph.d. Dissertation, VPI, 1987).

7-4. Repeat Problem 7-3 for an operation above the resonant frequency ($\omega_s > \omega_0$) but with the normalized output voltage of 0.9 and the normalized output current of 0.4. Use the design curves of Fig. 7-15 and Fig. P7-4.

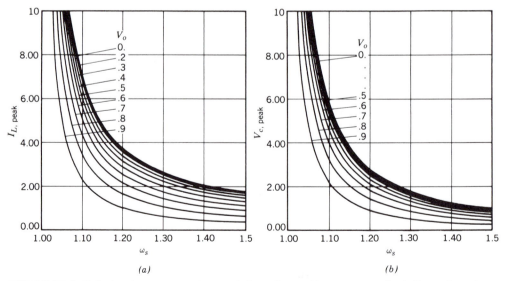

FIGURE P7-4: SLR dc – dc converter characteristics; all quantities are normalized (*Source*: Ramesh Oruganti, Ph.d. Dissertation, VPI, 1987).

7-5. Compare the values of S in Problems 7-2 through 7-4.

PLR dc–dc CONVERTERS

7-6. A PLR dc–dc converter of Fig. 7-17a with an isolation transformer of turns-ratio $n:1$ is operating in a discontinuous mode and the voltage and current waveforms are shown in Fig. 7-18.

Show that in a discontinuous mode,

$$V_{c,\text{peak}} = V_d$$

and

$$I_{L,\text{peak}} = \frac{I_o}{n} + \omega_0 c_r \frac{V_d}{2}$$

7-7. Design a half-bridge, transformer isolated PLR dc–dc converter. The input dc voltage $V_d = 155$ V, and the operating frequency $f_s = 300$ kHz. The output is at 5 V and 20 A. Obtain the transformer turns-ratio n, L_r, and C_r, assuming a discontinuous mode of operation, normalized operating frequency of 0.45, normalized $C_r = 1.2$ pu, and normalized $L_r = 0.833$ pu where,

$$(C_r)_{\text{base}} = \frac{I_o/n}{\omega_0 V_d/2} \quad \text{and} \quad (L_r)_{\text{base}} = \frac{V_d/2}{\omega_0 I_o/n}$$

Using the design curves of Fig. 7-21, and Problem 7-6, obtain the peak values of v_c and i_L. Calculate S, which was defined in Problem 7-2.

7-8. Design the converter of Problem 7-7, assuming a continuous mode below the resonant frequency. Let the normalized operating frequency be 0.8, and the values of normalized C_r and L_r as in Problem 7-7. Normalized I_o is 0.8.
(a) Calculate n, L_r, and C_r.
(b) Using the design curves of Figs. 7-21 and P7-8, obtain the peak values of v_c and i_L. Calculate S, as defined in Problem 7-2.

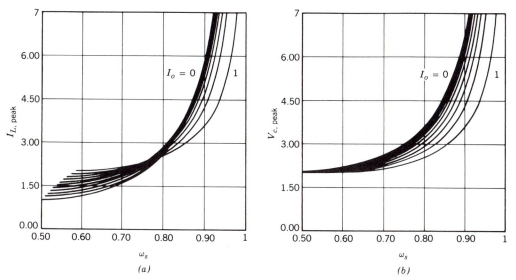

FIGURE P7-8: PLR dc–dc converter characteristics; all quantities are normalized (*Source:* Ramesh Oruganti, Ph.d. Dissertation, VPI, 1987).

7-9. Repeat Problem 7-8 for operation above the resonant frequency with the normalized frequency of 1.1. Use the design curves of Fig. 7-21 and P7-9.

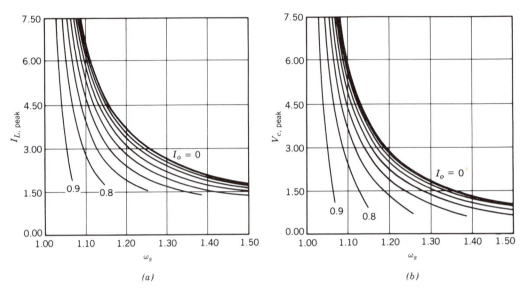

(a)   *(b)*

FIGURE P7-9: PLR dc – dc converter characteristics; all quantities are normalized (Source: Ramesh Oruganti, Ph.d. Dissertation, VPI, 1987).

7-10. Compare the values of S calculated in Problems 7-7 through 7-9.

ZCS RESONANT-SWITCH CONVERTERS

7-11. In the ZCS resonant-switch circuit of Fig. 7-30a, $f_0 = 1$ MHz, $Z_0 = 10 \ \Omega$, $P_{load} = 10$ W, $V_d = 15$ V, and $V_o = 10$ V. Assume L_2 to be quite large and all components to be ideal.
Calculate the i_L and v_c waveforms as a function of time. Sketch the waveforms for i_L and v_c and label the important transition points. Also label the peak values of i_L and v_c and the time instants at which they occur.

7-12. Repeat Problem 7-11 assuming a diode is connected in antiparallel with the switch in Fig. 7-30a.

ZVS RESONANT-SWITCH CONVERTERS

7-13. In a ZVS resonant-switch dc–dc converter of Fig. 7-31a, $V_d = 40$ V. I_o varies in a range of 4 A to 20 A. Calculate the theoretical minimum value of the switch voltage rating.

7-10   REFERENCES

Series-Loaded Resonant (SLR) DC – DC Converters

1. R. ORUGANTI, "State-Plane Analysis of Resonant Converters," Ph.D. Dissertation, Virginia Polytechnic Institute, 1987; available from University Microfilms International, P.O. Box 1764, Ann Arbor, MI 48106.
2. R. J. KING and T. A. STUART, "A Normalized Model for the Half-Bridge Series Resonant Converters," *IEEE Transactions on Aero-space and Electronics Systems*, Vol. AES-17, No. 2, pp. 190–198, March 1981.

3. R. J. KING and T. A. STUART, "Modelling the Full-Bridge Series-Resonant Power Converter," *IEEE Transactions on Aerospace and Electronic Systems*, Vol. AES-18, No. 4, pp. 449-459, July 1982.

4. R. J. KING and T. A. STUART, "Inherent Over-Load Protection for the Series-Resonant Converters," *IEEE Transactions on Aerospace and Electronics Systems*, Vol. AES-19, No. 6, pp. 820–830, Nov. 1983.

5. R. ORUGANTI and F. C. LEE, "Resonant Power Processors: Part 1—State Plane Analysis," *IEEE-IAS Annual Meeting Conference Records*, pp. 860–867, 1984.

6. A. F. WITULSKI and R. W. ERICKSON, "Design of the Series Resonant Converter for Minimum Component Stress," *IEEE Transactions on Aerospace and Electronic Systems*, Vol. AES-22, No. 4, pp. 356–363, July 1986.

7. J. G. HAYES, N. MOHAN and C. P. HENZE, "Zero-Voltage Switching in a Digitally Controlled Resonant DC–DC Power Converter," Proceedings of the 1988 IEEE Applied Power Electronics Conference.

Parallel-Loaded Resonant (PLR) DC–DC Converters

8. N. MAPHAM, "An SCR Inverter with Good Regulation and Sine-wave Output," *IEEE Transactions on Industry and General Applications*, Vol. IGA-3, pp. 176–187, March/April 1967.

9. V. T. RANGANATHAN, P. D. ZIOGAS, and V. R. STEFANOVIC, "A Regulated DC–DC Voltage Source Converter Using a High Frequency Link," *IEEE Transactions on Industry Applications*, Vol. IA-18, No. 3, pp. 279–287, May/June 1982.

10. R. ORUGANTI, "State-Plane Analysis of Resonant Converters," Ph.D. Dissertation, Virginia Polytechnic Institute, 1987.

11. M. C. W. LINDMARK, "Switch-Mode Power Supply," U.S. Patent #4,097,773, June 27, 1978.

12. I. J. PITEL, "Phase-Modulated Resonant Power Conversion Techniques for High Frequency Inverters," *IEEE-IAS Annual Meeting Proceedings*, 1985.

13. Y. G. KANG and A. K. UPADHYAY, "Analysis and Design of a Half-Bridge Parallel Resonant Converter," *1987 IEEE Power Electronics Specialists Conference*, pp. 231–243.

14. F. S. TSAI, P. MATERU, and F. C. LEE, "Constant-Frequency, Clamped-Mode Resonant Converters," *1987 IEEE Power Electronics Specialists Conference*, pp. 557–566.

Hybrid Resonant DC–DC Converters

15. D. V. JONES, "A New Resonant-Converter Topology," *Proceeding of the 1987 High Frequency Power Conversion Conference*, pp. 48–53.

Current-Source Parallel Resonant DC-to-AC Inverters for Induction Heating

16. K. THORBORG, *Power Electronics*, Prentice Hall International (UK) Ltd, 1988.

Class-E Converters

17. N. O. SOKAL and A. D. SOKAL, "Class-E, A New Class of High Efficiency Tuned Single-Ended Switching Power Amplifiers," *IEEE Journal of Solid State Circuits*, Vol. SC-10, pp. 168–176, June 1975.

18. F. H. RAAB, "Idealized Operation of Class-E Tuned Power Amplifier," *IEEE Transactions on Circuits and Systems*, Vol. CAS-24, No. 12, pp. 725–735, December 1977.

19. K. LÖHN, "On the Overall Efficiency of the Class-E Power Converters," *1986 IEEE Power Electronics Specialists Conference*, pp. 351–358.

20. M. Kazimierczuk and K. Puczko, "Control Circuit for Class-E Resonant DC/DC Converter," *Proceedings of the National Aerospace Electronics Conference 1987,* Vol. 2, pp. 416–423.

21. G. Lutteke and H. C. Raets, "220 V Mains 500 kHz Class-E Converter, *1985 IEEE Power Electronics Specialists Conference*, pp 127–135.

22. H. Omori, T. Iwai, et al., "Comparative Studies between Regenerative and Non-Regenerative Topologies of Single-Ended Resonant Inverters," *Proceedings of the 1987 High Frequency Power Conversion Conference.*

ZCS and ZVS Resonant-Switch Converters

23. P. Vinciarelli, "Forward Converter Switching at Zero Current," U.S. Patent #4,415,959, Nov. 1983.

24. R. Oruganti, "State-Plane Analysis of Resonant Converters," Ph.D. Dissertation, Virginia Polytechnic Institute, 1987.

25. K. H. Liu and F. C. Lee, "Resonant Switches—A Unified Approach to Improve Performances of Switching Converters," *IEEE INTELEC Conference Record,* pp. 344–351, 1984.

26. K. H. Liu, R. Oruganti, and F. C. Lee, "Resonant Switches—Topologies and Characteristics," *1986 IEEE Power Electronics Specialists Conference*, pp. 106–116.

27. K. H. Liu and F. C. Lee, "Zero Voltage Switching Technique in DC–DC Converters, *1986 IEEE Power Electronics Specialists Conference*, pp. 58–70.

28. K. D. T. Ngo, "Generalization of Resonant Switches and Quasi-Resonant DC–DC Converters," *1987 IEEE Power Electronics Specialists Conference,* pp. 395–403.

29. W. A. Tabisz and F. C. Lee, "Zero-Voltage Switching Multi-Resonant Technique—A Novel Approach To Improve Performance of High-Frequency Quasi-Resonant Converters", IEEE PESC Record, 1988.

30. M. F. Schlecht and L. F. Casey, "Comparison of the Square-Wave and Quasi-Resonant Topologies", Second Annual Applied Power Electronics Conference, San Diego, CA., pp. 124–134, 1987.

Zero-Voltage-Switching Clamped-Voltage (ZVS-CV) Converters

31. C. P. Henze, H. C. Martin, and D. W. Parsley, "Zero-Voltage Switching in High Frequency Power Converters Using Pulse Width Modulation," *Proceedings of the 1988 IEEE Applied Power Electronics Conference.*

32. R. Goldfarb, "A New Non-Dissipative Load-Line Shaping Technique Eliminates Switching Stress in Bridge Converters," *Proceedings of Powercon 8*, pp. D-4-1 to D-4-6, 1981.

33. T. M. Undeland, "Snubbers for Pulse Width Modulated Bridge Converters with Power Transistors or GTOs," *1983 International Power Electronics Conference*, Tokyo, Japan, pp. 313–323.

34. O. D. Patterson and D. M. Divan, "Pseudo-Resonant Full-Bridge DC/DC Converter" *1987 IEEE Power Electronics Specialists Conference,* pp. 424–430.

Resonant-DC-Link Converters

35. D. M. Divan, "The Resonant DC Link Converter—A New Concept in Static Power Conversion," *1986 IEEE-IAS Annual Meeting Record*, pp. 648–656.

36. M. Kheraluwala and D. M. Divan, "Delta Modulation Strategies for Resonant Link Inverters," *1987 IEEE Power Electronics Specialists Conference*, pp. 271–278.

37. K. S. Rajashekara, et al., "Resonant DC Link Inverter-Fed AC Machines Control," *1987 IEEE Power Electronics Specialists Conference*, pp. 491–496.

High-Frequency-Link Converters

38. P. K. Sood, T. A. Lipo, and I. G. Hansen, "A Versatile Power Converter for High Frequency Link Systems," *1987 IEEE Applied Power Electronics Conference*, pp. 249–256.
39. L. Gyugyi and F. Cibulka, "The High-Frequency Base Converter—A New Approach to Static High Frequency Conversion," *IEEE Transactions on Industry Applications*, Vol. IA-15, No. 4, pp. 420–429, July/August 1979.
40. P. M. Espelage and B. K. Bose, "High Frequency Link Power Conversion," *1975 IEEE-IAS Annual Meeting Record*, pp. 802-808.

Summary

41. N. Mohan, "Power Electronic Circuits: An Overview", *1988 IEEE Industrial Electronics Society Conference*, pp. 522–527.

PART

3

Power Supply Applications

8

Switching DC
Power Supplies

_____
_____
_____

8-1 INTRODUCTION

Regulated dc power supplies are needed for most analog and digital electronic systems. Most power supplies are designed to meet some or all of the following requirements:

- *Regulated output.* The output voltage must be held constant within a specified tolerance for changes within a specified range in the input voltage and the output loading.

- *Isolation.* The output may be required to be electrically isolated from the input.

- *Multiple outputs.* There may be multiple outputs (positive and negative) that may differ in their voltage and current ratings. Such outputs may be isolated from each other.

In addition to these requirements, common goals are to reduce power supply size and weight, and improve their efficiency. Traditionally, linear power supplies have been used. However, advances in the semiconductor technology have led to switching power supplies, which are smaller and much more efficient compared to linear power supplies. The cost comparison between linear and switching supplies depends on the power rating.

8-2 LINEAR POWER SUPPLIES

To appreciate the advantages of the switching supplies, it is desirable first to consider the linear power supplies. Figure 8-1a shows the schematic of a linear supply. In order to provide electrical isolation between the input and the output and to deliver

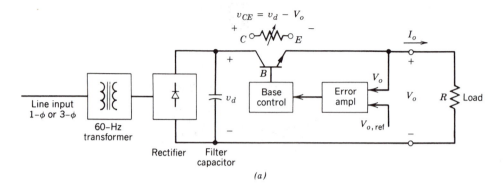

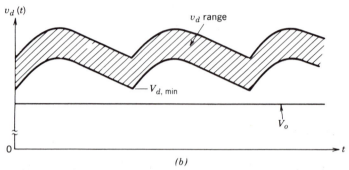

FIGURE 8-1: Linear power supply: (*a*) schematic, (*b*) selection of transformer turns-ratio so that $V_{d,min} > V_o$ by a small margin.

the output in the desired voltage range, a 60-Hz transformer is needed. A transistor is connected in series that operates in its active region.

Comparing V_o with a reference voltage V_{ref}, the control circuit in Fig. 8-1*a* adjusts the transistor base current such that $V_o (= v_d - v_{CE})$ equals $V_{o,ref}$. The transistor in a linear supply acts as an adjustable resistor where the voltage difference $(v_d - V_o)$ between the input and the desired output voltage appears across the transistor and causes power losses in it. For a given range of 60 Hz ac input voltage, the rectified and filtered output $v_d(t)$ may be as shown in Fig. 8-1*b*. To minimize the transistor power losses, the transformer turns ratio should be carefully selected such that $V_{d,min}$ in Fig. 8-1*b* is greater than V_o but does not exceed V_o by a large margin.

The preceding discussion points out two major shortcomings of a linear power supply:

1. A low-frequency (60-Hz) transformer is required. Such transformers are larger in size and weight compared to high-frequency transformers.

2. The transistor operates in its active region, incurring a significant amount of power loss. Therefore, the overall efficiencies of linear power supplies are usually in a range of 30 to 60%.

On the positive side, these supplies utilize simple circuitry and therefore may cost less in small power ratings ($<$ 25 W). Also, these supplies do not produce large electromagnetic interference (EMI) with other equipment.

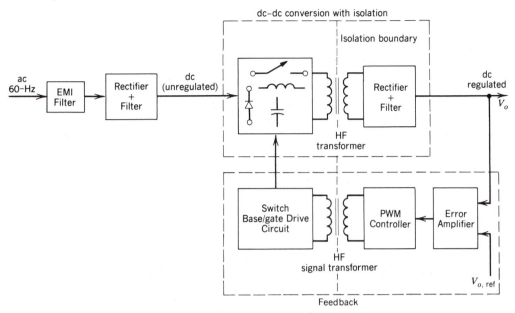

FIGURE 8-2: Schematic of a switch-mode dc power supply.

8-3   OVERVIEW OF SWITCHING POWER SUPPLIES

As opposed to linear power supplies, in switching power supplies, the transformation of dc voltage from one level to another is accomplished by using dc-to-dc converter circuits (or those derived from them), which were discussed in Chapters 5 and 7. These circuits employ solid-state devices (transistors, MOSFETs, etc.), which operate as a switch: either completely off or completely on. Since the power devices are not required to operate in their active region, this mode of operation results in a lower power dissipation. Increased switching speeds, higher voltage and current ratings, and a relatively lower cost of these devices are the factors that have contributed to the emergence of switching power supplies.

Figure 8-2 shows a switching supply with electrical isolation in a simplified block-diagram form. The input ac voltage is rectified into an unregulated dc voltage by means of a diode rectifier of the type discussed in Chapter 3. It should be noted that an EMI filter, as discussed in Chapter 17, is used at the input to prevent the conducted EMI. The dc–dc converter block in Fig. 8-2 converts the input dc voltage from one level to another dc level. This is accomplished by high-frequency switching, which produces high frequency ac across the isolation transformer. The secondary output of the transformer is rectified and filtered to produce V_o. The output of the dc supply in Fig. 8-2 is regulated by means of a feedback control that employs a PWM controller as discussed in Chapter 5, where the control voltage is compared with a sawtooth waveform at the switching frequency. The electrical isolation in the feedback loop is provided either through an isolation transformer as shown or through an optocoupler.

In many applications, multiple outputs (both positive and negative) are required. These outputs may be required to be electrically isolated from each other, depending on the application. Figure 8-3 shows the block diagram of a switching

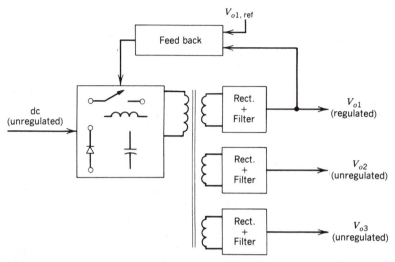

FIGURE 8-3: Multiple outputs.

supply where only one output V_{o1} is regulated and the other two are unregulated. If V_{o2} and/or V_{o3} need to be regulated, then linear regulator(s) can be used to regulate the other output(s).

Two major advantages of switching power supplies over linear power supplies are now apparent. These are as follows:

- The switching elements (power transistors or MOSFETs) operate as a switch: either completely off or completely on. By avoiding their operation in their active region, a significant reduction in power losses is achieved. This results in a higher energy efficiency in a 70 to 90% range. Moreover, a transistor operating in on/off mode has a much larger power-handling capability compared to its linear mode.

- Since a high frequency isolation transformer is used (as compared to a 50/60-Hz transformer in a linear power supply), the size and weight of switching supplies can be significantly reduced.

On the negative side, switching supplies are more complex, and proper measures must be taken to prevent electromagnetic interference (EMI) due to high-frequency switchings.

The above-mentioned advantages of switching supplies (over linear supplies) outweigh their shortcomings above a certain power rating. The power rating where this breakover occurs is steadily decreasing with time due to advances in semiconductor technology.

Switching dc power supplies, in general, utilize modifications of the following two classes of converter topologies:

1. Switch-mode dc–dc converters discussed in Chapter 5, where the switches operate in a switch mode.

2. Resonant converters discussed in Chapter 7, which utilize zero-voltage and/or zero-current switchings.

In this chapter, the switch-mode converter topologies are used to describe the operation of switching power supplies. Many of the basic principles discussed in this chapter also apply to the switching power supplies with resonant converters.

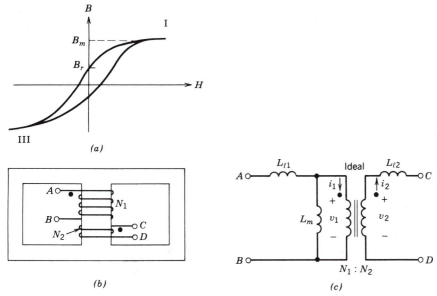

FIGURE 8-4: Transformer representation: (*a*) Typical B-H loop of transformer core, (*b*) two-winding transformer, (*c*) equivalent circuit.

8-4   DC – DC CONVERTERS WITH ELECTRICAL ISOLATION

8-4-1   Introduction to DC – DC Converters with Isolation

As seen by the block diagram of Fig. 8-2, the electrical isolation in switching dc power supplies is provided by a high-frequency isolation transformer. Figure 8-4*a* shows a typical transformer core characteristic in terms of its B-H (hysteresis) loop. B_m is the maximum flux density beyond which saturation occurs and B_r is the remnant flux density. Various types of dc–dc converters (with isolation) can be divided into two basic categories, based on the way they utilize the transformer core:

1. Unidirectional core excitation where only the positive part (quadrant 1) of the B-H loop is used, and

2. Bidirectional core excitation where both the positive (quadrant 1) and the negative (quadrant 3) part of the B-H loop are utilized alternatively.

8-4-1-1   UNIDIRECTIONAL CORE EXCITATION

Some of the dc–dc converters (without isolation) discussed in Chapter 5 can be modified to provide electrical isolation by means of unidirectional core excitation. Two such modifications are

- Flyback converter (derived from buck–boost converter)
- Forward converter (derived from step-down converter)

The output voltage of these converters is regulated by means of the PWM switching scheme discussed in Chapter 5.

8-4-1-2   BIDIRECTIONAL CORE EXCITATION

To provide electrical isolation by means of bidirectional core excitation, the single-phase switch-mode inverter topologies of Chapter 6 can be used to produce a square-wave ac at the input of the high-frequency isolation transformer in Fig. 8-2. We will discuss the following inverter topologies, which can constitute a switching dc power supply:

- Push–pull
- Half-bridge
- Full-bridge

As in Chapters 5 and 6, for analyzing the following circuits, the switches are treated as being ideal and the power losses in the inductive, capacitive, and the transformer elements are neglected. Some of these losses limit the operational capabilities of these circuits and are discussed separately.

All of the circuits are analyzed under a steady-state operating condition, and the filter capacitor at the output is assumed to be so large (as in Chapter 5) as to allow the assumption that the output voltage $v_o(t) \simeq V_o$ (i.e., essentially a pure dc). The analysis is presented only for the continuous-conduction mode and the analysis of the discontinuous-conduction mode is left as an exercise.

8-4-1-3 ISOLATION TRANSFORMER REPRESENTATION

A high-frequency transformer is required to provide electrical isolation. Neglecting the losses in the transformer of Fig. 8-4b, an approximate equivalent circuit for a two winding transformer is redrawn in Fig. 8-4c where $N_1 : N_2$ is the transformer winding turns ratio, L_m is the magnetizing inductance referred to the primary side, and L_{l1} and L_{l2} are the leakage inductances. In the ideal transformer, $v_1/v_2 = N_1/N_2$ and $N_1 i_1 = N_2 i_2$.

In a switch-mode dc–dc converter, it is desirable to minimize the leakage inductances L_{l1} and L_{l2} by providing a tight magnetic coupling between the two windings. The energy associated with the leakage inductances has to be absorbed by the switching elements and their snubber circuits, thus clearly indicating a need to minimize the leakage inductances. Similarly, in a switch-mode dc–dc converter, it is desirable to make the magnetizing inductance L_m in Fig. 8-4c as high as possible to minimize the magnetizing current i_m that flows through the switches and thus increases their current ratings.

It is important to consider the effect of the transformer leakage inductances in switch selection and snubber design. However, these inductances have a minor effect on the converter voltage transfer characteristics and therefore have been neglected in the converter analysis to follow.

In one of the converter topologies to be discussed, called the flyback converter, the transformer is in fact intended to be a two-winding inductor, which has dual functions of providing energy storage as in an inductor and electrical isolation as in a transformer. Therefore, the previous comments to make L_m high do not apply to this topology. However, the simplified transformer equivalent circuit can still be used for the analysis purposes.

The transformer design considerations in resonant power supplies are different than the ones discussed before for switch-mode power supplies. There, the leakage inductances and/or the magnetizing inductance may in fact be utilized to provide zero-voltage and/or zero-current switchings.

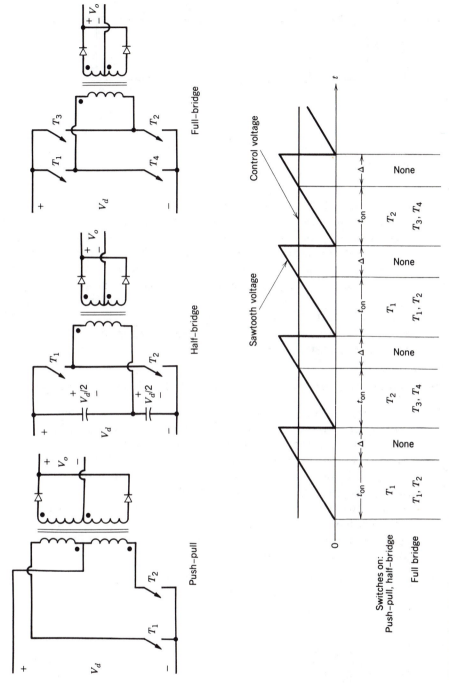

FIGURE 8-5: PWM scheme used in dc – dc converters, where the converter output is rectified to produce a dc output.

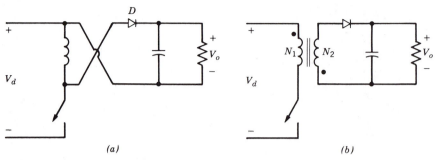

FIGURE 8-6: Flyback converter.

8-4-1-4 CONTROL OF DC-DC CONVERTERS WITH ISOLATION

In the single-switch topologies like the flyback and the forward converters, the output voltage V_o for a given input V_d is controlled by pulse-width modulation in a manner similar to that used for their nonisolated counterparts discussed in Chapter 5.

In the push-pull, half-bridge and the full-bridge dc-dc converters, where the converter output is rectified to produce a dc output, the dc output voltage V_o is controlled by using a PWM scheme shown in Fig. 8-5 which controls the interval Δ during which all the switches are off simultaneously. This is unlike the PWM schemes used in Chapter 5 to control full-bridge dc-dc converters and in Chapter 6 to control single-phase dc-to-ac inverters.

8-4-2 Flyback Converters (Derived from Buck – Boost Converters)

Flyback converters are derived from the buck–boost converter discussed in Chapter 5 and redrawn in Fig. 8-6a. By placing a second winding on the inductor, it is possible to achieve electrical isolation, as shown in Fig. 8-6b.

Figure 8-7a shows the converter circuit where the two-winding inductor is represented by its approximate equivalent circuit. When the switch is on, due to the winding polarities, the diode D in Fig. 8-7a becomes reverse biased. The continuous-current-conduction mode in a buck–boost converter corresponds to an incomplete demagnetization of the inductor core in the flyback converter. Therefore, as shown by the waveforms in Fig. 8-8, the inductor core flux increases linearly from its initial value $\phi(0)$, which is finite and positive:

$$\phi(t) = \phi(0) + \frac{V_d}{N_1}t \qquad 0 < t < t_{\text{on}} \tag{8-1}$$

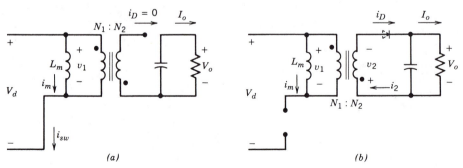

FIGURE 8-7: Flyback converter circuit states: (*a*) switch on, (*b*) switch off.

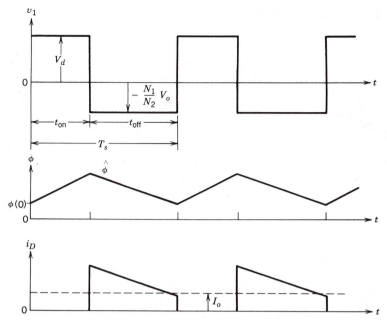

FIGURE 8-8: Flyback converter waveforms.

and, the peak flux $\hat{\phi}$ at the end of the on interval is given as

$$\hat{\phi} = \phi(t_{on}) = \phi(0) + \frac{V_d}{N_1}t_{on} \tag{8-2}$$

After t_{on}, the switch is turned off and the energy stored in the core causes the current to flow in the secondary winding through the diode D, as shown by Fig. 8-7b. The voltage across the secondary winding $v_2 = -V_0$ and therefore, the flux decreases linearly during t_{off}:

$$\phi(t) = \hat{\phi} - \frac{V_o}{N_2}(t - t_{on}) \qquad t_{on} < t < T_s \tag{8-3}$$

and

$$\phi(T_s) = \hat{\phi} - \frac{V_o}{N_2}(T_s - t_{on}) \tag{8-4}$$

$$= \phi(0) + \frac{V_d}{N_1}t_{on} - \frac{V_o}{N_2}(T_s - t_{on}) \quad \text{(using Eq. 8-2)} \tag{8-5}$$

Since the net change of flux through the core over one time period must be zero in steady state,

$$\phi(T_s) = \phi(0) \tag{8-6}$$

Therefore, from Eqs. 8-5 and 8-6

$$\frac{V_o}{V_d} = \frac{N_2}{N_1} \cdot \frac{D}{1-D} \tag{8-7}$$

where $D = t_{on}/T_s$ is the switch duty ratio. Equation 8-7 shows that the voltage transfer ratio in a flyback converter depends on D in an identical manner as the buck–boost converter.

The voltage and current waveforms shown in Fig. 8-8 can be obtained from the equations below. During the on interval, the transformer primary voltage $v_1 = V_d$. Therefore, the inductor current rises linearly from its initial value $I_m(0)$:

$$i_m(t) = i_{sw}(t) = I_m(0) + \frac{V_d}{L_m}t \qquad 0 < t < t_{on} \tag{8-8}$$

and

$$\hat{I}_m = \hat{I}_{sw} = I_m(0) + \frac{V_d}{L_m}t_{on} \tag{8-9}$$

During the off interval, the switch current goes to zero and $v_1 = -(N_1/N_2)V_o$. Therefore, i_m and the diode current i_D can be expressed during $t_{on} < t < T_s$ as

$$i_m(t) = \hat{I}_m - \frac{V_o(N_1/N_2)}{L_m}(t - t_{on}) \tag{8-10}$$

and

$$i_D(t) = \frac{N_1}{N_2}i_m(t) = \frac{N_1}{N_2}\left[\hat{I}_m - \frac{V_o(N_1/N_2)}{L_m}(t - t_{on})\right] \tag{8-11}$$

Since the average diode current equals I_o, from Eq. 8-11

$$\hat{I}_m = \hat{I}_{sw} = \frac{N_2}{N_1}\frac{1}{1-D}I_o + \frac{N_1}{N_2}\frac{(1-D)T_s}{2L_m}V_o \tag{8-12}$$

The voltage across the switch during the off interval equals

$$v_{sw} = V_d + \frac{N_1}{N_2}V_o = \frac{V_d}{1-D} \tag{8-13}$$

8-4-2-1 OTHER FLYBACK CONVERTER TOPOLOGIES

Two modifications of the flyback converter topology are shown in Fig. 8-9. Another flyback topology that is well suited for low output voltage applications is discussed in Reference 5.

Two-Transistor Flyback Converter. Figure 8-9a shows a two-transistor version of a flyback converter where T_1 and T_2 are turned on and off simultaneously. The advantage of such a topology over a single-transistor flyback converter, discussed earlier, is that voltage rating of the switches is one-half of the single-transistor version. Moreover, since a current path exists through the diodes connected to the primary

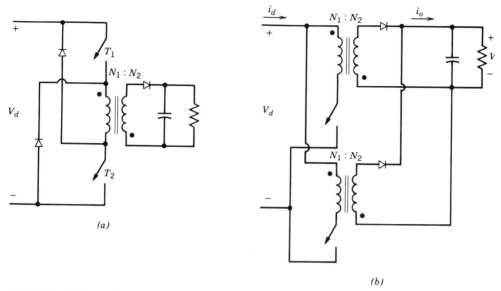

FIGURE 8-9: Other flyback topologies: (*a*) two transistor flyback converter, (*b*) parallelled flyback converters.

winding, a dissipative snubber across the primary winding is not needed to dissipate the energy associated with the transformer primary-winding leakage inductance (see Reference 14).

Paralleling Flyback Converters. At high power levels, it may be beneficial to parallel two or more flyback converters rather than using a single higher power unit. Some of the advantages of paralleling, which are not limited just to the flyback converter are (1) that it provides higher system reliability due to redundancy, (2) that it increases the effective switching frequency and hence decreases current pulsations at the input and/or the output and, (3) that it allows low power modules to be standardized where a number of these can be paralleled to provide a higher power capability.

The problem of current sharing among the parallel converters can be remedied by means of current-mode control, which is discussed later in this chapter.

Figure 8-9*b* shows two flyback converters in parallel; these operate at the same switching frequency, but the switches in the two converters are sequenced to turn on a half time period apart from one another. This results in improved input and output current waveforms (see Problem 8-4).

8-4-3 Forward Converter (Derived from Step-Down Converter)

Figure 8-10 shows an idealized forward converter. As will be discussed shortly, the transformer magnetizing current must be taken into account in these converters.

Initially, assuming a transformer to be ideal, when the switch is turned on, D_1 becomes forward biased and D_2 reverse biased. Therefore in Fig. 8-10,

$$v_L = \frac{N_2}{N_1} V_d - V_o \qquad 0 < t < t_{\mathrm{on}} \qquad (8\text{-}14)$$

which is positive. Therefore, i_L increases. When the switch is turned off, the inductor

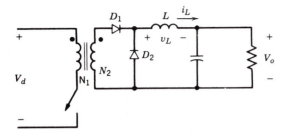

FIGURE 8-10: Idealized forward converter.

current i_L circulates through the diode D_2 and,

$$v_L = -V_o \qquad t_{on} < t < T_s \tag{8-15}$$

which is negative and therefore causes i_L to decrease linearly. Equating the integral of the inductor voltage over one time period to zero using Eqs. 8-14 and 8-15 yields

$$\frac{V_o}{V_d} = \frac{N_2}{N_1} D \tag{8-16}$$

Equation 8-16 shows that the voltage ratio in the forward converter is proportional to the switch duty ratio D, similar to the step-down converter. Another way to obtain the voltage ratio is to equate the average value of v_{oi} (defined in Fig. 8-11b) over one time period to V_o, recognizing that the average value of v_L is zero.

In a practical forward converter, the transformer magnetizing current must be taken into consideration for a proper converter operation. Otherwise, the stored energy in the transformer core would result in converter failure. An approach that allows the transformer magnetic energy to be recovered and fed back to the input supply is shown in Fig. 8-11a. It requires a third demagnetizing winding. Figure 8-11b shows the transformer in terms of its equivalent circuit, with the leakage inductances neglected. When the switch is on,

$$v_1 = V_d \qquad 0 < t < t_{on} \tag{8-17}$$

and i_m increases linearly from 0 to $\hat{I}_m$ as shown in Fig. 8-11c. When the switch is turned off, $i_1 = -i_m$. With the current directions shown in Fig. 8-11b, $N_1 i_1 + N_3 i_3 = N_2 i_2$. Because of D_1, i_2 equals zero and therefore

$$i_3 = \frac{N_1}{N_3} i_m \tag{8-18}$$

which flows through D_3 into the input dc supply. During the time interval t_m in Fig. 8-11c when i_3 is flowing, the voltage across the transformer primary as well as L_m is

$$v_1 = -\frac{N_1}{N_3} V_d \qquad t_{on} < t < (t_{on} + t_m) \tag{8-19}$$

Once the transformer demagnetizes, $i_m = 0$ and $v_1 = 0$. The time interval t_m can be obtained by recognizing that the time integral of voltage v_1 across L_m must be zero

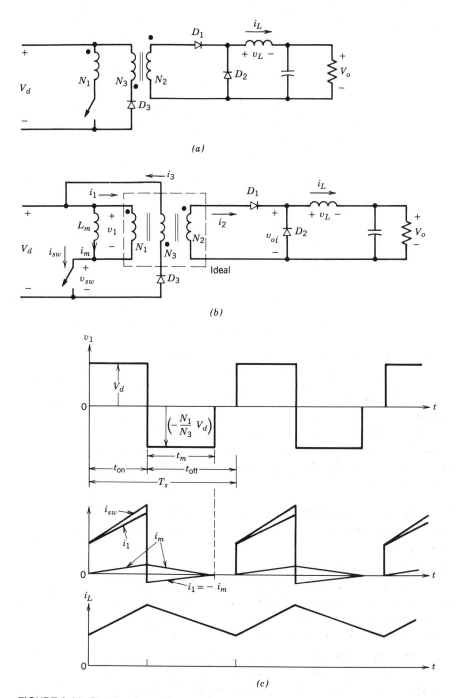

FIGURE 8-11: Practical forward converter.

over one time period. Using Eqs. 8-17 and 8-19

$$\frac{t_m}{T_s} = \frac{N_3}{N_1} D \qquad (8\text{-}20)$$

If the transformer is to be totally demagnetized before the next cycle begins, the maximum value (t_m/T_s) can attain is $(1 - D)$. Therefore, using Eq. 8-20, the maximum duty ratio D_{max} with a given turns ratio N_3/N_1 is

$$(1 - D_{max}) = \frac{N_3}{N_1} D_{max}$$

or

$$D_{max} = \frac{1}{1 + \dfrac{N_3}{N_1}} \qquad (8\text{-}21)$$

The foregoing analysis shows that with an equal number of turns for the primary and the demagnetizing windings ($N_1 = N_3$, a common practice), the maximum duty ratio in such converters is limited to 0.5.

Note that since a large voltage isolation requirement does not exist between the primary and the demagnetizing windings, these two can be wound bifilar, in order to minimize the leakage inductance between the two windings. The demagnetizing winding requires a much smaller size of wire, since it has to carry only the demagnetizing current. It should be noted that when the transformer magnetizing inductance is included, the voltage transformation ratio V_o/V_d remains the same as given by Eq. 8-16, which was derived by assuming an ideal transformer. Instead of using a third demagnetizing winding, the energy in the core can be dissipated in the Zener diode connected across the switch.

8-4-3-1 OTHER FORWARD CONVERTER TOPOLOGIES

Some of the common modifications of the forward converter topologies are shown in Fig. 8-12.

Two-Switch Forward Converter. As is shown in Fig. 8-12a, the two switches are turned on and off simultaneously. The voltage rating of each of the switches is one-half of that in a single-switch topology. More significantly, when the switches are off, the magnetizing current flows into the input supply through the diodes, thus eliminating the need for a separate demagnetizing winding or snubbers.

Paralleling Forward Converters. The same advantages can be gained by paralleling two or more forward converters as those discussed in the flyback converter section. Figure 8-12b shows two forward converters in parallel whose switches are sequenced to turn on a half time period apart from one another. At the output, a common filter can be used, thus significantly reducing the size of the output filter capacitor and inductor (see Problem 8-7).

8-4-4   Push–Pull Converter (Derived from Step-Down Converter)

Figure 8-13a shows the circuit arrangement for a push–pull dc–dc converter where the push–pull inverter of Chapter 6 is used to produce a square-wave ac at the

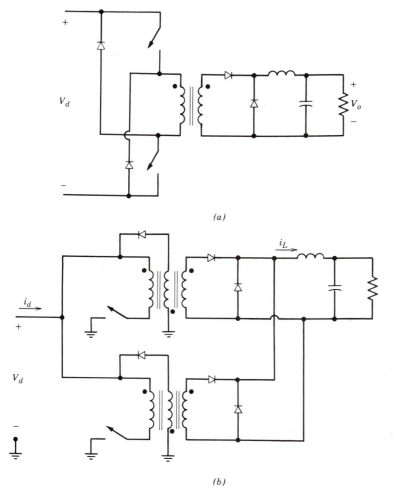

(a)

(b)

FIGURE 8-12: Other topologies of forward converter: (a) two-switch forward converter, (b) parallelled forward converters.

input of the high-frequency transformer. The PWM switching scheme described by Fig. 8-5 is used to regulate the output voltage. A center-tapped secondary is used, which results in only one diode voltage drop on the secondary side.

In Fig. 8-13a, when T_1 is on, D_1 conducts and D_2 gets reverse biased. This results in $v_{oi} = (N_2/N_1)V_d$ in Fig. 8-13b. Therefore, the voltage across the filter inductor is given as

$$v_L = \frac{N_2}{N_1}V_d - V_o \qquad 0 < t < t_{on} \tag{8-22}$$

and i_L through D_1 increases linearly as shown by Fig. 8-13b.

During the interval Δ when both the switches are off, the inductor current splits equally between the two secondary half-windings and $v_{oi} = 0$. Therefore, during $t_{on} < t < (t_{on} + \Delta)$

$$v_L = -V_o \tag{8-23}$$

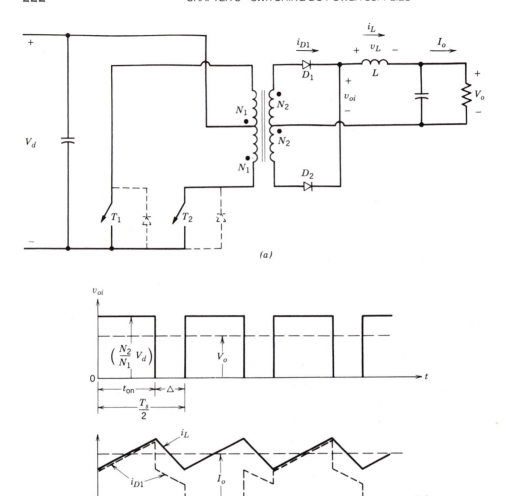

FIGURE 8-13: Push–pull converter.

and

$$i_{D1} = i_{D2} = \frac{i_L}{2} \tag{8-24}$$

The next half-cycle consists of t_{on} (during which T_2 is on), and the interval Δ. The waveforms repeat with a period $T_s/2$ and

$$t_{on} + \Delta = \frac{T_s}{2} \tag{8-25}$$

Equating the time integral of the inductor voltage during one repetition period $T_s/2$ to zero using Eqs. 8-22, 8-23, and 8-25 yields

$$\frac{V_o}{V_d} = 2\frac{N_2}{N_1}D \qquad 0 < D < 0.5 \tag{8-26}$$

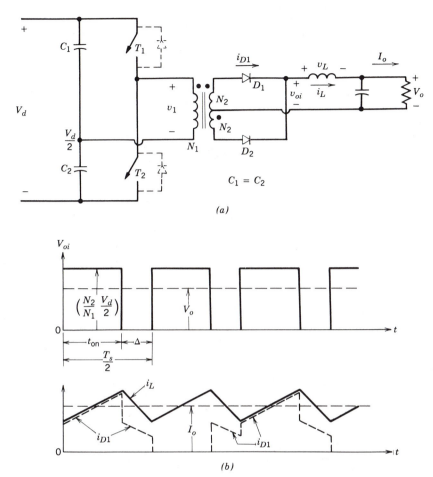

FIGURE 8-14: Half-bridge dc–dc converter.

where $D = t_{on}/T_s$ is the duty ratio of the switches 1 and 2 and the maximum value it can attain is 0.5 (in practice, to maintain a small blanking time to avoid turning both the switches on simultaneously, D should be kept smaller than 0.5). The average value of v_{oi} waveform in Fig. 8-13b equals V_o.

It should be noted that in the push–pull inverter of Chapter 6, the feedback diodes connected in antiparallel to the switches were required to carry the reactive current and their conduction interval depended inversely on the power factor of the output load. In the push–pull dc–dc converter, these antiparallel diodes shown dotted in Fig. 8-13a are needed to provide a path for the current required due to leakage flux of the transformer.

In push–pull circuits, due to a slight and unavoidable difference in the switching times of two switches T_1 and T_2, there is always an unbalance between the peak values of the two switch currents. This unbalance can be eliminated by means of current-mode control of the converter, which is discussed later in this chapter.

8-4-5 Half-Bridge Converter (Derived from Step-Down Converter)

Figure 8-14a shows a half-bridge dc–dc converter. As discussed in Chapter 6 in connection with the half-bridge inverters, the capacitors C_1 and C_2 establish a voltage midpoint between 0 and the input dc voltage. The switches T_1 and T_2 are turned on alternatively, each for an interval t_{on}. With T_1 on, $v_{oi} = (N_2/N_1)(V_d/2)$ as shown in Fig. 8-14b and, therefore,

$$v_L = \frac{N_2}{N_1}\left(\frac{V_d}{2}\right) - V_o \qquad 0 < t < t_{on} \tag{8-27}$$

During the interval Δ, when both switches are off, the inductor current splits equally between the two secondary halves. Assuming ideal diodes, $v_{oi} = 0$ and, therefore,

$$v_L = -V_o \qquad t_{on} < t < (t_{on} + \Delta) \tag{8-28}$$

In steady state, the waveforms repeat with a period $T_s/2$ and

$$t_{on} + \Delta = \frac{T_s}{2} \tag{8-29}$$

Equating the time integral of the inductor voltage during one repetition period to zero using Eqs. 8-27 through 8-29 yields

$$\frac{V_o}{V_d} = \frac{N_2}{N_1}D \tag{8-30}$$

where $D = t_{on}/T_s$ and $0 < D < 0.5$. The average value of v_{oi} in Fig. 8-14b equals V_o.

The diodes in antiparallel with the switches T_1 and T_2 are used for switch protection, as in a push–pull converter.

8-4-6   Full-Bridge Converter (Derived from Step-Down Converter)

Figure 8-15a shows a full-bridge converter where (T_1, T_2) and (T_3, T_4) are switched as pairs alternatively at the selected switching frequency. When (T_1, T_2) or (T_3, T_4) are on, $v_{oi} = (N_2/N_1)V_d$ as shown in Fig. 8-15b and therefore

$$v_L = \frac{N_2}{N_1}V_d - V_o \qquad 0 < t < t_{on} \tag{8-31}$$

When both the switch pairs are off, the inductor current splits equally between the two secondary halves. Assuming ideal diodes, $v_{oi} = 0$ and therefore

$$v_L = -V_o \qquad t_{on} < t < (t_{on} + \Delta) \tag{8-32}$$

Equating the time integral of the inductor voltage over one time period to zero in steady state and recognizing that $t_{on} + \Delta = T_s/2$

$$\frac{V_o}{V_d} = 2\frac{N_2}{N_1}D \tag{8-33}$$

where $D = t_{on}/T_s$ and $0 < D < 0.5$. In Fig. 8-15b, the average value of v_{oi} equals V_o.

The diodes connected in antiparallel to the switches (shown as dotted) provide a path to the current due to the energy associated with the primary winding leakage inductance.

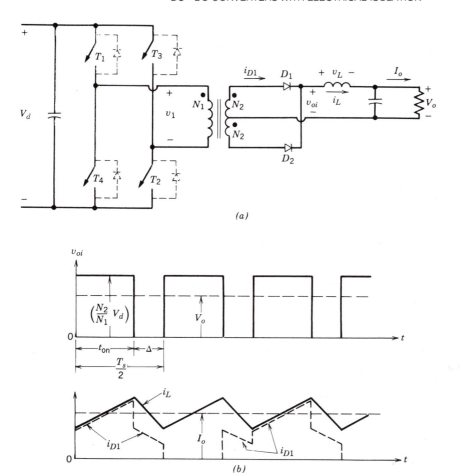

FIGURE 8-15: Full-bridge converter.

Comparison of the full-bridge (FB) converter with the half-bridge (HB) converter for identical input and output voltages and power ratings requires the following turns ratios:

$$\left(\frac{N_2}{N_1}\right)_{\text{HB}} = 2\left(\frac{N_2}{N_1}\right)_{\text{FB}} \tag{8-34}$$

Neglecting the ripple in the current through the filter inductor at the output and assuming the transformer magnetizing current to be negligible in both circuits, the switch currents I_{sw} are

$$\left(I_{sw}\right)_{\text{HB}} = 2\left(I_{sw}\right)_{\text{FB}} \tag{8-35}$$

In both converters, the input V_d appears across the switches; however, they are required to carry twice as much current in the half-bridge compared with the full-bridge converter. Therefore, in large power ratings, it may be advantageous to use a full-bridge over a half-bridge converter to reduce the number of paralleled devices in the switch.

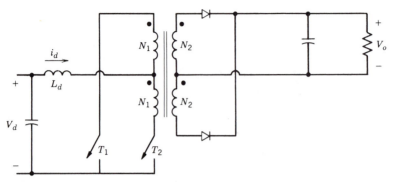

FIGURE 8-16: Current-source converter ($D > 0.5$).

8-4-7   Current-Source DC – DC Converters

The dc–dc converters (derived from the step-down converter topology) in the previous sections are supplied with a voltage at their input and, therefore, are voltage-source converters. By inserting an inductor at the input of a push–pull circuit, as shown in Fig. 8-16, and operating the switches at a duty ratio D of greater than 0.5, the converter is fed through a current source. D greater than 0.5 implies simultaneous conduction of the two switches, which was to be strictly avoided in the normal voltage-source push–pull converter.

When both switches are on, the voltage across each primary half-winding becomes zero. The input current i_d builds up linearly and the energy is stored in the input inductor. When only one of the two switches is conducting, the input voltage and the stored energy in the input inductor supply the output stage. Therefore, this circuit operates in a manner similar to the step-up converter of Chapter 5.

In the continuous-current-conduction mode, its voltage transfer ratio can be derived to be (see Problem 8-9)

$$\frac{V_o}{V_d} = \frac{N_2}{N_1}\frac{1}{2(1 - D)} \qquad D > 0.5 \tag{8-36}$$

which is similar to the voltage transfer ratio of a step-up converter.

Current-source converters have the disadvantage of having a low power-to-weight ratio compared to voltage-source converters.

8-4-8   Transformer Core Selection in DC – DC Converters with Electrical Isolation

It is desirable to have power transformers that are small in weight and size and have low power losses. The motivation for using high switching frequencies is to reduce the size of the power transformer and the filter components. If this benefit is to be realized, the power loss in the transformer core should remain low even at high frequencies.

Ferrite materials such as 3C8 are commonly used to build transformer cores. Similar to Fig. 8-4a, Fig. 8-17a shows a typical B-H loop for such a material, where the maximum flux density B_m beyond which the saturation occurs is in a range of 0.2 to 0.4 Wb/m^2 and the remnant flux density B_r is in a range of 0.1 to 0.2 Wb/m^2. In Fig. 8-17b, the core loss per unit weight for several switching frequencies is plotted as a

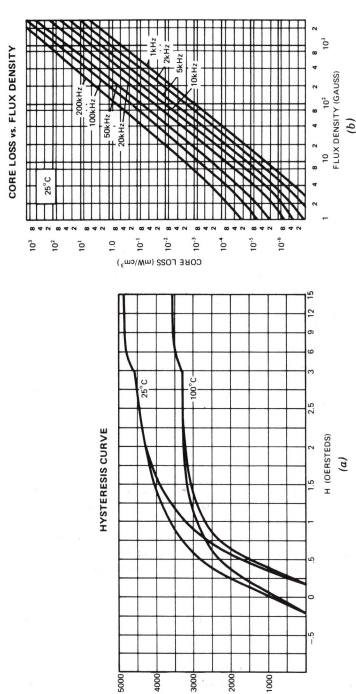

FIGURE 8-17: 3C8 Ferrite characteristic curves (*Source*: Ferroxcube Division of Amperex Electronic Corporation): (*a*) B-H loop, (*b*) core loss curves.

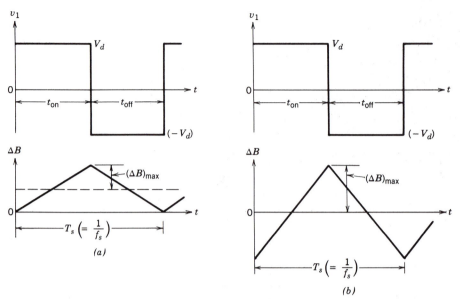

FIGURE 8-18: Core excitation: (*a*) forward converter with *D* = 0.5, (*b*) full-bridge converter, *D* = 0.5.

function of $(\Delta B)_{max}$, where $(\Delta B)_{max}$ is the peak swing in the flux density around its average value during each cycle of the switching frequency f_s. In general, the expression for the core loss per unit weight or per unit volume is given as

$$\text{Core loss density} = kf_s^a\left[(\Delta B)_{max}\right]^b \qquad (8\text{-}37)$$

where the exponents a and b, and the coefficient k depend on the type of material.

A forward converter (with $N_1 = N_3$ in Fig. 8-11*a*) is chosen as an example of the unidirectional core excitation, and a full-bridge converter is chosen to represent a bidirectional core excitation. With a switch duty ratio of 0.5, the peak flux density excursions are calculated by using the waveforms shown in Figs. 8-18*a* and 8-18*b*, where v_1 is voltage across the primary winding. In both the converters

$$(\Delta B)_{max} = \frac{V_d}{4N_1A_cf_s} \qquad \left(\text{at } D = 0.5\right) \qquad (8\text{-}38)$$

where A_c is the cross-sectional area of the core and N_1 is the number of turns in the primary winding. In the forward converter with a unidirectional core excitation, waveforms of Fig. 8-18*a* and the B-H loop in Fig. 8-4*a* dictate that

$$(\Delta B)_{max} < (B_m - B_r)/2 \qquad (8\text{-}39a)$$

In the full-bridge converter with a bidirectional core excitation

$$(\Delta B)_{max} < B_m \qquad (8\text{-}39b)$$

Based on the foregoing discussion, the following conclusions can be reached regarding the desired core properties:

1. A large value of maximum flux density B_m allows $(\Delta B)_{max}$ to be large and results in a small A_c in Eq. 8-38, and hence in a smaller core size.

2. At switching frequencies, for example below 100 kHz, $(\Delta B)_{max}$ is limited by B_m. Therefore, a higher switching frequency in Eq. 8-38 results in a smaller core area. However, at switching frequencies above 100 kHz, a smaller value of $(\Delta B)_{max}$ is chosen to limit the core losses given in Fig. 8-17b.

3. In a forward converter topology where the core is excited in only one direction, $(\Delta B)_{max}$ is limited by $(B_m - B_r)$. Therefore, it is important to use a core with a low remnant flux density B_r in such a topology unless a complex core resetting mechanism is used. In practice, a small air gap is introduced in the core that linearizes the core characteristic and significantly lowers B_r (see Problem 8-11).

In the converters with bidirectional core excitation topologies, the presence of an air gap prevents core saturation under start-up and transient conditions but does not prevent core saturation if there is a volt-second imbalance during the two half-cycles of operation (a volt-second imbalance implies that a dc voltage component is applied to the transformer core). In a practical implementation, there are several causes of such a volt-second imbalances such as unequal conduction voltage drops and unequal switching times of the switches. The preferable way to avoid core saturation due to these practical limitations is to monitor switch currents, as is done in the current-mode control discussed in a later section in this chapter. Use of an appropriate control IC also eliminates saturation under start-up and transient conditions. The other way to prevent core saturation due to voltage imbalance is to use a blocking capacitor in series with the primary winding of the half-bridge and the full-bridge inverters. The blocking capacitor should be chosen appropriately so that it is not too large as to be ineffective under transient conditions and not too small to cause a large ac voltage drop across it under steady-state operating condition. In the push-pull converters, the current-mode control is used to prevent the switch currents from becoming unequal.

In the core of a two winding inductor of a flyback converter, an air gap must be present to provide energy storage capability. In the presence of this air gap, which is larger than that in the previous topologies, the remnant flux density B_r is essentially zero and the B-H characteristic becomes essentially linear.

The amount of inductance needed to operate only in the discontinuous mode (complete demagnetization mode) can be calculated from the given converter voltages and the switching frequency (see Problems).

8-5   CONTROL OF SWITCH-MODE DC POWER SUPPLIES

The output voltages of dc power supplies are regulated to be within a specified tolerance band (e.g. $\pm 1\%$ around its nominal value) in response to changes in the output load and the input line voltages. This is accomplished by using a negative feedback control system shown in Fig. 8-19a, where the converter output v_o is compared with its reference value $V_{o,\text{ref}}$. The error amplifier produces the control voltage v_c, which is used to adjust the duty ratio d of the switch(es) in the converter.

If the power stage of the switch mode converter in Fig. 8-19a can be linearized, then the Nyquist stability criterion and the Bode plots can be used to determine the appropriate compensation in the feedback loop for the desired steady-state and

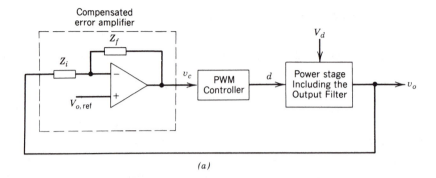

(a)

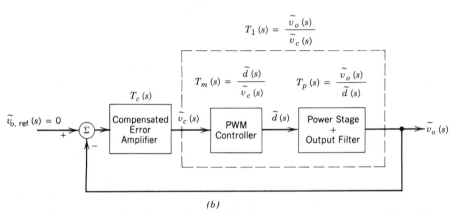

(b)

FIGURE 8-19: Voltage regulation: (a) feedback control system, (b) linearized feedback control system.

transient responses. Professors Middlebrook, Cúk, and their colleagues at the California Institute of Technology have developed a state-space averaging technique that results in a linear model of the power stage including the output filter in Fig. 8-19a for small ac signals, linearized around a steady-state dc operating point. Similarly, the PWM controller in Fig. 8-19a can be linearized around a steady-state operating point. Therefore, each block in Fig. 8-19a can be represented by a transfer function as shown in Fig. 8-19b, where the small ac signals are represented by " ~ ".

8-5-1   Linearization of the Power Stage Including the Output Filter, Using State-Space Averaging to Obtain $\tilde{v}_o(s) / \tilde{d}(s)$

The goal of the following analysis is to obtain a small signal transfer function $\tilde{v}_o(s)/\tilde{d}(s)$, where $\tilde{v}_o$ and $\tilde{d}$ are small perturbations in the output voltage v_o and the switch duty ratio d, respectively, around their steady-state dc operating values V_o and D. Only a converter operating in a continuous-conduction mode is discussed. The procedure is as follows:

STEP 1. *State-Variable Description for Each Circuit State.* In a converter operating in a continuous-conduction mode, there are two circuit states: one state corresponds to when the switch is on and the other when the switch is off. A third circuit state exists during the discontinuous interval, which is not considered in the following analysis because of the assumption of a continuous-conduction mode of operation.

During each circuit state, the linear circuit is described by means of the state-variable vector **x** consisting of the inductor current and the capacitor voltage. In the circuit description, the parasitic elements such as the resistance of the filter inductor and the equivalent series resistance (ESR) of the filter capacitor should also be included. V_d is the input voltage. A lowercase letter is used to represent a variable, which includes its steady-state dc value plus a small ac perturbation, for example, $v_o = V_o + \tilde{v}_o$. Therefore, during each circuit state, we can write the following state equations:

$$\dot{\mathbf{x}} = \mathbf{A}_1 \mathbf{x} + \mathbf{B}_1 v_d \qquad \text{during } d \cdot T_s \tag{8-40}$$

and

$$\dot{\mathbf{x}} = \mathbf{A}_2 \mathbf{x} + \mathbf{B}_2 v_d \qquad \text{during } (1 - d) \cdot T_s \tag{8-41}$$

where $\mathbf{A}_1$ and $\mathbf{A}_2$ are state matrices and $\mathbf{B}_1$, $\mathbf{B}_2$ are vectors.

The output v_o in all converters can be described in terms of their state variables alone as

$$v_o = \mathbf{C}_1 \mathbf{x} \qquad \text{during } d \cdot T_s \tag{8-42}$$

and

$$v_o = \mathbf{C}_2 \mathbf{x} \qquad \text{during } (1 - d) \cdot T_s \tag{8-43}$$

where $\mathbf{C}_1$ and $\mathbf{C}_2$ are transposed vectors.

STEP 2. *Averaging the State-Variable Description Using the Duty Ratio d.* To produce an average description of the circuit over a switching period, the equations corresponding to the two foregoing states are time weighted and averaged, resulting in the following equations:

$$\dot{\mathbf{x}} = \left[\mathbf{A}_1 d + \mathbf{A}_2(1 - d)\right]\mathbf{x} + \left[\mathbf{B}_1 d + \mathbf{B}_2(1 - d)\right]v_d \tag{8-44}$$

and

$$v_o = \left[\mathbf{C}_1 d + \mathbf{C}_2(1 - d)\right]\mathbf{x} \tag{8-45}$$

STEP 3. *Introducing Small AC Perturbations and Separation into AC and DC Components.* Small ac perturbations, represented by " $\sim$ ", are introduced in the dc steady-state quantities (which are represented by the uppercase letters). Therefore,

$$\mathbf{x} = \mathbf{X} + \tilde{\mathbf{x}} \tag{8-46}$$

$$v_o = V_o + \tilde{v}_o \tag{8-47}$$

and

$$d = D + \tilde{d} \tag{8-48}$$

In general, $v_d = V_d + \tilde{v}_d$. However, in view of our goal to obtain the transfer function between the output voltage $\tilde{v}_o$ and the duty ratio $\tilde{d}$, the perturbation $\tilde{v}_d$ is assumed to be zero in the input voltage to simplify our analysis. Therefore

$$v_d = V_d \tag{8-49}$$

By using Eqs. 8-46 through 8-49 into Eqs. 8-44 and recognizing that in steady state, $\dot{\mathbf{X}} = 0$,

$$\dot{\tilde{\mathbf{x}}} = \mathbf{AX} + \mathbf{B}V_d + \mathbf{A}\tilde{\mathbf{x}} + \left[(\mathbf{A}_1 - \mathbf{A}_2)\mathbf{X} + (\mathbf{B}_1 - \mathbf{B}_2)V_d\right]\tilde{d}$$

$$+ \text{terms containing products of } \tilde{\mathbf{x}} \text{ and } \tilde{d} \text{ (to be neglected)} \tag{8-50}$$

where

$$\mathbf{A} = \mathbf{A}_1 D + \mathbf{A}_2(1 - D) \tag{8-51}$$

and

$$\mathbf{B} = \mathbf{B}_1 D + \mathbf{B}_2(1 - D) \tag{8-52}$$

The steady-state equation can be obtained from Eq. 8-50 by setting all the time derivatives and the perturbation terms to zero. Therefore, the steady-state equation is

$$\mathbf{AX} + \mathbf{B}V_d = 0 \tag{8-53}$$

and therefore in Eq. 8-50

$$\dot{\tilde{\mathbf{x}}} = \mathbf{A}\tilde{\mathbf{x}} + \left[(\mathbf{A}_1 - \mathbf{A}_2)\mathbf{X} + (\mathbf{B}_1 - \mathbf{B}_2)V_d\right]\tilde{d} \tag{8-54}$$

Similarly, using Eqs. 8-46 through 8-49 into Eq. 8-45 results in

$$V_o + \tilde{v}_o = \mathbf{CX} + \mathbf{C}\tilde{\mathbf{x}} + \left[(\mathbf{C}_1 - \mathbf{C}_2)\mathbf{X}\right]\tilde{d} \tag{8-55}$$

where

$$\mathbf{C} = \mathbf{C}_1 D + \mathbf{C}_2(1 - D) \tag{8-56}$$

In Eq. 8-55, the steady-state output voltage

$$V_o = \mathbf{CX} \tag{8-57}$$

and therefore,

$$\tilde{v}_o = \mathbf{C}\tilde{\mathbf{x}} + \left[(\mathbf{C}_1 - \mathbf{C}_2)\mathbf{X}\right]\tilde{d} \tag{8-58}$$

Using Eqs. 8-53 and 8-57, the steady state dc voltage transfer function is

$$\frac{V_o}{V_d} = -\mathbf{CA}^{-1}\mathbf{B} \tag{8-59}$$

STEP 4. *Transformation of the AC Equations in to s-Domain to Solve for the Transfer Function.* Eqs. 8-54 and 8-58 consist of the ac perturbations. Using Laplace transformation in Eq. 8-54

$$s\tilde{\mathbf{x}}(s) = \mathbf{A}\tilde{\mathbf{x}}(s) + \left[(\mathbf{A}_1 - \mathbf{A}_2)\mathbf{X} + (\mathbf{B}_1 - \mathbf{B}_2)V_d\right]\tilde{d}(s) \quad (8\text{-}60)$$

or

$$\tilde{\mathbf{x}}(s) = [s\mathbf{I} - \mathbf{A}]^{-1}$$

$$\cdot \left[(\mathbf{A}_1 - \mathbf{A}_2)\mathbf{X} + (\mathbf{B}_1 - \mathbf{B}_2)V_d\right]\tilde{d}(s) \quad (8\text{-}61)$$

where **I** is a unity matrix.

Using Laplace transformation in Eq. 8-58 and expressing $\tilde{\mathbf{x}}(s)$ in terms of $\tilde{d}(s)$ from Eq. 8-61 results in the desired transfer function $T_p(s)$ of the power stages

$$T_p(s) = \frac{\tilde{v}_o(s)}{\tilde{d}(s)} = \mathbf{C}[s\mathbf{I} - \mathbf{A}]^{-1}$$

$$\cdot \left[(\mathbf{A}_1 - \mathbf{A}_2)\mathbf{X} + (\mathbf{B}_1 - \mathbf{B}_2)V_d\right] + (\mathbf{C}_1 - \mathbf{C}_2)\mathbf{X} \quad (8\text{-}62)$$

EXAMPLE 8-1

Obtain the transfer function $\tilde{v}_o(s)/\tilde{d}(s)$ in a forward converter operating in a continuous-conduction mode. Assume $N_1/N_2 = 1$ for simplicity.

SOLUTION

A forward converter is redrawn in Fig. 8-20a and the circuit states with the switch on and the switch off are shown in Figs. 8-20b and 8-20c, respectively. r_L is inductor resistance, r_c is the equivalent series resistance of the capacitor, and R is the load resistance.

Let x_1 and x_2 be as shown in Fig. 8-20. Then, in the circuit Fig. 8-20b with the switch on

$$-V_d + L\dot{x}_1 + r_L x_1 + R(x_1 - C\dot{x}_2) = 0 \quad (8\text{-}63)$$

and

$$-x_2 - Cr_c\dot{x}_2 + R(x_1 - C\dot{x}_2) = 0 \quad (8\text{-}64)$$

In a matrix form, these two equations can be written as

$$\begin{bmatrix} x_1 \\ x_2 \end{bmatrix} = \begin{bmatrix} -\dfrac{Rr_c + Rr_L + r_c r_L}{L(R + r_c)} & -\dfrac{R}{L(R + r_c)} \\[2ex] \dfrac{R}{C(R + r_c)} & -\dfrac{1}{C(R + r_c)} \end{bmatrix} \begin{bmatrix} x_1 \\ x_2 \end{bmatrix} + \begin{bmatrix} \dfrac{1}{L} \\[1ex] 0 \end{bmatrix} V_d \quad (8\text{-}65)$$

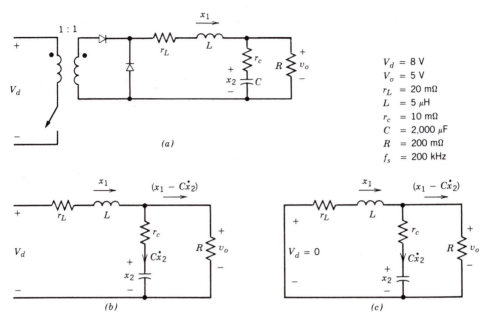

FIGURE 8-20: Forward converter: (a) circuit, (b) switch on, (c) switch off.

Comparing this equation with Eq. 8-40

$$\mathbf{A}_1 = \begin{bmatrix} -\dfrac{Rr_c + Rr_L + r_c r_L}{L(R + r_c)} & -\dfrac{R}{L(R + r_c)} \\ \dfrac{R}{C(R + r_c)} & -\dfrac{1}{C(R + r_c)} \end{bmatrix} \tag{8-66}$$

and

$$\mathbf{B}_1 = \begin{bmatrix} \dfrac{1}{L} \\ 0 \end{bmatrix} \tag{8-67}$$

The state equation for the circuit of Fig. 8-20c with the switch off can be written by observation, noting that the circuit of Fig. 8-20c is exactly the same as the circuit of Fig. 8-20b with V_d set to zero. Therefore, in Eq. 8-41

$$\mathbf{A}_2 = \mathbf{A}_1 \tag{8-68}$$

and

$$\mathbf{B}_2 = 0 \tag{8-69}$$

The output voltage in both the circuit states is given as

$$v_o = R(x_1 - C\dot{x}_2)$$

$$= \frac{Rr_c}{R + r_c}x_1 + \frac{R}{R + r_c}x_2 \qquad \text{(using } \dot{x}_2 \text{ from Eq. 8-64)}$$

$$= \left[\frac{Rr_c}{R + r_c} \quad \frac{R}{R + r_c} \right] \left[\begin{array}{c} x_1 \\ x_2 \end{array} \right] \qquad (8\text{-}70)$$

Therefore, in Eqs. 8-42 and 8-43

$$\mathbf{C}_1 = \mathbf{C}_2 = \left[\frac{Rr_c}{R + r_c} \quad \frac{R}{R + r_c} \right] \qquad (8\text{-}71)$$

Now, the following averaged matrices and vector can be obtained:

$$\mathbf{A} = \mathbf{A}_1 \qquad \text{(from Eqs. 8-51 and 8-68)} \qquad (8\text{-}72)$$

$$\mathbf{B} = \mathbf{B}_1 D \qquad \text{(from Eqs. 8-52 and 8-69)} \qquad (8\text{-}73)$$

and

$$\mathbf{C} = \mathbf{C}_1 \qquad \text{(from Eqs. 8-56 and 8-71)} \qquad (8\text{-}74)$$

Model Simplification
In all practical circuits,

$$R \gg (r_c + r_L) \qquad (8\text{-}75)$$

Therefore, **A** and **C** are simplified as

$$\mathbf{A} = \mathbf{A}_1 = \mathbf{A}_2 = \left[\begin{array}{cc} -\dfrac{r_c + r_L}{L} & -\dfrac{1}{L} \\ \dfrac{1}{C} & -\dfrac{1}{CR} \end{array} \right], \qquad (8\text{-}76)$$

$$\mathbf{C} = \mathbf{C}_1 = \mathbf{C}_2 \simeq [r_c \quad 1] \qquad (8\text{-}77)$$

and **B** remains unaffected as

$$\mathbf{B} = \mathbf{B}_1 D = \left[\begin{array}{c} 1/L \\ 0 \end{array} \right] D \qquad (8\text{-}78)$$

where $\mathbf{B}_2 = 0$. From Eq. 8-76,

$$\mathbf{A}^{-1} = \frac{LC}{1 + (r_c + r_L)/R} \left[\begin{array}{cc} -\dfrac{1}{CR} & \dfrac{1}{L} \\ -\dfrac{1}{C} & -\dfrac{r_c + r_L}{L} \end{array} \right] \qquad (8\text{-}79)$$

Using Eqs. 8-76 through 8-79 into Eq. 8-59, the steady state dc voltage transfer function is

$$\frac{V_o}{V_d} = D\frac{R + r_c}{R + (r_c + r_L)} \simeq D \qquad (8\text{-}80)$$

Similarly, using Eqs. 8-76 through 8-79 into Eq. 8-62

$$T_p(s) = \frac{\tilde{v}_0(s)}{\tilde{d}(s)} \simeq V_d\frac{1 + sr_cC}{LC\left[s^2 + s\left(\dfrac{1}{CR} + \dfrac{r_c + r_L}{L}\right) + \dfrac{1}{LC}\right]} \qquad (8\text{-}81)$$

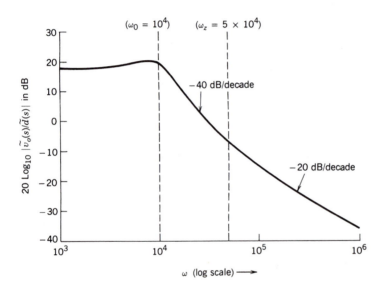

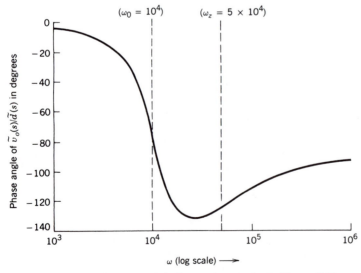

FIGURE 8-21: (a) Gain plot of the forward converter in Figure 8-20a.
(b) Phase plot of the forward converter in Figure 8-20a.

The terms in the square brackets in the denominator of Eq. 8-81 are of the form $s^2 + 2\xi\omega_0 s + \omega_0^2$, where

$$\omega_0 = \frac{1}{\sqrt{LC}} \qquad (8\text{-}82)$$

and

$$\xi = \frac{\left(\dfrac{1}{CR} + \dfrac{r_c + r_L}{L}\right)}{2\omega_0} \qquad (8\text{-}83)$$

Therefore, from Eq. 8-81 the transfer function $T_p(s)$ of the power stage and the output filter can be written as

$$T_p(s) = \frac{\tilde{v}_0(s)}{\tilde{d}(s)} = V_d \frac{\omega_0^2}{\omega_z} \cdot \frac{s + \omega_z}{s^2 + 2\xi\omega_0 s + \omega_0^2}, \qquad (8\text{-}84)$$

where a zero is introduced due to the equivalent series resistance of the output capacitor at the frequency

$$\omega_z = \frac{1}{r_c C} \qquad (8\text{-}85)$$

Figure 8-21 shows the Bode plot for the transfer function in Eq. 8-84, using the numerical values given in Fig. 8-20a. It shows that the transfer function has a fixed gain and a minimal phase shift at low frequencies. Beyond the resonant frequency $\omega_0 = \sqrt{1/LC}$ of the L-C output filter, the gain begins to fall with a slope of -40 dB/decade and the phase tends toward $-180°$. At frequencies beyond ω_z, the gain falls with a slope of -20 dB/decade and the phase angle tends toward $-90°$. The gain plot shifts vertically with V_d but the phase plot is not affected. ∎

In the flyback converter operating in a continuous mode, the transfer function is a nonlinear function $f(D)$ of the duty ratio D and is given as

$$\frac{\tilde{v}_0(s)}{\tilde{d}(s)} = V_d f(D) \frac{(1 + s/\omega_{z1})(1 - s/\omega_{z2})}{as^2 + bs + c} \qquad (8\text{-}86)$$

where the zero ω_{z2} in the transfer function appears in right half of the s-plane. The frequency of the right-half-plane zero depends on the load resistance and the effective value of the filter inductance, where the effective value of the filter inductance is the filter inductance times a nonlinear function of the steady-state dc duty ratio D. A Bode plot of such a transfer function is drawn in Figs. 8-22a and 8-22b. Unlike the transfer function of the converters derived from step-down converters and discussed earlier, the gain at low frequencies is a nonlinear function of the dc operating point (i.e., of V_d). Also the frequency at which the gain falls with a slope of -40 dB/decade depends on the dc operating point. The phase associated with this gain slope tends toward $-180°$. Assuming that $\omega_{z2} > \omega_{z1}$, at frequencies beyond the frequency ω_{z1} of the left-half-plane zero caused by the equivalent series resistance of the capacitor, the gain falls with a slope of -20 dB/decade and the phase angle tends toward $-90°$. At

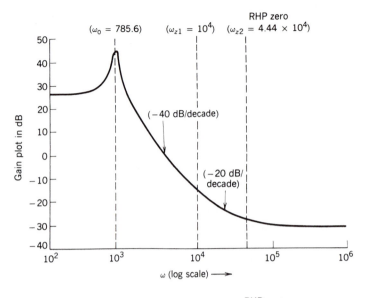

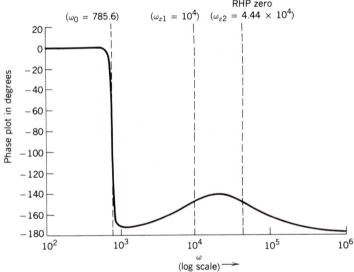

FIGURE 8-22: (a) Gain plot for a flyback converter. (b) Phase plot for a flyback converter.

$$\frac{\tilde{V}_o(s)}{\tilde{d}(s)} = 13.33 \times 10^6 \frac{\left(1 + \dfrac{s}{\omega_{z1}}\right)\left(1 - \dfrac{s}{\omega_{z2}}\right)}{s^2 + 2\xi\omega_0 s + \omega_0^2} \quad \text{where} \quad \begin{aligned} \omega_{z1} &= 10^4 \\ \omega_{z2} &= 4.44 \times 10^4 \\ \omega_0 &= 785.6 \\ \xi &= 0.012 \end{aligned}$$

frequencies beyond the frequency ω_{z2} of the right-half-plane zero, the gain curve flattens out but the phase angle begins to decrease again. The additional phase lag introduced by the right-half-plane zero must be considered in designing the compensation of such a system to provide enough gain and phase margins.

The presence of the right-half-plane zero can be explained by noting that in a flyback converter operating in a continuous mode, if the duty ratio d is increased instantaneously, the output voltage decreases momentarily because the inductor current has not had the time to increase, but the time interval $(1 - d)T_s$ during which the inductor transfers energy to the output stage has been suddenly decreased. This initial decline in the output voltage with the increase in d is opposite of what eventually takes place. This effect results in a zero in the right-half plane, which introduces phase lag in the transfer function $\tilde{v}_o(s)/\tilde{d}(s)$.

In a flyback converter operating in a discontinuous mode, the foregoing effect does not occur and the output voltage always increases with the increased duty ratio. Therefore, in the discontinuous mode of operation, the right-half-plane zero in the transfer function of Eq. 8-86 does not exist; thus, compensating the feedback loop to provide enough gain and phase margins is simpler.

8-5-2 Transfer Function $\tilde{d}(s) / \tilde{v}_c(s)$ of the Direct Duty Ratio Pulse-Width Modulator

In the direct duty ratio pulse-width modulator, the control voltage $v_c(t)$, which is the output of the error amplifier, is compared with a repetitive waveform $v_r(t)$, which establishes the switching frequency f_s, as shown in Fig. 8-23a. The control voltage $v_c(t)$ consists of a dc component and a small ac perturbation component

$$v_c(t) = V_c + \tilde{v}_c(t) \tag{8-87}$$

where $v_c(t)$ is in a range between 0 and $\hat{V}_r$, as shown in Fig. 8-23a. $\tilde{v}_c(t)$ is a sinusoidal ac perturbation in the control voltage at a frequency ω, where ω is much smaller than the switching frequency $\omega_s(= 2\pi f_s)$. The ac perturbation in the control voltage can be expressed as

$$\tilde{v}_c(t) = a \sin(\omega t - \phi) \tag{8-88}$$

by means of an amplitude a and an arbitrary phase angle ϕ.

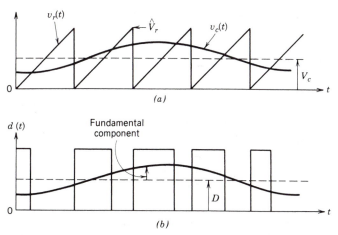

FIGURE 8-23: Pulse-width modulator.

In Fig. 8-23b, the instantaneous switch duty ratio $d(t)$ is as follows:

$$d(t) = 1.0 \quad \text{if} \quad v_c(t) \geq v_r(t) \tag{8-89}$$

$$= 0 \quad \text{if} \quad v_c(t) < v_r(t) \tag{8-90}$$

Similar to the analysis of sinusoidal PWM carried out in Chapter 6, $d(t)$ in Fig. 8-23b can be expressed in terms of the Fourier series as

$$d(t) = \frac{V_c}{\hat{V}_r} + \frac{a}{\hat{V}_r} \sin(\omega t - \phi) + \text{Other high frequency components} \tag{8-91}$$

The higher frequency components in the output voltage v_o due to the high frequency components in $d(t)$ are eliminated because of the low pass filter at the output of the converter. Therefore, the high-frequency components in Eq. 8-91 can be ignored. In terms of its dc value and its ac perturbation

$$d(t) = D + \tilde{d}(t) \tag{8-92}$$

Comparing Eqs. 8-91 and 8-92

$$D = \frac{V_c}{\hat{V}_r} \tag{8-93}$$

and

$$\tilde{d}(t) = \frac{a}{\hat{V}_r} \sin(\omega t - \phi) \tag{8-94}$$

From Eqs. 8-88 and 8-94, the transfer function $T_m(s)$ of the modulator is

$$T_m(s) = \frac{\tilde{d}(s)}{\tilde{v}_c(s)} = \frac{1}{\hat{V}_r} \tag{8-95}$$

Therefore, the theoretical transfer function of the pulse-width-modulator is surprisingly simple, without any time delays. However, the time delay associated with the comparator can lead to a delay in the modulator response.

EXAMPLE 8-2

In practice, the transfer function of the modulator may not have to be calculated from Eq. 8-95. Figure 8-24 shows the approximate transfer function of a commonly used PWM integrated circuit, supplied as a part of the data sheets, in terms of the duty ratio d as a function of the control voltage v_c, where v_c is the output of the error amplifier.

Calculate the transfer function $\tilde{d}(s)/\tilde{v}_c(s)$ for this PWM integrated circuit.

SOLUTION

For this particular modulator, the duty ratio d increases from 0 (at $v_c = 0.8$ V) to 0.95 (at $v_c = 3.6$ V). Therefore, the slope of the transfer function in Fig. 8-24 is equal to the

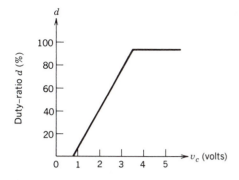

FIGURE 8-24: Pulse-width modulator transfer function.

transfer function of this modulator:

$$\frac{\tilde{d}(s)}{\tilde{v}_c(s)} = \frac{\Delta d}{\Delta v_c}$$

$$= \frac{0.95 - 0}{3.6 - 0.8} \simeq 0.34 \qquad (8\text{-}96)$$

With this modulator, the transfer function between v_o and the control voltage v_c can be obtained as

$$T_1(s) = \frac{\tilde{v}_o(s)}{\tilde{v}_c(s)} = \frac{\tilde{v}_o(s)}{\tilde{d}(s)} \cdot \frac{\tilde{d}(s)}{\tilde{v}_c(s)} = T_p(s) \cdot T_m(s) \qquad (8\text{-}97)$$

The gain plot of the transfer function $\tilde{v}_o(s)/\tilde{v}_c(s)$ can be obtained by adjusting the gain curve in the Bode plot of Fig. 8-21a or Fig. 8-22a to account for a constant gain of 0.34 ($= -9.37$ dB) of the modulator. Assuming zero delay in the modulator, the phase plot of $\tilde{v}_o(s)/\tilde{v}_c(s)$ is the same as that of $\tilde{v}_o(s)/\tilde{d}(s)$. ∎

8-5-3 Compensation of the Feedback System Using a Direct Duty Ratio Pulse-Width Modulator

In the switch-mode power supply shown in Fig. 8-19b, the overall open-loop transfer function is

$$T_{OL}(s) = T_1(s) \cdot T_c(s) \qquad (8\text{-}98)$$

where $T_1(s)$ is as given by Eq. 8-97 and

$$T_c(s) = \text{transfer function of the compensated error amplifier} \qquad (8\text{-}99)$$

For a given $T_1(s)$, the transfer function of the compensated error amplifier $T_c(s)$ must be properly tailored so that $T_{OL}(s)$ meets the performance requirements expected of

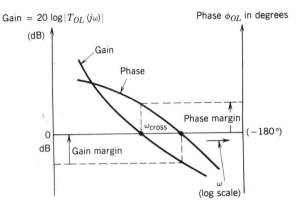

FIGURE 8-25: Gain and phase margins.

the power supply. Some of the desired characteristics of the open-loop transfer function $T_{OL}(s)$ are as follows:

1. The gain at low frequencies should be high to minimize the steady-state error in the power supply output.

2. The crossover frequency is the frequency at which the gain of $T_{OL}(s)$ falls to 1.0 (0 dB) as shown in Fig. 8-25. This crossover frequency ω_{cross} should be as high as possible, but approximately an order of magnitude below the switching frequency, to allow the power supply to respond quickly to the transients such as a sudden change of load.

3. The phase margin (PM) is defined by means of Fig. 8-25 as

$$\text{PM} = \phi_{OL} + 180° \qquad (8\text{-}100)$$

where ϕ_{OL} is the phase angle of $T_{OL}(s)$ at the crossover frequency and is negative. The phase margin, which should be a positive quantity in Eq. 8-100 determines the transient response of the output voltage in response to sudden changes in the load and the input voltage. A phase margin in a range of 45° to 60° is desirable.

To meet these requirements simultaneously, a general error amplifier is shown in Fig. 8-26, where the amplifier can be assumed to be ideal. One of the inputs to the amplifier is the output voltage v_0 of the converter; the other input is the desired (reference) value V_{ref} of v_0. The output of the error amplifier is the control voltage v_c. In terms of Z_i and Z_f in Fig. 8-26, the transfer function between the input and the

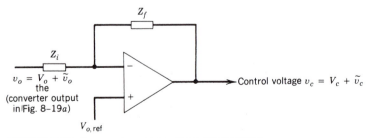

FIGURE 8-26: A general compensated error amplifier.

output perturbations can be obtained as

$$\frac{\tilde{v}_c(s)}{\tilde{v}_o(s)} = -\frac{Z_f(s)}{Z_i(s)} = -T_c(s) \tag{8-101}$$

where $T_c(s)$ is as defined in Fig. 8-19b.

One of the options in the selection of $T_c(s)$ is to introduce a pole-zero pair, in addition to a pole at the origin so that $T_c(s)$ is of the form

$$T_c(s) = \frac{A\ (s + \omega_z)}{s\ (s + \omega_p)} \tag{8-102}$$

where A is positive and $\omega_z < \omega_p$. In Eq. 8-102, due to the pole at the origin, the phase of $T_c(s)$ starts with $-90°$ as shown in Fig. 8-27a. The presence of the zero causes the phase angle to increase (or in other words provides a "boost") to be something greater than $-90°$. Eventually, because of the pole at ω_p, the phase angle of $T_c(s)$ comes back down to $-90°$. The gain plot is also shown in Fig. 8-27a. The parameters in Eq. 8-102 can be chosen such that the minimum phase lag in $T_c(s)$ occurs at the specified (desired) crossover frequency of the overall open-loop transfer function $T_{OL}(s)$.

The transfer function in Eq. 8-102 can be realized by means of an amplifier network shown in Fig. 8-27b, where

$$T_c(s) = \frac{1}{R_1 C_2} \frac{s + \omega_z}{s(s + \omega_p)} \tag{8-103}$$

$$\omega_z = 1/(R_2 C_1) \tag{8-104}$$

and

$$\omega_p = \frac{C_1 + C_2}{R_2 C_1 C_2} \tag{8-105}$$

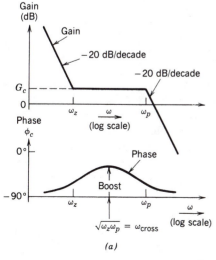

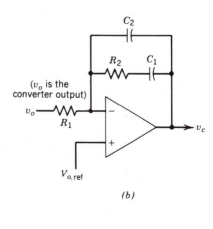

(b)

(a)

FIGURE 8-27: Error amplifier $T_c(s) = \dfrac{1}{R_1 C_2} \dfrac{s + \omega_z}{s(s + \omega_p)}$.

A step-by-step explanation that is easy to follow in the selection of the foregoing parameters has been provided in Reference 13, using a K-factor approach. This procedure suggests that as a first-step, the crossover frequency ω_{cross}, where $|T_{OL}(s)|$ would equal 0 dB, should be selected. This crossover frequency also defines the frequency in Fig. 8-27a, where the minimum phase lag occurs in the transfer function of $T_c(s)$. The K-factor is used such that in the transfer function $T_c(s)$ of Eq. 8-103

$$\omega_z = \omega_{\text{cross}}/K \tag{8-106}$$

and

$$\omega_p = K\omega_{\text{cross}} \tag{8-107}$$

It is shown in Reference 13 that K in Eqs. 8-106 and 8-107 is related to the boost (defined in Fig. 8-27a) in the following manner:

$$K = \tan\left(45° + \frac{\text{Boost}}{2}\right) \tag{8-108}$$

Therefore, the next step is to define the phase margin (PM) and, hence, the boost needed from the error amplifier at the crossover frequency to calculate K in Eqs. 8-106 and 8-107. From the definition of phase margin in Eq. 8-100

$$\text{PM} = 180° + \phi_1 + \phi_c \tag{8-109}$$

where ϕ_c is the phase angle of $T_c(s)$ at the crossover frequency. From Eq. 8-97

$$\phi_1 = \phi_p(s) + \phi_m(s) \tag{8-110}$$

where ϕ_1 is the phase angle of $T_1(s)$, $\phi_p(s)$ is the phase angle of the power stage $T_p(s)$, and $\phi_m(s)$ is the phase angle (if any) of the modulator $T_m(s)$. From the phase plot of the transfer function $T_c(s)$ shown in Fig. 8-27a

$$\phi_c = -90° + \text{Boost} \tag{8-111}$$

From Eqs. 8-109 and 8-111,

$$\text{Boost} = \text{PM} - \phi_1 - 90° \tag{8-112}$$

Therefore, once the phase margin (usually in a range of 45° to 60°) is chosen, the boost is defined from Eq. 8-112 where ϕ_1 (assuming ϕ_m to be zero) can be obtained from Fig. 8-21b or Fig. 8-22b at the frequency chosen to be the crossover frequency. Knowing the boost, K can be calculated from Eq. 8-108.

The next step in the design procedure is to ensure that the gain G_{OL} of the overall open-loop is equal to 1 (i.e., $G_{OL} = |T_{OL}(s)| = 1$) at the chosen crossover frequency. This requires that from Eq. 8-98, the gain G_c of the compensated error amplifier at ω_{cross} be as follows:

$$G_c(\text{at } \omega_{\text{cross}}) = \frac{1}{G_1(\text{at } \omega_{\text{cross}})} \tag{8-113}$$

where G_1 is the magnitude $|T_1(j\omega_{\text{cross}})|$ of the transfer function $T_1(s) = T_p(s) \cdot T_m(s)$

at ω_{cross}. Therefore, at $\omega = \omega_{\text{cross}}$, from Eq. 8-113 and by the substitution of Eqs. 8-104 through 8-107 into Eq. 8-103:

$$G_c = \frac{1}{KC_2 R_1 \omega_{\text{cross}}} = \frac{1}{G_1} \tag{8-114}$$

R_1 in the circuit of Fig. 8-27b is chosen arbitrarily and the rest of the circuit parameters can be calculated as follows from the K-factor procedure outlined before using Eqs. 8-104 through 8-107 and Eq. 8-114,

$$C_2 = \frac{G_1}{KR_1 \omega_{\text{cross}}} \tag{8-115}$$

$$C_1 = C_2(K^2 - 1) \tag{8-116}$$

and

$$R_2 = K/(C_1 \omega_{\text{cross}}) \tag{8-117}$$

For the converters, such as a flyback converter operating in a continuous mode, it may be necessary to use an error amplifier that has two pairs of poles and zeros in addition to the pole at the origin for a proper compensation.

8-5-4 Voltage Feed-Forward PWM Control

In the direct duty ratio PWM control discussed in the previous two subsections, if the input voltage changes, an error is produced in the output voltage, which eventually gets corrected by the feedback control. This results in a slow dynamic performance in regulating the output in response to the changes in input voltage.

If the duty ratio could be adjusted directly to accommodate the change in the input voltage, then the converter output would remain unchanged. This can be accomplished by feeding the input voltage level to the PWM IC. The PWM switching strategy here is very similar to the one discussed in connection with the direct duty ratio PWM control except for one difference: the ramp (and, hence, the peak) of the sawtooth waveform does not stay constant but varies in direct proportion to the input voltage as shown in Fig. 8-28. This shows how an increased input voltage (hence,

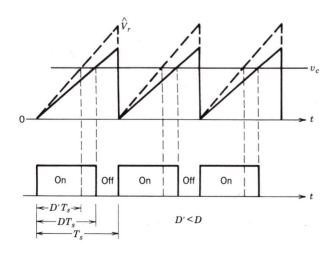

FIGURE 8-28: Voltage feedforward: effect on duty ratio.

increased $\hat{V}_r$) results in a decreased duty ratio, shown by dotted curves in Fig. 8-28. This type of control in step-down derived converters (e.g., forward converters) results in $\tilde{v}_o(s)/\tilde{v}_d(s)$ equal to zero, and hence in an excellent inherent regulation for the changes in the input voltage. The same is true for a flyback converter operating in a complete demagnetization mode.

If this voltage feedforward is implemented in a double-ended power supply (like push–pull, half-bridge, full-bridge), then care must be taken to provide a dynamic volt–time balance so that the on times of the two switches are kept equal on a dynamic basis to prevent the saturation of the high-frequency isolation transformer.

8-5-5 Current-Mode Control

The PWM direct duty-ratio control discussed so far is shown in Fig. 8-29a, where the control voltage v_c (amplified error signal between the actual output and the reference) controls the duty ratio of the switch by comparing the control voltage with a fixed frequency sawtooth waveform. This control of the switch duty ratio adjusts the voltage across the inductor and hence the inductor current (which feeds the output stage), and eventually brings the output voltage to its reference value.

In a current-mode control, an additional inner control loop is used as shown in Fig. 8-29b, where the control voltage v_c directly controls the output inductor current that feeds the output-stage, and thus the output voltage. Ideally, the control voltage

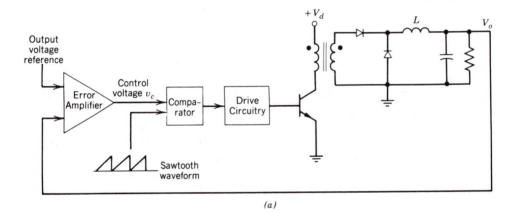

(a)

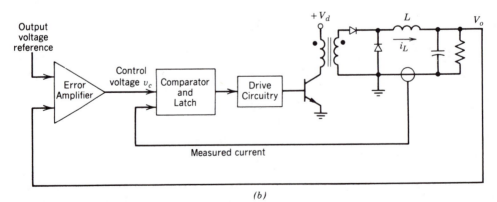

(b)

FIGURE 8-29: PWM duty ratio versus current-mode control. (a) PWM duty-ratio control, (b) current-mode control.

should act to directly control the *average* value of the inductor current for the fastest response, though, as we will see later, various types of current-mode controls tend to accomplish this differently. The fact that the current feeding the output stage is controlled directly in a current-mode control has a profound effect on the dynamic behavior of the negative feedback control loop.

There are three basic types of current-mode controls:

1. Tolerance-band control,

2. Constant-"off"-time control, and

3. Constant-frequency control with turn-on at clock time.

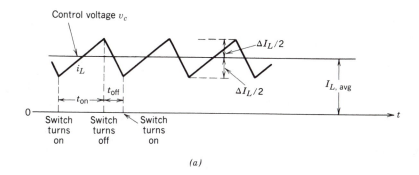

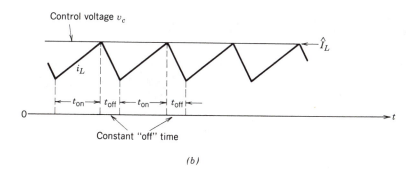

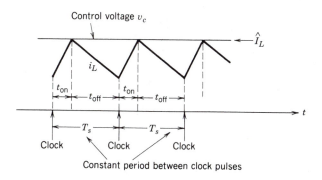

FIGURE 8-30: Three types of current-mode control: (*a*) tolerance-band control, (*b*) constant "off"-time control, (*c*) constant frequency with turn-on at clock time.

In all these types of controls, either the inductor current or the switch current, which is proportional to the output inductor current is measured and compared with the control voltage.

In the *tolerance-band control*, the control voltage v_c dictates the average value of the inductor current as shown in Fig. 8-30a. ΔI_L is a design parameter. The switching frequency depends on ΔI_L, the converter parameters, and the operating conditions. This direct control over the average value of i_L is a very desirable feature of this type of control. However, this scheme works well only in the continuous-current conduction mode. Otherwise, in the discontinuous-current conduction mode, the inductor current becomes zero (though $\Delta I_L/2$ would actually be demanding a negative i_L, which is not possible). If the controller is not designed to handle this discontinuous current when i_L is zero and i_L being demanded by the controller is negative, the switch will never turn on and the inductor current will decay to zero.

In the *constant-"off"-time control*, the control voltage dictates $\hat{I}_L$, as is shown in Fig. 8-30b. Once $\hat{I}_L$ is reached, the switch turns off for a fixed (constant) off time, which is a design parameter. Here also, the switching frequency is not fixed and depends on the converter parameters and its operating condition.

The *constant-frequency control with a turn-on at clock time* is thus far the most common type of current-mode control. Here, the switch is turned on at the beginning of each constant frequency switching time period. The control voltage dictates $\hat{I}_L$ and the instant at which the switch is turned off, as shown in Fig. 8-30c. The switch remains off until the beginning of the next switching cycle. A constant switching frequency makes it easier to design the output filter.

In the current-mode control in practice, a slope compensation is added to the control voltage, as shown in Fig. 8-31 to provide stability, to prevent subharmonic oscillations, and to provide a feedforward property. Figure 8-31 shows the waveforms for a forward converter of the type shown in Fig. 8-29b, where the slope of the slope-compensation waveform is one-half of the slope of the inductor current when the switch is off. With given input and output voltages, the duty ratio is D_1 and the waveform of the inductor current i_L is shown by the solid lines. If the input voltage is increased but the output voltage is to remain unaffected, the duty ratio decreases to D_2

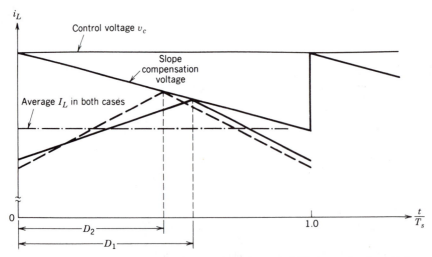

FIGURE 8-31: Slope compensation in current-mode control (D_2 is smaller for a higher input voltage with a constant V_o).

and the inductor current waveform is shown by dotted lines. The average value of the inductor current, which equals the load current, remains the same in both cases in spite of a change in the input voltage. This shows the voltage feedforward property of the current-mode control with a proper slope compensation.

The current-mode control has several advantages over the conventional direct duty ratio PWM control:

1. *It limits peak switch current.* Since either the switch current is directly measured or the current is measured somewhere in the circuit (like through the output inductor) where it represents the switch current without delay, the peak value of the switch current can be limited by simply putting an upper limit on the control voltage. This can be easily accomplished in the controllers that control $\hat{I}_L$.

2. *It removes one pole* (corresponding to the output filter-inductor) from the control-to-output transfer function $\tilde{v}_o(s)/\tilde{v}_c(s)$, thus simplifying the compensation in the negative feedback system, especially in the presence of the right-half-plane zero.

3. It allows a *modular design* of power supplies by equal current sharing where several power supplies can be operated in parallel and provide equal currents, if the same control voltage is fed to all the modules.

4. It results in a *symmetrical flux excursion* in a push–pull converter, thus eliminating the problem of transformer core saturation.

5. *It provides input voltage feedforward.* As shown by Fig. 8-31, an input voltage feedforward is automatically accomplished, resulting in an excellent rejection of input line transients.

8-5-6 Digital Pulse-Width Modulation Control

In recent years, there has been an ongoing attempt to implement the foregoing types of controls by means of digital controllers. The main advantages of a digital approach over its analog counterpart are (1) its lower sensitivity to the changes in the environment such as temperature, supply voltage fluctuations, aging of components, and so on and (2) possibility of a lower parts count, thus improving the supply reliability. A proportional-integral-differential (PID) control can be implemented in either hardware, software, or both to minimize the steady-state error and to yield a satisfactory transient response.

8-6 POWER SUPPLY PROTECTION

In addition to a stable control that provides appropriate steady-state and transient performance, it is important that the control of the power supply also provide its protection against abnormal operating conditions. These protective control features are explained by means of a direct duty ratio PWM-IC belonging to the 1524 family, which has been used in a large number of power supplies.

The modulator UC1524A is an enhanced version of the original modulator. It can be used for switching frequencies of up to 500 kHz. The block diagram of the UC1524A is shown in Fig. 8-32a. The internal reference circuit provides a regulated 5-V output (pin 16) for the input supply voltage variations in a range of 8 to 40 V (pin 15).

BLOCK DIAGRAM

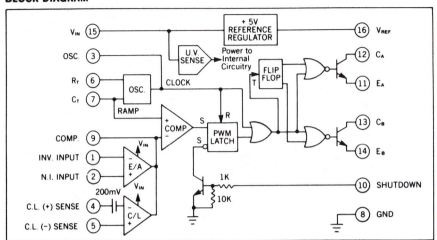

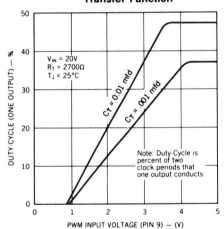

**Pulse Width Modulator
Transfer Function**

FIGURE 8-32: Pulse-width modulator UC1524A (Source: Unitrode Integrated Circuits Corp.): (*a*) block diagram, (*b*) transfer function.

An error amplifier (a transconductance type) allows the measured output voltage of the power supply (connected to pin 1) to be compared to the reference or the desired output voltage (connected to pin 2) of the amplifier. An appropriate feedback network to provide compensation and the loop gain can be connected from the error amplifier output (pin 9) to the inverting input (pin 1). If a separate error amplifier with compensation, as discussed in Section 8-5-3 is used, then this error amplifier can be wired to be a noninverting amplifier with a unity gain. R_T and C_T (between pins 6 and 7 to ground), determine the frequency of the oscillator, which produces a sawtooth waveform at pin 7. The oscillator frequency is determined as

$$\text{Oscillator frequency (kHz)} = \frac{1.15}{R_T(k\Omega)C_T(\mu F)} \tag{8-118}$$

The sawtooth waveform is compared with the error amplifier output in a comparator to determine the duty ratio of the switches. The output of the oscillator (pin 3) is a narrow 3.5-V clock pulse of a 0.5 μs pulse width at a frequency determined by Eq. 8-118.

This IC allows PWM control of push–pull and bridge (half and full) converters where two switches (or switch pairs) need to be controlled alternatively. Comparison of the sawtooth waveform with the error amplifier output, along with the flip-flop (triggered by the oscillator clock pulse) and the logic gates, provide positive base drives alternatively to output switches A and B, which can be used to drive the power switches. One complete switching period requires two oscillator frequency cycles and hence the switch operating frequency is one-half the oscillator frequency. The PWM latch ensures that only a single pulse is allowed to reach the appropriate output stage within each period.

There is another significant function of the oscillator output clock pulse. As the switch duty-ratio begins to approach 0.5, this narrow pulse ensures a blanking time between the turning off and the turning on of the converter switches (or switch pairs). The selection of C_T determines the blanking time from 0.5 μs to as large as 4 μs. The oscillator frequency is determined by Eq. 8-118, by selecting R_T, whereas C_T is selected to yield a desired blanking time.

The transfer function between the duty ratio of the one of the two outputs, as a function of the input voltage at pin 9 (which is the output of the error amplifier internal to this IC) is shown in Fig. 8-32b. To control the single-switch converters (such as forward and flyback), both the IC output switches A and B can be connected in parallel and the switch duty ratio can be as high as 0.95.

Some of the protective features are explained in the following sections.

8-6-1 Soft Start

A soft start in switch mode dc power supplies is provided by increasing the duty ratio and hence the output voltage slowly, subsequent to the input voltage switch-on. This can be provided by connecting a simple circuit to pin 9.

8-6-2 Voltage Protection

Overvoltage and undervoltage protection can be incorporated by adding a few external components to the shut-down pin 10.

8-6-3 Current Limiting

For protection against overcurrent at the output, the circuit output current can be sensed by measuring the voltage across a sensing resistor. This voltage is applied across pins 4 and 5. When this sensed voltage exceeds a temperature compensated threshold of 200 mV, the output of the error amplifier is pulled toward ground and linearly decreases the output pulse width.

8-6-3-1 FOLDBACK CURRENT LIMITING

In a constant current limited power supply, if the gain of the current-limiting stage is high, the supply V_o–I_o characteristic can be as shown in Fig. 8-33a, where once a critical value of current I_{limit} is reached, I_o is not allowed to increase any more and the output voltage V_o depends on the load line. Therefore, a load resistance R_1 yields an output voltage V_{o1} and a load resistance R_2 yields V_{o2}, as shown in Fig. 8-33a. Even with a complete short circuit across the output supply, the output current does not

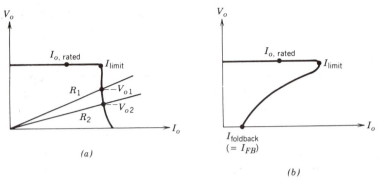

FIGURE 8-33: Current limiting: (*a*) constant current limiting, (*b*) foldback current limiting.

exceed the current limit (a design parameter) by any appreciable amount. This may be a requirement in a dc power supply, which may be used to supply a constant current and hold the output current to a specified value, once the output load resistance decreases below a certain value.

However, in many applications, the output current exceeding the critical value represents an abnormal load condition, and a foldback current limit is introduced where, as the load resistance decreases, the output current also decreases (along with the decreasing output voltage V_o) as shown in Fig. 8-33b. Here, in case of a short circuit across the output, the current will have a much smaller value I_{FB} in comparison to I_{limit}. The motivation for this foldback current limit is to reduce the current flowing through the supply unnecessarily and to bring it to a much smaller value under an abnormal load condition. Once the load recovers to its normal value, the supply once again begins to regulate V_o to its reference value. This foldback current limit can be implemented using a PWM controller such as that shown in Fig. 8-32a.

8-7   ELECTRICAL ISOLATION IN THE FEEDBACK LOOP

In an electrically isolated switching power supply, it is necessary to provide electrical isolation in the feedback path where the output voltage on the secondary side of the high-frequency power transformer is measured, to control the power switches which are on the primary side of the high-frequency power transformer. Two options are presented in Fig. 8-34a and 8-34b.

In the secondary-side control shown in Fig. 8-34a, the PWM controller such as the UC1524A discussed earlier, is on the secondary side of the power transformer. Its supply (or bias) voltage is provided by a bias supply through an isolation transformer from the primary side. The signals to the switch-driver circuit are provided through small signal transformers, thus maintaining isolation in the feedback loop.

As an alternative, the primary-side control is shown in Fig. 8-34b, where the PWM controller is on the primary side along with the power switches. This requires electrical isolation between the output voltage error amplifier and the PWM controller. The advantages of having the PWM controller on the same side of the switches are that it simplifies the interface with the switch-driver circuit and it is possible to implement the input voltage feedforward control.

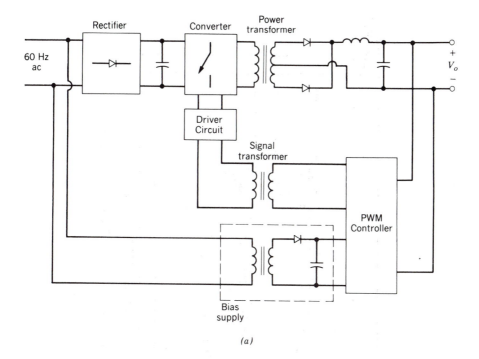

(a)

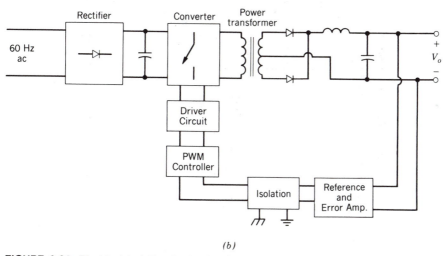

(b)

FIGURE 8-34: Electrical isolation in the feedback loop: (*a*) secondary-side control, (*b*) primary-side control.

One way to implement isolation in the control of Fig. 8-34*b* is to use an optocoupler between the dc output of the error amplifier and the PWM controller. However, the optocoupler approach suffers from several drawbacks, namely, the stability of the optocoupler gain with temperature and time, thus making it difficult to guarantee the stability and the performance of the power supply.

The other alternative in the primary-side control is to use an amplitude-modulated oscillator such as UC1901 shown in Fig. 8-35. The high-frequency oscillator output is coupled through a small high-frequency signal transformer to a demodulator

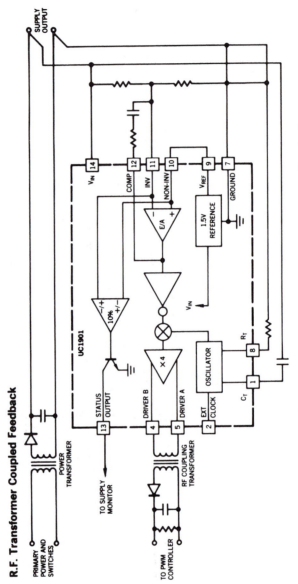

FIGURE 8-35: Isolated Feedback Generator UC1901 (Source: Unitrode Integrated Circuits Corp.).

that supplies the dc error voltage to the PWM controller. Reference 21, provides a detailed description of using such a feedback isolation technique.

8-8 DESIGNING TO MEET THE POWER SUPPLY SPECIFICATIONS

Power supplies have to meet several specifications. The considerations for meeting some of these specifications are discussed in the following sections.

8-8-1 Input Filter

A simple low-pass filter such as a single-stage filter consisting of L and C as shown in Fig. 8-36 may be used at the input to the switch-mode supply to improve its power factor of operation and to reduce the conducted EMI. From the energy efficiency standpoint, this filter should have as little power loss as possible. However, one must consider the possibility of oscillations in the presence of such a low-loss filter.

A regulated switch-mode power supply appears as a negative resistance across the input filter capacitor. The reason for a negative resistance is the fact that with increasing input voltage, the input current decreases, since the output voltage is regulated and, hence, the output power and the input power do not change. Decreasing input current with an increasing input voltage implies a negative input resistance.

If an adequate damping is not provided, a possibility of sustained oscillation exists. A useful design criterion requires that the resonant frequency of the input filter be a decade lower than the resonant frequency of the output filter to avoid interaction between the two. The input filter capacitor should be chosen to be as large as possible, and additional damping elements should be included.

It is also possible to provide active wave-shaping of the input current as described in Chapter 17, which results in an essentially harmonic-free current at a unity power factor.

8-8-2 Input Rectifier Bridge

To be able to operate either from a nominal rms ac voltage of 115 V or 230 V, it is possible to use a voltage-doubler circuit such as that of Fig. 3-8, which was discussed in Chapter 3.

8-8-3 Bulk Capacitor and the Hold-up Time

The dc link capacitor C_d, usually referred to as the bulk capacitor, reduces the voltage ripple in the input to the dc–dc converter. In addition, it also provides a hold-up time during which the regulated supply keeps on providing the regulated

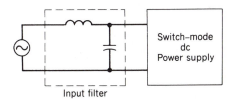

Input filter

FIGURE 8-36: Input filter.

voltage output in the absence of the ac input voltage caused by a momentary power outage. The bulk capacitor C_d can be calculated as a function of the desired hold-up time:

$$C_d \simeq 2 \times \frac{\text{Rated power output} \times \text{Hold-up time}}{\left(V_{d,\text{nominal}}^2 - V_{d,\text{min}}^2\right) \times \eta} \tag{8-119}$$

where, $V_{d,\text{min}}$ is chosen to be in a range of 60% to 75% of the nominal input voltage $V_{d,\text{nominal}}$, and η is the energy efficiency of the power supply.

It should be noted that for a given capacitance, the capacitor volume is roughly proportional to the voltage rating and the maximum energy storage capability is proportional to the square of the voltage rating. This points out the significant advantage of switch-mode power supplies over linear power supplies, since energy storage in switch-mode supplies is at a much higher voltage compared with linear power supplies.

8-8-4 Limiting Inrush (Surge) Current at Initial Turn-on

When the power switch to the supply is initially turned on, the bulk capacitor C_d initially appears effectively as a short circuit across the ac source, which may result in an unacceptably large inrush current. To limit this inrush current, a series element between the dc side of the rectifier bridge and C_d can be used. This series element can be a thermister, which initially has a large resistance when it is cold, thus limiting the inrush current at switch on. As it heats up, its resistance goes down to a reasonably low value to yield a reasonable efficiency. However, it has a long thermal time constant and therefore if a short-term power outage occurs that is long enough to discharge the bulk capacitor but not long enough to allow the thermistor to cool down, a large inrush current can result when the line power comes back on.

Another option is to use a current-limiting resistor, with a thyristor in parallel to make up the series element. Initially the thyristor is off and the current-limiting resistor limits the inrush current at turn-on. When the bulk-capacitor voltage charges up, the thyristor is turned on, thus bypassing the current-limiting resistor. It is also possible to design the series element by a device such as a MOSFET or an IGBT. The device is slowly turned on, thus limiting the peak inrush current.

8-8-5 Equivalent Series Resistance (ESR) of the Output Filter Capacitor

The equivalent series resistance (ESR) of the output filter capacitor in Fig. 8-37 is required to be as low as possible. In high switching frequency applications, the ESR significantly contributes to the peak-to-peak and the rms values of the ripple in the output voltage (see Problems 8-16). The peak deviation in the output voltage from its steady-state value, following a step change in load, also depends on the capacitor ESR. For a step change of load, the output filter inductor in Fig. 8-37 acts as a source of constant current during this load transient, and the change in load current as a

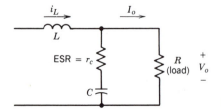

FIGURE 8-37: ESR in the output capacitor.

transient is supplied by the filter capacitor. Hence, following a load transient,

$$\Delta V_o = - \text{ESR} \times \Delta I_o \qquad (8\text{-}120)$$

8-8-6 Synchronous Rectifier to Improve Energy Efficiency

There is an increasing requirement in equipment such as computers, for power supplies with voltages of even lower than 5 V, for example in a 2 to 3 V range, as a consequence of increasing integration of logic gates on a single monolithic substrate. In switching power supplies with a low output voltage, the diodes in the output rectifier stage can be the biggest source of power loss. Even the commonly used Schottky diodes have a relatively large voltage drop and, hence, a large power loss in such low-output voltage applications. As a remedy, low voltage MOSFET with a very low on-state resistance $r_{\text{DS(on)}}$ and low voltage bipolar junction transistors with a very low on-state voltage $V_{\text{CE(sat)}}$ can be used to replace the diodes in the output stage. These devices in this application are commonly referred to as synchronous rectifiers.

8-8-7 Multiple Outputs

For a multiple output power supply, dynamic cross-regulation refers to how well the power supply can regulate the voltage of its regulated output, if a load change occurs on one of its unregulated outputs. For separate filter inductors used for each of the multiple outputs, the dynamic cross-regulation is very poor. This is because a load change on one of the unregulated outputs takes a relatively long time for its effect to show up at the regulated output and for the controller to take the corrective action. If the output filter inductors are coupled (i.e., wound on a common core), the dynamic cross-regulation is much better, since the change in the unregulated output voltage is immediately propagated to the regulated output, thus forcing the feedback controller to act.

8-8-8 EMI Considerations

Switching power supplies must meet the conducted and the radiated EMI specifications. These specifications and the EMI filters are discussed in Chapter 17.

8-9 SUMMARY

In this chapter, principles behind a successful switching power supply design are discussed. Various topics unique to the design of switching power supply are covered. These topics include converter topologies, transformer core excitation, various types of controls and the compensation of the feedback loop in regulated supplies, power supply protection, providing isolation in the feedback loop, and design consideration to meet the power supply specifications.

PROBLEMS

LINEAR POWER SUPPLIES

8-1. A 12-V regulated linear power supply of the type shown in Fig. 8-1a is designed to operate with a 60-Hz ac voltage in a range of 120 V ($+10\%$, -25%). At the maximum load, the peak-to-peak ripple in the capacitor voltage is 1.0 V. The power supply is designed such that $V_{d,\,\text{min}} - V_o = 0.5\,\text{V}$ in Fig. 8-1b.

Calculate the loss of efficiency due to power losses in the transistor at full load, with the input voltage at its maximum (*Hint*: Approximate the capacitor voltage waveform with straight-line segments).

FLYBACK CONVERTER

8-2. A flyback converter is operating in a complete demagnetization mode. Derive the voltage transfer ratio V_o/V_d in terms of the load resistance R, switching frequency f_s, transformer inductance L_m, and the duty ratio D.

8-3. In a regulated flyback converter with a 1:1 turns ratio, $V_o = 12\,\text{V}$, $V_d = 12\,\text{V}$ to $24\,\text{V}$, $P_{\text{load}} = 6\,\text{W}$ to $60\,\text{W}$, and the switching frequency $f_s = 200\,\text{kHz}$. Calculate the maximum value of the magnetizing inductance L_m that can be used if the converter is always required to operate in a complete demagnetization (equivalent to a discontinuous conduction) mode. Assume ideal components.

8-4. A flyback converter is operating in an incomplete demagnetization mode with a duty ratio of 0.4. In the same application, another option may be to parallel two half-size flyback converters as shown in Fig. 8-9b. Compare the ripple in the input current i_d and the output stage current i_o waveforms in these two options, assuming a very large output capacitor such that $v_o(t) \simeq V_o$.

FORWARD CONVERTER

8-5. A switch-mode power supply is to be designed with the following specifications:

$$V_d = 48\,\text{V} \pm 10\%$$

$$V_o = 5\,\text{V (regulated)}$$

$$f_s = 100\,\text{kHz}$$

$$P_{\text{load}} = 15\,\text{W to } 50\,\text{W}$$

A forward converter operating in a *continuous-conduction mode* with a demagnetizing winding ($N_3 = N_1$) is chosen. Assume all components to be ideal except for the presence of transformer magnetization inductance.
(a) Calculate (N_2/N_1) if this turns ratio is desired to be as small as possible.
(b) Calculate the minimum value of the filter inductance.

8-6. A forward converter with a demagnetizing winding is designed to operate with a maximum duty ratio $D_{\max}$ of 0.7. Calculate the voltage rating of the switch in terms of the input voltage V_d.

8-7. In the circuit of Fig. 8-12b with two parallel forward converters, draw the input current i_d and i_L waveforms, if each converter is operating at a duty ratio of 0.6 in a continuous-conduction mode. Compare these two waveforms with those if a single forward converter (with twice the power rating but with the same value of the output filter inductance as in Fig. 8-12b) is used. Assume $v_o(t) \simeq V_o$.

PUSH-PULL CONVERTERS

8-8. In a push-pull converter of Fig. 8-13a, assume the losses to be zero and each switch duty ratio to be 0.25. The transformer has a finite magnetizing inductance and i_m is the magnetizing current.
(a) At a large load where $i_L(N_2/N_1) \gg i_m$, draw the i_m, i_{D1} and i_{D2} waveforms.
(b) At essentially no-load, draw the i_m waveform and show that the peak value of i_m is higher than in part (a).

CURRENT-SOURCE CONVERTERS

8-9. Derive the voltage transfer-ratio given by Eq. 8-36 in the current-source converter of Fig. 8-16.

TRANSFORMER CORE

8-10. A transformer for a full-bridge converter is built with a ferrite material with properties similar to those shown in Figs. 8-17a and 8-17b. $V_d = 170$ V, $f_s = 50$ kHz, and $(\Delta B)_{max} = 0.2$ Wb/m^2 with a switch duty ratio of 0.5. The peak magnetizing current is measured to be 1.0 A. Estimate the core losses in watts in the transformer at 25°C, under the operating conditions described.

8-11. A toroidal transformer core is built with a material whose B-H loop is shown in Fig. 8-17a. An air gap is included whose length is one-hundredth of the length of the flux path in the core. Plot the B-H loop and calculate the remnant flux in the gapped core.

DIRECT DUTY RATIO CONTROL

8-12. The forward converter of Fig. 8-20a is to have a gain crossover frequency $\omega_{cross} = 10^5$ rad/s with a phase margin of 30°. Use the Bode plot of Figs. 8-21a and 8-21b for the transfer function $\tilde{v}_o(s)/\tilde{d}(s)$. The PWM transfer function is given by Fig. 8-24.

Calculate the values for R_2, C_1, and C_2 in the compensated error amplifier of Fig. 8-27, assuming $R_1 = 1$ kΩ.

8-13. Repeat Problem 8-12 for a flyback converter, assuming that the Bode plots in Figs. 8-22a and 8-22b are for its transfer function $\tilde{v}_o(s)/\tilde{v}_c(s)$. The crossover frequency ω_{cross} and the phase margin are required to be 5×10^3 rad/s and 30°, respectively.

CURRENT-MODE CONTROL

8-14. In a forward converter with $N_1/N_2 = 1$, the output voltage is regulated to be 6.0 V by means of a current-mode control, where the slope of the slope-compensation ramp is one-half of the slope of the inductor current with the switch off.

Draw the waveforms as in Fig. 8-31 to show that the average inductor current remains the same if V_d changes from 10 V to 12 V.

CAPACITOR HOLD-UP TIME

8-15. A 100-W power supply with a full-load efficiency of 85% has a hold-up time of 40 ms at full load, when it is supplied with a nominal input voltage of 120 V at 60 Hz. A full-bridge rectifier is used at the input. If the power supply can operate only if the average dc voltage V_d is above 100 V, calculate the required value of the input capacitor C_d. (Hint: Assume that the capacitor voltage is charged approximately to the peak of the ac input voltage).

ESR OF THE OUTPUT FILTER CAPACITOR

8-16. In the forward converter shown in Fig. 8-20a, use the numerical values given, except assume r_L to be zero. Under a steady-state operating condition, plot i_L, voltage across r_c, voltage across C, and the ripple in v_o. Compare the peak-to-peak ripple in the following three voltages: v_o, voltage across C, and voltage across r_c.

8-10 REFERENCES

Books on Switch Mode DC Power Supplies

1. R. P. SEVERNS and G. E. BLOOM, *Modern DC-to-DC Switch Mode Power Converter Circuits*, Van Nostrand Reinhold, New York, 1985.
2. K. KIT SUM, *Switch Mode Power Conversion — Basic Theory and Design*, Marcel Dekker, New York and Basel, 1984.
3. R. E. TARTER, *Principles of Solid State Power Conversion*, H. W. Sams Co., Indianapolis, IN, 1985.
4. G. CHRYSSIS, *High Frequency Switching Power Supplies: Theory and Design*, McGraw Hill, New York, 1984.

Flyback Converters

5. H. C. MARTIN, "Miniature Power Supply Topology for Low Voltage Low Ripple Requirements," 1986, U.S. Patent 4,618,919.

Forward Converters

6. B. BRAKUS, "100 Amp Switched Mode Charging Rectifier for Three-Phase Mains," *IEEE / INTELEC 1984*, pp. 72–78.

Push – Pull Converters

7. R. REDL, M. DOMB, and N. SOKAL, "How to Predict and Limit Volt-Second Unbalance in Voltage-Fed Push–Pull Power Converters," *PCI Proceedings*, pp. 314–330, April 1983.

Current-Source Converters

8. REFERENCES 1 AND 3.

Transformer Core

9. FERROXCUBE CORPORATION, "Ferroxcube Linear Ferrite Materials and Components," Saugerties, NY, 1988.

Control Linearization

10. R. D. MIDDLEBROOK and S. CÚK, "A General Unified Approach to Modelling Switching—Converter Power Stages," *1976 IEEE Power Electronics Specialists Conference Record*, pp. 18–34.
11. REFERENCE 1.
11. R. D. MIDDLEBROOK, "Predicting Modulator Phase Lag in PWM Converter Feedback Loops," 8th International Solid-State Power Electronics Conference, April 27–30, 1981, Dallas, TX.

Control — Feedback Compensation

12. K. OGATA, *Modern Control Engineering*, Prentice-Hall, Englewood Cliffs, NJ, 1970.
13. H. DEAN VENABLE, "The *k*-Factor: A New Mathematical Tool for Stability Analysis and Synthesis," *Proceedings of Powercon 10*, March 22–24, 1983, San Diego, CA.

Feedforward Control

14. UNITRODE, "Switching Regulated Power Supply Design Seminar Manual," Unitrode Corporation, 1986.

Current-Mode Control

15. B. HOLLAND, "Modeling, Analysis and Compensation of the Current-Mode Converter," *Proceedings of the Powercon 11*, 1984, pp. I-2-1 through I-2-6.
16. R. REDL and N. SOKAL, "Current-Mode Control, Five Different Types, Used with the Three Basic Classes of Power Converters," *1985 IEEE Power Electronics Specialists Conference Record*, pp. 771–785.
17. REFERENCE 14.

Digital Control

18. C. P. HENZE and N. MOHAN, "Modelling and Implementation of a Digitally Controlled Power Converter Using Duty-Ratio Quantization," *Proceedings of ESA (European Space Agency) Sessions at the 1985 IEEE Power Electronics Specialists Conference*, pp. 245–255.

Hold-up Time and Capacitor ESR

19. B. LANDON, "Myth—Holdup is Free with SMPS," *Powerconversion International Magazine*, pp. 72–80, October 1981.
20. W. CHASE, "Capacitors for Switching Regulator Filters," *Powerconversion International Magazine*, pp. 57–60, May 1981.

Electrical Isolation in the Feedback Loop

21. UNITRODE, *Unitrode Applications Handbook 1987–1988*.

Limiting Inrush Currents

22. R. ADAIR, "Limiting Inrush Current to a Switching Power Supply Improves Reliability, Efficiency," *Electronic Design News* (EDN), May 20, 1980.

EMI

23. D. L. INGRAM, "Designing Switch-Mode Converter Systems for Compliance with FCC Proposed EMI Requirements," *Power Concepts*, pp. G1-1 through G1-11, 1977.

9

Power Conditioners and Uninterruptible Power Supplies

———————
———————
———————

9-1 INTRODUCTION

In the previous chapters, it was mentioned that power electronics converters produce electromagnetic interference (EMI) and inject current harmonics into the utility system. An interface between a power electronic system and the utility source, which can minimize these potential problems is discussed in Chapter 17. In this chapter, the focus is on powerline disturbances and how power electronic converters can be utilized to prevent the powerline disturbances from disrupting the operation of critical loads such as computers used for controlling important processes, medical equipment, and the like.

9-2 POWERLINE DISTURBANCES

Ideally, the voltage supplied by the utility system should be a perfect sinewave without any harmonics, at its nominal frequency of 60 Hz and at its nominal magnitude. For a three-phase system, the voltages should form a balanced set, with each phase displaced by 120° with respect to the others.

9-2-1 Types of Disturbances

In practice, however, voltages can significantly depart from the ideal condition due to the powerline disturbances listed below:

- *Overvoltage.* The voltage magnitude is substantially higher than its nominal value for a sustained period of a few cycles.

- *Undervoltage (Brownout).* The voltage is substantially lower than its nominal value for a few cycles.

262

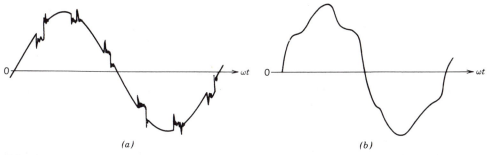

FIGURE 9-1: Possible distortions in input voltage: (*a*) chopped voltage waveform, (*b*) distorted voltage waveform due to harmonics.

- *Outage (Blackout).* The utility system voltage collapses for a few cycles or more.
- *Voltage Spikes.* These are superimposed on the normal 60-Hz waveforms and occur occasionally (not on a repetitive basis). These can be either of a line-mode (differential-mode) or a common-mode type.
- *Chopped Voltage Waveform.* This refers to a repetitive chopping of the voltage waveform and the associated ringing as shown in Fig. 9-1*a*.
- *Harmonics.* A distorted voltage waveform as shown in Fig. 9-1*b* contains harmonic voltage components at harmonic frequencies (usually low-order multiples of the line frequency). These harmonics exist on a sustained basis.
- *EMI.* This refers to high-frequency noise, which may be conducted on the power line or radiated from its source.

9-2-2 Sources of Disturbances

Sources that produce these disturbances are very diverse. Overvoltages may be caused by sudden decreases in the system load, thus causing the utility voltage to go up. Undervoltages may be caused by overload conditions, by start of induction motors, or for many other reasons. Occasional large voltage spikes may be a result of switching in or out of power-factor-correction capacitors, powerlines, or even such things as pump/compressor motors in the vicinity. Chopping of the voltage waveform may be caused by ac to dc line-frequency thyristor converters of the type discussed in Chapter 4, if such converters are used to interface the power electronic equipment with the utility system. These converters produce a short circuit on the ac voltage source through the ac system impedance on a repetitive basis. The voltage harmonics may be caused by a variety of sources. These include magnetic saturation of power system transformers as well as the harmonic currents injected by power electronic loads. These harmonic currents flowing through the ac system impedances result in harmonic voltages. EMI is produced by most power electronics equipment due to rapid switching of voltages and currents, as will be discussed in Chapter 17.

9-2-3 Effect on Sensitive Equipment

The effect of such powerline disturbances on the sensitive equipment depends on the following factors: (1) type and magnitude of the powerline disturbance, (2) type of equipment and how well it is designed, and (3) if any power conditioning equipment is used. Sustained overvoltages and undervoltages may cause the equipment to trip out, which is highly undesirable in certain applications. Large voltage spikes may cause a hardware failure in the equipment. Manufacturers of critical equipment often provide a certain degree of protection by incorporating surge arrestors such as metal oxide varistors (MOVs) at the input to guard against such failures. However, spikes of very large magnitude in combination with a higher frequency of occurrence can still result in a hardware failure. Chopped voltage waveforms and voltage harmonics have the potential of interfering with the equipment if it is not designed to be immune from such effects. Power conditioners consisting of filters and an isolation transformer can correct such problems.

The effect of power system outage depends on the duration of the outage and the equipment design. As an example, a personal computer power supply may be designed such that for an outage of less than 100 ms, the power supply outputs to the digital circuitry ride through and hold their nominal values, and no effect is felt on the computer operation. For an outage of a longer duration, after a 100-ms interval a logic signal within the computer allows the computer CPU an additional 50 ms to backup the existing information, beyond which time all power supply output voltages decrease rapidly. Figure 9-2 shows the tolerance of large mainframe computers to powerline disturbances, beyond which a backup procedure may be initiated and a shutdown occurs for some time. Typical power quality specified by major computer manufacturers is listed in Table 9-1. In case of critical applications where such a shutdown is unacceptable, the backup to the utility grid is provided by means of

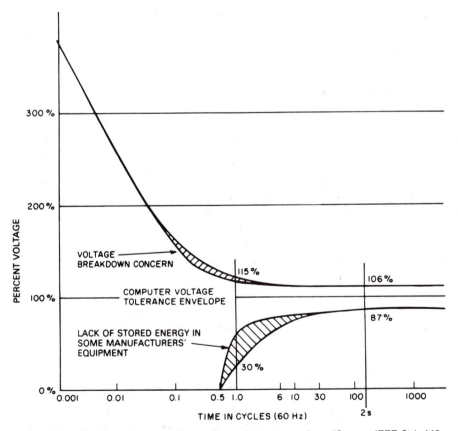

FIGURE 9-2: Typical computer system voltage tolerance envelope (*Source*: IEEE Std. 446 "Recommended Practice for Emergency and Standby Power Systems for Industrial and Commercial Applications").

TABLE 9-1

Typical Range of Input Power Quality and Load Parameters of
Major Computer Manufacturers

Parameters[a]	Range or Maximum
(1) Voltage regulation, steady state	$+5, -10$ to $+10\%, -15\%$ (ANSI C84.1–1970 is $+6, -13\%$)
(2) Voltage disturbances	
Momentary undervoltage	-25 to -30% for less than 0.5 s with -100% acceptable for 4 to 20 ms
Transient overvoltage	$+150$ to 200% for less than 0.2 ms
(3) Voltage harmonic distortion[b]	3–5% (with linear load)
(4) Noise	No standard
(5) Frequency variation	60 Hz $\pm$ 0.5 Hz to ± 1 Hz
(6) Frequency rate of change	1 Hz/s (slew rate)
(7) 3ϕ, Phase voltage unbalance[c]	2.5 to 5%
(8) 3ϕ, Load unbalance[d]	5 to 20% maximum for any one phase
(9) Power factor	0.8 to 0.9
(10) Load demand	0.75 to 0.85 (of connected load)

[a]Parameters (1), (2), (5), and (6) depend on the power source, while parameters (3), (4), and (7) are the product of an interaction of source and load, and parameters (8), (9), and (10) depend on the computer load alone.
[b]Computed as the sum of all harmonic voltages added vectorially.
[c]Computed as follows:

$$\% \text{ phase voltage unbalance} = \frac{3(V_{max} - V_{min})}{V_a + V_b + V_c} \cdot 100$$

[d]Computed as difference from average single-phase load.
Source: IEEE Std. 446, "Recommended Practice for Emergency and Standby Power Systems for Industrial and Commercial Applications."

uninterruptible power supplies (UPS). Both the power conditioners and the uninterruptible power supplies are discussed in the following sections.

9-3 POWER CONDITIONERS

Power conditioners provide an effective way of suppressing some or all of the electrical disturbances other than the power outages and frequency deviations from 60 Hz (frequency deviation is not a problem in an interconnected ac power system). Some of these power conditioners are listed:

- Metal oxide varistors (MOVs) provide protection against line-mode voltage spikes.

- EMI filters help to prevent the effect of the chopped waveform on the equipment, as well as to prevent the equipment from conducting high-frequency noise into the utility grid.

- Isolation transformers with electrostatic shields not only provide galvanic isolation but also filter the line-mode and the common-mode voltage spikes.

- Ferroresonant transformers provide voltage regulation as well as filtering of the line-mode spikes. They are also partially effective in filtering the common-mode noise.

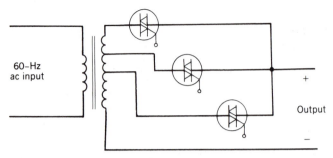

FIGURE 9-3: Electronic tap-changer.

- Linear conditioners are used in many sensitive applications to supply clean power.

Since none of these power conditioners employs switch-mode or resonant-mode power electronics, they are not discussed here any further.

For voltage regulation, an electronic tap changing scheme using triacs as shown in Fig. 9-3 can be used, where triacs or back-to-back connected thyristors replace a mechanical contact and allow a bidirectional current flow.

9-4   UNINTERRUPTIBLE POWER SUPPLIES (UPS)

For supplying very critical loads such as computers used for controlling important processes, some medical equipment, and the like, it may be necessary to use UPS. These provide protection against power outages, as well as voltage regulation during powerline overvoltage and undervoltage conditions. They are also excellent in terms of suppressing incoming line transient and harmonic disturbances.

UPS in their block-diagram form are shown in Fig. 9-4. A rectifier is used for converting single-phase or three-phase ac input into dc, which supplies power to the inverter as well as to the battery bank to keep it charged.

In the normal mode of operation, the power to the inverter is provided by the rectifier. In case of a line outage, power comes from the battery bank. The inverter produces either a single-phase or a three-phase sinusoidal waveform depending on the UPS. The output voltage of the inverter is filtered, prior to being applied to the load.

9-4-1   Rectifier

For supplying power to the inverter and for keeping the battery bank charged, two rectifier arrangements are shown in Fig. 9-5. In a conventional arrangement shown in Fig. 9-5a, a phase-controlled rectifier as in Chapter 4 is used. It is also

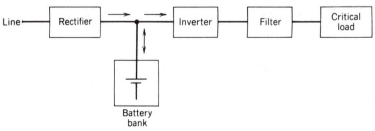

FIGURE 9-4: UPS block diagram.

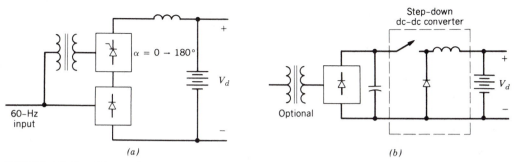

FIGURE 9-5: Possible rectifier arrangements.

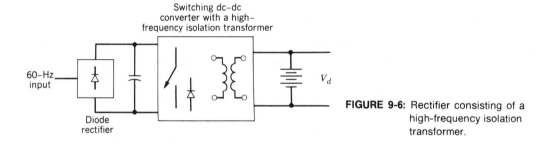

FIGURE 9-6: Rectifier consisting of a high-frequency isolation transformer.

possible to use a diode rectifier bridge in cascade with a step-down dc–dc converter as in Chapter 5, as shown in Fig. 9-5b.

When an electrical isolation from the mains is required, it is possible to use a dc–dc converter with a high-frequency isolation transformer as shown in Fig. 9-6. The dc–dc converter with electrical isolation may be similar to the ones used in the switch-mode dc power supplies of Chapter 8 or may utilize resonant converter concepts discussed in Chapter 7.

Another rectifier arrangement is shown in Fig. 9-7, where the bulk of the power (supplied to the inverter) flows through the diode bridge and only the power required for charging of the battery bank flows through a single-phase phase-controlled thyristor rectifier. V_{charge} can be controlled in magnitude and polarity for proper charging of the battery bank. Thyristor T_1 normally remains off; it is turned on in the event of a power outage.

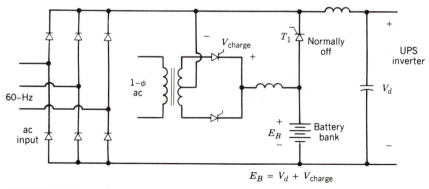

$$E_B = V_d + V_{\text{charge}}$$

FIGURE 9-7: A rectifier with a separate battery-charger circuit.

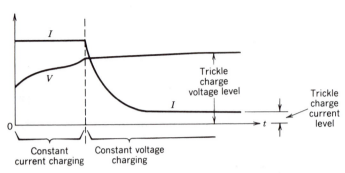

FIGURE 9-8: Charging of a battery after a line-outage causes battery discharge.

9-4-2   Batteries

There are many different types of battery systems. Of these, the conventional lead–acid batteries are commonly used for the UPS applications.

In the normal mode when the line voltage is present, the battery is trickle charged to offset the slight self-discharge by the battery. This requires that a constant trickle charge voltage be applied across the battery and the battery continuously draws a small amount of current, thus maintaining itself in a fully charged state.

In the event of a line outage, the battery supplies the load. The capacity of a battery is expressed in ampere-hours, which is the product of a constant discharge current and the duration beyond which the battery voltage falls below a voltage level called the final discharge voltage. The battery voltage should not be allowed to fall below the final discharge voltage level; otherwise the battery life is shortened. Typically, a 10-hour current is defined as the current in amperes that causes the fully charged battery to discharge in 10 hours to its final voltage level. Discharge currents in excess of the 10-hour current cause the final discharge voltage to be reached sooner than their magnitudes would suggest. Therefore, the higher discharge currents reduce the effective battery capacity.

Once the line voltage is restored, the battery bank in a UPS is brought back to its fully charged state. This is accomplished by initially charging the battery at a constant charging current rate as shown in Fig. 9-8. This causes the battery terminal voltage to increase to its trickle charge voltage level. Once the trickle charge voltage level is reached, the voltage applied is kept constant, as shown in Fig. 9-8, and the charging current finally decreases to the trickle charge current and stays at that level. It is possible to program the battery charging characteristic to bring it to a full-charge state more quickly.

9-4-3   Inverters

The filtered output of the inverter is normally specified to contain very little harmonic distortion, even though most loads are highly nonlinear and, hence, inject large harmonic currents into the UPS. Therefore, the inverter must allow almost instantaneous control over its output ac waveform. The output voltage harmonic content is specified by means of a term called total harmonic distortion (THD), which was defined in Chapter 3 as

$$\%\text{THD} = 100 \times \frac{\left(\sum_{h=2}^{\infty} V_h^2 \right)^{1/2}}{V_1} \tag{9-1}$$

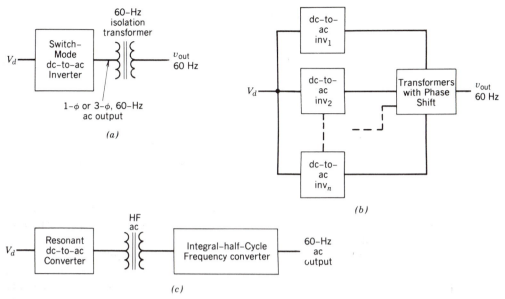

FIGURE 9-9: Various inverter arrangements.

where V_1 is the fundamental frequency rms value of the output voltage and V_h is the rms magnitude at the harmonic of order h. Typically, THD is specified to be less than 5%; each harmonic voltage as a ratio of V_1 is specified to be less than 3%.

Modern UPS normally use PWM dc-to-ac inverters of Chapter 6, with either a single-phase or three-phase ac output. A schematic is shown in Fig. 9-9a. An isolation transformer is generally used at the output. Large UPS may employ a scheme where the outputs of two or more such inverters are paralleled through transformers with phase shift, as shown schematically in Fig. 9-9b. This allows the inverters to operate at a relatively lower switching frequency, utilizing either a low-frequency PWM, selective harmonic cancellation, or a square-wave switching scheme. As shown schematically in Fig. 9-9c, it is also possible to use resonant converters, high-frequency isolation transformers, and integral-half-cycle frequency converter concepts discussed in Chapter 7.

It is important to minimize the harmonics content of the inverter output. This decreases the filter size, which not only results in cost savings but also results in an improved dynamic response of the UPS as the load changes. A feedback control is

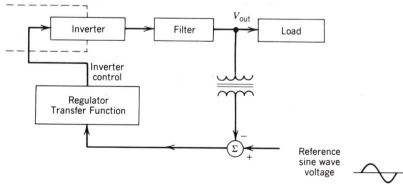

FIGURE 9-10: UPS control.

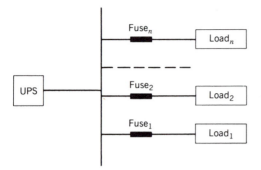

FIGURE 9-11: A UPS supplying several loads.

shown in Fig. 9-10, where the actual output waveform is compared with the sinusoidal reference. The error is used to modify the inverter switching. A control loop with a fast response is needed for a good dynamic performance.

Above a few kilowatts, most UPS provide power to several loads connected in parallel. As shown in Fig. 9-11, each load is supplied through a fuse. In the event of a short circuit in one of the loads, it is important for the UPS to blow that particular fuse and to keep on supplying the rest of the loads. Therefore, the current rating of the

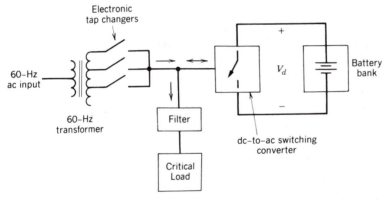

FIGURE 9-12: UPS arrangement where the functions of battery charging and inverter are combined.

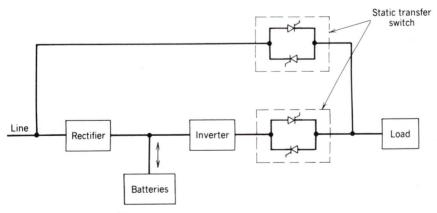

FIGURE 9-13: Line as back-up.

UPS under a sustained short-circuit condition should be sufficient to blow the fuse of the faulted load. In this respect, a rotating-type UPS with a large short-circuit current capacity is far superior to the power electronics type UPS.

An alternative scheme, where the functions of battery charging and the inverter are combined, is shown in Fig. 9-12. In the normal mode, the switching converter operates as a rectifier, charging the battery bank. In addition, it can draw inductive or capacitive currents from the mains, thus providing a fine regulation of the voltage supplied to the load. In case of a utility outage, the utility is isolated and the switching converter operates as an inverter, supplying power to the load from the battery bank. This arrangement is usually referred to as the "standby power supply," as we discuss in the next section.

9-4-4 Static Transfer Switch

For additional reliability, the powerline itself is used as a backup to the UPS and a static transfer switch transfers the load from the UPS to the line, as shown in Fig. 9-13 by means of a block diagram.

As an alternative, in the normal mode the load is supplied by the line in Fig. 9-13. In the event of a line outage, the static transfer switch transfers the load to the UPS. This arrangement is usually referred to as the "standby power supply." When a static transfer switch is used, the inverter output should be synchronized to the line voltage. Therefore, transferring the load from one source to the other results in the least amount of disturbance seen by the load.

9-5   SUMMARY

There are many types of disturbances associated with the mains input. Power conditioners provide an effective way to protect sensitive electronic loads from these disturbances except for the power outages and frequency deviations.

For very critical loads, uninterruptible power supplies (UPS) are used so that the power flow to the load is uninterrupted even in the event of a power outage. The storage capacity of the battery bank is sized based on the likelihood of an outage of a specified duration.

PROBLEMS

9-1. A UPS with transformer-coupled inverters is shown in Fig. P9-1. A programmed harmonic elimination switching scheme is used to eliminate the eleventh and the thirteenth harmonics, which also provides a control over the magnitude of the fundamental.

Show that third, fifth, and seventh harmonics are neutralized by the transformers, assuming that the voltage waveforms of $inverter_2$ lag that of $inverter_1$ by 30°.

9-2. What is the minimum switching frequency in Problem 9-1?

9-3. A UPS arrangement shown in Fig. 9-12 consists of taps to yield 95%, 100%, and 105% of the input voltage at no load. The transformer is rated at 120 V, 60 Hz, and 1 kVA. It has a leakage reactance (resistance can be neglected) of 6%.

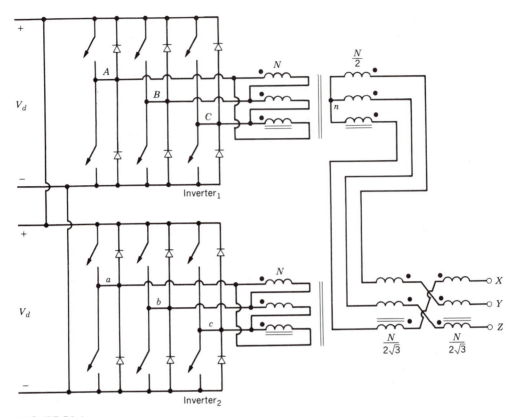

FIGURE P9-1:

Calculate the reactive power that the switch-mode converter must draw to bring the load voltage to 100% of nominal value of 120 V, if the utility voltage is 128 V. Assume that the critical load draws a sinusoidal current at a unity power factor.

9-6   REFERENCES

1. T. S. KEY, "Diagnosing Power Quality-Related Computer Problems," *IEEE Transactions on Industry Applications*, Vol. IA-15, No. 4, pp. 381–393, July/August 1979.
2. "IEEE Recommended Practice for Emergency and Standby Power Systems for Industrial and Commercial Applications," ANSI/IEEE Std. 446–1987.
3. K. THORBORG, "Power Electronics," Prentice Hall International (UK) Ltd, 1988.
4. H. GUMHALTER, *Power Supply Systems in Communications Engineering—Part I Principles,* John Wiley & Sons, 1984.
5. T. KAWABATA, S. DOI, T. MORIKAWA, T. NAKAMURA, and M. SHIGENOBU, "Large Capacity Parallel Redundant Transistor UPS," *1983 IPEC-Tokyo Conference Record*, Vol. 1, pp 660–671.
6. A. SKJELLNES, "A UPS with Inverter Specially Designed for Nonlinear Loads," *IEEE/ INTELEC 1987.*
7. S. MANIAS, P. D. ZIOGAS, and G. OLIVIER, "Bilateral DC to AC Converter Employing a High Frequency Link," *1985 IEEE/IAS Conference Records*, pp. 1156–1162.

PART

4

Motor Drive Applications

10

Introduction to Motor Drives

10-1 INTRODUCTION

Motor drives are used in a very wide power range, from a few watts to many thousands of kilowatts, in applications ranging from very precise, high-performance position controlled drives in robotics to variable-speed drives for adjusting flow rates in pumps. In all drives where the speed and position are controlled, a power electronic converter is needed as an interface between the input power and the motor.

Primarily, there are the following four types of motor drives, which are discussed in Chapters 11 through 14:

- DC-motor drives
- Induction-motor drives
- Synchronous-motor drives
- Step-motor drives

A general block diagram for the control of motor drives is shown in Fig. 10-1. The process determines the requirements on the motor drive; for example, a servo-quality drive (called the servo drive) is needed in robotics, whereas only an adjustable-speed drive may be required in an air conditioning system, as explained further.

In servo applications of motor drives, the response time and the accuracy with which the motor follows the speed and position commands are extremely important. These servo systems, using one of these motor drives, require speed or position feedback for a precise control as shown in Fig. 10-2. In addition, if an ac-motor drive is used, the controller must incorporate sophistication, such as field-oriented control, to make the ac motor (through the power electronic converter) meet the servo-drive requirements.

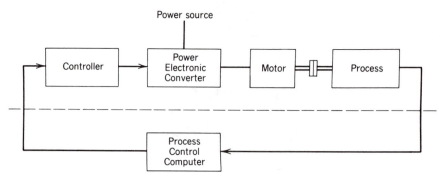

FIGURE 10-1: Control of motor drives.

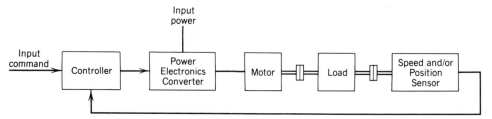

FIGURE 10-2: Servo drives.

However, in a large number of applications, the accuracy and the response time of the motor to follow the speed command is not critical. As shown in Fig. 10-1, there is a feedback loop to control the process, outside of the motor drive. Because of the large time constants associated with the process-control feedback loop, the motor drive's accuracy and the time of response to speed commands are not critical. An example of such an adjustable-speed drive is shown in Fig. 10-3 for an air conditioning system.

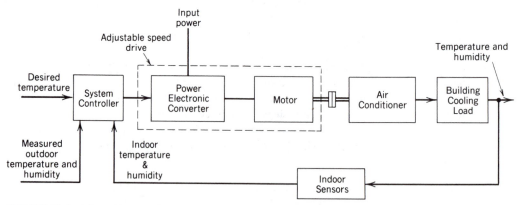

FIGURE 10-3: Adjustable speed drive in an air conditioning system.

10-2   CRITERIA FOR SELECTING DRIVE COMPONENTS

As shown in Figs. 10-1 through 10-3, a motor drive consists of an electric motor, a power electronic converter, and possibly a speed and/or position sensor. In this section, criteria for optimum match between the mechanical load and the drive components are discussed in general terms.

10-2-1 Match Between the Motor and the Load

Prior to selecting the drive components, the load parameters and requirements such as the load inertia, maximum speed, speed range, and direction of motion must be available. The motion profile as a function of time, for example as shown in Fig. 10-4a, must also be specified. By means of modeling the mechanical system, it is possible to obtain a load–torque profile. Assuming a primarily inertial load with a negligible damping, the torque profile, corresponding to the speed profile in Fig. 10-4a, is shown in Fig. 10-4b. The torque required by the load peaks during the acceleration and deceleration.

One way to drive a rotating load is to couple it directly to the motor. In such a direct coupling, the problems and the losses associated with a gearing mechanism are avoided. But the motor must be able to provide peak torques at specified speeds. The

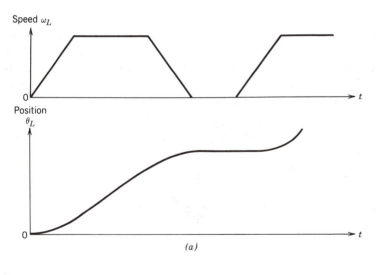

(a)

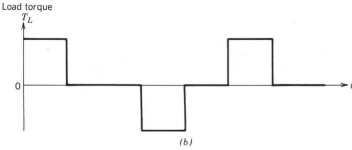

(b)

FIGURE 10-4: Load profile: (a) load-motion profile, (b) load-torque profile (assuming a purely inertial load).

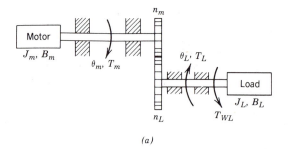

(a)

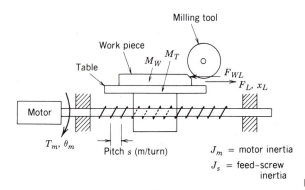

(b)

FIGURE 10-5: Coupling mechanisms: (a) gear, (b) feed screw.

other option for a rotating load is to use a gearing mechanism. A coupling mechanism such as rack-and-pinion, belt-and-pulley, or feed-screw must be used to couple a load with a linear motion to a rotating motor. A gear and a feed-screw drive are shown in Figs. 10-5a and 10-5b, respectively. Assuming the energy efficiency of the gear in Fig. 10-5a to be 100%, the torques on the two sides of the gear are related as

$$\frac{T_m}{T_L} = \frac{\omega_L}{\omega_m} = \frac{\theta_L}{\theta_m} = \frac{n_m}{n_L} = a \qquad (10\text{-}1)$$

where the angular speed $\omega = \dot{\theta}$, n_m and n_L are the number of teeth, and a is the coupling ratio.

In a feed-screw drive of Fig. 10-5b, the torque and the force are related as

$$\frac{T_m}{F_L} = \frac{v_L}{\omega_m} = \frac{x_L}{\theta_m} = \frac{s}{2\pi} = a \qquad (10\text{-}2)$$

where the linear velocity $v_L = \dot{x}_L$, s is the pitch of the feed-screw in m/turn, and a is the coupling ratio.

The electromagnetic torque T_{em} required from the motor can be calculated on the basis of energy considerations in terms of the inertias, required load acceleration, coupling ratio a, and the working torque or force. In Fig. 10-5a, T_{WL} is the working torque of the load and $\dot{\omega}_L$ is the load acceleration. Therefore,

$$T_{em} = \frac{\dot{\omega}_L}{a}\left[J_m + a^2 J_L\right] + a T_{WL} + \frac{\omega_L}{a}\left(B_m + a^2 B_L\right) \qquad (10\text{-}3a)$$

This equation can be written in terms of the motor speed (recognizing that $\omega_m = \omega_L/a$), the equivalent total inertia $J_{eq} = J_m + a^2 J_L$, the equivalent total damping $B_{eq} = B_m +$

$a^2 B_L$, and the equivalent working torque of the load $T_{Weq} = a T_{WL}$:

$$T_{em} = J_{eq} \dot{\omega}_m + B_{eq} \omega_m + T_{Weq} \tag{10-3b}$$

Similarly, for the feed-screw system in Fig. 10-5b with F_{WL} as the working or the machining force and a as the coupling-ratio calculated in Eq. 10-2 in terms of pitch s, T_{em} can be calculated as (see Problem 10-3)

$$T_{em} = \frac{\dot{v}_L}{a} \left[J_m + J_s + a^2 (M_T + M_W) \right] + a F_{WL} \tag{10-4}$$

where $\dot{v}_L$ is the linear acceleration of the load.

As indicated by Eqs. 10-1 and 10-2, the choice of the coupling ratio a affects the motor speed. At the same time, the value of a affects the peak electromagnetic torque T_{em} required from the motor, as is indicated by Eqs. 10-3a and 10-4. In selecting the optimum value of the coupling ratio a, the cost and losses associated with the coupling mechanism must also be included.

10-2-2 Thermal Considerations in Selecting the Motor

In the previous section, the match between the load and the motor is discussed that establishes the peak torque and the maximum speed required from the motor. This matching also establishes the motor–torque profile, which, for example, has the same form (but different magnitudes) as the load–torque profile of Fig. 10-4b.

As another example, the electromagnetic torque required from the motor as a function of time is obtained as shown in Fig. 10-6a. In electric machines, the electromagnetic torque produced by the motor is proportional to the motor current i,

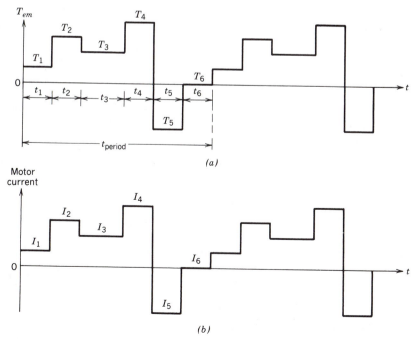

FIGURE 10-6: Motor torque and current.

provided the flux in the air gap of the motor is kept constant. Therefore, the motor current profile is identical to the motor–torque profile, as shown in Fig. 10-6b. The motor current in Fig. 10-6b during various time intervals is a dc current for a dc motor. For an ac motor, the motor current shown is approximately the rms value of the ac current drawn during various time intervals. The power loss P_R in the winding resistance R_M due to the motor current is a large part of the total motor losses, which get converted into heat. This resistive loss is proportional to the square of the motor current and, hence, proportional to T_{em}^2 during various time intervals in Figs. 10-6a and 10-6b. If the time period t_{period} in Fig. 10-6, with which the waveforms repeat, is short compared with the motor thermal time constant, then the motor heating and the maximum temperature rise can be calculated based on the resistive power loss P_R averaged over the time period t_{period}. Therefore, in Fig. 10-6, the rms value of the current over the period of repetition can be obtained as

$$P_R = R_M I_{\text{rms}}^2 \tag{10-5}$$

where

$$I_{\text{rms}}^2 = \left(\sum_{k=1}^{m} I_k^2 t_k \right) \Big/ t_{\text{period}} \tag{10-6}$$

and $m = 6$ in this example.

Because of the motor current being linearly proportional to the motor torque, the rms value of the motor torque over t_{period} from Fig. 10-6 and Eq. 10-6 is

$$T_{em,\text{rms}}^2 = k_1 \left(\sum_{k=1}^{m} I_k^2 t_k \right) \Big/ t_{\text{period}} \tag{10-7}$$

and therefore,

$$T_{em,\text{rms}}^2 = k_1 I_{\text{rms}}^2 \tag{10-8}$$

where k_1 is a constant of proportionality.

From Eqs. 10-5 and 10-8, the average resistive power loss P_R is given as

$$P_R = k_2 T_{em,\text{rms}}^2 \tag{10-9}$$

where k_2 is a constant of proportionality.

In addition to P_R, there are other losses within the motor that contribute to its heating. These are P_{FW} due to friction and windage, P_{EH} due to eddy currents and hysteresis within the motor laminations, and P_s due to switching frequency ripple in the motor current, since it is supplied by a switching power electronic converter rather than an ideal source. There are always some power losses called stray power losses P_{stray} that are not included with the foregoing losses. Therefore, the total power loss within the motor is

$$P_{\text{loss}} = P_R + P_{\text{FW}} + P_{\text{EH}} + P_s + P_{\text{stray}} \tag{10-10}$$

Under a steady-state condition, the motor temperature rise $\Delta\Theta$ in degrees centigrade is given as

$$\Delta\Theta = \frac{P_{\text{loss}}}{R_{TH}} \tag{10-11}$$

where P_{loss} is in watts and the thermal resistance R_{TH} of the motor is in degrees centigrade per watt.

For a maximum allowable temperature rise $\Delta\Theta$, the maximum permissible value of P_{loss} in steady state depends on the thermal resistance R_{TH} in Eq. 10-11. In general, the loss components other than P_R in the right side of Eq. 10-10 increase with the motor speed. Therefore, the maximum allowable P_R and, hence, the maximum continuous motor–torque output from Eq. 10-9 would decrease at higher speed, if R_{TH} remains constant. However, in self-cooled motors with the fan connected to the motor shaft, for example, R_{TH} decreases at higher speeds due to increased air circulation at higher motor speeds. Therefore, the maximum safe operating area in terms of the maximum rms torque available from a motor at various speeds depends on the motor design and is specified in the motor data sheets (specially in case of servo motors). For a motor–torque profile like that shown in Fig. 10-6a, the motor should be chosen such that the rms value of the torque required from the motor remains within the motor's safe operating area in the speed range of operation.

10-2-3 Match Between the Motor and the Power Electronic Converter

A match between the load and the various characteristics of the motor, such as its inertia, and the peak and the rms torque capability, have been discussed in the previous two sections. Depending on the power rating, speed of operation, operating environment, reliability, various other performance requirements by the load, and the cost of the overall drive, one of the following four types of motor drive is selected: dc-motor drive, induction-motor drive, synchronous-motor drive, and the step-motor drive. The advantages and the disadvantages associated with each of these motor drives are discussed in Chapters 11 through 14.

The power electronic converter topology and its control depend on the type of motor drive selected. In general, the power electronic converter provides a controlled voltage to the motor in order to control the motor current and, hence, the electromagnetic torque produced by the motor. Some of the considerations in matching the power electronic converter to the motor are discussed in the following subsections.

10-2-3-1   CURRENT RATING

As we discussed previously, the rms value of the torque that a motor can supply depends on its thermal characteristics. However, a motor can supply substantially larger peak torques (as much as four times the continuous maximum torque) provided that the duration of the peak torque is small compared with the thermal time constant of the motor. Since T_{em} is proportional to i, a peak torque requires a corresponding peak current from the power electronic converter. The current capability of the power semiconductor devices used in the converter is limited by the maximum junction temperature within the devices and other considerations. A higher current results in a higher junction temperature due to power losses within the power semiconductor device. The thermal time constants associated with the power semiconductor devices are in general much smaller than the thermal time constants of various motors. Therefore, the current rating of the power electronic converter must be selected based on both the rms and the peak values of the torque that the motor is required to supply.

10-2-3-2   VOLTAGE RATING

In both dc and ac motors, the motor produces a counter-emf e that opposes the voltage v applied to it, as shown by a simplified generic circuit of Fig. 10-7. The rate

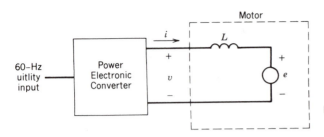

FIGURE 10-7: Simplified circuit of a motor drive.

at which the motor current and, hence, the torque can be controlled is given by

$$\frac{di}{dt} = \frac{v - e}{L} \tag{10-12}$$

where L is the inductance presented by the motor to the converter.

To be able to quickly control the motor current and, hence, its torque, the output voltage v of the power electronic converter must be reasonably greater than the counter-emf e. The magnitude of e in a motor increases linearly with the motor speed, with a constant flux in the air gap of the motor. Therefore, the voltage rating of the power electronic converter depends on the maximum motor speed with a constant air-gap flux.

10-2-3-3   SWITCHING FREQUENCY AND THE MOTOR INDUCTANCE

In a servo drive, the motor current should be able to respond quickly to the load demand, thus requiring L to be small in Eq. 10-12. Also, the steady-state ripple in the motor current should be as small as possible to minimize the motor loss P_s in Eq. 10-10, and the ripple in the motor torque. A small current ripple requires the motor inductance L in Eq. 10-12 to be large. Because of the conflicting requirements on the value of L, the ripple in the motor current can be reduced by increasing the converter switching frequency. However, the switching losses in the power electronic converter increase linearly with the switching frequency. Therefore, a reasonable compromise must be made in selecting the motor inductance L and the switching frequency.

10-2-4 Selection of Speed and Position Sensors

In selecting the speed and position sensors, the following items must be considered: direct or indirect coupling, sensor inertia, possibility and avoidance of torsional resonance, and the maximum sensor speed.

To control the instantaneous speed within a specified range, the ripple in the speed sensor should be small. This can be understood in terms of incremental position encoders, which are often used for measuring speed as well as position. If such a sensor is used at very low speeds, the number of pulse outputs per revolution must be large to provide instantaneous speed measurement with sufficient accuracy. Similarly, an accurate position information will require an incremental position encoder with a large number of pulse outputs per revolution.

10-2-5 Servo-Drive Control and Current Limiting

A block diagram of a servo drive was shown in Fig. 10-2. In most practical applications, a very fast response to a sudden change in position or speed command

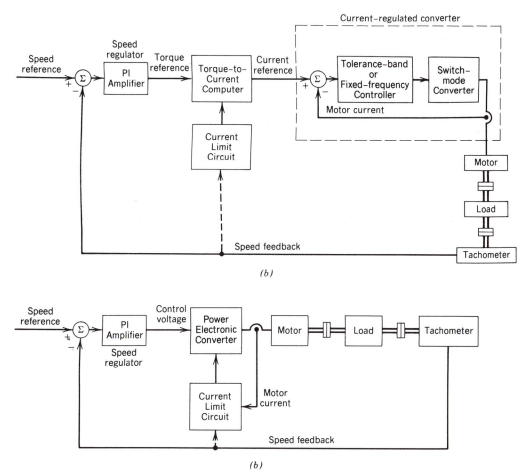

FIGURE 10-8: Control of servo drives: (*a*) inner current loop, (*b*) no inner current loop.

would require a large peak torque, which would result in a large peak current. This may be prohibitive in terms of the cost of the converter. Therefore, the converter current (same as the motor current) is limited by the controller. Figures 10-8*a* and 10-8*b* show two ways of implementing the current limit.

In Fig. 10-8*a*, an inner current loop is used where the actual current is measured, and the error between the reference and the actual current controls the converter output current by means of a current-regulated modulation similar to that discussed in Section 6-6-3. Here, the power electronic converter operates as a current-regulated voltage-source converter. An inner control loop improves the response time of the drive. As shown, the limit on the reference current may be dependent on the speed.

In the other control scheme shown in Fig. 10-8*b*, the error between the speed reference and the actual speed controls the converter through a PI amplifier. The output of the PI amplifier, which controls the converter, is suppressed only if the converter current exceeds the current limit. The current limit can be made to be speed dependent.

In a position control system, the speed reference signal in Figs. 10-8*a* and 10-8*b* is obtained from the position regulator. The input to such a position regulator will be the error between the reference position and the actual position.

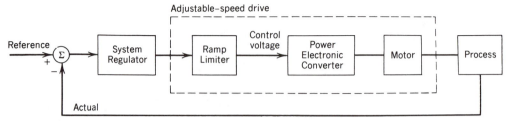

FIGURE 10-9: Ramp limiter to limit motor current.

10-2-6 Current Limiting in Adjustable Speed Drives

In adjustable-speed drives such as that shown in Fig. 10-3, the current is kept from exceeding its limit by means of limiting the rate of change of control voltage with time in the block diagram of Fig. 10-9.

10-3 SUMMARY

1. Primary types of motor drives are dc-motor drives, induction-motor drives, synchronous-motor drives, and the step-motor drives.

2. Most of the applications of motor drives belong to one of the two categories: servo-drives or adjustable-speed drives. In servo drive applications, the response time and the accuracy with which the motor follows the speed and/or position commands are extremely important. In adjustable-speed drive applications, response time to changes in speed command is not as critical; in fact, in many applications, it is not necessary to control the speed accurately where the process feedback loop has large time constants, relative to the response time of the drive.

3. A modeling of the mechanical system is necessary to determine the dynamics of the overall system and to select the motor and the power electronic converter of the appropriate ratings.

4. Servo drives require speed and/or position sensors to close the feedback loop. It is possible to operate with or without an inner current feedback loop. In an adjustabled-speed drive, the current is kept within its limit by limiting the rate of change of control voltage to the power electronic converter with time.

PROBLEMS

10-1. In the system shown in Fig. 10-5a, the gear ratio $n_L/n_m = 2$, $J_L = 10$ kg-m^2, and $J_m = 2.5$ kg-m^2. Damping can be neglected. For a load-speed profile in Fig. P10-1, draw the torque profile and the rms value of the electromagnetic torque required from the motor.

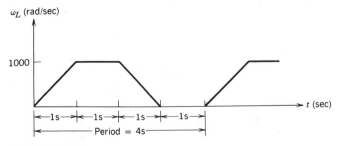

FIGURE P10-1:

10-2. Consider a belt-and-pulley system shown in Fig. P10-2:

$$J_m = \text{Motor inertia}$$

$$M = \text{Mass of the load}$$

$$r = \text{Pulley radius}$$

Other inertias are negligible.

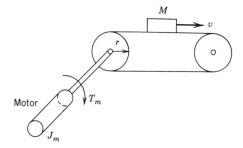

FIGURE P10-2:

(a) Calculate the torque T_{em} required from the motor to accelerate a load of 0.5 kg from rest to a velocity 1 m/s in a time of 3 s. Assuming the motor torque to be constant during this interval.

(b) Calculate the electromagnetic torque T_{em} required from the motor in part (a) if pulley radius $r = 0.1$ m and the motor inertia $J_m = 0.006$ kg-m^2.

10-3. Derive Eq. 10-4.

10-4. In the system of Fig. 10-5a, assume a triangular velocity profile with equal acceleration and deceleration rates. The system is purely inertial and B_m, B_L, and T_{WL} can be neglected.

Assuming a gear efficiency of 100% and an optimum gear ratio (such that the reflected load inertia equals the motor inertia), calculate the time needed to rotate the load by an angle θ_L in terms of J_m, J_L, and the peak torque T_{em} that the motor must be capable of developing.

10-4 REFERENCES

1. H. GROSS, EDITOR, *Electrical Feed Drives for Machine Tools,* Siemens and John Wiley & Sons, New York, 1983.
2. *DC Motors Speed Controls ServoSystem—An Engineering Handbook.* 5th Ed., Electro-Craft Corporation, 1600 Second Street South, Hopkins, MN, 1980.
3. A. E. FITZGERALD, C. KINGSLEY, JR., and S. D. UMANS, *Electric Machinery,* 4th Ed. McGraw-Hill, New York, 1983.
4. G. R. SLEMON and A. STRAUGHEN, *Electric Machines,* Addison-Wesley Publishing Co., Reading, MA, 1980.

11

DC-Motor Drives

11-1 INTRODUCTION

Traditionally, dc-motor drives have been used for speed and position control applications. In the past few years, the use of ac-motor servo drives in these applications is increasing. In spite of that, in applications where an extremely low maintenance is not required, dc drives continue to be used because of their low initial cost and excellent drive performance.

11-2 EQUIVALENT CIRCUIT OF DC MOTORS

In a dc motor, the field-flux ϕ_f is established by the stator, either by means of permanent magnets as shown in Fig. 11-1a, where ϕ_f stays constant, or by means of a field winding as shown in Fig. 11-1b, where the field current I_f controls ϕ_f. If the magnetic saturation in the flux path can be neglected, then

$$\phi_f = k_f I_f \qquad (11\text{-}1)$$

where k_f is a field constant of proportionality.

The rotor carries in its slots the so-called armature winding, which handles the electrical power. This is in contrast to most ac motors, where the power handling winding is on the stator for ease of handling the larger amount of power. However, the armature winding in a dc machine has to be on the rotor to provide a "mechanical" rectification of voltages and currents (which alternate direction as the conductors rotate from the influence of one stator pole to the next) in the armature-winding conductors, thus producing a dc voltage and a dc current at the terminals of the armature winding. The armature winding, in fact, is a continuous winding, without any beginning or end, and it is connected to the commutator segments. These commutator segments, usually made up of copper, are insulated from each other and rotate with the shaft. At least one pair of stationary carbon brushes is used to make

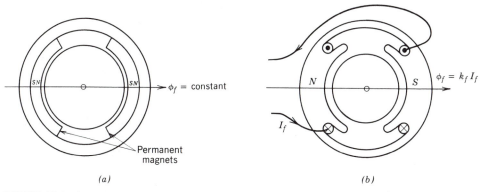

FIGURE 11-1: dc motor: (*a*) permanent-magnet motor, (*b*) dc motor with a field winding.

contact between the commutator segments (and, hence, the armature conductors), and the stationary terminals of the armature winding that supply the dc voltage and current.

In a dc motor, the electromagnetic torque is produced by the interaction of the field-flux ϕ_f and the armature current i_a:

$$T_{em} = k_t \phi_f i_a \tag{11-2}$$

where k_t is the torque constant of the motor. In the armature circuit, a back-emf is produced by the rotation of armature conductors at a speed ω_m in the presence of a field-flux ϕ_f:

$$e_a = k_e \phi_f \omega_m \tag{11-3}$$

where k_e is the voltage constant of the motor.

In SI units, k_t and k_e are equal (both numerically and dimensionally), which can be shown by equating the electrical power $e_a i_a$ and the mechanical power $\omega_m T_{em}$. The electrical power is

$$P_e = e_a i_a = k_e \phi_f \omega_m i_a \quad \text{(using Eq. 11-3)} \tag{11-4}$$

and the mechanical power is

$$P_m = \omega_m T_{em} = k_t \phi_f \omega_m i_a \quad \text{(using Eq. 11-2)} \tag{11-5}$$

In steady state,

$$P_e = P_m \tag{11-6}$$

Therefore, from the foregoing equations

$$k_t \left[\frac{N - m}{A \cdot Wb} \right] = k_e \left[\frac{V}{Wb \cdot rad/s} \right] \tag{11-7}$$

In practice, a controllable voltage source v_t is applied to the armature terminals to establish i_a. Therefore, the current i_a in the armature circuit is determined by v_t, the induced back-emf e_a, the armature winding resistance R_a, and the armature-wind-

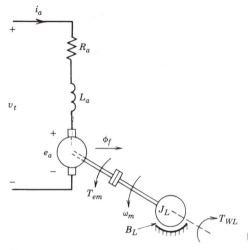

FIGURE 11-2: dc motor equivalent circuit.

ing inductance L_a:

$$v_t = e_a + R_a i_a + L_a \frac{di_a}{dt} \tag{11-8}$$

Equation 11-8 is illustrated by an equivalent circuit in Fig. 11-2.

The interaction of T_{em} with the load torque, as given by Eq. 10-3b of Chapter 10, determines how the motor speed builds up:

$$T_{em} = J \frac{d\omega_m}{dt} + B\omega_m + T_{WL}(t) \tag{11-9}$$

where J and B are the total equivalent inertia and damping, respectively of the motor-load combination and T_{WL} is the equivalent working torque of the load.

DC machines are seldom used as generators. However, they act as generators while braking, where their speed is being reduced. Therefore, it is important to consider dc machines in their generator mode of operation. In order to consider braking, we will assume that the flux ϕ_f is kept constant and the motor is initially driving a load at a speed of ω_m. To reduce the motor speed, if v_t is reduced below e_a in Fig. 11-2, then the current i_a will reverse in direction. The electromagnetic torque T_{em} given by Eq. 11-2 now reverses in direction and the kinetic energy associated with the motor-load inertia is converted into electrical energy by the dc machine, which now acts as a generator. This energy must be somehow absorbed by the source of v_t or dissipated in a resistor.

During the braking operation, the polarity of e_a does not change, since the direction of rotation has not changed. Equation 11-3 still determines the magnitude of the induced emf. As the rotor slows down, e_a decreases in magnitude (assuming that ϕ_f is constant). Ultimately, the generation stops when the rotor comes to a standstill and all the inertial energy is extracted. If the terminal-voltage polarity is also reversed, the direction of rotation of the motor will reverse. Therefore, a dc motor can be operated in either direction and its electromagnetic torque can be reversed for braking, as shown by the four quadrants of torque-speed plane in Fig. 11-3.

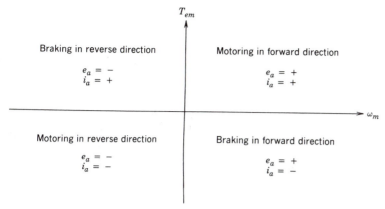

FIGURE 11-3: Four-quadrant operation of a dc motor.

11-3 PERMANENT-MAGNET DC MOTORS

Often in small dc motors, permanent magnets on the stator as shown in Fig. 11-1a produce a constant field-flux ϕ_f. In steady state, assuming a constant field-flux ϕ_f, Eqs. 11-2, 11-3, and 11-8 result in

$$T_{em} = k_T I_a \tag{11-10}$$

$$E_a = k_E \omega_m \tag{11-11}$$

and

$$V_t = E_a + R_a I_a \tag{11-12}$$

where $k_T = k_t \phi_f$ and $k_E = k_e \phi_f$. Equations 11-10 through 11-12 correspond to the equivalent circuit of Fig. 11-4a. From the above equations, it is possible to obtain the steady-state speed ω_m as a function of T_{em} for a given V_t:

$$\omega_m = \frac{1}{k_E}\left(V_t - \frac{R_a}{k_T}T_{em}\right) \tag{11-13}$$

The plot of this equation in Fig. 11-4b shows that as the torque is increased, the torque-speed characteristic at a given V_t is essentially vertical, except for the droop due to the voltage drop $I_a R_a$ across the armature-winding resistance. This droop in speed is quite small in integral HP dc motors, but may be substantial in small servo motors. More importantly, however, the torque-speed characteristics can be shifted horizontally in Fig. 11-4b by controlling the applied terminal voltage V_t. Therefore, the speed of a load with an arbitrary torque-speed characteristic can be controlled by controlling V_t in a permanent-magnet dc motor with a constant ϕ_f.

In a continuous steady-state, the armature-current I_a should not exceed its rated value and, therefore, the torque should not exceed the rated torque. Therefore, the characteristics beyond the rated torque are shown as dotted in Fig. 11-4b. Similarly, the characteristic beyond the rated speed is shown as dotted, because increasing the speed beyond the rated speed would require the terminal voltage V_t to exceed its rated value, which is not desirable. This is a limitation of a permanent-magnet dc motor, where the maximum speed is limited to the rated speed of the motor. The torque

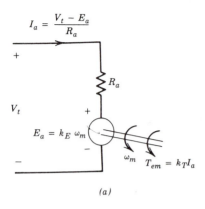

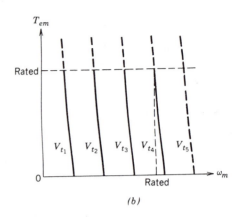

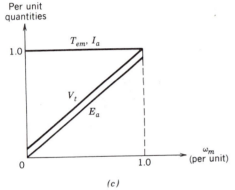

FIGURE 11-4: Permanent-magnet dc motor: (a) equivalent circuit, (b) torque-speed
characteristics; $V_{t5} > V_{t4} > V_{t3} > V_{t2} > V_{t1}$, where V_{t4} is the rated voltage,
(c) continuous torque-speed capability.

capability as a function of speed is plotted in Fig. 11-4c. It shows the steady-state
operating limits of the torque and current; it is possible to significantly exceed current
and torque limits on a short-term basis. Figure 11-4c also shows the terminal voltage
required as a function of speed and the corresponding E_a.

11-4   DC MOTORS WITH A SEPARATELY EXCITED FIELD WINDING

Permanent-magnet dc motors are limited to ratings of a few horsepower and
also have a maximum speed limitation. These limitations can be overcome if ϕ_f is
produced by means of a field winding on the stator, which is supplied by a dc current
I_f, as shown in Fig. 11-1b. To offer the most flexibility in controlling the dc motor, the
field winding is excited by a separately controlled dc source v_f, as shown in Fig. 11-5a.
As indicated by Eq. 11-1, the steady-state value of ϕ_f is controlled by $I_f (= V_f/R_f)$,
where R_f is the resistance of the field winding.

Since ϕ_f is controllable, Eq. 11-13 can be written as follows:

$$\omega_m = \frac{1}{k_e\phi_f}\left(V_t - \frac{R_a}{k_t\phi_f}T_{em}\right) \tag{11-14}$$

recognizing that $k_E = k_e\phi_f$ and $k_T = k_t\phi_f$. Equation 11-14 shows that in a dc motor

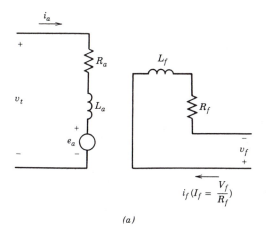

(a)

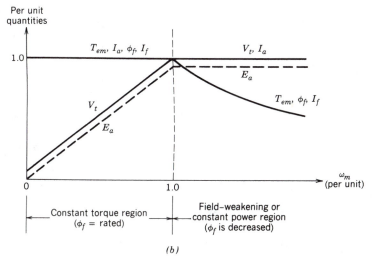

(b)

FIGURE 11-5: Separately excited dc motor: (a) equivalent circuit,
(b) continuous torque-speed capability.

with a separately excited field winding, both V_t and ϕ_f can be controlled to yield the desired torque and speed. As a general practice, to maximize the motor-torque capability, ϕ_f (hence, I_f) is kept at its rated value for speeds less than the rated speed. With ϕ_f at its rated value, the relationships are the same as given by Eqs. 11-10 through 11-13 of a permanent-magnet dc motor. Therefore, the torque-speed characteristics are also the same as those for a permanent-magnet dc motor that were shown in Fig. 11-4b. With ϕ_f constant and equal to its rated value, the motor torque-speed capability is as shown in Fig. 11-5b, where this region of constant ϕ_f is often called the constant torque region. The required terminal voltage V_t in this region increases linearly from approximately 0 to its rated value, as the speed increases from 0 to its rated value. V_t and the corresponding E_a are shown in Fig. 11-5b.

To obtain speeds beyond its rated value, V_t is kept constant at its rated value and ϕ_f is decreased by decreasing I_f. Since I_a is not allowed to exceed its rated value on a continuous basis, the torque capability declines, since ϕ_f is reduced in Eq. 11-2. In this so called field-weakening region, the maximum power $E_a I_a$ (equal to $\omega_m T_{em}$) into the motor is not allowed to exceed its rated value on a continuous basis. This region, also called the constant power region, is shown in Fig. 11-5b, where T_{em} declines with ω_m, and V_t, E_a, and I_a stay constant at their rated values. It should be emphasized that Fig. 11-5b is the plot of the maximum continuous capability of the motor in steady state. Any operating point within the regions shown is, of course, permissible. In the field-weakening region, the speed may be exceeded by 50% to 100% of its rated value, depending on the motor specifications.

11-5   EFFECT OF ARMATURE CURRENT WAVEFORM

In dc–motor drives, the output voltage of the power electronic converter contains an ac ripple voltage superimposed on the desired dc voltage. Ripple in the terminal voltage can lead to a ripple in the armature current with the following consequences that must be recognized:

11-5-1 Form Factor

The Form factor for the dc-motor armature current is defined as

$$\text{Form factor} = \frac{I_a(\text{rms})}{I_a(\text{average})} \tag{11-15}$$

The form factor will be unity only if i_a is a pure dc. The more i_a deviates from a pure dc, the higher will be the value of the form factor. The power input to the motor (and hence the power output) varies proportionally with the average value of i_a, whereas the losses in the resistance of the armature winding depend on $I_a^2(\text{rms})$. Therefore, the higher the form factor of the armature current, the higher the losses in the motor (i.e., higher heating) and, hence, the lower the motor efficiency.

Moreover, a form factor much higher than unity implies a much larger value of the peak armature current compared to its average value, which may result in excessive arcing in the commutator and brushes. To avoid serious damage to the motor that is caused by large peak currents, the motor may have to be derated (i.e., the maximum power or torque would have to be kept well below its rating) to keep the motor temperature from exceeding its specified limit and to protect the commutator and brushes. Therefore, it is desirable to improve the form factor of the armature current as much as possible.

11-5-2 Torque Pulsations

Since the instantaneous electromagnetic torque $T_{em}(t)$ developed by the motor is proportional to the instantaneous armature current $i_a(t)$, a ripple in i_a results in a ripple in the torque and hence in speed, if the inertia is not large. This is another reason to minimize the ripple in the armature current. It should be noted that a high-frequency torque ripple will result in smaller speed fluctuations, as compared with a low-frequency torque ripple of the same magnitude.

11-6   DC SERVO DRIVES

In servo applications, the speed and accuracy of response is important. In spite of the increasing popularity of ac servo drives, dc servo drives are still widely used. If it were not for the disadvantages of having a commutator and brushes, the dc motors

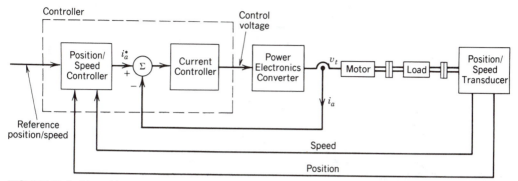

FIGURE 11-6: Closed-loop position / speed dc servo drive.

would be ideally suited for servo drives. The reason is that the instantaneous torque T_{em} in Eq. 11-2 can be controlled linearly by controlling the armature-current i_a of the motor.

11-6-1 Transfer Function Model for Small-Signal Dynamic Performance

Figure 11-6 shows a dc motor operating in a closed loop to deliver controlled speed or controlled position. To design the proper controller that will result in high performance (high speed of response, low steady-state error, and high degree of stability), it is important to know the transfer function of the motor. It is then combined with the transfer function of the rest of the system in order to determine the dynamic response of the drive for changes in the desired speed and position, or for a change in load. As we will explain later on, the linear model is valid only for small changes where the motor current is not limited by the converter supplying the motor.

For analyzing small-signal dynamic performance of the motor-load combination around a steady-state operating point, the following equations can be written in terms of small deviations around their steady-state values:

$$\Delta v_t = \Delta e_a + R_a \, \Delta i_a + L_a \frac{d}{dt}(\Delta i_a) \tag{11-16}$$

$$\Delta e_a = k_E \, \Delta \omega_m \tag{11-17}$$

$$\Delta T_{em} = k_T \, \Delta i_a \tag{11-18}$$

$$\Delta T_{em} = \Delta T_{WL} + B\Delta \omega_m + J\frac{d(\Delta \omega_m)}{dt} \qquad \text{(from Eq. 11-9)} \tag{11-19}$$

If we take the Laplace transform of these equations, where the Laplace variables represent only the small signal Δ values in Eqs. 11-16 through 11-19,

$$V_t(s) = E_a(s) + (R_a + sL_a)I_a(s)$$

$$E_a(s) = k_E \omega_m(s)$$

$$T_{em}(s) = k_T I_a(s)$$

$$T_{em}(s) = T_{WL}(s) + (B + sJ)\omega_m(s) \tag{11-20}$$

and

$$\omega_m(s) = s\theta_m(s)$$

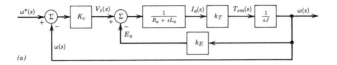

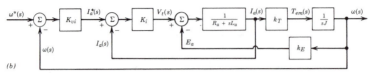

(b)

FIGURE 11-7: Block diagram representation of the motor and load (without any feedback).

These equations for the motor–load combination can be represented by transfer-function blocks, as shown in Fig. 11-7. The inputs to the motor–load combination in Fig. 11-7 are the armature terminal voltage $V_t(s)$ and the load torque $T_{WL}(s)$. Applying one input at a time by setting the other input to zero, the superposition principle yields (note that this is a linearized system)

$$\omega_m(s) = \frac{k_T}{(R_a + sL_a)(sJ + B) + k_T k_E} V_t(s) - \frac{R_a + sL_a}{(R_a + sL_a)(sJ + B) + k_T k_E} T_{WL}(s)$$

(11-21)

This equation results in two closed-loop transfer functions:

$$G_1(s) = \left. \frac{\omega_m(s)}{V_t(s)} \right|_{T_{WL}(s)=0} = \frac{k_T}{(R_a + sL_a)(sJ + B) + k_T k_E}$$

(11-22)

and

$$G_2(s) = \left. \frac{\omega_m(s)}{T_{WL}(s)} \right|_{V_t(s)=0} = -\frac{R_a + sL_a}{(R_a + sL_a)(sJ + B) + k_T k_E}$$

(11-23)

As a simplification to gain better insight into the dc-motor behavior, the friction term, which is usually small, will be neglected by setting $B = 0$ in Eq. 11-22. Moreover, considering just the motor without the load, J in Eq. 11-22 is then the motor inertia J_m. Therefore

$$G_1(s) = \frac{k_T}{sJ_m(R_a + sL_a) + k_T k_E} = \frac{1}{k_E \left(s^2 \dfrac{L_a J_m}{k_T k_E} + s \dfrac{R_a J_m}{k_T k_E} + 1 \right)}$$

(11-24)

We will define the following constants:

$$\tau_m = \frac{R_a J_m}{k_T k_E} = \text{Mechanical time constant}$$

(11-25)

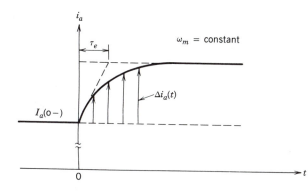

FIGURE 11-8: Electrical time constant τ_e; speed ω_m is assumed to be constant.

and

$$\tau_e = \frac{L_a}{R_a} = \text{Electrical time constant} \qquad (11\text{-}26)$$

Using τ_m and τ_e in the expression for $G_1(s)$ yields

$$G_1(s) = \frac{1}{k_E\left(s^2\tau_m\tau_e + s\tau_m + 1\right)} \qquad (11\text{-}27)$$

Since in general $\tau_m \gg \tau_e$, it is a reasonable approximation to replace $s\tau_m$ by $s(\tau_m + \tau_e)$ in the foregoing expression. Therefore

$$G_1(s) = \frac{\omega_m(s)}{V_t(s)} \simeq \frac{1}{k_E(s\tau_m + 1)(s\tau_e + 1)} \qquad (11\text{-}28)$$

The physical significance of the electrical and the mechanical time constants of the motor should also be understood. The electrical time constant τ_e determines how quickly the armature current builds up, as shown in Fig. 11-8, in response to a step change Δv_t in the terminal voltage, where the rotor speed is assumed to be constant.

The mechanical time constant τ_m determines how quickly the speed builds up in response to a step change Δv_t in the terminal voltage, provided that the electrical time constant τ_e is assumed to be negligible and, hence, the armature current can change instantaneously. Neglecting τ_e in Eq. 11-28, the change in speed from the steady-state condition can be obtained as

$$\omega_m(s) = \frac{V_t(s)}{k_E(s\tau_m + 1)} = \frac{\Delta v_t}{k_E s(s\tau_m + 1)} = \frac{\Delta v_t}{k_E} \frac{1/\tau_m}{s(s + 1/\tau_m)} \qquad (11\text{-}29)$$

recognizing that $V_t(s) = \Delta v_t/s$. From Eq. 11-29

$$\Delta\omega_m(t) = \frac{\Delta v_t}{k_E}\left(1 - e^{-t/\tau_m}\right) \qquad (11\text{-}30)$$

where τ_m is the mechanical time constant with which the speed changes in response to a step change in the terminal voltage, as shown in Fig. 11-9a. The corresponding change in the armature current is plotted in Fig. 11-9b. Note that if the motor current is limited by the converter during large transients, the torque produced by the motor is simply $k_T I_{a,\,max}$.

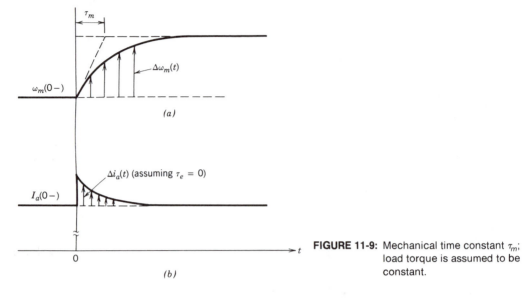

FIGURE 11-9: Mechanical time constant τ_m; load torque is assumed to be constant.

11-6-2 Power Electronic Converter

Based on the previous discussion, a power electronic converter supplying a dc motor should have the following capabilities:

1. The converter should allow both its output voltage and current to reverse, in order to yield a four-quadrant operation as shown in Fig. 11-3.

2. The converter should be able to operate in a current-controlled mode by holding the current at its maximum acceptable value during fast acceleration and deceleration. The dynamic current limit is generally several times higher than the continuous steady-state current rating of the motor.

3. For accurate control of position, the average voltage output of the converter should vary linearly with its control input, independent of the load on the motor. This item is further discussed in Section 11-6-5.

4. The converter should produce an armature current with a good form factor and should minimize the fluctuations in torque and speed of the motor.

5. The converter output should respond as quickly as possible to its control input, thus allowing the converter to be represented essentially by a constant gain without a dead time in the overall servo drive transfer function model.

A linear power amplifier satisfies all the requirements listed above. However, because of its low energy efficiency, this choice is limited to a very low power range. Therefore, the choice must be made between switch-mode dc–dc converters of the type discussed in Chapter 5 or the line-frequency controlled converters discussed in Chapter 4. Here, only the switch-mode dc–dc converters are described. Drives with line-frequency converters can be analyzed in the same manner.

A full-bridge switch-mode dc–dc converter produces a four-quadrant controllable dc output. This full-bridge dc–dc converter (also called H-bridge) was discussed in Chapter 5. The overall system is shown in Fig. 11-10, where the line-frequency ac input is rectified into dc by means of a diode rectifier of the type discussed in Chapter 3 and filtered by means of a filter capacitor. An energy dissipation circuit is included to prevent the filter capacitor voltage from becoming large in case of braking of the dc motor.

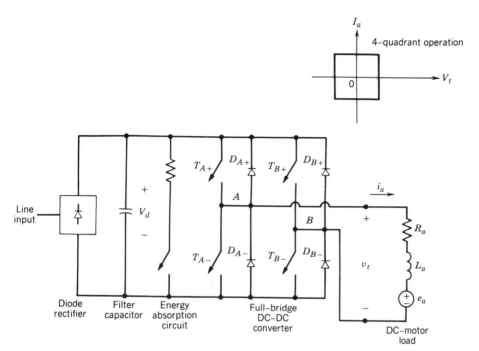

FIGURE 11-10: dc-motor servo drive; four-quadrant operation.

As discussed in Chapter 5, all four switches in the converter of Fig. 11-10 are switched during each cycle of the switching frequency. This results in a true four-quadrant operation with a continuous-current conduction, where both V_t and I_a can smoothly reverse, independent of each other. Ignoring the effect of blanking time, the average voltage output of the converter varies linearly with the input control voltage $v_{control}$, independent of the load:

$$V_t = k_c v_{control} \qquad (11\text{-}31)$$

where k_c is the gain of the converter.

As discussed in Sections 5-7-1 and 5-7-2 of Chapter 5, either a PWM bipolar voltage switching or a PWM unipolar voltage switching scheme can be used. Thus, the converter in Fig. 11-6 can be replaced by an amplifier gain k_c given by Eq. 11-31.

11-6-3 Ripple in the Armature Current i_a

In Chapter 5, it was mentioned that the current through a PWM full-bridge dc–dc converter supplying a dc-motor load flows continuously even at small values of I_a. However, it is important to consider the peak-to-peak ripple in the armature current because of its impact on the torque pulsations and heating of the motor. Moreover, a larger current ripple requires a larger peak current rating of the converter switches.

In the system of Fig. 11-10 under a steady-state operating condition, the instantaneous speed ω_m can be assumed to be constant if there is sufficient inertia, and therefore $e_a(t) = E_a$. The terminal voltage and the armature current can be expressed in terms of their dc and the ripple components as

$$v_t(t) = V_t + v_r(t) \qquad (11\text{-}32)$$

and

$$i_a(t) = I_a + i_r(t) \qquad (11\text{-}33)$$

where $v_r(t)$ and $i_r(t)$ are the ripple components in v_t and i_a, respectively. Therefore, in the armature circuit, from Eq. 11-8

$$V_t + v_r(t) = E_a + R_a[I_a + i_r(t)] + L_a\frac{di_r(t)}{dt} \qquad (11\text{-}34)$$

where

$$V_t = E_a + R_aI_a \qquad (11\text{-}35)$$

and

$$v_r(t) = R_ai_r(t) + L_a\frac{di_r(t)}{dt} \qquad (11\text{-}36)$$

Assuming that the ripple current is primarily determined by the armature-inductance L_a, and R_a has a negligible effect, from Eq. 11-36

$$v_r(t) \simeq L_a\frac{di_r(t)}{dt} \qquad (11\text{-}37)$$

The additional heating in the motor is approximately $R_aI_r^2$, where I_r is the rms value of the ripple current i_r.

By means of an example and Fig. 5-30, it was shown in Chapter 5 that for a PWM bipolar voltage switching, the ripple voltage is maximum when the average output voltage is zero and all switches operate at equal duty ratios. Applying these results to the dc-motor drive, Fig. 11-11a shows the voltage ripple $v_r(t)$ and the resulting ripple current $i_r(t)$, using Eq. 11-37. From these waveforms, the maximum

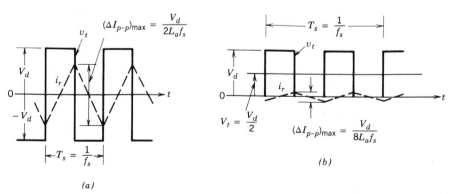

FIGURE 11-11: Ripple i_r in the armature current: (a) PWM bipolar-voltage switching; $V_t = 0$, (b) PWM unipolar-voltage switching; $V_t = V_d / 2$.

peak-to-peak ripple can be calculated as

$$\left(\Delta I_{p-p}\right)_{max} = \frac{V_d}{2L_a f_s} \tag{11-38}$$

where, V_d is input dc voltage to the full-bridge converter.

The ripple voltage for a PWM unipolar voltage switching is shown to be maximum when the average output voltage is equal to $V_d/2$. Applying this result to a dc-motor drive, Fig. 11-11b shows $i_r(t)$ waveform, where

$$\left(\Delta I_{p-p}\right)_{max} = \frac{V_d}{8L_a f_s} \tag{11-39}$$

Equations 11-38 and 11-39 show that the maximum peak-to-peak ripple current is inversely proportional to L_a and f_s. Therefore, careful consideration must be given to the selection of f_s and L_a, where L_a can be increased by adding an external inductor in the series with the motor armature.

11-6-4 Control of Servo Drives

A servo system where the speed error directly controls the power electronic converter is shown in Fig. 11-12a. The current-limiting circuit comes into operation only when the drive current tries to exceed an acceptable limit $I_{a,max}$ during fast accelerations and decelerations. During these intervals, the output of the speed regulator is suppressed and the current is held at its limit until the speed and position approach their desired values.

To improve the dynamic response in high-performance servo drives, an internal current loop is used as shown in Fig. 11-12b, where the armature current and, hence, the torque is controlled. The current control is accomplished by comparing the actual measured armature current i_a with its reference value i_a^* produced by the speed regulator. The current i_a is inherently controlled from exceeding the current rating of the drive, by limiting the reference current i_a^* to $I_{a,max}$.

The armature current provided by the dc–dc converter in Fig. 11-12b can be controlled in a similar manner as the current-regulated modulation in a dc-to-ac inverter, discussed in Section 6-6-3. The only difference is that the reference current in steady state in a dc–dc converter is a dc rather than a sinusoidal waveform as in Section 6-6-3. Either a variable-frequency tolerance-band control discussed in Section 6-6-3-1 or a fixed-frequency control discussed in Section 6-6-3-2 can be used for current control.

11-6-5 Nonlinearity due to Blanking Time

In a practical full-bridge dc–dc converter, where the possibility of a short circuit across the input dc bus exists, a blanking time is introduced between the instant at which a switch turns off and the instant at which the other switch in the same leg turns on. The effect of the blanking time on the output of dc-to-ac full-bridge PWM inverters was discussed in detail in Section 6-5. That analysis is also valid for PWM full-bridge dc–dc converters for dc-servo drives. Equation 6-77 and Fig. 6-32b show the effect of blanking time on the output voltage magnitude. Recognizing that the output voltage of the converter is proportional to the motor speed ω_m and the output

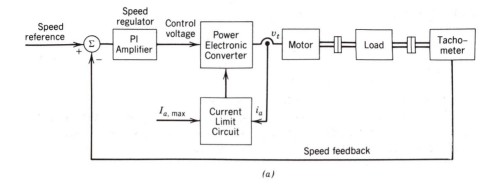

(a)

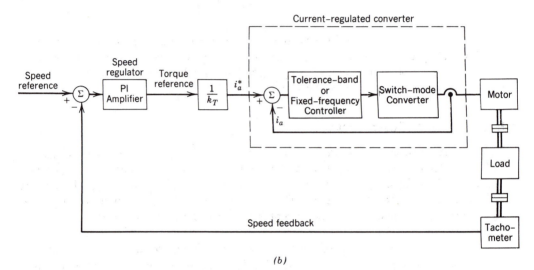

(b)

FIGURE 11-12: Control of servo drives: (a) no internal current-control loop, (b) internal current-control loop.

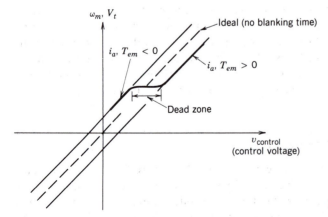

FIGURE 11-13: Effect of blanking time.

current i_a is proportional to the torque T_{em}, Fig. 6-32b is redrawn as in Fig. 11-13. If at an arbitrary speed ω_m, the torque and, hence, i_a are to be reversed, there is a dead zone in $v_{control}$ as shown in Fig. 11-13 during which i_a and T_{em} remain small. The effect of this nonlinearity due to blanking time on the performance of the servo system is minimized by means of current-controlled mode of operation discussed in the block diagram of Fig. 11-12b, where an internal current loop directly controls i_a.

11-6-6 Selection of Servo-Drive Parameters

Based on the foregoing discussion, the effects of armature inductance L_a, switching frequency f_s, blanking time t_Δ, and the switching times t_c of the solid-state devices in the dc–dc converter can be summarized as follows:

1. The ripple in the armature current, which causes torque ripple and additional armature heating, is proportional to (L_a/f_s).
2. The dead-zone in the transfer function of the converter, which degrades the servo performance, is proportional to $(f_s \cdot t_\Delta)$.
3. Switching losses in the converter are proportional to $(f_s \cdot t_c)$.

All these factors need to be considered simultaneously in the selection of the appropriate motor and the power electronic converter.

11-7 ADJUSTABLE-SPEED DC DRIVES

Unlike servo drives, the response time to speed and torque commands is not as critical in adjustable-speed drives. Therefore, either switch-mode dc–dc converters as discussed for servo drives or the line-frequency controlled converters discussed in Chapter 4 can be used for speed control.

11-7-1 Switch-Mode DC – DC Converter

If a four-quadrant operation is needed and a switch-mode converter is utilized, then a full-bridge converter as shown in Fig. 11-10 is used.

If the speed does not have to reverse but braking is needed, then a two-quadrant converter as shown in Fig. 11-14a can be used. It consists of two switches where one of the switches is on at any time, to keep the output voltage independent of the direction of i_a. The armature current can reverse, and a negative value of I_a corresponds to the braking mode of operation, where the power flows from the dc motor to V_d. The output voltage V_t can be controlled in magnitude but it always remains unipolar. Since i_a can flow in both directions, unlike in the single-switch step-down and step-up dc–dc converters discussed in Chapter 5, i_a in the circuit of Fig. 11-14a will not become discontinuous.

For a single-quadrant operation where the speed remains unidirectional and braking is not required, a step-down converter as shown in Fig. 11-14b can be used.

11-7-2 Line-Frequency Controlled Converters

In many adjustable-speed dc drives, especially in large power ratings, it may be economical to utilize a line-frequency controlled converter of the type discussed in Chapter 4. Two of these converters are repeated in Fig. 11-15 for single-phase and

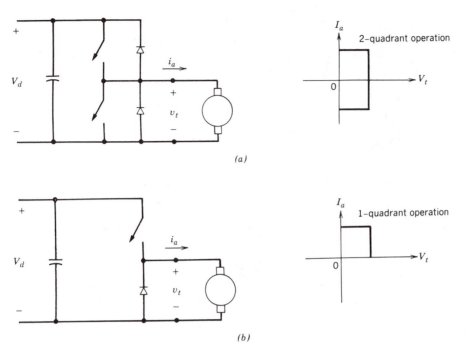

FIGURE 11-14: (*a*) Two-quadrant operation. (*b*) Single-quadrant operation.

three-phase ac inputs. The output of these line-frequency converters, also called the phase-controlled converters, contains an ac ripple that is a multiple of the 60-Hz line frequency. Because of this low frequency ripple, an inductance in series with the motor armature may be required to keep the ripple in i_a low, to minimize its effect on armature heating and the ripple in torque and speed.

A disadvantage of the line-frequency converters is the longer dead time in responding to the changes in the speed control signal, compared to high-frequency switch-mode dc–dc converters. Once a thyristor or a pair of thyristors is triggered on in the circuits of Fig. 11-15, the delay angle α that controls the converter output voltage applied to the motor terminals can not be increased for a portion of the 60-Hz cycle. This may not be a problem in adjustable-speed drives where the response time to speed and torque commands is not too critical. But it clearly shows the limitation of line-frequency converters in servo-drive applications.

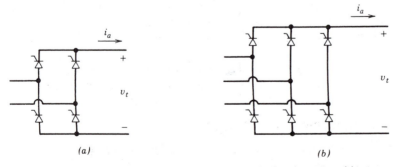

FIGURE 11-15: Line-frequency controlled converters for dc motor drives:
(*a*) single-phase input, (*b*) three-phase input.

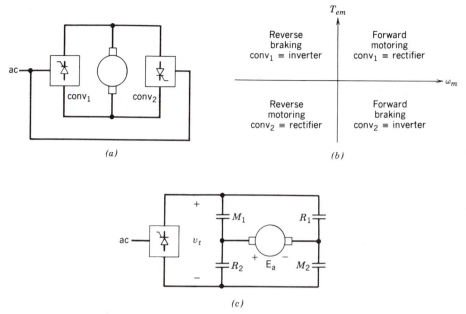

FIGURE 11-16: Line-frequency controlled converters for four-quadrant operation: (*a*) back-to-back converters for four-quadrant operation (without circulating current), (*b*) converter operation modes, (*c*) contactors for four-quadrant operation.

The current through these line-frequency controlled converters is unidirectional, but the output voltage can reverse polarity. The two-quadrant operation with the reversible voltage is not suited for dc motor braking, which requires the voltage to be unidirectional but the current to be reversible. Therefore, if regenerative braking is required, two back-to-back connected thyristor converters can be used, as shown in Fig. 11-16*a*. This, in fact, gives a capability to operate in all four-quadrants, as depicted in Fig. 11-16*b*.

An alternative to using two converters is to use one phase-controlled converter together with two pairs of contactors, as shown in Fig. 11-16*c*. When the machine is to be operated as a motor, the contactors M_1 and M_2 are closed. During braking where the motor speed is to be reduced rapidly, since the direction of rotation remains the same, E_a is of the same polarity as in the motoring mode. Therefore, to let the converter go into an inverter mode, contactors M_1 and M_2 are opened and R_f and R_2 are closed. It should be noted that the contactors switch at zero current when the current through them is brought to zero by the converter.

11-7-3 Effect of Discontinuous Armature Current

In line-frequency phase-controlled converters and single-quadrant step-down switch-mode dc–dc converters, the output current can become discontinuous at light loads on the motor. For a fixed control voltage $v_{control}$ or the delay angle α, the discontinuous current causes the output voltage to go up. This voltage rise causes the motor speed to increase at low values of I_a (which correspond to low torque load), as shown generically by Fig. 11-17. With a continuously flowing i_a, the drop in speed at higher torques is due to the voltage drop $R_a I_a$ across the armature resistance; additional drop in speed occurs in the phase-controlled-converter driven motors due to

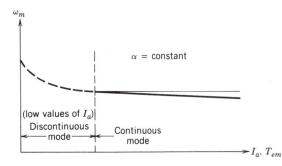

FIGURE 11-17: Effect of discontinuous i_a on ω_m.

commutation voltage drops across the ac-side inductance L_s, which approximately equal $(2\omega L_s/\pi)I_a$ in single-phase converters and $(3\omega L_s/\pi)I_a$ in three-phase converters, as discussed in Chapter 4. These effects results in poor speed regulation under an open-loop operation.

11-7-4 Control of Adjustable-Speed Drives

The type of control used depends on the drive requirements. An open-loop control is shown in Fig. 11-18 where the speed command ω^* is generated by comparing the drive output with its desired value (which, for example, may be temperature in case of a capacity-modulated heat pump). A d/dt limiter allows the speed command to change slowly, thus preventing the rotor current from exceeding its rating. The slope of the d/dt limiter can be adjusted to match the motor–load inertia. The current limiter in such drives may be just a protective measure, whereby if the measured current exceeds its rated value, the controller shuts the drive off. A manual restart may be required. A closed-loop control as discussed in Section 11-6 can also be implemented.

11-7-5 Field Weakening in Adjustable-Speed DC-Motor Drives

In a dc motor with a separately excited field winding, the drive can be operated at higher than the rated speed of the motor by reducing the field-flux ϕ_f. Since many adjustable-speed drives, especially at higher power ratings, employ a motor with a wound field, this capability can be exploited by controlling the field current and ϕ_f. A

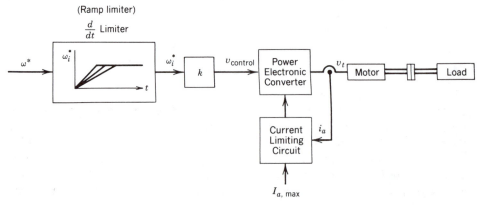

FIGURE 11-18: Open-loop speed control.

simple line-frequency phase-controlled converter as shown in Fig. 11-15 is normally used to control I_f through the field winding, where the current is controlled in magnitude but it always flows in only one direction. If a converter topology consisting of only thyristors (such as in Fig. 11-15) is chosen, where the converter output voltage is reversible, the field current can be decreased rapidly.

11-7-6 Power Factor of the Line Current in Adjustable-Speed Drives

The motor operation at its torque limit is shown in Fig. 11-19a in the constant torque region below the rated speed and in the field-weakening region above the rated speed. In a switch-mode drive, which consists of a diode rectifier bridge and a PWM

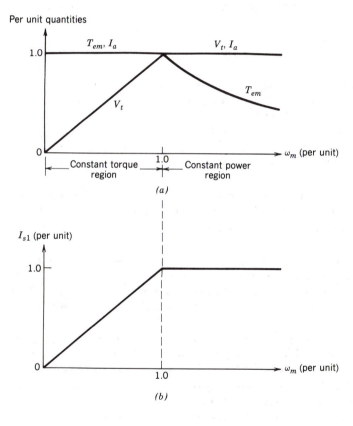

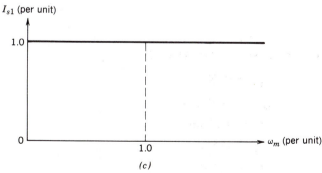

FIGURE 11-19: Line current in adjustable-speed dc drives: (a) drive capability, (b) switch-mode converter drive, (c) line-frequency thyristor converter drive.

dc–dc converter, the fundamental frequency component I_{s1} of the line current as a function of speed is shown in Fig. 11-19b. Figure 11-19c shows I_{s1} for a line-frequency phase-controlled thyristor drive. Assuming the load torque to be constant, I_{s1} decreases with decreasing speed in a switch-mode drive. Therefore, the switch-mode drive results in a good displacement power factor. On the other hand, in a phase-controlled thyristor drive, I_{s1} remains essentially constant as speed decreases, thus resulting in a very poor displacement power factor at low speeds.

As discussed in Chapters 3 and 4, both the diode rectifiers and the phase-controlled rectifiers draw line currents that consist of large harmonics in addition to the fundamental. These harmonics cause the power factor of operation to be poor in both types of drives. The circuits described in Chapter 17 can be used to remedy the harmonics problem in the switch-mode drives, thus resulting in a high power factor of operation.

11-8   SUMMARY

1. Because of mechanical contact between the commutator segments and brushes, dc motors require periodic maintenance. Because of arcing between these two surfaces, dc motors are not suitable for certain environments.

2. In a dc motor, the field flux is established by either a field winding supplied through a dc current or by permanent magnets located on the stator. The magnitude of the electromagnetic torque is directly proportional to the field-flux and the armature current magnitude. This makes a dc motor ideal for servo-drive applications.

3. The induced back-emf across the armature winding terminals is directly proportional to the field-flux magnitude and the rotational speed of the rotor.

4. A simple transfer function model can be obtained for a dc motor to obtain its dynamic performance.

5. Form factor of the armature current is defined as the ratio of its rms value to its average value. A poor armature current waveform with a high form factor results in excessive armature heating, arcing across commutator segments and brushes, and large torque pulsations. Therefore, an appropriate remedial action should be taken to avoid damage to the dc motor.

6. Dc-motor drives utilize either the line-frequency controlled converters or the dc–dc switch-mode converters. By field weakening in a wound field dc motor, the speed can be controlled beyond its rated value, without exceeding the rated armature voltage.

7. The power factor at which a dc-motor drive operates from the utility grid and the current harmonics injected into the utility grid depend on the type of the converter used: line-frequency controlled converter or a switch-mode dc–dc converter.

PROBLEMS

11-1. Consider a permanent-magnet dc servo motor with the following parameters:

$$T_{\text{rated}} = 10 \text{ N-m}$$

$$n_{\text{rated}} = 3700 \text{ rpm}$$

$$k_T = 0.5 \text{ N-m/A}$$

$$k_E = 53 \text{ V/krpm}$$

$$R_a = 0.37 \text{ }\Omega$$

$$\tau_e = 4.05 \text{ ms}$$

$$\tau_m = 11.7 \text{ ms}$$

Calculate the terminal voltage V_t in steady state if the motor is required to deliver a torque of 5 N-m at a speed of 1500 rpm.

11-2. $G_1(s) = [\omega_m(s)/V_t(s)]$ is the transfer function of an unloaded and uncontrolled dc motor. Express $G_1(s)$ given by Eq. 11-27 in the following form:

$$G_1(s) = \frac{1/k_E}{1 + 2sD/\omega_n + s^2/\omega_n^2}$$

Calculate D and ω_n for the servomotor parameters given in Problem 11-1. Plot the magnitude and the phase of $G_1(s)$ by means of a Bode plot.

11-3. Using the servomotor parameters given in Problem 11-1, calculate and plot the change in ω_m as a function of time for a step increase of 10 V in the terminal voltage of that uncontrolled, unloaded servomotor.

11-4. The servomotor of Problem 11-1 is driven by a full-bridge dc–dc converter operating from a 200-V dc bus. Calculate the peak-to-peak ripple in the motor current if a PWM bipolar voltage switching scheme is used. The motor is delivering a torque of 5 N-m at a speed of 1500 rpm. The switching frequency is 20 kHz.

11-5. Repeat Problem 11-4 if a unipolar-voltage-switching scheme is used.

11-6. In the servo drive of Problem 11-1, a PI regulator is used in the speed loop to obtain a transfer function of the following form in Fig. P11-6

$$F_\omega(s) = \frac{\omega(s)}{\omega^*(s)} = \frac{1}{1 + s\dfrac{2D}{\omega_n} + s^2/\omega_n^2}$$

where $D = 0.5$ and $\omega_n = 300$ rad/sec.
(a) Draw the Bode plot of the closed-loop transfer function $F_\theta(s) = [\theta(s)/\theta^*(s)]$ if a gain $k_p = 60$ $(s)^{-1}$ is used for the proportional position regulator in Fig. P11-6.
(b) What is bandwidth of the above closed loop system?

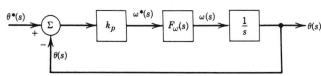

FIGURE P11-6:

11-7. Consider the servomotor of Problem 11-1 in a speed-control loop. If an internal current loop is not used, the block diagram is as shown in Fig. P11-7a, where only a proportional control is used. If an internal current loop is used, the block diagram without the current limits is as shown in Fig. P11-7b, where ω_n is 10 times that in part a.

Design the controllers (K_v, K_{vi}, K_i) to yield a control loop with slightly underdamped response $(D = 0.7)$. Compare the two control schemes in terms of bandwidth and transient performance, assuming that the current limit is not reached in either of them.

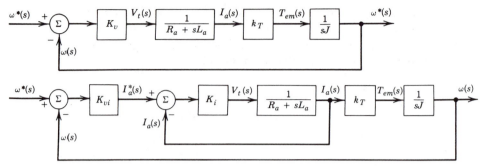

FIGURE P11-7:

11-9 REFERENCES

1. A. E. FITZGERALD, C. KINGSLEY, JR., and S. D. UMANS, *Electric Machinery*, 4th Ed. McGraw-Hill, New York, 1983.

2. P. C. SEN, *Thyristor DC Drives*, John Wiley & Sons, New York, 1981.

3. G. R. SLEMON and A. STRAUGHEN, *Electric Machines*, Addison-Wesley Publishing Co., Reading, MA, 1980.

4. T. KENJO and S. NAGAMORI, *Permanent-Magnet and Brushless DC Motors*, Clarendon Press, Oxford, 1985.

5. *DC Motors · Speed Controls · Servo System—An Engineering Handbook*, 5th Ed., Electro-Craft Corporation; 1600 Second Street South, Hopkins, MN, 1980.

12

Induction Motor Drives

12-1 INTRODUCTION

Induction motors with squirrel-cage rotors are the workhorse of industry because of their low cost and rugged construction. When operated directly from the line voltages (60 Hz utility input at essentially a constant voltage), an induction motor operates at a nearly constant speed. However, by means of power electronic converters, it is possible to vary the speed of an induction motor. The induction-motor drives can be classified into two broad categories based on their applications:

1. *Adjustable-Speed Drives.* One important application of these drives is in process control by controlling the speed of fans, compressors, pumps, blowers and the like.
2. *Servo Drives.* By means of sophisticated control, induction motors can be used as servo drives in computer peripherals, machine tools, and robotics.

The emphasis in this chapter is on understanding the behavior of induction motors and how it is possible to control their speed where the dynamics of speed control need not be very fast and precise. This is the case in most process control applications where induction motor drives are used. As a side benefit, use of induction-motor drives results in energy conservation, as discussed below.

Consider a simple example of an induction motor driving a centrifugal pump as shown in Fig. 12-1a, where the motor and the pump operate at a nearly constant speed. To reduce the flow rate, the throttling valve is partially closed. This causes loss of energy across the throttling valve. This energy loss can be avoided by eliminating the throttling valve and driving the pump at a speed which results in the desired flow rate, as in Fig. 12-1b.

In the system of Fig. 12-1b, the input power decreases significantly as the speed is decreased to reduce flow rate. This decrease in power requirement can be calculated by recognizing that in a centrifugal pump:

$$\text{Torque} \simeq k_1(\text{speed})^2 \qquad (12\text{-}1)$$

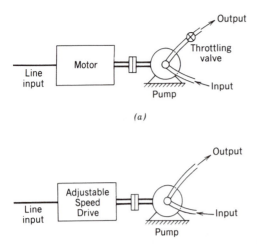

FIGURE 12-1: Centrifugal pump: (a) constant-speed drive, (b) adjustable-speed drive.

and, therefore, the power required by the pump from the motor is

$$\text{Power} \simeq k_2(\text{speed})^3 \tag{12-2}$$

where k_1 and k_2 are the constants of proportionality.

If the motor and the pump energy efficiencies can be assumed to be constant as their speed and loadings change, then the input power required by the induction motor would also vary as the $(\text{speed})^3$. Therefore, in comparison with a throttling valve to control the flow rate, the variable-speed driven pump can result in significant energy conservation, where reduced flow rates are required for long periods of time. Moreover, pump systems are usually designed to provide a flow margin of 20% to 30% over the maximum values of their actual flow. Therefore, an adjustable-speed pump can result in substantial energy conservation. This conclusion is valid only if it is possible to adjust the motor speed in an energy-efficient manner. As we will see in this chapter, energy efficiency associated with power electronic inverters (described in Chapter 6) used for controlling induction motor speeds is high over wide speed and load ranges.

12-2 BASIC PRINCIPLES OF INDUCTION MOTOR OPERATION

In a large majority of applications, induction-motor drives incorporate a three-phase, squirrel-cage motor. Therefore, the discussion here also assumes a three-phase, squirrel-cage induction motor. The stator of an induction motor consists of three phase windings distributed in the stator slots. These three windings are displaced by $120°$ in space, with respect to each other. The squirrel-cage rotor consists of a stack of insulated laminations. It has electrically conducting bars inserted through it, close to the periphery in the axial direction, which are electrically shorted at each end of the rotor by end rings, thus producing a cagelike structure. This also illustrates the simple, low-cost, and rugged nature of the rotor.

The objective of the following analysis is to explain as simply as possible the interaction between the induction motor and the power electronic converter. With this objective in mind, the details of proportionalities between various motor variables are

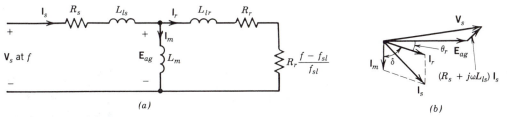

FIGURE 12-2: Per-phase representation: (*a*) equivalent circuit, (*b*) phasor diagram.

simply expressed as k_j (where the subscript j is assigned arbitrary numeric values). Moreover, the motor is assumed to operate without any magnetic saturation.

If a balanced set of three-phase sinusoidal voltages at a frequency $f = \omega/2\pi$ are applied to the stator, it results in a balanced set of currents, which establishes a flux-density distribution B_{ag} in the air gap with the following properties: (1) it has a constant amplitude and (2) it rotates with a constant speed, also called the synchronous speed, of ω_s radians per second. The synchronous speed in a p-pole motor, supplied by frequency f, can be obtained as

$$\omega_s = \frac{2\pi/(p/2)}{1/f} = \frac{2}{p}(2\pi f) = \frac{2}{p}\omega \qquad [\text{rad}/s] \tag{12-3}$$

which is synchronized to the frequency f of the applied voltages and currents to the stator windings. In terms of the revolutions per minute (rpm), the synchronous speed is

$$n_s = 60 \cdot \frac{\omega_s}{2\pi} = \frac{120}{p}f \tag{12-4}$$

The air-gap flux ϕ_{ag} (due to the flux-density distribution B_{ag}) rotates at a synchronous speed relative to the stationary stator windings. As a consequence, a counter-emf often called the air-gap voltage E_{ag} is induced in each of the stator phases at frequency f. This can be illustrated by means of a per-phase equivalent circuit shown in Fig. 12-2a, where $\mathbf{V}_s$ is the per-phase voltage (equal to the line-line rms voltage V_{LL} divided by $\sqrt{3}$) and $\mathbf{E}_{ag}$ is the air-gap voltage. R_s is the resistance of the stator winding and L_{ls} is the leakage inductance of the stator winding. The magnetizing component $\mathbf{I}_m$ of the stator current $\mathbf{I}_s$ establishes the air-gap flux. From the magnetic-circuit analysis, it can be seen that

$$N_s\phi_{ag} = L_m i_m \tag{12-5}$$

where N_s is an equivalent number of turns per-phase of the stator winding and L_m is the magnetizing inductance as shown in Fig. 12-2a.

From Faraday's law

$$e_{ag} = N_s\frac{d\phi_{ag}}{dt} \tag{12-6}$$

With the air-gap flux linking the stator phase winding to be $\phi_{ag}(t) = \phi_{ag}\sin\omega t$, Eq. 12-6 results in

$$e_{ag} = N_s\omega\phi_{ag}\cos\omega t \tag{12-7}$$

which has an rms value of

$$E_{ag} = k_3 f \phi_{ag} \qquad (12\text{-}8)$$

where k_3 is a constant.

The torque in an induction motor is produced by the interaction of air-gap flux and the rotor currents. If the rotor is rotating at the synchronous speed, there will be no relative motion between ϕ_{ag} and the rotor, and hence there will be no induced rotor voltages, rotor currents, and torque. At any other speed ω_r of the rotor in the same direction of the air-gap flux rotation, the rotor is "slipping" with respect to the air-gap flux at a relative speed called the slip speed ω_{sl}, where

$$\omega_{sl} = (\omega_s - \omega_r) \qquad (12\text{-}9)$$

This slip speed, normalized by the synchronous speed, is simply called the "slip" s

$$\underset{\text{(in per unit)}}{\text{Slip } s} = \frac{\text{Slip speed}}{\text{Synchronous speed}} = \frac{\omega_s - \omega_r}{\omega_s} \qquad (12\text{-}10)$$

Therefore, the speed of the air-gap flux with respect to the rotor equals

$$\text{Slip speed } \omega_{sl} = \omega_s - \omega_r = s\omega_s \qquad (12\text{-}11)$$

From Faraday's law, the induced voltages in the rotor circuit are at a slip frequency f_{sl}, which is proportional to the slip speed:

$$f_{sl} = \frac{\omega_{sl}}{\omega_s} \cdot f = sf \qquad (12\text{-}12)$$

The magnitude E_r of this slip-frequency voltage that is induced in any of the rotor conductors can be obtained in a similar manner as the induced voltages in the stator phases. The same air-gap flux ϕ_{ag} links the rotor conductors as the one that links the stator windings. However, the flux-density distribution in the air-gap rotates at a slip speed ω_{sl} with respect to the rotor conductors. Therefore, the induced emf E_r in the rotor conductors can be obtained by replacing f in Eq. 12-8 by the slip frequency f_{sl}. By assuming the squirrel-cage rotor to be represented by a three-phase short-circuited winding with the same equivalent number of turns N_s per phase as on the stator

$$E_r = k_3 f_{sl} \phi_{ag} \qquad (12\text{-}13)$$

where k_3 is the same as in Eq. 12-8.

Since the rotor squirrel-cage winding is short-circuited by the end-rings, these induced voltages at the slip frequency result in rotor currents I_r at the slip frequency f_{sl}:

$$\mathbf{E}_r = R_r \mathbf{I}_r + j2\pi f_{sl} L_{lr} \mathbf{I}_r \qquad (12\text{-}14)$$

where R_r and L_{lr} are the resistance and the leakage inductance of the per-phase equivalent rotor winding. The slip-frequency rotor currents produce a field that rotates at the slip speed with respect to the rotor and, hence, at the synchronous speed with respect to the stator (since $\omega_{sl} + \omega_r = \omega_s$). The interaction of ϕ_{ag} and the field produced by the rotor currents results in an electromagnetic torque. Losses in the rotor winding resistance are

$$P_r = 3R_r I_r^2 \tag{12-15}$$

Multiplying both sides of Eq. 12-14 by f/f_{sl} and using Eqs. 12-8 and 12-13

$$\mathbf{E}_{ag} = \frac{f}{f_{sl}}\mathbf{E}_r = f\frac{R_r}{f_{sl}}\mathbf{I}_r + j2\pi f L_{lr}\mathbf{I}_r \tag{12-16}$$

as shown in Fig. 12-2a, where fR_r/f_{sl} is represented as a sum of R_r and $R_r(f - f_{sl})/f_{sl}$. In Eq. 12-16, all rotor quantities are referred to N_s (the stator number of turns). By multiplying both sides of the Eq. 12-16 by $\mathbf{I}_R$, the power crossing the air gap, often called the air-gap power P_{ag}, can be obtained as

$$P_{ag} = 3\frac{f}{f_{sl}}R_r I_r^2 \tag{12-17}$$

From Eqs. 12-17 and 12-15, the electromechanical power P_{em} is

$$P_{em} = P_{ag} - P_r = 3R_r\frac{f - f_{sl}}{f_{sl}}I_r^2 \tag{12-18a}$$

and

$$T_{em} = P_{em}/\omega_r \tag{12-18b}$$

From Eqs. 12-9, 12-17, 12-18a, and 12-18b

$$T_{em} = P_{ag}/\omega_s \tag{12-18c}$$

In the equivalent circuit of Fig. 12-2a, the loss in the rotor resistance and the per-phase electromechanical power are shown by splitting the resistance $f(R_r/f_{sl})$ in Eq. 12-16 into R_r and $R_r(f - f_{sl})/f_{sl}$.

The total current $\mathbf{I}_s$ drawn by the stator is the sum of the magnetizing current $\mathbf{I}_m$ and the equivalent rotor current $\mathbf{I}_r$ ($\mathbf{I}_r$ here is the component of the stator current that cancels out the ampere-turns produced by the actual rotor current):

$$\mathbf{I}_s = \mathbf{I}_m + \mathbf{I}_r \tag{12-19}$$

The phasor diagram for the stator voltages and currents is shown in Fig. 12-2b. The magnetizing current $\mathbf{I}_m$, which produces ϕ_{ag}, lags the air-gap voltage by 90°. $\mathbf{I}_r$, which is responsible for producing the electromagnetic torque, lags $\mathbf{E}_{ag}$ by the power factor angle θ_r of the rotor circuit:

$$\theta_r = \tan\frac{2\pi f_{sl}L_{lr}}{R_r} = \tan\frac{2\pi f L_{lr}}{R_r f/f_{sl}} \tag{12-20}$$

From electromagnetic theory, the torque produced is

$$T_{em} = k_4 \phi_{ag} I_r \sin \delta \qquad (12\text{-}21)$$

where

$$\delta = 90° - \theta_r \qquad (12\text{-}22)$$

is the torque angle between the magnetizing current $\mathbf{I}_m$, which produces ϕ_{ag}, and $\mathbf{I}_r$, which represents the rotor field. The applied per-phase stator voltage $\mathbf{V}_s$ is

$$\mathbf{V}_s = \mathbf{E}_{ag} + (R_s + j2\pi f L_{ls})\mathbf{I}_s \qquad (12\text{-}23)$$

In induction motors of normal design, the following condition is true in the rotor circuit at low values of f_{sl} corresponding to normal operation:

$$2\pi f_{sl} L_{sl} \ll R_r \qquad (12\text{-}24)$$

Therefore, θ_r in Eq. 12-20 approximately equals zero and the torque angle δ in Eq. 12-22 equals 90°. Therefore, in Eq. 12-21

$$T_{em} \simeq k_4 \phi_{ag} I_r \qquad (12\text{-}25)$$

From Eqs. 12-13 and 12-14, using the approximation in Eq. 12-24

$$I_r \simeq k_5 \phi_{ag} f_{sl} \qquad (12\text{-}26)$$

Combining Eqs. 12-25 and 12-26

$$T_{em} \simeq k_6 \phi_{ag}^2 f_{sl} \qquad (12\text{-}27)$$

The approximation in Eq. 12-24 also allows the relationship in Eq. 12-19 to result in

$$I_s \simeq \sqrt{I_m^2 + I_r^2} \qquad (12\text{-}28)$$

For normal motor parameters, except at low values of operating frequency f as will be discussed later, in Eq. 12-23

$$V_s \approx E_{ag} \qquad (12\text{-}29)$$

Using Eq. 12-8 in Eq. 12-29

$$V_s \simeq k_3 \phi_{ag} f \qquad (12\text{-}30)$$

From Eqs. 12-15 and 12-18a, the ratio of the power loss in the rotor to the electromechanical output power P_{em} is

$$\%P_r = \frac{P_r}{P_{em}} = \frac{f_{sl}}{f - f_{sl}} \qquad (12\text{-}31)$$

TABLE 12-1
Important Relationships

$$\omega_s = k_7 f$$

$$s = \frac{\omega_s - \omega_r}{\omega_s}$$

$$f_{sl} = sf$$

$$\%P_r = \frac{f_{sl}}{f - f_{sl}}$$

$$V_s \simeq k_3 \phi_{ag} f$$

$$I_r \simeq k_5 \phi_{ag} f_{sl}$$

$$T_{em} \simeq k_6 \phi_{ag}^2 f_{sl}$$

$$I_m = k_8 \phi_{ag} \qquad \text{(from Eq. 12-5)}$$

$$I_s \simeq \sqrt{I_m^2 + I_r^2}$$

The important equations for a frequency-controlled induction motor are summarized in Table 12-1, some of which assume that the condition in Eq. 12-24 is valid. The following important observations can be drawn from these relationships:

1. The synchronous speed can be varied by varying the frequency f of the applied voltages.

2. Except at low values of f, the percentage power loss in the rotor resistance is small, provided f_{sl} is small. Therefore, in steady state, the slip frequency f_{sl} should not exceed its rated value (corresponding to the motor operation at the rated conditions listed on its name-plate).

3. With small f_{sl}, except at low values of f, the slip s is small and the motor speed varies approximately linearly with the frequency f of the applied voltages.

4. For the torque capability to equal the rated torque at any frequency, ϕ_{ag} should be kept constant and equal to its rated value. This requires that the V_s must vary proportionately to f (the voltage boost needed at low values of f is discussed later on).

5. Since I_r is proportional to f_{sl}, to restrict the motor current I_s from exceeding its rated value, the steady-state slip frequency f_{sl} should not exceed its rated value.

Based on the preceding observations, it can be concluded that the motor speed can be varied by controlling the applied frequency f, and the air-gap flux should be kept constant at its rated value by controlling the magnitude of the applied voltages in proportion to f. If an induction motor is controlled in such a way, then the motor is capable of supplying its rated torque while f_{sl}, I_r, I_s, and the percentage losses in the rotor circuit all remain within their respective rated values.

12-3 INDUCTION MOTOR CHARACTERISTICS AT RATED (LINE) FREQUENCY AND RATED VOLTAGE

Typical characteristics of an induction motor under nameplate values of frequency and voltage are shown in Figs. 12-3 and 12-4 where T_{em} and I_r, respectively, are plotted as functions of rotor speed and f_{sl}. At low values of f_{sl}, T_{em} and I_r vary

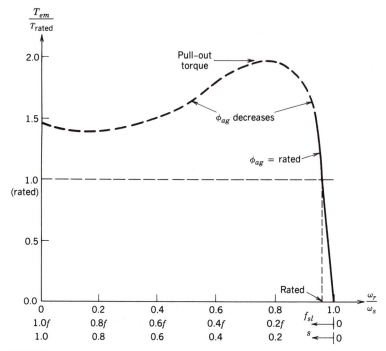

FIGURE 12-3: A typical torque-speed characteristic; V_s and f are constant at their rated values.

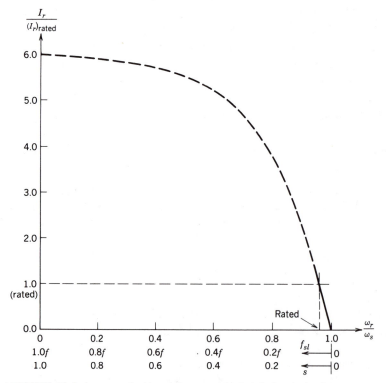

FIGURE 12-4: I_r versus f_{sl}; V_s and f are constant at their rated values.

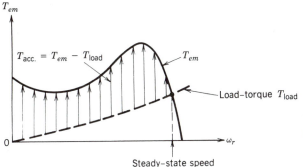

FIGURE 12-5: Motor start-up; V_s and f are constant at their rated values.

linearly with f_{sl}. As f_{sl} becomes larger, T_{em} and I_r no longer increase linearly with f_{sl} for the following reasons: (1) rotor-circuit inductive reactance term is no longer negligible compared to R_r in Eq. 12-14, (2) θ_r in Eq. 12-20 becomes significant, thus causing δ to depart from its optimum value of $90°$ and, (3) large values of $\mathbf{I}_r$, and hence $\mathbf{I}_s$ cause significant voltage drop across the stator winding impedance in Eq. 12-23 and hence cause ϕ_{ag} ($= E_{ag}/f$) to decline for a fixed supply input V_s at frequency f. All of these effects take place simultaneously, and the resulting torque and current characteristics for large f_{sl} are shown as dotted in Figs. 12-3 and 12-4. The maximum torque that the motor can produce is called the pull-out torque.

It should be emphasized that in the commonly used induction-motor drives, which are discussed in detail in this chapter, f_{sl} is kept small and, hence, the dotted portions of the torque and the current characteristics of Figs. 12-3 and 12-4 are not used. However, if an induction motor is started from the line-voltage supply without a power electronic controller, it would draw 6 to 8 times its rated current at start-up as shown in Fig. 12-4. Figure 12-5 shows the available acceleration torque ($T_{em} - T_{\text{load}}$) for the motor to accelerate from standstill. Here, an arbitrary torque–speed characteristic of the load is assumed and the intersection of the motor and the load characteristics determines the steady-state point of operation.

12-4 SPEED CONTROL BY VARYING STATOR FREQUENCY AND VOLTAGE

The discussion in Section 12-2 suggested that the speed can be controlled by varying f, which controls the synchronous speed (and, hence, the motor speed, if the slip is kept small), keeping ϕ_{ag} constant by varying V_s in a linear proportion to f. We will examine other speed control techniques later on, but varying the stator frequency and voltage is the preferred technique in most variable-speed induction motor drive applications and, hence, we will discuss it in detail.

12-4-1 Torque-Speed Characteristics

From the relationships in Table 12-1 for small values of f_{sl}, keeping ϕ_{ag} constant results in a linear relationship between T_{em} and f_{sl} at any value of f:

$$T_{em} \simeq k_9 f_{sl} \qquad (12\text{-}32)$$

which represents the solid portion of the torque-speed characteristic in Fig. 12-3. Since

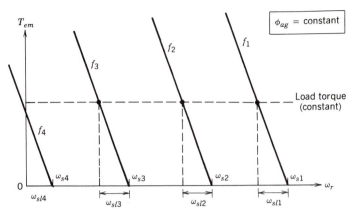

FIGURE 12-6: Torque-speed characteristics at small slip with a constant ϕ_{ag}; constant load torque.

f is varied, it is preferable to express T_{em} as a function of the slip speed ω_{sl}. From Eqs. 12-3 and 12-12

$$\omega_{sl} = \frac{f_{sl}}{f}\omega_s = \frac{4\pi}{p}f_{sl} \tag{12-33}$$

From Eqs. 12-32 and 12-33

$$T_{em} \simeq k_{10}\omega_{sl} \tag{12-34}$$

Such a characteristic is shown in Fig. 12-6 for frequency f equal to f_1 with a corresponding synchronous speed ω_{s1}.

The torque-speed characteristics shift horizontally in parallel, as shown in Fig. 12-6 for four different values of f. To explain this, consider two frequencies f_1 and f_2. The synchronous speeds ω_{s1} and ω_{s2} are in proportion to f_1 and f_2. If an equal load-torque is to be delivered at both these frequencies, from Eq. 12-34, $\omega_{sl1} = \omega_{sl2}$. Therefore, in the torque-speed plane of Fig. 12-6, equal torques and equal slip speeds (at f_1 and f_2) result in parallel but horizontally shifted characteristics.

Note that at a constant load-torque, the slip frequency (which is the frequency of the induced voltages and currents in the rotor circuit in hertz) is constant, but from Eq. 12-12 the slip s goes up as frequency f goes down. From Eq. 12-31, the percentage

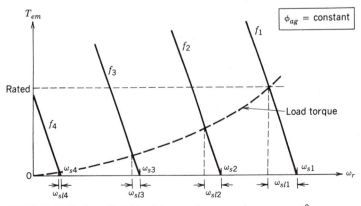

FIGURE 12-7: Centrifugal load torque; torque varies as speed^2.

power loss in the rotor increases as f is decreased to reduce the motor speed. However, in many loads such as the centrifugal pumps, compressors, and fans, the load-torque varies by the square of the speed, as given by Eq. 12-1. In such cases, f_{sl} as well as s declines with decreasing frequency, as shown in Fig. 12-7. Hence, the rotor losses remain small.

EXAMPLE 12-1

A four-pole, 10-hp, 460-V motor is supplying its rated power to a centrifugal load at a 60-Hz frequency. Its rated speed is 1746 rpm.

   Calculate its speed, slip frequency, and slip when it is supplied by a 230-V, 30-Hz source.

   Solution

   At 60 Hz:

$$n_s = 1800 \text{ rpm (four-pole)}$$

$$s_{\text{rated}} = \frac{1800 - 1746}{1800} = 3\%$$

$$(f_{sl})_{\text{rated}} = s_{\text{rated}} \cdot f = 0.03 \times 60 = 1.8 \text{ Hz}$$

$$(n_{sl})_{\text{rated}} = 1800 - 1746 = 54 \text{ rpm}$$

   At 30 Hz, keeping (V_s/f) constant:

$$T_{em} \simeq \frac{T_{\text{rated}}}{4} \text{ (centrifugal load; using Eq. 12-1)}$$

$$f_{sl} = \frac{(f_{sl})_{\text{rated}}}{4} = \frac{1.8}{4} = 0.45 \text{ Hz} \qquad \text{(using Eq. 12-32)}$$

$$n_{sl} = \frac{120}{\text{Poles}} f_{sl} = \frac{120}{4} \times 0.45 = 13.5 \text{ rpm}$$

$$n_s = 900 \text{ rpm}$$

$$\therefore n_r = n_s - n_{sl} = 900 - 13.5 = 886.5 \text{ rpm}$$

$$s = \frac{f_{sl}}{f} = \frac{0.45}{30} = 1.5\%$$ ∎

12-4-2 Start-up Considerations

For a solid-state inverter-driven induction motor, it is an important considera-tion to keep the current draw from becoming large during start-up. This can be achieved by considering the following relationship: for a constant ϕ_{ag} from Eq. 12-26

$$I_r \simeq k_{11} f_{sl} \tag{12-35}$$

Using Eqs. 12-32 and 12-35, T_{em} and I_r are plotted in Fig. 12-8 to show how the motor can be started at a small applied frequency f $(= f_{\text{start}})$. Since at start-up f_{sl} equals f_{start}, I_r can be limited by selecting an appropriate f_{start}. With a constant I_m due to a constant ϕ_{ag}, the stator current I_s is therefore kept from becoming large.

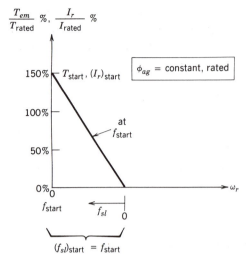

FIGURE 12-8: Frequency at start-up.

For example, if the starting torque is required to be 150% of the rated torque and the solid-state drive can withstand a current overload of 150% on a short-term basis, the starting frequency f_{start} can be determined from the motor nameplate ratings. For the motor in Example 12-1, the starting frequency for 150% torque (and, hence, current) based on the rated speed of 1746 rpm at 60 Hz is calculated by using Fig. 12-8 as

$$f_{start} = \frac{T_{start}}{T_{rated}} \left(f_{sl} \right)_{rated} \tag{12-36}$$

$$= 1.5 \times 1.8 = 2.7 \text{ Hz}$$

In practice, the stator frequency f is increased continuously at a preset rate as shown in Fig. 12-9, which does not let the current I_s exceed a specified limit (like 150% of the rated) until the final desired speed has been achieved. This rate is decreased for higher inertia loads to allow the rotor speed to catch up.

12-4-3   Voltage Boost Required at Low Frequencies

The effect of R_s at low values of operating frequency f cannot be neglected, even if f_{sl} is small. This can be easily seen if the following observation is made: in induction motors of normal design, $(2\pi f L_{lr})$ is negligible in comparison to $[R_r(f/f_{sl})]$ in the equivalent circuit of Fig. 12-2a. Therefore, $\mathbf{I}_r$ will be in phase with $\mathbf{E}_{ag}$. Using

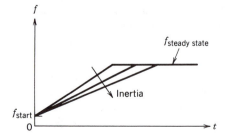

FIGURE 12-9: Ramping of frequency f at start-up.

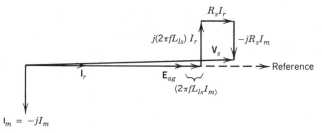

FIGURE 12-10: Phasor diagram at a small value of f_{sl}.

$\mathbf{E}_{ag}$ as the reference phasor, $\mathbf{I}_s = I_r - jI_m$. Therefore, Eq. 12-23 can be written as

$$\mathbf{V}_s \simeq \left[E_{ag} + (2\pi f L_{ls})I_m + R_s I_r \right] + j\left[(2\pi f L_{ls})I_r - R_s I_m \right] \qquad (12\text{-}37)$$

and represented by the phasor diagram of Fig. 12-10. As shown by Fig. 12-10, the second term in the right-hand side of the Eq. 12-37 corresponds to a phasor that is almost perpendicular to $\mathbf{V}_s$ and, therefore, its influence on the magnitude of V_s can be neglected

$$V_s \approx E_{ag} + (2\pi f L_{ls})I_m + R_s I_r \qquad (12\text{-}38a)$$

If ϕ_{ag} is kept constant, E_{ag} varies linearly with f. If ϕ_{ag} is kept constant, I_m is also constant. Therefore, the additional voltage required due to L_{ls} in Eq. 12-38a is also proportional to the operating frequency f. Therefore, for a constant ϕ_{ag}, Eq. 12-38a can be written as

$$V_s \approx k_{12} f + R_s I_r \qquad (12\text{-}38b)$$

Equation 12-38b shows that the additional voltage required to compensate for the voltage drop across R_s to keep ϕ_{ag} constant does not depend on f but depends on I_r. Recognizing that I_r is proportional to T_{em}, the terminal voltage V_s required to keep ϕ_{ag} constant at the rated torque is shown by a solid line in Fig. 12-11. A voltage proportional to f, with the rated voltage at the rated frequency, is indicated by a dotted line in Fig. 12-11. The voltage boost required to maintain a constant air-gap

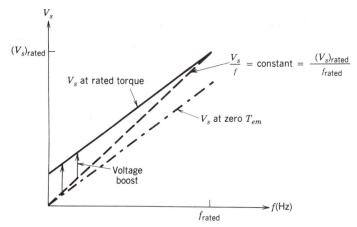

FIGURE 12-11: Voltage boost required to keep ϕ_{ag} constant.

flux for a given T_{em} can be obtained from Eq. 12-38b and Fig. 12-11. Figure 12-11 shows that to keep ϕ_{ag} constant, a much higher percentage voltage boost is required at low operating frequencies due to the voltage drop across R_s, whereas at large values of f, the voltage drop across R_s can be neglected in comparison to E_{ag}. The voltage required at no-load is shown by a dashed line.

12-4-4   Induction Motor Capability — Below and Above the Rated Speed

Speed control by means of frequency (and voltage) variation also allows the capability to operate the motor not only at speeds below the rated speed, but also at above the rated speed. This capability is very attractive in many applications, since most induction motors, because of their rugged construction, can be operated up to twice the rated speed without mechanical problems. However, the torque and power capabilities as a function of rotor speed need to be clearly established.

The motor torque-speed characteristics are shown in Fig. 12-12a. V_s, I_r, I_m, and T_{em} as functions of the normalized rotor speed are plotted in Fig. 12-12b. f_{sl} and s are plotted in Fig. 12-12c. It should be noted that in large motors at the limit of motor capability, $I_s \approx I_r$ since the contribution of I_m to I_s in Eq. 12-28 is small.

12-4-4-1   BELOW THE RATED SPEED — CONSTANT TORQUE REGION

In the region of speed below its rated value, the solid curves in Fig. 12-12a show the motor torque-speed characteristics at low values of f_{sl} where ϕ_{ag} is kept constant by controlling V_s/f. The stator voltage magnitude is decreased approximately in proportion to the frequency from its rated value down to very low values, as shown in Fig. 12-12b. If ϕ_{ag} is maintained constant, the motor can deliver its rated torque (on a continuous basis) by drawing its rated current at a constant f_{sl} as shown in Fig. 12-12. Therefore, this region (below the rated speed) is called the constant torque region.

In this region, f_{sl} remains constant at its full-load (rated) value while delivering the rated torque. f_{sl} and s are shown in Fig. 12-12c.

At the constant rated torque, the power loss $P_r = 3R_r I_r^2$ in the rotor resistances is also constant, where I_r stays constant. However, in practice, getting rid of this rotor heat due to P_r becomes a problem at low speeds due to reduced cooling. Therefore, unless the motor has a constant-speed fan or is designed to be totally enclosed and nonventilated, the torque capability drops off at very low speeds. It should be noted that this is of no concern in centrifugal loads where the torque requirement is very low at low speeds.

12-4-4-2   BEYOND THE RATED SPEED — CONSTANT POWER REGION

By increasing the stator frequency above its nominal (rated) value, it is possible to increase the motor speed beyond the rated speed. In most adjustable-speed drive applications, the motor voltage is not exceeded beyond its rated value, unlike that explained in Section 12-4-4-4. Therefore, by keeping V_s at its rated value, increasing the frequency f results in a reduced V_s/f and, hence, a reduced ϕ_{ag}. From Eqs. 12-27, 12-30, and 12-33 in this region

$$T_{em} \simeq \frac{k_{13}}{f^2} \omega_{sl} \tag{12-39}$$

which results in torque–speed curves, whose slopes are proportional to $(1/f)^2$, as shown in Fig. 12-12a for higher than rated frequencies.

At the limit of the motor capability in this region, I_r equals its rated value, similar to the previous region. This corresponds to a constant $s = f_{sl}/f$ in this region,

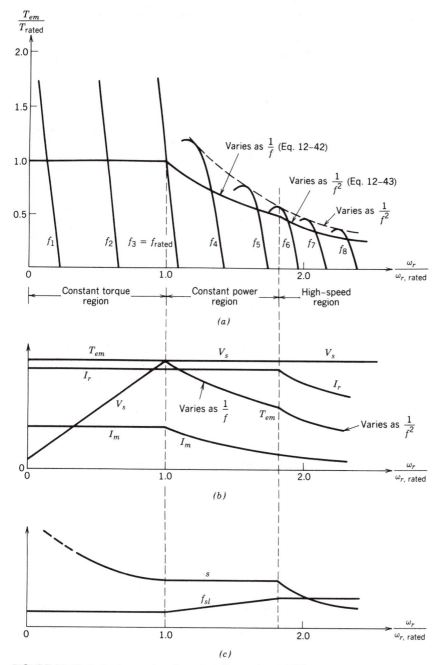

FIGURE 12-12: Induction motor characteristics and capabilities.

which can be shown by using Eqs. 12-12, 12-26 and 12-30

$$I_r \simeq k_{14}\frac{f_{sl}}{f} \simeq k_{14}s = \text{constant} \tag{12-40}$$

The slip frequency f_{sl} now increases with f, as shown in Fig. 12-12c. At a constant slip

$$\omega_r = (1 - s)\omega_s = k_{15}f \tag{12-41}$$

Using both V_s and f_{sl}/f as constants, the maximum torque in this region can be calculated from Eqs. 12-27 and 12-30 in terms of the rated torque and the rated frequency:

$$T_{em,max} = \frac{f_{rated}}{f} \cdot T_{rated} \qquad (12\text{-}42)$$

Therefore $P_{em,max} = \omega_r T_{em,max}$ can be held constant at its rated value, recognizing from Eq. 12-41 that ω_r is proportional to f. Hence, this region of operation is called the constant power region. V_s, I_r, I_m, and the maximum steady-state T_{em} are plotted in Fig. 12-12b.

In practice, the motor can deliver higher than its rated power by noting that (1) I_m goes down as a result of decreased ϕ_{ag} and, therefore, I_s equal to its rated value allows a higher value of I_r and, hence, higher torque and power, and (2) since I_m is decreased, the core losses are reduced and, at the same time, there is better cooling at higher speeds.

12-4-4-3   HIGH-SPEED OPERATION—CONSTANT f_{sl} REGION

With V_s equal to its rated value, depending on the motor design, beyond a speed somewhere in a range of 1.5 to 2 times the rated speed, ϕ_{ag} is reduced so much that the motor approaches its pull-out torque as is shown in Fig. 12-12a. At still higher speeds, the motor can deliver only a fixed percentage of the pull-out torques, as shown graphically by Fig. 12-12a, and $\omega_{sl}(f_{sl})$ becomes constant. Therefore, the torque capability declines as

$$T_{em,max} \simeq k_{16}\frac{1}{f^2} \qquad (12\text{-}43)$$

Both the torque and the motor current decline with speed, as is shown in Fig. 12-12b. Keeping V_s constant, the motor torque in this region is not limited by the current handling capability of the motor, since the current at the limit is less than its rated value and declines with speed, as is shown in Fig. 12-12b; rather, it is limited by the maximum torque produced by the motor.

12-4-4-4   HIGHER VOLTAGE OPERATION

In most motors, the voltage insulation level is much higher than the specified rated voltage of the motor. Therefore, by means of a proper solid-state power source, it is possible to apply a higher than rated voltage at speeds above the rated speed of the motor. This is particularly easy to see in the case of a dual voltage motor, for example, a 230/460-V motor. With the motor connected for a 230-V operation, if 460 V are applied at twice the rated frequency, the motor operates at the rated air-gap flux and, hence, can deliver its rated torque without exceeding its current rating. Since the motor under these conditions will run at twice its rated speed, it will also deliver twice its rated power. Before using a motor in this manner, the motor manufacturer ought to be consulted.

12-4-5   Braking in Induction Motors

In many applications, it is repeatedly required to quickly reduce the motor speed or to bring it to a halt. One of the advantages of using a variable-frequency controller for speed control is that it can accomplish this in a controlled manner. This provides a controlled alternative to the following ways of speed reduction, which are

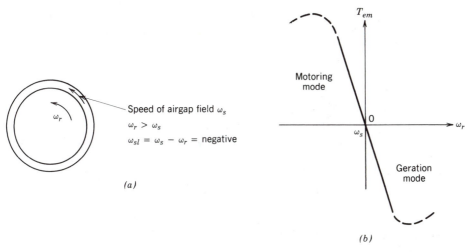

FIGURE 12-13: Generation mode.

not as attractive: mechanical brakes, which waste energy associated with load–motor inertia and whose brake pads wear out with repeated use; letting the motor coast to a halt, which could take a long time; "plugging" where the phase sequence of the utility supply to the motor is suddenly reversed, causing large currents to flow into the utility source and bringing the motor to a halt in an uncontrolled manner.

To understand braking in induction motors, it should be realized that it is possible to operate an induction machine as a generator by mechanically driving it above the synchronous speed (which is related to the supply-voltage frequency). As shown in Fig. 12-13a, a rotor speed higher than ω_s results in a negative slip speed ω_{sl} and a negative slip s. The torque-speed characteristic corresponding to $\omega_r > \omega_s$ is shown in Fig. 12-13b, where the electromagnetic torque developed in this mode is negative and acts in an opposite direction to the direction of rotating magnetic field.

Note that for the induction machine to operate in a generator mode, the ac voltages must be present at the stator terminals, that is, the machine will not generate, for example, if only a resistor bank is connected to the stator terminals and the shaft is turned; there is no source to establish the rotating magnetic field in the air gap.

The generation mechanism discussed before is used to provide braking in variable-frequency induction motor drives. Figure 12-14 shows the motor torque-speed

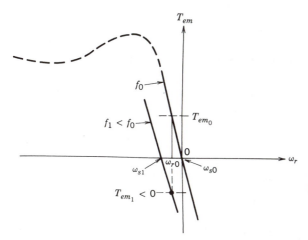

FIGURE 12-14: Braking (initial motor speed is ω_{r0} and the applied frequency is instantaneously decreased from f_0 to f_1).

characteristics at two frequencies, assuming a constant ϕ_{ag}. These curves are extended beyond the corresponding synchronous speeds. Consider that the motor is initially operating with a stator frequency f_0 at a rotor speed of ω_{r0} below ω_{s0}. If the stator frequency is decreased to f_1, the new synchronous speed is ω_{s1}. The slip speed becomes negative and thus T_{em} becomes negative, as is shown in Fig. 12-14. This negative T_{em} causes the motor speed to decrease quickly and some of the energy associated with the motor–load inertia is fed into the source connected to the stator.

In practice, the stator frequency (keeping ϕ_{ag} constant) is reduced slowly to avoid large currents through the variable frequency controller. This procedure, if used to bring the motor to a halt, can be viewed as opposite of the start-up procedure. It should be noted that the variable-frequency controller must be capable of handling the energy supplied by the motor in the braking mode.

12-5 IMPACT OF NONSINUSOIDAL EXCITATION ON INDUCTION MOTORS

In the preceding section, it was assumed that the induction motors are supplied from a three-phase, balanced, and sinusoidal set of voltages. In practice, the inverters used in variable-frequency controllers produce three phase voltages or currents that are identical in each phase, except for the 120° phase displacement. Unfortunately, these are not purely sinusoidal and contain higher frequency components that are harmonics of the fundamental frequency, as discussed in Chapter 6. In the following analysis, we will assume that the motor is supplied by three phase voltage sources as in the case of a voltage-source inverter. This analysis can be easily modified to three phase current sources as in the case of a current-source inverter.

12-5-1 Harmonic Motor Currents

As a first-order approximation, the motor currents in the presence of harmonic voltage components can be found by calculating each harmonic current component i_h (at harmonic h) from the per-phase equivalent circuit of Fig. 12-2a. Then the motor currents can be obtained by using the principle of superposition and adding the fundamental and all other harmonic current components.

At a harmonic h (which, in practice, will be odd and not a multiple of 3), the flux produced by the voltage components (v_{ah}, v_{bh}, v_{ch}) rotates in the air gap at a speed of

$$\omega_{sh} = h\omega_s \qquad (12\text{-}44)$$

where the direction may be the same or in opposition to the rotor's direction of rotation. It can be easily verified that the flux produced by the harmonics $h = 6n - 1$ (where, $n = 1, 2, 3, \dots$) have an opposite phase rotation compared with the fundamental. Therefore, these harmonics result in a flux rotation opposite to that of the rotor. Harmonics $h = 6n + 1$ (where $n = 1, 2, 3, \dots$) produce a flux rotation in the same direction as the rotor.

Under a variable-frequency operation for speed control, the motor rotates at a reasonably small value of slip; thus, to a first-order approximation, the rotor speed can be assumed to be the same as the fundamental frequency, synchronous speed:

$$\omega_r \simeq \omega_s \qquad (12\text{-}45)$$

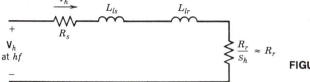

FIGURE 12-15: Per-phase harmonic equivalent circuit.

Therefore, at a harmonic h in the equivalent circuit of Fig. 12-2a, using Eqs. 12-44 and 12-45, the rotor slip relative to the synchronous speed at the harmonic frequency is

$$\text{Slip } s_h = \frac{\omega_{sh} \pm \omega_r}{\omega_{sh}} \simeq \frac{h \pm 1}{h} \approx 1 \qquad (12\text{-}46)$$

where $+$ or $-$ sign corresponds to the direction of air-gap flux rotation in opposition to or the same as the rotor's direction of rotation, respectively. Recognizing that $\omega_r \approx \omega_s$ and $s_h \simeq 1$, an approximate equivalent circuit at the frequency of harmonic h is shown in Fig. 12-15, which is obtained by the equivalent circuit of Fig. 12-2a by neglecting L_m. If the motor is excited from a voltage source and the harmonic components of the stator voltage v_s are known, the corresponding harmonic components in the motor current i_s can be obtained by using the principle of superposition and the harmonic equivalent circuit of Fig. 12-15 for each harmonic, one at a time.

For calculating the harmonic current components, the magnetizing components can generally be neglected and the harmonic current magnitude is primarily determined by the leakage reactances at the harmonic frequency, which dominate over R_s and R_r:

$$I_h \approx \frac{V_h}{h\omega(L_{ls} + L_{lr})} \qquad (12\text{-}47)$$

Equation 12-47 for I_h shows that by increasing the frequencies at which the harmonic voltages occur in the converter output (which is accomplished by increasing the switching frequency, as discussed in Chapter 6), the magnitudes of harmonic currents can be reduced. Note that the foregoing procedure for calculating harmonic currents is at best a first-order approximation, since the motor leakage reactances and resistances vary with frequency.

12-5-2 Harmonic Losses

The per-phase additional power loss in the copper of stator and rotor windings due to these harmonic currents can be approximated as

$$\Delta P_{\text{cu}} = \sum_{h=2}^{\infty} (R_s + R_r)I_h^2 \qquad (12\text{-}48)$$

where R_s and R_r increase in a nonlinear manner with harmonic frequencies. It is tedious to estimate the additional core losses due to harmonic frequency eddy currents and hysteresis. These and the additional stray losses depend on the motor geometry, magnetic material used, lamination thickness, which may have been optimized for the 60-Hz frequency, and so on. These additional losses, which may be significant, can be measured, and the estimation procedures have been discussed in the literature. In general, these additional losses are in a range of 10% to 20% of the total power losses at the rated load.

12-5-3   Torque Pulsations

Presence of harmonics in the stator excitation results in a pulsating torque component. If the pulsating torques are at low frequencies, they can cause troublesome speed fluctuations and shaft fatigue.

Considering the lowest harmonic frequencies, which are fifth and seventh in a three-phase square-wave inverter, the generation of pulsating torque component can best be explained by considering these harmonic excitations one at a time.

In Fig. 12-16a, the seventh harmonic excitation results in an air-gap flux component rotating at a speed $7 \omega_s$, in the same direction as the fundamental air-gap flux and the rotor. Assuming the rotor speed to be approximately equal to ω_s, it is easy to see that the rotor field produced consists of the fundamental component B_{r1} at a speed of ω_s and the seventh harmonic component B_{r7} at a speed of $7 \omega_s$, as is shown in Fig. 12-16a. Fields ϕ_{ag1} and B_{r1} rotate at the same speed and, hence, result in a nonpulsating torque. The same is true for the interaction of ϕ_{ag7} and B_{r7}, which rotate at the same speed. However, the relative speed between ϕ_{ag7} and B_{r1} is $6 \omega_s$. Similarly, the relative speed between ϕ_{ag1} and B_{r7} is $6 \omega_s$. Therefore, both these interactions produce torque components that pulsate at a sixth harmonic frequency.

In Fig. 12-16b, the fifth harmonic excitation results in an air-gap flux that rotates at a speed of $5 \omega_s$ in a direction opposite to the rotor. The induced rotor fields are shown in Fig. 12-16b. ϕ_{ag5} interacts with B_{r1} and, ϕ_{ag1} interacts with B_{r5} to produce torque components, both of which pulsate at a sixth harmonic frequency.

The above discussion shows that both fifth and seventh harmonic excitations combine to produce a torque that pulsates at the sixth harmonic frequency. Similar calculations can be made for other harmonic frequency excitations.

The effect of torque ripple on the ripple in the rotor speed can be written as follows, assuming no resonance occurs:

$$\text{Amplitude of speed ripple} = k_{17} \frac{\text{Amplitude of torque ripple}}{\text{Ripple frequency} \cdot \text{Inertia}} \qquad (12\text{-}49)$$

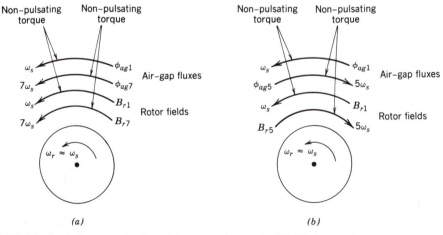

FIGURE 12-16: Torque pulsations: (a) seventh harmonic, (b) fifth harmonic.

which shows that a given amplitude of torque ripple may result in negligible speed ripple at high ripple frequencies.

12-6 VARIABLE-FREQUENCY CONVERTER CLASSIFICATIONS

Based on the discussion in the previous section, the variable-frequency converters, which act as an interface between the utility power system and the induction motor, must satisfy the following basic requirements:

1. Ability to adjust the frequency according to the desired output speed.
2. Ability to adjust the output voltage so as to maintain a constant air-gap flux in the constant torque region.
3. Ability to supply a rated current on a continuous basis at any frequency.

Except for a few special cases of very-high-power applications where cycloconverters are used (these are briefly discussed in Chapter 13), variable-frequency drives employ inverters with a dc input as discussed in Chapter 6. Figure 12-17 illustrates the basic concept where the utility input is converted into dc by means of either a controlled or an uncontrolled rectifier and then inverted to provide three phase voltages and currents to the motor, adjustable in magnitude and frequency. These converters can be classified based on the type of rectifier and inverter used in Fig. 12-17:

1. Pulse-width-modulated voltage-source inverter (PWM-VSI) with a diode rectifier.
2. Square-wave voltage-source inverter (square-wave-VSI) with a thyristor rectifier.
3. Current-source inverter (CSI) with a thyristor rectifier.

As the names imply, the basic difference between the voltage-source inverters (VSI) and the current-source inverters (CSI) is the following: In VSI, the dc input appears as a dc voltage source (ideally with no internal impedance) to the inverter. On the other hand, in the CSI, the dc input appears as a dc current source (ideally with the internal impedance approaching infinity) to the inverter.

Figure 12-18a shows the schematic of a PWM-VSI with a diode rectifier. In the square-wave-VSI of Fig. 12-18b, a controlled rectifier is used at the front end and the inverter operates in a square-wave mode (also called the six-step). The line voltage may be single-phase or three-phase. In both VSI controllers, a large dc-bus capacitor is used to make the input to the inverter appear as a voltage source with a very small internal impedance at the inverter switching frequency.

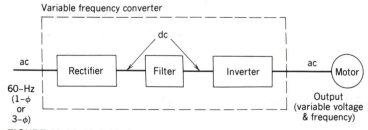

FIGURE 12-17: Variable-frequency converter.

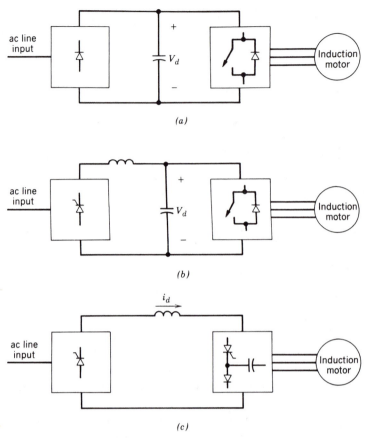

FIGURE 12-18: Classification of variable-frequency converters: (a) PWM-VSI with a diode rectifier, (b) square-wave VSI with a controlled rectifier, (c) CSI with a controlled rectifier.

From the schematic of VSI converters shown in Figs. 12-18a and 12-18b, it is recognized that the switch-mode, dc-to-ac voltage source inverters have been discussed previously in Chapter 6 in both square-wave and PWM modes of operation. It should be noted that in practice, only three-phase motors are controlled by means of variable frequency. Therefore, only the dc to three-phase ac inverters are applicable here. Also, the controlled and uncontrolled (diode) rectification of single-phase and three-phase ac inputs to dc has been discussed in detail in Chapters 3 and 4. Therefore, the main emphasis in this chapter will be on the interaction of voltage-source inverters with induction motor type of loads.

Figure 12-18c shows the schematic of a current-source inverter (CSI) drive where a line-voltage-commutated controlled converter (discussed in Chapter 4) is used at the front end. Because of a large inductor in the dc link, the input to the inverter appears as a dc current source. The inverter utilizes thyristors, diodes, and capacitors for forced commutation.

12-7 VARIABLE-FREQUENCY PWM-VSI DRIVES

Figure 12-19a shows the schematic of a pulse-width-modulated PWM-VSI drive, assuming a three-phase utility input. As a brief review of what has already been

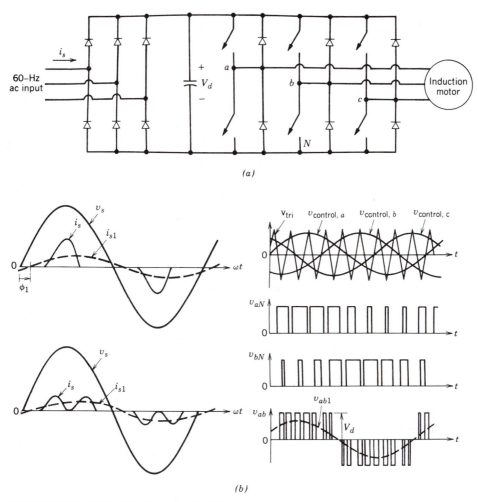

FIGURE 12-19: PWM-VSI: (a) schematic, (b) waveforms.

covered in Chapter 6, a PWM inverter controls both the frequency and the magnitude of the voltage output. Therefore, at the input, an uncontrolled diode bridge rectifier is generally used. One possible method of generating the inverter switch control signals is by comparing three sinusoidal control voltages (at the desired output frequency and proportional to the output voltage magnitude) with a triangular waveform at a selected switching frequency, as shown in Fig 12-19b.

As discussed in Chapter 6, in a PWM inverter, the harmonics in the output voltage appear as sidebands of the switching frequency and its multiples. Therefore, a high switching frequency results in an essentially sinusoidal current (plus a superimposed small ripple at a high frequency) in the motor.

Since the ripple current through the dc-bus capacitor is at the switching frequency, the "input dc source" impedance seen by the inverter would be smaller at higher switching frequencies. Therefore, a small value of capacitance suffices in PWM inverters, but this capacitor must be able to carry the ripple current. A small capacitance across the diode rectifier also results in a better input current waveform drawn from the utility source. However, care should be taken in not letting the voltage ripple in the dc-bus voltage become too large, which would cause additional harmonics in the voltage applied to the motor.

12-7-1   Impact of PWM-VSI Harmonics

In a PWM inverter output voltage, since the harmonics are at a high frequency, the ripple in the motor current is usually small due to high leakage reactances at these frequencies. Since these high-frequency voltage harmonics can have as high or even higher amplitude compared to the fundamental frequency component, the iron losses (eddy current and hysteresis in the stator and the rotor iron) dominate. In fact, the total losses due to harmonics may even be higher with a PWM than with a square-wave inverter. This comparison would of course depend on the motor design class, magnetic material property, and switching frequency. Because of these additional harmonic losses, it is generally recommended that a standard motor with a 5% to 10% higher power rating be used.

In a PWM drive, the pulsating torques developed are small in amplitude and are at high frequencies (compared to the fundamental). Therefore, as shown by Eq. 12-49, they produce little speed pulsations because of the motor inertia.

12-7-2   Input Power Factor and Current Waveform

The input ac current drawn by the rectifier of a PWM-VSI drive contains a large amount of harmonics similar to that discussed in Chapter 3. Its waveform is shown in Fig. 12-19b for a single-phase and a three-phase input. As discussed in Chapter 3, the input inductance L_s improves the input ac current waveform somewhat. Also, a small dc-link capacitance will result in a better waveform.

The power factor at which the drive operates from the utility system is essentially independent of the motor power factor and the drive speed. It is only a slight function of the load power, improving slightly at a higher power. The displacement power factor (DPF) is approximately 100%, as can be observed from the input current waveforms of Fig. 12-19b.

12-7-3   Electromagnetic Braking

As we discussed in Section 12-4-5, the power flow during electromagnetic braking is from the motor to the variable-frequency controller. During braking, the voltage polarity across the dc-bus capacitor remains the same as in the motoring mode. Therefore, the direction of the dc-bus current to the inverter gets reversed. Since the current direction through the diode rectifier bridge normally used in PWM-VSI drives cannot reverse, some mechanism must be implemented to handle this energy during braking; otherwise the dc-bus voltage can reach destructive levels.

One way to accomplish this goal is to switch on a resistor in parallel with the dc-bus capacitor, as is shown in Fig. 12-20a, if the capacitor voltage exceeds a preset level, in order to dissipate the braking energy.

An energy-efficient technique is to use a four-quadrant converter (switch-mode or a back-to-back connected thyristor converter) at the front-end in place of the diode bridge rectifier. This would allow the energy recovered from the motor-load inertia to be fed back to the utility supply as shown in Fig. 12-20b, since the current through the four-quadrant converter used for interfacing with the utility source can reverse in direction. This is called regenerative braking since the recovered energy is not wasted. The decision to employ regenerative braking over dissipative braking depends on the additional equipment cost versus the savings on energy recovered, and the desirability of sinusoidal currents and unity power factor operation from the utility source.

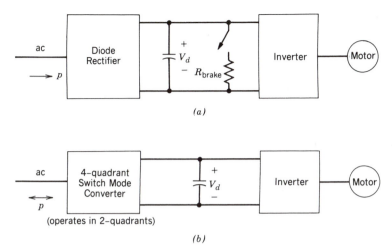

(a)

(operates in 2–quadrants)

(b)

FIGURE 12-20: Electromagnetic braking in PWM-VSI: (*a*) dissipative braking, (*b*) regenerative braking.

12-7-4 Adjustable-Speed Control of PWM-VSI Drives

In VSI drives (both PWM and square-wave type), the speed can be controlled without a speed feedback loop, where there may be a slower acting feedback loop through the processor controller, as explained in Chapter 10. Figure 12-21 shows such a control. The frequency f of the inverter output voltages is controlled by the input speed reference signal ω_{ref}. The input command ω_{ref} is modified, for protection and improved performance as will be discussed shortly, and the required control inputs (ω_s or f and V_s signals) to the PWM controller in Fig. 12-21 are calculated. The PWM controller can be realized by analog components, as discussed in Chapter 6 and indicated by Fig. 12-19*b*. The control signals $v_{a,\,control}$, and so on, can be calculated from the f and V_s signals and by knowing V_d and $\hat{V}_{tri}$.

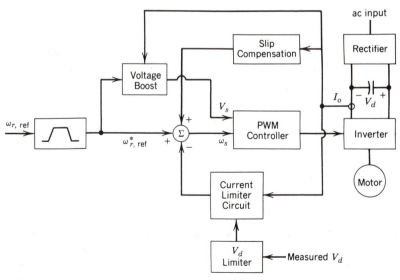

FIGURE 12-21: Speed control circuit. Motor speed is not measured.

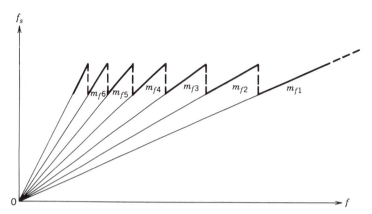

FIGURE 12-22: Switching frequency versus the fundamental frequency.

As discussed in Chapter 6, a synchronous PWM must be used. This requires that the switching frequency vary in proportion to f. To keep the switching frequency close to its maximum value, there are jumps in m_f and, hence, in f_s as f decreases, as shown in Fig. 12-22. To prevent jittering at frequencies where jumps occur, a hysteresis must be provided. Digital ICs such as HEF5752V are commercially available, which incorporate many of the functions of the PWM controller described earlier.

For protection and better speed accuracy, current and voltage feedback may be employed. These signals are required anyway for starting/stopping of the drive, to limit the maximum current through the drive during acceleration/deceleration or under heavy load conditions, and to limit the maximum dc link voltage during braking of the induction motor. Because of slip, the induction motor operates at a speed lower than the synchronous speed. It is possible to approximately compensate for this slip speed, which increases with torque, without measuring the actual speed. Moreover a voltage boost is required at lower speeds. To meet these objectives, the motor currents and the dc link voltage V_d across the capacitor are measured. To represent the instantaneous three-phase ac-motor currents, a current i_o at the inverter input, as shown in Fig. 12-21, is measured. The following control options are described:

1. *Speed-Control Circuit.* As shown in Fig. 12-21, a speed-control circuit accepts the speed reference signal $\omega_{r,\text{ref}}$ as the input that controls the frequency of the inverter output voltages. By the ramp limiter, the maximum acceleration/deceleration rates can be specified by the user through potentiometers that adjust the rate-of-change allowed to the speed reference signal. During the acceleration/deceleration condition, it is necessary to keep the motor current i_o and the dc link voltage V_d within limits.

   If the speed regulation is to be improved, to be more independent of the load-torque, it also accepts an input from the slip-compensation subcircuit, as shown in Fig. 12-21 and explained in item 3 below.

2. *Current-Limiting Circuit.* A current-limiting circuit is necessary if a speed ramp limiter as in Fig. 12-21 is not used. In the motoring mode, if ω_s is increased too fast compared to the motor speed, then ω_{sl} and, hence, i_o would increase. To limit the maximum rate of acceleration so that the motor currents stays below the current limit, the actual motor current is compared with the current limit,

and the error, through a controller, acts on the speed control circuit by reducing the acceleration rate (i.e., by reducing ω_s).

In the braking mode, if ω_s is reduced too fast, the negative slip would become large in magnitude and would result in a large braking current through the motor and the inverter. To restrict this current to the current limit during the braking, the actual current is compared with the current limit, and the error, fed through a controller, acts on the speed-control circuit by decreasing the deceleration rate (i.e., by increasing ω_s). During braking, the dc-bus capacitor voltage must be kept within a maximum limit. If there is no regenerative braking, a dissipation resistor is switched on in parallel with the dc-bus capacitor to provide a dynamic braking capability. If the energy recovered is larger than that lost through various losses, the capacitor voltage could become excessive. Therefore, if the voltage limit is exceeded, the control circuit decreases the deceleration rate (by increasing ω_s).

3. *Compensation for Slip.* To keep the rotor speed constant, a term must be added to the applied stator frequency, which is proportional to the motor torque T_{em}, as can be seen from Fig. 12-6

$$\omega_s = \omega_{r,\text{ref}} + k_{18}T_{em} \tag{12-50}$$

The second term in Eq. 12-50 is calculated by the slip-compensation block of Fig. 12-21. One option is to estimate T_{em}. This can be done by measuring the dc power to the motor and subtracting the losses in the inverter and in the stator of the motor to get the air-gap power P_{ag}. From Eqs. 12-3 and 12-18c, T_{em} can be calculated.

4. *Voltage Boost.* To keep the air-gap flux ϕ_{ag} constant, the motor voltage must be (as found by combining Eqs. 12-38b and 12-25)

$$V_s = k_{19}\omega_s + k_{20}T_{em} \tag{12-51}$$

Using T_{em} as calculated in item 3 above and knowing ω_s, the required voltage can be calculated from Eq. 12-51. This provides the necessary voltage boost in Fig. 12-21.

It should be noted that, if needed, the speed can be precisely controlled by measuring the actual speed and thereby using the actual slip in the block diagram of Fig. 12-21. By knowing the slip, the actual torque can be calculated from Eq. 12-27, thereby allowing the voltage boost to be calculated more accurately.

12-7-5 Induction-Motor Servo Drives

In the previous sections, the emphasis has been on controlling the speed of induction motors. Recently, because of the ready availability of the digital signal processors (DSPs), induction motors are beginning to be used for servo drives. In servo drives, the torque developed by the motor should respond quickly and precisely to the torque command without oscillations, at all speeds including at rest, since these drives are used for position control.

The control of induction-motor servo drives is normally done by a field-oriented space-vector-based calculations of what the stator currents of the induction motor should be to provide an electromagnetic torque T_{em} equal to the torque-command specified by the speed-regulator. In these calculations, a model of the induction motor

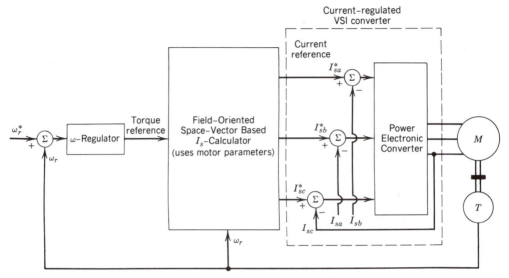

FIGURE 12-23: Field-oriented control for induction motor servo drive.

is needed; therefore, the motor parameters must be plugged into the model in the I_s-calculator block shown in Fig. 12-23. Many of these models rely on an accurate knowledge of the rotor resistance, R_r. As resistance of copper varies 40% when temperature varies by 100°C, such knowledge is not easy to come by during motor operation. Most of these models also need the actual speed information; this is, however, not a serious restriction, since speed is measured in servo drives anyway. Adaptive control with parameter estimation is often utilized.

As shown in Fig. 12-23, the field-oriented controller calculates the three-phase reference currents that must be delivered by the power converter to the motor. A current-regulated VSI inverter (CR-VSI), as discussed in Chapter 6, can be utilized where the inverter switch control signals are obtained by comparing the reference currents with the actual phase currents measured.

12-8   VARIABLE-FREQUENCY SQUARE-WAVE-VSI DRIVES

The schematic of such a drive was shown in Fig. 12-18b. The inverter operates in a square-wave mode which results in phase-to-motor-neutral voltage as shown in Fig. 12-24a. With the square-wave inverter operation, described in Chapter 6, each inverter switch is on for 180° and a total of three switches are on at any instant of time. The resulting motor current waveform is also shown in Fig. 12-24b. Because of the inverter operating in a square-wave mode, the magnitude of the motor voltages is controlled by controlling V_d in Fig. 12-18b by means of a line-frequency phase-controlled converter.

Voltage harmonics in the inverter output decrease as V_1/h with $h = 5, 6, 11, 13$, etc., where V_1 is the fundamental frequency phase-to-neutral voltage. Because of substantial magnitudes of low-order harmonics, harmonic currents calculated from Eq. 12-47 are significant. These harmonic currents result in large torque ripple, which can produce troublesome speed ripple at low operating speeds.

The line-rectifier in Fig. 12-18b is similar to the line-frequency phase-controlled converters described in Chapter 4. Assuming a continuously flowing current through

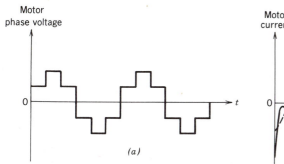

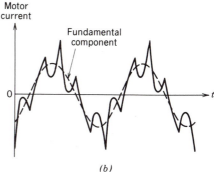

FIGURE 12-24: Square-wave VSI waveforms.

the rectifier and for simplicity ignoring the line–side inductances

$$V_d = 1.35V_{LL}\cos\alpha \qquad (12\text{-}52)$$

where V_{LL} is the line–line rms line voltage. From Eq. 6-58, the motor line–line voltage for a given V_d is

$$V_{LL1}^{\text{motor}} = 0.78V_d \qquad (12\text{-}53)$$

From Eqs. 12-52 and 12-53,

$$V_{LL1}^{\text{motor}} = 1.05V_{LL}\cos\alpha \simeq V_{LL}\cos\alpha \qquad (12\text{-}54)$$

which shows that the maximum line–line fundamental frequency motor voltage (at $\alpha = 0$) is approximately equal to V_{LL}. Note that the same maximum motor voltage (equal to the line voltage) can be approached in PWM-VSI drives only by overmodulation as described in Chapter 6. Therefore, in both PWM and square-wave-VSI drives, the maximum available motor voltage in Fig. 12-12b is approximately equal to the line voltage. This allows the use of standard 60-Hz motors, since the inverter is able to supply the rated voltage of the motor at its rated frequency of 60 Hz.

In a square-wave drive, from Eq. 12-54 and assuming V_s/f ratio to be a constant,

$$\frac{\omega_r}{\omega_{r,\text{rated}}} \approx \frac{V_{LL1}^{\text{motor}}}{V_{LL}} \approx \cos\alpha \qquad (12\text{-}55)$$

From Eqs. 4-30 and 12-55, the drive operates at the following power factor from the line (assuming that a sufficiently large filter inductor is present in Fig. 12-18b at the rectifier output)

$$\text{Line power factor} \approx 0.955\cos\alpha \approx 0.955\,\frac{\omega_r}{\omega_{r,\text{rated}}} \qquad (12\text{-}56)$$

which shows that the line power factor at the rated speed is better than that of an induction motor supplied directly by the line. At low speed, however, the line power factor of a square-wave drive can become quite low, as seen from Eq. 12-56. This can

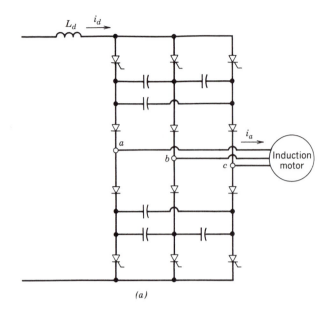

(a)

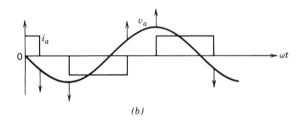

(b)

FIGURE 12-25: CSI-drive: (a) inverter, (b) idealized phase waveforms.

be remedied by replacing the thyristor rectifier by a diode-rectifier bridge in combination with a step-down dc–dc converter.

12-9 VARIABLE-FREQUENCY CURRENT-SOURCE INVERTER (CSI) DRIVES

Figure 12-18c shows the schematic of a CSI drive. Basically it consists of a phase-controlled rectifier, a large inductor, and a dc-to-ac inverter. A large inductor is used in the dc link, which makes the input appear as a current source to the inverter.

Since the induction motor operates at a lagging power factor, circuits for forced commutation of the inverter thyristors are needed, as shown in Fig. 12-25a. These forced-commutation circuits consist of diodes, capacitors, and the motor leakage inductances. This requires that the inverter be used with the specific motor for which it is designed. At any time, only two thyristors conduct: one of the thyristors connected to the positive dc bus and the other connected to the negative dc bus. The motor current and the resulting phase voltage waveforms are shown in Fig. 12-25b. In a CSI drive, the regenerative braking can be easily provided without any additional circuits.

In the past, the fact that line-frequency thyristors with simple commutation circuits act as the inverter switches, was a very important asset of CSI drives. With the availability of controllable switches in ever-increasing power ratings, nowadays CSI drives are used mostly in very large horsepower applications.

TABLE 12-2
Comparison of Adjustable Frequency Drives

	PWM	Square-Wave	CSI
Input power factor	+	−	− −
Torque pulsations	+ +	−	−
Multimotor capability	+	+	−
Regeneration	−	−	+ +
Short-circuit protection	−	−	+ +
Open-circuit protection	+	+	−
Handle undersized motor	+	+	−
Handle oversized motor	−	−	−
Efficiency at low speeds	−	+	+
Size and weight	+	+	− −
Ride-through capability	+	−	−

12-10 COMPARISON OF VARIABLE-FREQUENCY DRIVES

It is possible to use all three types of drives (PWM-VSI, square-wave-VSI, and CSI) with general-purpose induction motors. All three can provide a constant torque capability, from the rated speed down to some small speed where the reduced cooling in the motor dictates that the torque capability will decline. Motor derating as a percentage of the nameplate hp rating is essentially independent of the drive type.

In spite of the somewhat similar nature of these three drive types, there are certain basic differences, which are compared in Table 12-2 where " + " is a positive and " − " is a negative attribute. It should be kept in mind that this comparison addresses the inherent capability of each drive. By means of additional circuits, most of the limitations can be overcome.

Some general comments can be made about the trend in applying these drives. For retrofit applications, PWM-VSI is preferred over CSI, which requires a better match between the inverter and the motor. In sizes below a few hundred horsepower, thee is an increasing trend to use PWM-VSI.

To make these solid-state controllers more reliable, a host of other protective features are incorporated. These include instantaneous overcurrent trip, input circuit-breakers, current-limiting fuses, line reactors or isolation transformers at the input, output disconnect switch between the voltage-source inverter and the motor, motor thermal protection incorporated with the controller, trips in case of overvoltage, undervoltage or loss of a phase, and so on.

12-11 LINE-FREQUENCY VARIABLE-VOLTAGE DRIVES

The variable-frequency variable-voltage drives we described earlier are the most energy efficient and versatile way to control the speed of squirrel-cage induction motors. However, in some applications, it may be cheaper to use line-frequency variable-voltage drives as discussed in this chapter.

In the equivalent circuit of Fig. 12-2a, with f equal to the line frequency and a fixed value of f_{sl}, the power in any resistive element is proportional to V_s^2. Therefore, using Eqs. 12-17 and 12-18, the torque T_{em} will be proportional to V_s^2 for a value of

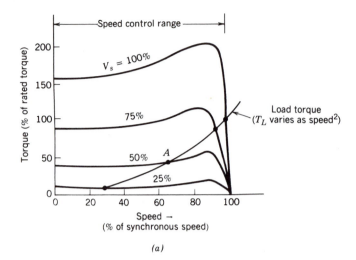

(a)

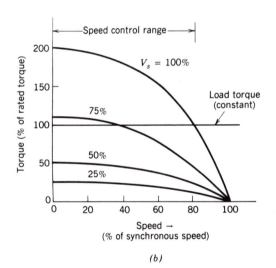

(b)

FIGURE 12-26: Speed control by stator voltage control: (a) motor with a low value of s_{rated}; fan-type load, (b) motor with a large s_{rated}; a constant torque load.

rotor speed ω_r determined by f and f_{sl}:

$$T_{em} = k_{21}V_s^2 \tag{12-57}$$

Based on Eq. 12-57, Fig. 12-26a shows the motor torque-speed curves at various values of V_s for a normal induction motor with a small value of the rated slip. The load-torque of a fan- or a pump-type load varies approximately as the square of speed. Therefore, only a small torque is required at low speeds, and as Fig. 12-26a shows, the speed can be controlled over a wide range. Because of the heavy dependence of the load torque on speed, the motor operating point, for example A (intersection of the load and the motor torque–speed curves) in Fig. 12-26a, is stable. The operating point would not be stable if the load torque remains constant with speed.

For a load requiring a constant torque with speed, it is necessary to use a motor with a higher motor resistance whose torque-speed characteristics are shown in Fig. 12-26*b*. A motor with a high rotor resistance has a large value of slip at which the pull-out torque is developed. Such a motor allows the speed to be controlled over a wide range even when supplying a constant-torque load.

Speed control by controlling the stator voltage results in a very poor energy efficiency at low speeds because of high rotor losses caused by large slips. In a motor, the actual rotor loss must be below its rated rotor loss (which occurs while the motor is supplying the rated torque, supplied from the rated voltage). As seen from Figs. 12-26*a* and 12-26*b*, speed control by adjusting the stator voltage results in a large slip and, hence, in a large rotor power loss at reduced speeds. Therefore, the motor selected for this application must have a high enough rating so its rated rotor loss is larger than the maximum rotor loss encountered by using this technique. This technique is widely used in fractional horsepower fan or pump drives. These fractional horsepower motors are generally single-phase motors, but the analysis presented earlier for the three-phase motor is applicable. This speed-control technique is also used for cranes and hoists (which have speed-independent load torque) where the high slip, high power-loss operation is required for only a small portion of the load duty cycle.

A practical circuit for controlling the stator voltage of a three-phase induction motor is shown in Fig. 12-27*a*. It consists of three pairs of back-to-back connected thyristors.

Because of the interaction between the phases, it is not possible to analyze this circuit on a per-phase basis. It has been shown in the literature that for analysis purposes, each phase of the motor can be represented by a sinusoidal back-emf in series with an inductance. Figure 12-27*b* shows one of the motor phase voltages, v_{an}, and the phase current i_a. As shown in Fig. 12-27*b*, the motor currents are no longer sinusoidal and their harmonic components result in a pulsating torque and in a higher power loss compared with a sinusoidal supply. These losses are in addition to the rotor circuit losses due to a high slip operation at low speeds. These nonsinusoidal currents also flow into the utility system. Because of high rotor losses, this technique for controlling the speed is limited to low-horsepower or intermittent-load applications.

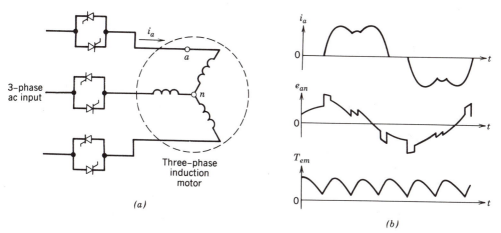

FIGURE 12-27: Stator voltage control: (*a*) circuit, (*b*) waveforms.

12-12 REDUCED VOLTAGE STARTING ("SOFT START") OF INDUCTION MOTORS

The circuit of Fig. 12-27a can also be used in constant-speed drives to reduce the motor voltages at start-up, thereby reducing the starting currents. In normal (low-slip) induction motors, the starting currents can be as large as 6 to 8 times the full-load current. To reduce these large starting currents, the motor can be started at reduced voltages obtained from the circuit of Fig. 12-27a. Provided the torque developed at reduced voltage is sufficient to overcome the load, the motor accelerates (slip s will decrease) and the motor current decreases. During the steady-state operation, each thyristor conducts for an entire half-cycle. Then, these thyristors can be shorted out by mechanical contactors connected in parallel with the back-to-back connected thyristor pairs, to eliminate the power losses in the thyristors due to a finite $(1-2 \text{ V})$ conduction voltage drop across the thyristors.

The circuit of Fig. 12-27a can also be used in constant-speed drives to minimize motor losses. In an induction motor (single-phase and three-phase) at a given torque output, the motor losses vary with the stator voltage V_s. The stator voltage at which the minimum power loss occurs decreases with decreasing load. Therefore, it is possible to use the circuit of Fig. 12-27a to reduce V_s at reduced loads and, hence, save energy.

The amount of energy saved is significant (compared with the extra losses in the motors due to current harmonics and in the thyristors due to a finite conduction voltage drop) only if the motor operates at very light loads for substantial periods of time. In applications where reduced voltage starting (soft start) is required, the power switches are already implemented and only the control for the minimum power loss needs to be added. In such cases, this concept may be economical.

12-13 SPEED CONTROL BY STATIC SLIP-POWER RECOVERY

From the induction-motor equivalent circuit, it is possible to obtain torque-speed curves for various values of rotor resistances. In the equivalent circuit of Fig. 12-2a, if R_r/s is kept constant (i.e., R_r and s are increased by the same factor), I_r and, hence, T_{em} remain constant. This results in characteristics as shown in Fig. 12-28 for various

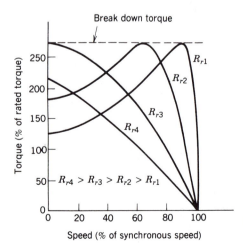

FIGURE 12-28: Torque-speed curves for a wound-rotor induction motor.

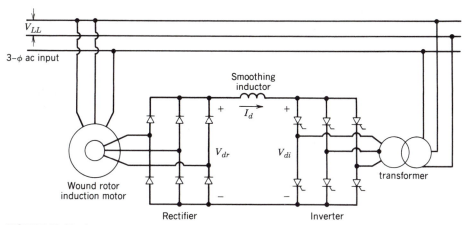

FIGURE 12-29: Static slip recovery.

values of rotor resistance R_r. In a wound rotor induction motor, the total resistance R_r in the rotor phases can be varied by adding an external resistance through the slip rings.

For the load-torque versus speed curve shown in Fig. 12-28, it is quite apparent that the speed of operation can be continuously varied by controlling the external resistance in the rotor circuit (the steady-state speed is given by the intersection of the load and the motor torque-speed curves as in all motor drives). However, high rotor losses (due to high slips) may be unacceptable.

The static slip power recovery scheme provides an alternative to the historical Scherbius and Kramer drives, both of which require a second rotating machine to recover the rotor circuit electrical power. In the static slip-power recovery system, rather than dissipating the slip power in the rotor external resistances, these resistances are simulated by means of a diode rectifier and the energy recovered is fed back to the ac source by a means of a line-voltage-commutated inverter, as is shown in Fig. 12-29.

This scheme requires a wound-rotor motor with slip rings. Such a motor is not as inexpensive and as maintenance-free as its squirrel-cage counterpart. However, in very large power ratings, this scheme may compete with the adjustable-frequency drive, if the speed needs to be controlled only in a small range around its nominal value. A small speed range results in a smaller rating of the solid-state converter required, thus making this scheme competitive.

12-14 SUMMARY

1. Induction motors are the workhorse of industry because of their low cost and rugged construction. When operated directly from the line voltages, an induction motor operates at nearly a constant speed. By means of power electronic converters, induction motors can be used for adjustable-speed and servo-drive applications. A major application of adjustable-speed induction motor drives is for improving the energy efficiency in various residential, industrial, and electric utility systems.

2. In a three-phase induction motor, the resultant field distribution in the air-gap is sinusoidal and rotates at a synchronous speed $\omega_s = (2\pi f)2/p$ radians per second for a p-pole winding when it is excited by three-phase voltages and currents at a frequency f.

3. The speed of an induction motor can be controlled by varying the stator frequency f, which controls the synchronous speed and, hence, the motor speed, since the slip s is kept

small. The air-gap flux ϕ_{ag} is kept constant by V_s in linear proportion to f. This technique allows the induction motor to deliver its rated torque at speeds up to its rated speed. Beyond the rated speed, the motor torque capability declines, though the motor can deliver its rated output power up to a certain speed.

4. For braking in an induction motor to reduce its speed, the stator frequency f is decreased so that the synchronous speed at which the air-gap magnetic field rotates is less than the rotor speed.

5. Switch-mode dc-to-ac inverters, as discussed in Chapter 6, are used to supply adjustable-frequency, adjustable-magnitude three-phase ac voltages for induction motor speed control. The harmonics in the inverter output voltages result in harmonics in the motor current, harmonic losses in the motor, and possibly the motor-torque pulsations. Therefore, care must be taken in selecting the inverter and the inverter switching frequency.

6. The inverters used for the induction motor speed control can be classified as pulse-width-modulated voltage-source inverters, square-wave voltage-source inverters, and the current-source inverters. The comparative advantages and disadvantages of these inverters are given in Table 12-2.

7. By means of field-oriented vector control, induction motor drives can be used for servo applications.

8. There are other means of controlling the speed of induction motors. Some of these are (a) stator pole changing, (b) pole-amplitude modulation, (c) stator voltage control at the line frequency and, (d) static slip-power recovery. In certain applications, one of these techniques rather than the stator-frequency control may be preferable.

PROBLEMS

12-1. A three-phase, 60-Hz, four-pole, 10-hp, 460-V (LL, rms) induction motor has a full-load speed of 1746 rpm. Assume the torque-speed characteristic in a range of 0 to 150% rated torque to be linear. It is driven by an adjustable-frequency sinusoidal supply such that the air-gap flux is held constant. Plot its torque-speed characteristics at the following values of frequency f: 60 Hz, 45 Hz, 30 Hz, and 15 Hz.

12-2. The drive in Problem 12-1 is supplying a centrifugal pump load, which at the full-load speed of the motor requires the rated torque of the motor. Calculate and plot speed, frequency f, slip frequency f_{sl}, and slip s at the following percentage values of pump rated torque: 100, 75, 50, and 25.

12-3. A 460-V, 60-Hz, four-pole induction motor develops its rated torque by drawing 10 A at a power factor of 0.866. The other parameters are as follows:

$$R_s = 1.53 \ \Omega \qquad X_{ls} = 2.2 \ \Omega \qquad X_m = 69.0 \ \Omega$$

If such a motor is to produce a rated torque at frequencies below 60 Hz while maintaining a constant air-gap flux, calculate and plot the required line-to-line voltage as a function of frequency.

12-4. The motor in Problem 12-3 has a full-load speed of 1750 rpm. Calculate f_{start}, I_{start}, and $(V_{LL})_{start}$ if the motor is to develop a starting torque equal to 1.5 times its rated torque. Assume the effect of L_{lr} to be negligibly small, and the air-gap flux to be at its rated value.

12-5. The idealized motor of Problem 12-1 is initially operating at its rated conditions at 60 Hz. If the supply frequency is suddenly decreased by 5% while maintaining a constant air-gap flux, calculate the braking torque developed as a percentage of its rated torque.

12-6. In a three-phase 60-Hz, 460-V induction motor, $(R_s + R_r) = 3.0\ \Omega$ and $(X_{ls} + X_{lr}) = 5.0\ \Omega$.

The motor is driven by a square-wave voltage-source inverter that supplies a 460-V line-line voltage at the frequency of 60 Hz. Estimate the harmonic currents and the additional copper losses due to these harmonic currents by including fifth, seventh, eleventh, and the thirteenth harmonics.

12-7. For harmonic frequency analysis, an induction motor can be represented by a per-phase equivalent circuit as shown in Fig. P12-7, which includes a fundamental frequency counter-emf or thevein voltage $\mathbf{E}_{TH}$. Also, $R_{TH} = 3.0\ \Omega$ and $X_{TH} = 5.0\ \Omega$. It is supplied by a voltage-source inverter, which produces a 60-Hz line-line voltage component of 460 V. The load on the motor is such that the fundamental frequency current drawn by the motor is 10 A, which lags the fundamental frequency voltage by an angle of 30°.

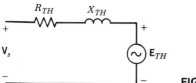

FIGURE P12-7:

Obtain and plot the current drawn by the motor as a function of time, if it is driven by a square-wave-VSI. What is the peak current that the inverter switches must carry?

12-8. Repeat Problem 12-7 if the induction motor is driven by a PWM-VSI with an amplitude modulation ratio $m_a = 1.0$ and the frequency modulation ratio $m_f = 15$. Compare the peak switch currents with those in Problem 12-7.

12-9. A square-wave-VSI drive supplies a 460-V line-line at a frequency of 60 Hz to an induction motor that develops a rated torque of 50 Nm at 1750 rpm. The motor and the inverter efficiencies can be assumed to be constant at 90% and 95%, respectively, while operating at the rated torque of the motor.

If the motor is operated at its rated torque and the rated air-gap flux, determine the equivalent resistance R_{eq} that can represent the inverter-motor combination in the Fig. P12-9 at the motor frequency of 60 Hz, 45 Hz, 30 Hz, and 15 Hz.

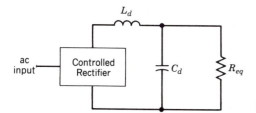

FIGURE P12-9:

12-10. Repeat Problem 12-9 if a PWM-VSI drive with an uncontrolled rectifier is used, where $m_a = 1.0$ at the 60-Hz output.

12-11. A CSI-driven induction motor is supplying a constant torque load equal to the rated torque of the motor. The CSI drive is supplied from a 460-V, three-phase, 60-Hz input. It supplies the motor a 460 V (L-L) voltage at a 60 Hz frequency with a fundamental frequency current of 100 A that lags behind the fundamental frequency voltage by an angle of 30°.

If the motor displacement power factor angle remains constant at 30°, estimate and plot the input power factor and the displacement power factor at the motor frequencies of 60 Hz, 45 Hz, 30 Hz, and 15 Hz. Idealize the motor current waveforms to be as shown in Fig. 12-25b and assume a constant air-gap flux in the motor. Neglect losses in the motor and the inverter.

12-12. Show that in a voltage-controlled induction motor supplying a constant load torque, the power loss in the rotor circuit at a voltage V_s, as a ratio of the power loss at the rated voltage condition can be approximated as

$$\frac{P_r}{(P_r)_{\text{rated voltage}}} \approx \left(\frac{V_{s,\text{rated}}}{V_s} \right)^2 ,$$

for reasonably small values of slip.

12-15 REFERENCES

1. N. Mohan and R. J. Ferraro, "Techniques for Energy Conservation in AC Motor Driven Systems," EPRI Final Report EM-2037, Project 1201-13, September 1981.
2. A. E. Fitzgerald, C. Kingsley, Jr., and S. D. Umans, *Electric Machinery*, 4th Ed., McGraw-Hill, New York, 1983.
3. G. R. Slemon and A. Straughen, *Electrical Machines*, Addison-Wesley, Reading, MA, 1980.
4. B. K. Bose, *Power Electronics and AC Drives*, Prentice-Hall, Englewood Cliffs, NJ, 1986.
5. W. Leonard, *Control of Electrical Drives*, Springer-Verlag, New York, 1985.

13

Synchronous-Motor Drives

13-1 INTRODUCTION

Synchronous motors are used as servo drives in applications such as computer peripheral equipment, robotics, and as adjustable-speed drives in a variety of applications such as load-proportional capacity-modulated heat pumps, large fans, and compressors. In low-power applications up to a few kilowatts, permanent-magnet synchronous motors are used (see Fig 13-1a). Such motors are often referred to as "brushless-dc" motors or electronically-commutated motors. Synchronous motors with wound rotor field are used in large power ratings (see Fig. 13-1b).

13-2 BASIC PRINCIPLES OF SYNCHRONOUS MOTOR OPERATION

The field winding on the rotor produces a flux ϕ_f in the air gap. This flux rotates at a synchronous speed ω_s rad/s, which is the same as the rotor speed. The flux ϕ_{fa} linking one of the stator phase windings, for example phase a, varies sinusoidally with time:

$$\phi_{fa}(t) = \phi_f \sin \omega t \tag{13-1}$$

where

$$\omega = 2\pi f = \frac{p}{2}\omega_s \tag{13-2}$$

and p is the number of poles in the motor. If we assume N_s as an equivalent number of turns in each stator phase winding, the emf induced in phase a from Eq. 13-1 is

$$e_{fa}(t) = N_s\frac{d\phi_{fa}}{dt} = \omega N_s\phi_f \cos \omega t \tag{13-3}$$

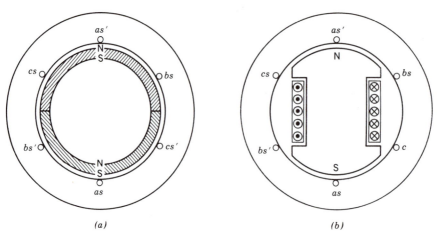

FIGURE 13-1: Structure of synchronous motors: (*a*) permanent magnet rotor (two-pole),
(*b*) salient-pole wound rotor (two-pole).

This induced voltage in the stator winding is called the excitation voltage whose rms
value is

$$E_{fa} = \frac{\omega N_s}{\sqrt{2}} \phi_f \qquad (13\text{-}4)$$

In accordance with the normal convention, the amplitudes of voltage and
current phasors are represented by their rms values; the amplitudes of flux phasors are
represented by their peak values. e_{fa} and ϕ_{fa}, being sinusoidal with time, can be
represented as phasors at $\omega t = 0$ where $\mathbf{E}_{fa} = E_{fa}$ is the reference phasor in Fig.
13-2*a*, and from Eq. 13-1

$$\phi_{fa} = -j\phi_f \qquad (13\text{-}5)$$

From Eqs. 13-3 through 13-5 and Fig. 13-2*a*

$$\mathbf{E}_{fa} = j\frac{\omega N_s}{\sqrt{2}} \phi_{fa} = E_{fa} \qquad (13\text{-}6)$$

In synchronous-motor drives, the stator is supplied with a set of balanced
three-phase currents, whose frequency is controlled to be f, which from Eq. 13-2 is

$$f = \frac{p}{4\pi} \omega_s \qquad (13\text{-}7)$$

The fundamental frequency components of these stator currents produce a constant
amplitude flux ϕ_s in the air gap, which rotates at the synchronous speed ω_s. The
amplitude of ϕ_s is proportional to the amplitudes of the fundamental frequency
components in the stator currents.

In this three-phase motor, the flux linking with phase a due to ϕ_s produced by
all three stator currents is $\phi_{sa}(t)$. As shown in Reference 1, $\phi_{sa}(t)$ is proportional to the
phase a current $i_a(t)$:

$$N_s\phi_{sa}(t) = L_a i_a(t) \qquad (13\text{-}8)$$

where the armature inductance L_a is $3/2$ times the self-inductance of phase a. Therefore, the voltage induced in phase a due to $\phi_{sa}(t)$, from Eq. 13-8, is

$$e_{sa}(t) = N_s \frac{d\phi_{sa}}{dt} = L_a \frac{di_a}{dt} \tag{13-9}$$

Assuming the fundamental component of the supplied current to the stator phase a to be

$$i_a(t) = \sqrt{2} I_a \sin(\omega t + \delta), \tag{13-10}$$

$$e_{sa}(t) = \sqrt{2} \omega L_a I_a \cos(\omega t + \delta) \tag{13-11}$$

from Eq. 13-9, where δ is defined later on to be the torque angle. i_a and e_{sa} can be represented as phasors, which at $\omega t = 0$

$$\mathbf{I}_a = I_a e^{j(\delta - \frac{\pi}{2})} \tag{13-12}$$

and as shown in Fig. 13-2a

$$\mathbf{E}_{sa} = j\omega L_a \mathbf{I}_a = \omega L_a I_a e^{+j\delta} \tag{13-13}$$

The resultant air-gap flux $\phi_{ag,\,a}(t)$ linking the stator phase a is the sum of $\phi_{fa}(t)$ and $\phi_{sa}(t)$:

$$\phi_{ag,\,a}(t) = \phi_{fa}(t) + \phi_{sa}(t) \tag{13-14}$$

which can be represented as a phasor

$$\phi_{ag,\,a} = \phi_{fa} + \phi_{sa} \tag{13-15}$$

The air-gap voltage $e_{ag,\,a}(t)$ due to the resultant air-gap flux linking phase a is

$$e_{ag,\,a}(t) = N_s \frac{d\phi_{ag,\,a}}{dt} = e_{fa}(t) + e_{sa}(t) \tag{13-16}$$

from Eqs. 13-14, 13-3, and 13-9. Equations 13-6 and 13-13 combined with Eq. 13-16 result in

$$\mathbf{E}_{ag,\,a} = \mathbf{E}_{fa} + \mathbf{E}_{sa} = E_{fa} + j\omega L_a \mathbf{I}_a \tag{13-17}$$

All of these phasors are drawn in Fig. 13-2a. Based on Eq. 13-17 and the phasor diagram, a per-phase equivalent circuit of a synchronous motor is shown in Fig. 13-2b, where R_s and L_{ls} are the stator winding resistance and leakage inductance, respectively. Including the voltage drop across R_s and L_{ls}, the per-phase terminal voltage in phase a is

$$\mathbf{V}_a = \mathbf{E}_{ag,\,a} + (R_a + j\omega L_s)\mathbf{I}_a \tag{13-18}$$

The phasor diagram corresponding to Eq. 13-18 is shown in Fig. 13-2c, where θ_a is the angle between the current and the terminal voltage phasors.

From the per-phase equivalent circuit of Fig. 13-2b and the phasor diagram of Fig. 13-2a, the electromagnetic torque T_{em} can be obtained as follows: the electrical

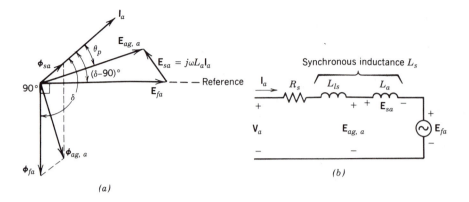

(a)

(b)

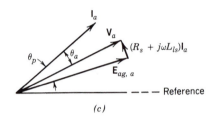

(c)

FIGURE 13-2: Per-phase representation: (a) phasor diagram, (b) equivalent circuit, (c) terminal voltage.

power that gets converted into the mechanical power P_{em} is

$$P_{em} = 3E_{fa}I_a \cos\left(\delta - \frac{\pi}{2}\right) \tag{13-19}$$

and

$$T_{em} = P_{em}/\omega_s \tag{13-20}$$

Using Eqs. 13-19, 13-20, and 13-4

$$T_{em} = k_t \phi_f I_a \sin\delta \tag{13-21}$$

where the angle δ between ϕ_{fa} and I_a is called the torque angle, and k_t the constant of proportionality.

In the phasor diagram of Fig. 13-2c, I_a leads V_a. This leading power factor operation is required if the synchronous motor is supplied by a drive where the current through the inverter thyristors is commutated by the synchronous-motor voltages.

A torque angle δ equal to 90° results in a decoupling of the field-flux ϕ_f and the field due to the stator currents, which is important in high-performance servo drives. With $\delta = 90°$, a constant field-flux ϕ_f and the amplitudes of the stator phase currents equal to I_s, Eq. 13-21 can be written as

$$T_{em} = k_T I_s \tag{13-22}$$

where k_T is the motor torque constant. A phasor diagram corresponding $\delta = 90°$ is shown in Fig. 13-3, where I_a must lead ϕ_{fa} by 90°. This condition for servo drives implies that the current i_a must become positive maximum, $\omega t = 90°$ or $t =$

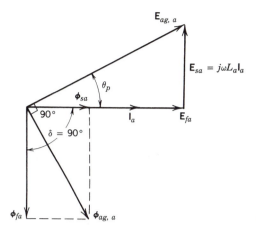

FIGURE 13-3: Phasor diagram with $\delta = 90°$.

$(\pi/2)/(p/2)\omega_s$ seconds before ϕ_{fa} reaches its positive maximum value. Another observation from the phasor diagram of a servo drive is that $\mathbf{I}_a$ is at a lagging power factor. Therefore, the inverter of the drive must consist of self-controlled switches.

In the previous analysis, the rotor saliency was ignored. The effect of rotor saliency cannot be represented by a per-phase equivalent circuit because of a different magnetic permeance along the rotor-pole axis (called the d-axis) and along the axis midway between two rotor poles (called the q-axis). This d- and q-axis analysis is beyond the scope of this book, but in qualitative terms, an additional reluctance torque component is present because of the difference in reactances in the d- and q-axes. This component is usually small, but not negligible, compared with the electromagnetic torque component discussed above, assuming a nonsalient round rotor.

13-3 SYNCHRONOUS SERVOMOTOR DRIVES WITH SINUSOIDAL WAVEFORMS

The air-gap flux-density distribution and the induced excitation voltages in the stator phase windings in such a motor are nearly sinusoidal. In this regard, the description of this motor is identical to that presented in the previous section. Moreover, the torque angle δ is maintained at 90 degrees. For controlling such a synchronous servo drive, the rotor field position is measured by means of an absolute position sensor, with respect to a stationary axis, for example as shown in Fig. 13-4 for a two-pole motor. Recognizing that at $\theta = 0$ in a 2-pole motor of Fig. 13-4, i_a should be at its positive peak

$$i_a(t) = I_s \cos\left[\theta(t)\right] \tag{13-23}$$

where the amplitude I_s is obtained from Eq. 13-22. For a p-pole motor, in general, if θ is the mechanical angle measured, then the electrical angle θ_e is

$$\theta_e(t) = \frac{p}{2}\theta(t) \tag{13-24}$$

If we use Eqs. 13-23 and 13-24 and recognize that $i_b(t)$ and $i_c(t)$ are delayed by 120°

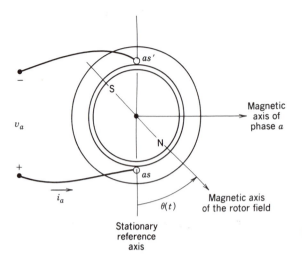

FIGURE 13-4: Measured rotor position θ at time t.

and 240°, respectively

$$i_a(t) = I_s \cos\left[\theta_e(t)\right] \tag{13-25}$$

$$i_b(t) = I_s \cos\left[\theta_e(t) - 120°\right] \tag{13-26}$$

$$i_c(t) = I_s \cos\left[\theta_e(t) - 240°\right] \tag{13-27}$$

This control strategy can also be used for induction motor drives as described in Reference 10.

With the frequency of the stator currents "locked" or synchronized to the rotor position, which is continuously measured, there is no possibility of losing synchronism, and the torque angle δ remains at its optimal value of 90°. If a holding torque is required at zero speed to overcome the load torque and, hence, to keep the load from moving from a position where θ is constant, as is often the case in servo drives, a synchronous servomotor drive provides this torque by applying dc currents to the stator as given by Eqs. 13-24 to 13-27.

Figure 13-5 shows the overall block diagram of a synchronous servomotor drive with sinusoidal waveforms. The absolute rotor field position is sensed by means of an absolute position sensor such as a high-accuracy resolver, which is mechanically prealigned to measure the rotor field position θ with respect to a known axis, for

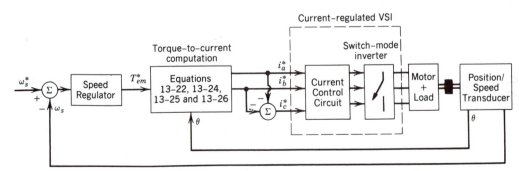

FIGURE 13-5: Synchronous motor servo drive.

example as indicated in Fig. 13-4. By using prestored cosine tables in read-only memory (ROM), cosine functions required in Eqs. 13-25 to 13-27 are generated for two of the three phase currents, for example, a and b. The stator current amplitude I_s is determined by the torque/speed loop, using Eq. 13-22. Once the reference currents i_a^* and i_b^* are defined for phases a and b, $i_c^* = -i_a^* - i_b^*$ in a three-wire motor. As discussed in Chapter ó, a current-regulated voltage-source inverter is used to force the motor currents to equal the reference currents.

13-4 SYNCHRONOUS MOTOR DRIVES WITH TRAPEZOIDAL WAVEFORMS

The motors described in the previous section are designed such that the induced excitation emfs in the stator due to the field flux are sinusoidal and the stator currents produce a sinusoidal field. The motors described in this section are designed with concentrated coils, and the magnetic structure is shaped such that the flux density of the field because of the permanent magnets and the induced excitation voltages have trapezoidal waveforms.

Figure 13-6a shows the induced emf $e_{fa}(t)$ in phase a, where the rotor is rotating in a counter-clockwise direction at a speed of ω_s rad/s and θ is measured with respect to the stator as shown in Fig. 13-4. The electrical angle θ_e is defined by Eq. 13-24 for a p-pole motor. This emf waveform has a flat-portion, which occurs for at least 120° (electrical) during each half-cycle. The amplitude $\hat{E}_f$ is proportional to the rotor speed

$$\hat{E}_f = k_E \omega_s \tag{13-28}$$

where k_E is the motor voltage constant. Similar voltage waveforms are induced in phases b and c, displaced by 120° and 240°, respectively.

To produce as ripple-free torque as possible in such a motor, the phase currents supplied should have rectangular waveforms as is shown in Fig. 13-6b, which result in instantaneous electrical power $p_a(t) = e_{fa}(t) \cdot i_a(t)$, and so on, as shown in Fig. 13-6c. Since $p_{\text{total}}(t) = p_a(t) + p_b(t) + p_c(t) = P_{em}$ is independent of time, the instantaneous electromagnetic torque is also independent of time and depends only on the current amplitude I_s:

$$T_{em}(t) = \frac{P_{em}}{\omega_s} = k_T I_s \tag{13-29}$$

where k_T and k_E of Eqs. 13-28 and 13-29 are related (see Problem 13-5). In practice, because of the finite time required for the phase currents to change, T_{em} contains ripple.

A current-regulated VSI, similar to that shown in Fig. 13-5, is used where the sinusoidal reference currents are replaced by rectangular current references, which are shown in Fig. 13-6b. One complete cycle is divided into six intervals of 60 electrical degrees each. In each interval, the current through two of the phases is constant and proportional to the torque command. To obtain these current references, the rotor position is usually measured by Hall-effect sensors that indicate the six current-commutation instants per electrical cycle of waveforms. In non-servo applications, it is possible to use the three-phase induced emfs to determine the current-commutation instants, thereby eliminating the need for any rotor position sensor.

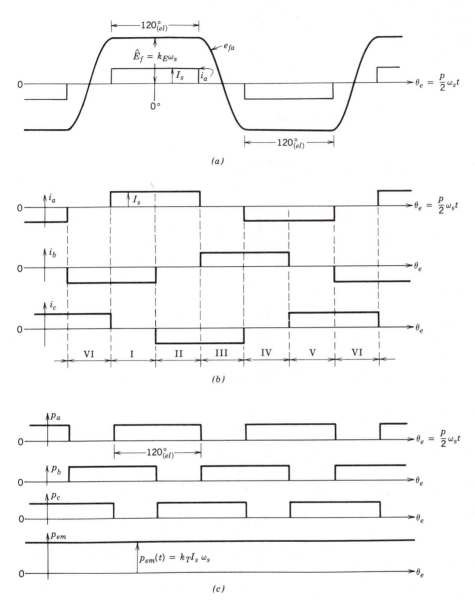

FIGURE 13-6: Trapezoidal waveform synchronous motor drive.

13-5 LOAD-COMMUTATED-INVERTER (LCI) DRIVES

In very large power ratings in excess of 1000 hp, LCI-synchronous motor drives become competitive with the induction motor drives in adjustable-speed applications. The circuit diagram of a LCI drive is shown in Fig. 13-7a. Each phase of the synchronous motor is represented by an internal voltage in series with the motor inductance as discussed in the previous sections, assuming a nonsalient-pole motor.

As an overview, the LCI drive is the source of three-phase currents to the motor. The frequency and the phase of these currents are synchronized to the rotor position. The current commutation in the load inverter to supply currents to the motor phases in an appropriate sequence is provided by the induced emfs in the motor. The

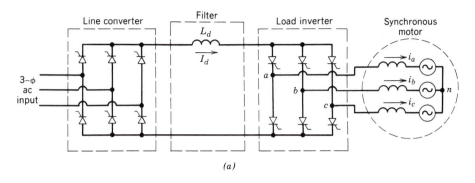

(a)

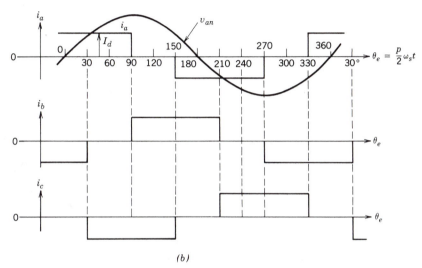

(b)

FIGURE 13-7: LCI drive: (a) circuit, (b) idealized waveforms.

amplitude of the currents supplied to the motor is controlled by the phase-controlled line converter through a filter inductance L_d. The filter inductance reduces the current harmonics and ensures that the input to the load inverter and, hence, to the motor appears as a current source.

Note that the line converter in Fig. 13-7a is identical to the phase-controlled, line-frequency converters discussed in Chapter 4. By controlling the firing angle of the converter, its dc output voltage and, hence, the current magnitude (in the dc link as well as to the motor) can be controlled. Normally, it operates in a rectifier mode.

The load inverter is identical to the line converter, that is, it also consists of only thyristors, but it normally operates in an inverter mode. The current commutation is provided by the internally induced emfs in the synchronous motor. The presence of these three-phase emfs facilitates the current commutation in the load converter in an identical manner as in a line-frequency thyristor converter operating in an inverter mode. The idealized motor current waveforms are shown in Fig. 13-7b.

At start-up and at low speeds (less than 10% of the full speed) the induced emf in the synchronous motor is not sufficient to provide current commutation in the load converter. Under this condition, the current commutation is provided by the line converter by going into an inverter mode and forcing I_d to become zero, thus providing turn-off of thyristors in the load inverter.

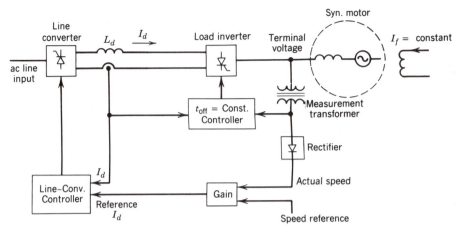

FIGURE 13-8: LCI drive controller.

There are many control possibilities. Considering first the range below the rated speed, one possibility is to keep the field excitation current I_f constant. Also, the turn-off time t_{off} available to the thyristors in the load converter is kept constant. The dc-link current, which is proportional to the motor current, is varied with torque at a given speed. The voltage waveforms at the motor terminals are measured to calculate the rotor-field position as a function of time. The measured three-phase voltages are rectified to provide a dc signal proportional to the instantaneous rotor speed. Keeping I_f and t_{off} constant, the actual speed is compared with the reference speed, as is shown in Fig. 13-8. The amplified error signal determines the I_d reference. If the actual I_d is less than its reference value, the line-converter increases the dc voltage applied to the link thus increasing I_d and, hence, the torque produced by the motor. In response to increased T_{em}, the motor speed goes up. Based on I_d and the information obtained from the measured terminal voltage waveforms, the firing pulses to the thyristor gates of the load inverter are provided such as to keep t_{off} constant. In practice, I_f is not kept constant. Rather, it is controlled as a function of torque and speed to result in the rated air-gap flux in the motor.

For speeds above the rated speed, the motor torque capability declines but the drive can supply the rated power. For operation above the rated speed, the field flux needs to be reduced by reducing I_f. Therefore, this region is also called the flux-weakening region.

Some of the other important properties of LCI drives are described as follows:

1. Use of synchronous motors in very large horsepower ratings (> 1000 hp) results in an overall drive efficiencies exceeding 95% at the rated power, a few percentage points higher than what can be accomplished in the induction motor drives.

2. The load-commutated inverter is much simpler and has lower losses compared to the inverter used in CSI-induction motor drives. Eliminating the requirement for self-controlled switches is a distinct advantage at high voltage and current ratings.

3. As in any power-electronic motor drive, there is no inrush current at start-up, unlike line-started motors. By designing the pole-face (or the damper-cage) winding to provide a sufficient torque, a synchronous motor can be line-started on an induction motor principle. Once it reaches a speed close to the syn-

chronous speed, the rotor field is excited, thus making it operate as a synchronous motor. Being able to line-start and operate at the line frequency provides additional reliability in case of the inverter failure, where the line voltages act as backup, as in the induction motor drive (though the drive would operate only at one speed).

4. LCI drives have an inherent capability to provide regenerative braking by making the synchronous motor operate as a generator, rectifying the motor voltages by means of the load converter and feeding the power into the utility grid by operating the line converter in an inverter mode.

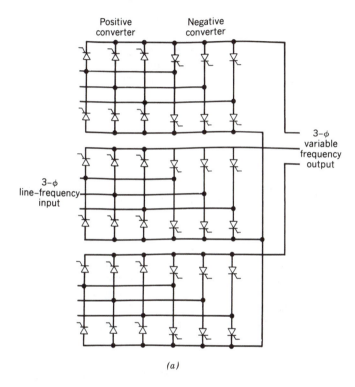

(a)

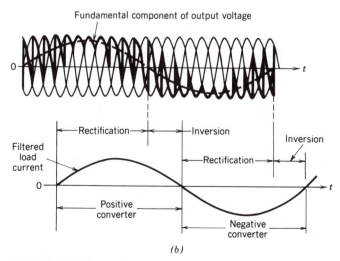

(b)

FIGURE 13-9: Three-phase cycloconverter.

13-6   CYCLOCONVERTERS

In low speed and very large horsepower applications, it is possible to use cycloconverters to control the speed of synchronous and induction motors. A basic cycloconverter circuit that utilizes line-frequency-commutated converters is shown in Fig. 13-9a. The three-phase 60-Hz input is through isolation transformers. Each phase consists of two back-to-back connected line-frequency thyristor converters as discussed in Chapter 4. The firing (or delay) angles of the two converters in each phase are cyclically controlled to yield a low-frequency sinusoidal output. One of the phase outputs is shown in Fig. 13-9b where the operating mode (rectification or inversion) of the positive and the negative converters depend on the direction of the output load current.

The cycloconverter output is derived directly from the line frequency input, without an intermediate dc link. The maximum output frequency is limited to about one-third of the input ac frequency to maintain an acceptable waveform with a low harmonic content.

13-7   SUMMARY

1. In synchronous motors, the flux is produced by the rotor either by means of permanent magnets or by a field winding excited by a dc current. This feature of synchronous motors allows them to offer higher efficiencies compared to induction motors of similar ratings. In large horsepower ratings, a wound-field construction is used, whereas permanent magnet rotors are used at smaller power ratings.

2. The synchronous motor drives can be categorized as (a) precision servo drives for computer peripheral equipment and robotics and (b) adjustable-speed drives for controlling the speed such as in load-proportional capacity-modulated air conditioners and heat pumps. In very large (> 1000 hp) power ratings, the load-commutated-inverter (LCI) synchronous motor drives may be used due to their higher efficiency and a simpler inverter compared with the induction motor drives.

3. A per-phase equivalent circuit can be drawn in terms of the induced emf $\mathbf{E}_{fa}$, $\mathbf{I}_a$ and the per-phase inductances as shown in Fig. 13-2 for phase a.

4. For an optimal torque condition in servo drives, the torque angle δ is kept at $90°$. This results in a decoupling between the stator field and the rotor field. Thereby, the torque T_{em} required by the load is supplied by controlling the stator current amplitude I_s, without affecting the field flux ϕ_f. This makes T_{em} linearly proportional to I_s.

5. Synchronous motors used for servo drives can be broadly classified as (a) sinusoidal waveform motors and (b) trapezoidal waveform motors.

6. In the sinusoidal waveform synchronous motor drives, the rotor position θ with respect to a stationary axis is accurately measured by means of an absolute position encoder, for example a resolver. The three stator phase currents i_a, i_b, and i_c are calculated based on θ as given by Eqs. 13-25 to 13-27 where the amplitude I_s is determined by the torque requirement. Figure 13-5 shows the overall block diagram of such a servo drive.

7. The trapezoidal waveform synchronous motor drives are used for both the servo and the adjustable-speed applications. Here, the induced emfs, for example e_{fa}, in phase a has a trapezoidal waveform with a $120°$ long flat portion during each half-cycle. The stator currents have a rectangular waveform as shown in Fig. 13-6a. Since the stator currents are rectangular, their on and off instants are determined by the rotor position, which is determined by three Hall-effect sensors in servo drives, or by measuring the three-phase terminal voltage waveforms in non-servo drives.

8. In the LCI drives in very large horsepower ratings, a current-source thyristor inverter is used. This inverter is similar to a line-frequency inverter as discussed in Chapter 4 since

the current commutation from one phase to the next is provided by the induced emfs in the motor.

9. In low speed and very large horsepower applications, cycloconverters can be used.

PROBLEMS

13-1. A brushless, permanent-magnet, four-pole, three-phase motor has the following parameters:

$$\text{Torque constant} = 0.229 \, N\text{m/A}$$

$$\text{Voltage constant} = 24.0 \, \text{V/krpm}$$

$$\text{Phase-to-phase resistance} = 8.4 \, \Omega$$

$$\text{Phase-to-phase winding inductance} = 16.8 \, \text{mH}$$

The above motor produces a trapezoidal back-emf. The torque constant is obtained as a ratio of the maximum torque produced to the current flowing through two of the phases. The voltage constant is the ratio of the peak phase-to-phase voltage to the rotational speed. If the motor is operating at a speed of 3000 rpm and delivering a torque of 0.25 Nm, plot the idealized phase current waveforms.

13-2. In a sinusoidal waveform, three-phase, two-pole brushless dc motor with a permanent magnet rotor, k_T equals 0.5 Nm/A where k_T is defined by Eq. 13-22. Calculate i_a, i_b, and i_c if the motor is required to supply a holding torque of 0.75 Nm (to keep the load from moving) at the rotor position of $\theta = 30°$, where θ is defined in Fig. 13-4.

13-3. Estimate the minimum dc input voltage to the switch-mode converter required to supply the motor in Problem 13-1, if the maximum speed is 5000 rpm.

13-4. In a sinusoidal-waveform permanent-magnet brushless servo motor, phase-to-phase resistance is 8.0 Ω and the phase-to-phase inductance is 16.0 mH. The voltage constant, which is the ratio of the peak phase voltage induced to the rotational speed is 25 V/krpm. p = 2 and n = 10,000 rpm. Calculate the terminal voltage, if the load is such that the motor draws 10 A rms per phase. Calculate the power factor of operation.

13-5. Show the relationship between k_E and k_T in Eqs. 13-28 and 13-29 for a trapezoidal waveform brushless motor. Compare the result with the ratio of the torque-constant to the voltage-constant of the motor specified in Problem 13-1.

13-8 REFERENCES

1. A. E. FITZGERALD, C. KINGSLEY and S. D. UMANS, *Electrical Machinery*, 4th Ed., McGraw-Hill, New York, 1983.
2. B. K. BOSE, *Power Electronics and AC Drives*, Prentice-Hall, Englewood Cliffs, NJ, 1986.
3. T. KENJO and S. NAGAMORI, *Permanent-Magnet and Brush-less DC Motors*, Claredon Press, Oxford, 1985.
4. *DC Motors · Speed Controls · Servo Systems—An Engineering Handbook*, 5th Ed., Electro-Craft Corporation, 1600 Second Street South, Hopkins, MN, 1980.
5. D. M. ERDMAN, H. B. HARMS, and J. L. OLDENKAMP, "Electronically commutated DC Motors for the Appliance Industry," *IEEE/IAS 1984 Annual Meeting Record*, pp. 1339–1345.

6. S. Meshkat and E. K. Persson, "Optimum Current Vector Control of a Brushless Servo Amplifier Using Microprocessors," *IEEE/IAS 1984 Annual Meeting Record*, pp. 451–457.

7. R. H. Comstock, "Trends in Brushless Permanent Magnet and Induction Motor Servo Drives," *Motion Magazine*, Second Quarter, pp. 4–12, 1985.

8. P. Zimmerman, "Electronically Commutated DC Feed Drives for Machine Tools," *Drives and Controls International*, pp. 13–19, Oct./Nov. 1982.

9. L. Gyugyi and B. R. Pelly, *Static Power Frequency Changers*, John Wiley & Sons, New York, 1975.

10. T. Undeland, S. Midttveit, and R. Nilssen, *Phasor-Applied Control (PAC) of Induction Motors: A New Concept for Servo-Quality Dynamic Performance*, 1986 Conference on Applied Motion Control, Minneapolis, MN, pp. 1–8.

14

Step-motor Drives

14-1 INTRODUCTION

Step-motors are used for position control in computer peripherals, the textile industry, integrated-circuit fabrication, robotics and many other applications. A step-motor can be considered as a digital electromechanical device where each electrical pulse input results in a movement of the rotor by a discrete angle called the step-angle of the motor. Therefore, a step-motor can be used to control position without the position feedback, by keeping count of the number of electrical pulse inputs.

There are a large variety of step-motors. These can be divided into three basic categories: variable-reluctance motors, permanent-magnet motors, and hybrid motors. Each of these is briefly discussed.

14-2 VARIABLE-RELUCTANCE STEP-MOTOR

In a variable-reluctance step-motor, both the stator and the rotor have different magnetic reluctance along various radial axes. The basic principle of operation of a variable-reluctance step-motor can be explained by considering an example shown in Fig. 14-1. The cross-sectional view of the motor chosen as an example shows that there are six stator teeth ($N_s = 6$). A phase winding is placed on two diametrically opposite teeth. Therefore, the number of stator phases $q = 3$. The number of teeth N_r on the rotor are not equal to N_s. Generally, $N_r = N_s \pm (N_s/q)$. In Fig. 14-1, $N_r = 4$.

In Fig. 14-1a, phase A is excited by a current i_A. This results in a torque T_{em} on the rotor that acts in a direction to minimize the magnetic reluctance to the flux produced by the current in phase A. This torque causes the rotor to align with phase A at $\theta = 0$ as shown in Fig. 14-1a, if there is no load torque on the rotor. For small deviations in θ, it can be shown from energy considerations that

$$T_{em} = \tfrac{1}{2}i_A^2\frac{dL}{d\theta} \qquad (14\text{-}1)$$

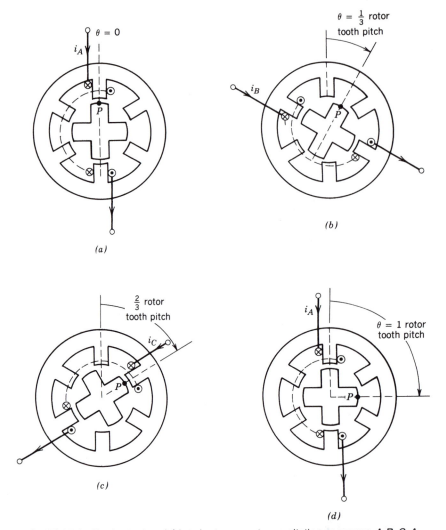

FIGURE 14-1: Single-stack variable-reluctance motor; excitation sequence *A-B-C-A—* (rotor positions shown for a zero load torque): (*a*) phase *A* excited; (*b*) phase *B* excited; (*c*) phase *C* excited; (*d*) phase *A* excited.

where the phase inductance L is a function of θ since it equals the square of the number of turns of the phase winding divided by the magnetic reluctance $R(\theta)$, which depends on the rotor position θ. The electromagnetic torque T_{em} at the equilibrium position $\theta = 0$ is zero. T_{em} is proportional to i_A^2 and is independent of the current direction. Defining T_{em} and θ to be positive in a clockwise direction, Fig. 14-2 shows a plot of T_{em} as a function of θ with a constant i_A. For a load torque in a counterclockwise direction, the rotor steady-state position will be $-\theta_1$ as shown in Fig. 14-2, where T_{em} is equal but opposite to the load torque.

With i_A equal to zero and phase B excited by i_B, the no-load equilibrium position is as shown in Fig. 14-1*b*. The point P on the rotor moves by the step-angle of the motor (if load was finite but constant, the rotor will still move by the same motor step-angle). Figures 14-1*c* and 14-1*d* show the rotor position for the next two excitations. For the excitation sequence A-B-C-A—, three changes of excitations shown in Fig. 14-1 result in the rotor movement by one rotor tooth pitch. The rotor

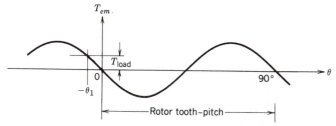

FIGURE 14-2: T_{em} versus θ with a constant i_A.

tooth-pitch equals $(360/N_r)$ degrees. Since all step-angles are equal, the step-angle of the three-phase motor in Fig.14-1 is $(360/N_r)/3 = 30°$. In general for a q-phase motor with N_r rotor teeth

$$\text{Step-angle} = \frac{360}{qN_r}\text{degrees} \qquad (14\text{-}2)$$

The direction of rotation can be controlled by the phase excitation sequence. For an excitation sequence A-C-B-A—, the rotor will move counterclockwise.

The motor described above is a single-stack variable-reluctance motor. There are multiple-stack variable-reluctance step-motors that operate on the same basic principle as described before.

14-3 PERMANENT-MAGNET STEP-MOTORS

The operating principle of a permanent-magnet step-motor can be illustrated by means of a cross section of a simple motor shown in Fig. 14-3, where the rotor is magnetized to consist of four permanent-magnet poles and the stator contains two phase windings that can be excited with either polarity currents (i_{A^+} refers to a positive current in phase A and i_{A^-} refers to a negative current). Each phase winding produces the same number of poles as the rotor. The magnetic poles produced by the stator currents cause the rotor to move as shown in Fig. 14-3, for the excitation sequence $i_{A^+}, i_{B^+}, i_{A^-}, i_{B^-}, i_{A^+}, \ldots$. By tracking the movement of the point P on the rotor, it is clear that the step-angle for the motor shown here is $45°$. It is important to be able to control the direction of phase currents in a sequence

$$i_{A^+}, i_{B^+}, i_{A^-}, i_{B^-}, i_{A^+}, \ldots \quad \text{for a clockwise rotation}$$

and

$$i_{A^+}, i_{B^-}, i_{A^-}, i_{B^+}, i_{A^+}, \ldots \quad \text{for a counterclockwise rotation.}$$

The electromagnetic torque T_{em} is produced by the interaction of the stator and rotor magnetic fields. Therefore, T_{em} is proportional to the phase current and a function of the small deviation in θ from its equilibrium position.

The permanent-magnet motors suffer from the disadvantage of higher inertia-to-torque ratio and the difficulty in manufacturing motors with a fairly small step-angle. They have the advantage that there is some torque to maintain position, if the drive fails.

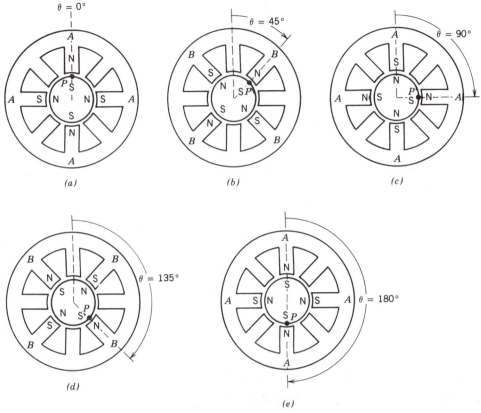

FIGURE 14-3: Two-phase permanent-magnet step motor; excitation sequence i_{A^+}, i_{B^+}, i_{A^-}, i_{B^-}, i_{A^+}, —: (a) i_{A^+}; (b) i_{B^+}; (c) i_{A^-}; (d) i_{B^-}; (e) i_{A^+}.

14-4 HYBRID STEP-MOTORS

Also called the synchronous-inductor motors, these commonly used step-motors combine the principles of the variable-reluctance and the permanent-magnet type of step-motors. An axial view of such a motor is shown in Fig. 14-4. The rotor consists of a permanent magnet that is magnetized parallel to the shaft axis to create a pair of poles. On this magnet, two end-caps are fitted at both ends. These end-caps consist of equal number of teeth N_r. The cross-sectional views, perpendicular to the shaft along X-X' and Y-Y' axes in Fig. 14.4 are shown in Fig. 14-5. The rotor teeth along X-X' act like S poles and the rotor teeth along Y-Y' act like N poles. The flux produced by the permanent magnets is shown dotted in Fig. 14-4. The stator is one single unit (laminated) where the winding slots are parallel to the motor shaft.

In the motor of Fig. 14-5, the stator consists of two phase windings, and each phase winding produces four stator poles. Excitation of a stator phase creates flux in a radial direction through the two air-gaps. Therefore, for example, in Fig. 14-5a, the flux through the stator pole number 1 is from the stator into the rotor in the cross section X-X' as well as in Y-Y', for a positive current i_{A^+} in phase A. In this example, each rotor end-cap consists of 10 teeth ($N_r = 10$). The rotor end-caps are misaligned by $\frac{1}{2}$ rotor-tooth-pitch on purpose, which is easily done since the two end-caps are independently fitted on the magnet. The stator teeth are, of course, aligned between sections X-X' and Y-Y', since the stator is a single unit.

The stator poles 1, 3, 5, 7 are excited by phase A and poles 2, 4, 6, 8 by phase B winding. Each phase winding produces adjacent poles of opposite polarity, that is,

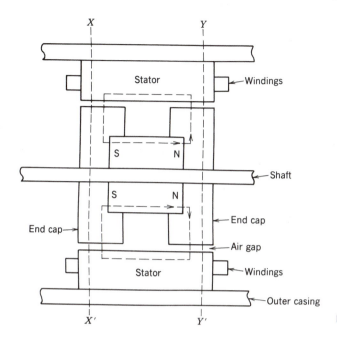

FIGURE 14-4: Hybrid step-motor: axial view.

- Positive $i_A(i_{A^+})$ causes the stator poles 1 and 5 to be north and 3 and 7 to be south. Therefore, because of i_{A^+} alone, the flux lines would be radially into-the-rotor at poles 1 and 5 and radially out-of-the-rotor at poles 3 and 7. This will be the same at cross section X-X' as well as Y-Y'. The opposite poles are created by i_{A^-}.

- i_{B^+} creates flux lines that are radially into the rotor at poles 2 and 6, and radially out of the rotor at poles 4 and 8.

When a phase winding is excited, a torque is exerted on the rotor because of its variable reluctance, to bring it into a position to maximize the flux linkage λ of the excited phase. In Fig. 14-5a, where the phase A winding is excited by a positive current i_{A^+}, the interaction of the rotor poles and the stator poles due to i_{A^+} results in the rotor position shown. In this position, under the stator poles where the rotor and the stator flux lines aid each other, the teeth are in alignment in both sections X-X' and Y-Y'. The opposite is true where these flux lines oppose each other. These two conditions exist simultaneously since the rotor teeth in the two end-caps are misaligned by $\frac{1}{2}$ rotor-tooth-pitch. The net effect is that with i_{A^+}, the equilibrium position of the rotor shown in Fig. 14-5a results in maximizing the flux linkage with the A phase winding.

Figure 14-5b shows that the rotor moves clockwise when the phase B is excited by a positive current i_{B^+}. This change of excitation from i_{A^+} to i_{B^+} causes the rotor to move by $\frac{1}{4}$ of the rotor-tooth-pitch in a clockwise direction, as indicated by the movement of point P on the rotor in Fig. 14-5. For an excitation sequence $i_{A^+}, i_{B^+}, i_{A^-}, i_{B^-}, i_{A^+}, \ldots$, the rotor rotates by one rotor-tooth-pitch for four changes in excitation. Therefore,

$$\begin{array}{c} \text{Step-angle} \\ \text{(2-phase motor)} \end{array} = \frac{\text{Rotor-tooth-pitch}}{4} = \frac{(360°/N_r)}{4} \qquad (14\text{-}3)$$

$$= 9° \ (\text{for } N_r = 10)$$

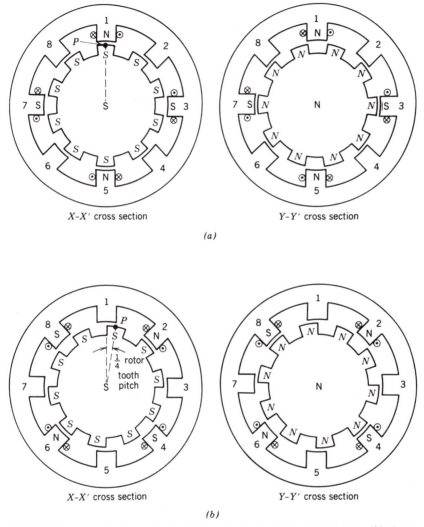

FIGURE 14-5: Hybrid step-motor excitation: (*a*) phase *A* is excited with i_{A+}; (*b*) phase *B* is excited with i_{B+}.

Torque in a hybrid step-motor is produced by interaction of the rotor- and the stator-produced fields. The rotor field is produced by a permanent magnet and hence stays constant. The stator field and therefore T_{em} is proportional to the phase current. Commonly available hybrid step-motors have a step angle of 1.8° (200 steps/revolution).

14-5   MODES OF EXCITATION IN STEP-MOTORS

In the previous sections, a single-phase excitation was assumed, where each of the stator phases is excited one at a time and the rotor moves by a step-angle for each change in excitation. In practice, there are various other possibilities, some of which are described here. It is advantageous to derive an equivalent circuit prior to describing these excitation modes.

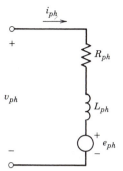

FIGURE 14-6: Equivalent circuit — per phase.

14-5-1 Equivalent-Circuit of a Step-Motor

The equivalent circuit for a step-motor phase winding is shown in Fig. 14-6, where in addition to the winding resistance and inductance, a back-emf e_{ph} of the polarity indicated is induced in the stator phase winding. In permanent-magnet and hybrid motors, as the rotor rotates, an emf e_{ph} is induced in the phase winding. The polarity of this induced voltage is such as to absorb power from the electrical source v_{ph} and its magnitude is proportional to the rotational speed ω

$$e_{ph} = k_E \omega \qquad (14\text{-}4)$$

where k_E is a voltage constant. The equivalent circuit of Fig. 14-6 is also valid for the variable-reluctance step-motors, where

$$v_{ph} - R_{ph} i_{ph} = \frac{d\lambda_{ph}}{dt} \qquad (14\text{-}5)$$

and the phase flux linkage

$$\lambda_{ph} = L_{ph} i_{ph}. \qquad (14\text{-}6)$$

By combining Eqs. 14-5 and 14-6, we find that

$$v_{ph} - R_{ph} i_{ph} = L_{ph}\frac{di_{ph}}{dt} + i_{ph}\frac{dL_{ph}}{dt} = L_{ph}\frac{di_{ph}}{dt} + \omega i_{ph}\frac{dL_{ph}}{d\theta} \qquad (14\text{-}7)$$

where the rotational angle $\theta = \omega t$. The second term on the right side of Eq. 14-7 is proportional to ω and hence is equivalent to e_{ph}. Therefore, in a variable-reluctance step-motor, the magnitude of the induced back-emf is

$$e_{ph} = \omega i_{ph}\frac{dL_{ph}}{d\theta} \qquad (14\text{-}8)$$

14-5-2 Two-Phase Excitation

In a step-motor operating in a step-mode where the rotor comes to rest after each step move, a change in excitation results in a step-response, where the rotor oscillates prior to settling down to its next equilibrium position. Quite often, two of the stator phases are simultaneously excited to provide a better damping of the rotor oscillations compared to a single-phase excitation.

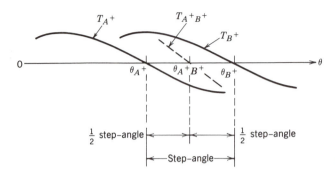

FIGURE 14-7: Two-phase excitation.

Figure 14-7 shows T_{A^+} and T_{B^+} as function θ in a two-phase hybrid motor similar to that in Fig. 14-5 with i_{A^+} and i_{B^+}, respectively. With both i_{A^+} and i_{B^+}, the resultant torque $T_{A^+B^+}$ is the sum of T_{A^+} and T_{B^+}, and is shown as dotted in Fig. 14-7. Figure 14-7 shows that the equilibrium position with two-phase excitation is midway between the equilibrium positions corresponding to i_{A^+} and i_{B^+} applied one at a time. A clockwise rotation in a hybrid motor of Fig. 14-5 requires the following two-phase excitation sequence: $(i_{A^+}i_{B^+}), (i_{B^+}i_{A^-}), (i_{A^-}i_{B^-}), (i_{B^-}i_{A^+}), (i_{A^+}i_{B^+}), \ldots$. Similarly a two-phase excitation scheme can be obtained for a variable-reluctance and a permanent-magnet step-motor (see Problems 14-1 and 14-2).

14-5-3 Half-Step Mode of Excitation

In the previous sections, the single-phase and the two-phase excitation schemes are discussed that cause the rotor to move by one step-angle for each change in excitation. Combining the single-phase and two-phase excitations makes it possible to achieve half-step rotations for each change of excitation. The clue to this is provided by the Fig. 14-7, which shows that the equilibrium positions for a single-phase excitation (either with i_{A^+} or i_{B^+}) are one-half step away from the position with a simultaneous two-phase excitation ($i_{A^+}i_{B^+}$). Therefore, to achieve a clockwise rotation in a hybrid motor of Fig. 14-5, the single-phase and the two-phase excitations can be combined to result in a half-step rotation by means of the following excitation sequence: $i_{A^+}, (i_{A^+}\, i_{B^+}), i_{B^+}, (i_{B^+}i_{A^-}), i_{A^-}, (i_{A^-}i_{B^-}), i_{B^-}, (i_{B^-}i_{A^+}), i_{A^+}, (i_{A^+}i_{B^+}), \ldots$. Similar sequences can be obtained for the permanent-magnet and the variable-reluctance motors (see Problems 14-3 and 14-4).

14-5-4 Micro-Stepping Mode of Operation

The procedure for half-stepping can be extended to subdivide the motor step-angle into very small steps called micro-steps. This requires that the magnitudes of phase currents be precisely controlled. Assuming the torque characteristics for a two-phase hybrid motor to be sinusoidal as in Fig. 14-8 and the equilibrium position for i_{A^+} excitation as $\theta = 0$, T_A and T_B with i_A and i_B respectively can be expressed as

$$T_A = -k\,i_A \sin \theta \tag{14-9}$$

$$T_B = +k\,i_B \cos \theta \tag{14-10}$$

where k is the motor torque constant and the rotor position θ is measured in electrical degrees (i.e., 360 degrees equal one rotor tooth-pitch). Note that i_A and i_B in Eqs. 14-9 and 14-10 can be positive or negative.

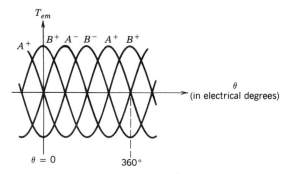

FIGURE 14-8: Torques in a two-phase hybrid motor.

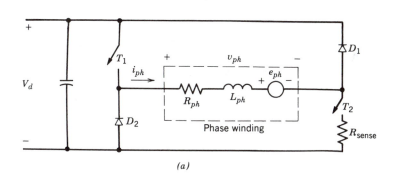

(a)

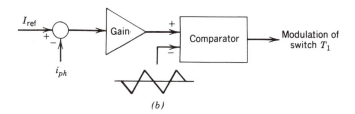

(b)

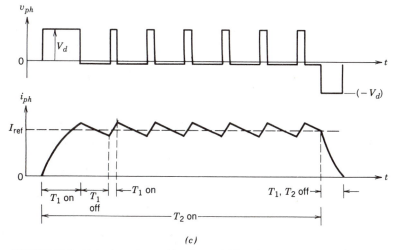

(c)

FIGURE 14-9: Unipolar voltage drive for variable-reluctance motor: (a) circuit; (b) current-mode control; (c) waveforms.

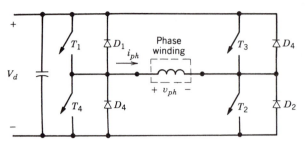

FIGURE 14-10: Bipolar voltage drive — per phase.

When both phases are excited simultaneously with i_A and i_B, the total torque developed is

$$T_{AB} = k(-i_A \sin \theta + i_B \cos \theta) \qquad (14\text{-}11)$$

To change the rotor angle θ by a micro-step angle μ, the two-phase currents are precisely controlled to two discrete levels such that

$$i_A = I_R \cos \mu$$

and $(14\text{-}12)$

$$i_B = I_R \sin \mu$$

where I_R is the rated phase current. Therefore, from Eqs. 14-11 and 14-12,

$$T_{AB} = kI_R(-\cos \mu \sin \theta + \sin \mu \cos \theta) = -kI_R \sin(\theta - \mu) \qquad (14\text{-}13)$$

The no-load stable equilibrium position corresponds to $T_{AB} = 0$, which occurs at $\theta = \mu$ in Eq. 14-13, with the currents i_A and i_B given by Eq. 14-12. Therefore, by varying i_A and i_B sinusoidally in discrete steps, the step-motor can be made to move in microsteps.

In commercially available micro-step controllers, a step-angle can be divided into as high as 125 micro-steps. With a two-phase hybrid motor with a step-angle of 1.8° or 200 steps/revolution, this controller results in 25,000 micro-steps/revolution.

14-6   DRIVE CIRCUITS FOR STEP-MOTORS

In order to maximize the torque capability of a step-motor, the drive circuit should be such that the turn-on current transition from 0 to a desired level and the turn-off transition, where the current is forced to zero, occur in as short a duration as possible. This is specially important at high stepping rates where the on and off transition times may correspond to a significant fraction of the time-interval for a step-angle rotation, thus resulting in a reduced torque output.

After a phase turn-on, the current should be held at its desired value (usually the rated value). At turn-off, as the current is forced to zero, it is desirable to recover the energy associated with the phase winding inductance, rather than dissipating it as heat.

In this section, only the appropriate drive circuits for energy-efficient operation of relatively high-power step-motors are discussed for the three types of step-motors.

14-6-1 Unipolar Voltage Drive for Variable-Reluctance Motors

In a variable-reluctance step-motor, it is sufficient to have unidirectional phase currents. A unipolar dc voltage V_d can be used to supply a phase current through a two-switch converter, as is shown in Fig. 14-9a. The motor phase is replaced by its equivalent circuit of Fig. 14-6. The magnitude of V_d is selected such that it can provide a rapid current buildup even at high stepping rates where the motor back-emf e_{ph} is large (e_{ph} opposes V_d during the buildup of phase current).

To energize a phase, both T_1 and T_2 in Fig. 14-9a are turned on. This applies V_d to the phase winding and i_{ph} builds up as shown in Fig. 14-9c. T_2 is kept closed so long as the phase is to remain energized. To maintain the phase current at I_{ref}, the actual phase current is sensed by means of a sense resistor R_{sense} in Fig. 14-9a. A fixed-frequency current-mode PWM control as discussed in Section 6-6-3-2 of Chapter 6 can be used. It is shown in a block diagram form in Fig. 14-9b. The current-mode control modulates the switch T_1. When T_1 is on, $v_{ph} = V_d$ and i_{ph} increases. When T_1 is turned off, i_{ph} circulates through (T_2, D_2) and decreases. The corresponding phase voltage and i_{ph} waveforms are shown in Fig. 14-9c.

When the phase is to be de-energized, both T_1 and T_2 are turned off. Now the phase current flows through diodes D_1 and D_2 and $v_{ph} = -V_d$. This causes the phase current to decrease rapidly to zero. The inductive energy is fed back to the dc supply.

14-6-2 Bipolar Voltage Drive for Permanent-Magnet and Hybrid Step-Motors

In permanent-magnet and hybrid step-motors, each phase current needs to be bidirectional—positive or negative. Therefore, the polarity of the dc voltage applied to a phase winding shold be reversible, that is, a bipolar voltage drive is required. A full-bridge dc–dc converter, as shown in Fig. 14-10 and previously discussed in Chapter 5, is commonly used for this purpose. T_1 and T_2 are switched as a pair; T_3 and T_4 are switched as the other pair.

When describing the circuit operation, a positive current excitation is assumed. Therefore, T_1 and T_2 are turned on to build up the phase current in the positive direction. Once the current reaches its reference value, it is maintained at that value by means of a current mode control, as discussed for the unipolar voltage drive in the previous section. When the pair (T_1, T_2) is on, i_{ph} increases. To regulate the current, the controller turns (T_1, T_2) off and, after a short blanking time, turns (T_3, T_4) on. Now i_{ph} circulates through D_3 and D_4 and flows into V_d. This causes i_{ph} to decrease. Therefore, the current-mode regulator maintains the average value of i_{ph} at its reference value by switching between the two switch pairs. When the phase is to be deenergized, all switches are turned off and the phase current quickly reduces to zero, since $v_{ph} = -V_d$. The inductive energy is recovered and fed back to the dc supply.

A similar operation, with the roles of (T_1, T_2) and (T_3, T_4) reversed, occurs when i_{ph} is to be in a negative direction.

To avoid the need for a bipolar-voltage drive, the phase windings can be bifilar wound. As shown in Fig. 14-11a, it is possible to use a unipolar-voltage drive with only two switches per phase. The direction of magnetic field produced by the stator phase depends on which of the two switches in Fig. 14-11a is on. In such a motor, the phase windings are not efficiently utilized, since only one of the bifilar windings is energized at a given time. Moreover, snubber circuits are needed in the unipolar-voltage drive of Fig. 14-11a, to prevent excessive voltages across the switches due to inductive energy when a switch turns off. Because of these shortcomings, it is preferable to use a

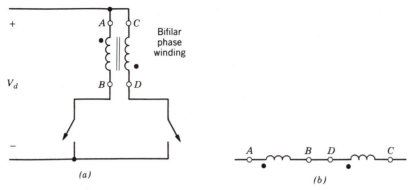

FIGURE 14-11: Bifilar-wound phase.

full-bridge bipolar-voltage drive by connecting the two bifilar-halves of such a motor, as shown in Fig. 14-11*b*.

14-7 OPEN-LOOP OPERATION OF STEP-MOTORS

In view of the discussion in the previous section, when moving an object from an initial to a target position, it is possible to establish a velocity (steps or pulses per second) profile as a function of time, as is shown in Fig. 14-12. Generally, the objective is to minimize the move time without missing any steps, with an acceptable settling time.

If the control is open-loop, any velocity profile selected must be tested out with the actual load (under all of its possible variations) to make sure that the operation takes place without missing any steps.

As is shown in Fig. 14-13, it is possible to keep track of the number of motor steps in the open-loop of operation. If a step is missed, it is immediately sensed and a corrective action may be required.

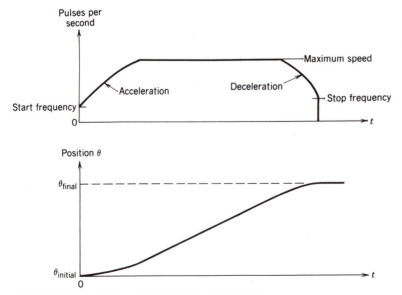

FIGURE 14-12: Velocity profile for open-loop control.

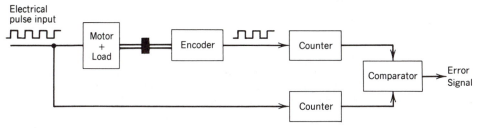

FIGURE 14-13: Detection of missed step in an open-loop operation.

14-8 CLOSED-LOOP CONTROL OF STEP-MOTORS

In a closed-loop control of step-motor, a position encoder is connected to the motor shaft. The next excitation pulse is applied only after the previous step movement is satisfactorily executed, as indicated by the position encoder. Therefore, there is no possibility of missing a step or losing synchronism. This control technique is desirable in applications where it is critical that not a single step be missed by the step-motor. This cannot be absolutely guaranteed in an open-loop operation, especially if there is a possibility of a sudden load change.

The price of this guaranteed synchronous operation is the added complexity of the position encoder and the closed-loop control. In such applications, servo-drives such as dc, synchronous, and the vector-controlled induction motor drives must also be considered along with a step-motor drive.

14-9 SWITCHED-RELUCTANCE MOTOR DRIVES

The switched-reluctance motors are basically variable-reluctance step-motors that are operated in a closed-loop manner. Based on the rotor position, the appropriate phases are energized or deenergized. These have the potential of competing with other servo and adjustable-speed drives.

14-10 SUMMARY

1. A step-motor can be considered a digital electromechanical device where each electrical pulse input results in a discrete output, in terms of a fixed step-angle by the motor. This presents an opportunity for an open-loop operation, whereby keeping count of the number of input pulses, it is possible to know the change in the rotor position.

2. The following three commonly used types of step-motors are discussed: variable-reluctance, permanent-magnet, and the hybrid step-motors.

3. The basic operating principle of a variable-reluctance motor is that when excited, a torque is produced in a direction to minimize the magnetic-circuit reluctance. The reluctance torque produced depends on the square of the phase current, and its direction is independent of the phase current direction.

4. There are two types of variable-reluctance step-motors: multiple-stack and single-stack. Both operate on the same basic principle. The step-angle equals $360°/(qN_r)$, where N_r is the number of rotor teeth; q is the number of stacks in the multistack motor and the number of phases in the single-stack motor, respectively.

5. A permanent-magnet step-motor consists of a permanent magnet on the rotor. The magnitude of the torque in this motor depends on the magnitude of the phase current, and the current direction should be reversible.

6. A hybrid step-motor combines the principles of the variable-reluctance and the permanent-magnet step-motors. The magnitude of the torque produced depends on the magnitude of the phase current. The direction of phase currents should be reversible. In a two-phase hybrid motor, the step-angle equals $90°/N_r$, where N_r is the number of rotor teeth.

7. Two modes of phase excitations are discussed that cause the rotor to move by one full-step for each change of excitation. These are the single-phase and the two-phase excitations. The two-phase excitation, where two phases are excited simultaneously, provides an electromagnetic damping of the mechanical oscillations.

8. By alternating between the single-phase and the two-phase excitations, it is possible to obtain a half-step mode of operation.

9. By precisely controlling the magnitudes of the phase currents, it is possible to obtain micro-stepping.

10. A unipolar voltage drive for a variable-reluctance motor is discussed. Once the phase current builds up to its rated value, it is maintained at that value until deenergized by a current-mode control similar to that discussed in Chapter 6.

11. A bipolar voltage drive is discussed for the permanent-magnet and the hybrid step-motors.

12. For an open-loop operation, a velocity profile should be defined. A closed-loop operation is also possible to ensure a synchronous operation without missing any steps.

13. Switched-reluctance motors are basically variable-reluctance step-motors where the rotor position determines the energizing and deenergizing of appropriate phases.

PROBLEMS

14-1. Determine the two-phase excitation sequence for clockwise rotation of a variable-reluctance step-motor of Fig. 14-1.

14-2. Repeat Problem 14-1 for a permanent-magnet step-motor of Fig. 14-3.

14-3. For half-step clockwise rotation in a variable-reluctance step-motor of Fig. 14-1, determine the excitation sequence.

14-4. Repeat Problem 14-3 for a permanent-magnet step-motor of Fig. 14-3.

14-5. In a 1.8° step-angle hybrid step-motor with both phases energized with rated currents, the peak electormagnetic torque is 0.75 Nm. Assuming the sinusoidal torque curves of Fig. 14-7, calculate the steady-state rotor position with respect to the no-load equilibrium position, for a load torque of 0.5 Nm.

14-11 REFERENCES

1. TAKASHI KENJO, *Stepping Motors and Their Microprocessor Controls*, Oxford Science Publications, Claredon Press, Oxford, 1985.
2. P. P. ACARNLEY, *Stepping Motors: A Guide to Modern Theory and Practice*, IEE Control Engineering Series 19, Revised 2nd Ed., 1984.
3. BENJAMIN C. KUO, (Ed.), *Incremental Motion Control: Step Motors and Control Systems*, SRL Publishing Company, Champaign, IL, 1979.
4. P. J. LAWRENSON et al., "Variable-Speed Switched-Reluctance Motors," *IEE Proceedings*, Vol. 127, Pt. B, No. 4, pp. 253–265, July 1980.

P A R T

5

Other Applications

15

Residential and Industrial Applications

15-1 INTRODUCTION

Power electronic converters are described in a generic manner in Chapters 1 to 7. Their applications in dc and ac power supplies are described in Chapters 8 and 9, respectively, and in motor drives in Chapters 10 to 14. The objectives of this chapter are twofold: (1) to give a brief overview of various residential power electronic applications and (2) to describe some additional industrial applications of power electronics such as welding and induction heating.

15-2 RESIDENTIAL APPLICATIONS

Residential homes and buildings are the endpoint of approximately 35% of the total electricity generated in the United States, which corresponds to approximately 8.5% of the total primary energy usage. The residential applications include space heating and air conditioning, refrigeration and freezer, water heating, lighting, cooking, television, clothes washer and dryer, and many other miscellaneous appliances.

The role of power electronics in residential applications is to provide energy conservation, reduced operating cost, increased safety, and greater comfort. Benefits of incorporating power electronics into some of the dominant residential applications are discussed here.

15-2-1 Space Heating and Air Conditioning

Approximately 25 to 30% of the electric energy in an all-electric home is used for space heating and air conditioning. Heatpumps are now being used in one out of every three new homes. Incorporating load-proportional capacity modulation can increase the heatpump efficiency by as much as 30% over the conventional single-speed

377

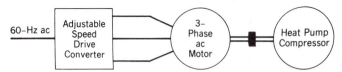

FIGURE 15-1: Load-proportional capacity modulated heat pump.

heatpumps. In a conventional heatpump, the compressor operates essentially at a constant speed when the motor is running. The compressor output in this system is matched to the building heating or cooling load by cycling the compressor on or off. In a load-proportional capacity modulated heatpump (shown in Fig. 15-1), the speed of the compressor motor and hence the compressor output is adjusted to match the building heating or cooling load, thus eliminating the on or off cycling of the compressor. Either an induction motor drive discussed in Chapter 12 or a self-synchronous motor drive discussed in Chapter 13 is used to adjust the compressor speed, in proportion to the building load.

The benefits of eliminating on or off cycling of the compressor are discussed by means of Fig. 15-2. In a conventional heatpump system in the cooling mode, if the sensed temperature of the building exceeds the upper limit of the thermostat setting, the compressor motor is turned on. The electric motor almost immediately begins to draw its maximum electrical power, but the compressor output increases slowly, as shown in Fig. 15-2. Therefore, the shaded area in the plot of the compressor output represents a loss in the compressor output and hence a loss in the energy efficiency of the system. When the building temperature reaches the lower limit of the thermostat temperature setting, the motor and the compressor are turned off. By this on/off cycling, the average compressor output, shown by a dotted line in the plot of the compressor output in Fig. 15-2, is matched to the building load, and the building temperature is maintained within a tolerance band around the thermostat temperature setting.

The loss in the compressor output due to on/off cycling is eliminated in the load-proportional capacity modulated heatpump, where the speed of the compressor and, hence, the compressor output is adjusted to equal the building load. In spite of some losses in the power electronic converter used in this system, the overall electric energy consumed can be reduced by as much as 30% compared with the conventional single-speed heatpumps. Moreover, the building temperature can be maintained in a narrower band, thus resulting in increased comfort.

15-2-2   High-Frequency Fluorescent Lighting

Lighting consumes approximately 15% of the energy in residential buildings and 30% in commercial buildings. Fluorescent lamps are 3 to 4 times more energy efficient compared with the incandescent lamps. The energy efficiency of fluorescent lamps can

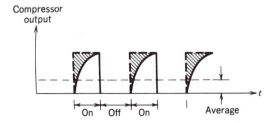

FIGURE 15-2: Conventional heat pump waveforms.

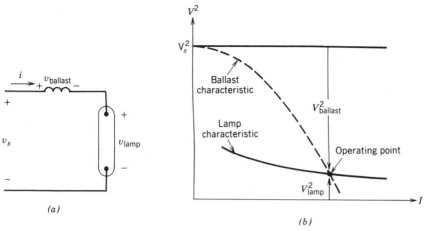

FIGURE 15-3: Fluorescent lamp with an inductive ballast.

be further increased by 20 to 30% by operating them at a high frequency (> 25 kHz), compared to the conventional 60-Hz fluorescent lamps.

Fluorescent lamps exhibit a negative resistance characteristic. This requires that an inductive ballast (also called a choke) be used in series for stable operation, as shown in the simplified schematic of Fig. 15-3a. Since the lamp impedance is essentially resistive, the three voltages in the circuit of Fig. 15-3a are related as

$$V_{ballast}^2 + V_{lamp}^2 = V_s^2 \qquad (15\text{-}1)$$

The lamp and the ballast characteristics are plotted in Fig. 15-3b in terms of V^2 and I. The intersection of the two characteristics provides a stable operating point.

Figure 15-4a shows a circuit schematic for the conventional 60-Hz rapid-start system consisting of two lamps in series. In this system, the lamp cathodes are continuously heated by the cathode heater windings A, B, and C. The circuit is redrawn in Fig. 15-4b without the heating windings to explain the basic operation. The input voltage is boosted by the autotransformer (primary in series with the secondary). The leakage inductances of the primary and the secondary transformer windings provide the ballast inductance needed for a stable operation. The starting capacitor has a low impedance compared to an unignited lamp and a high impedance compared to an ignited lamp. Therefore at start-up, the starting capacitor provides a shunt across lamp B and nearly all of the input voltage appears across lamp A thus striking an arc. Once the arc discharge is established in lamp A, a high voltage appears across lamp B, which ignites an arc in lamp B. Then, the series combination of lamps A and B is in series with a power actor correction capacitor C_{pf}, which is used to correct an otherwise poor power factor of operation.

The high-frequency fluorescent lighting system is shown in a block diagram form in Fig. 15-5a. The high-frequency electronics ballast converts the 60-Hz ac input into a high-frequency output, usually in a range of 25 to 40 kHz. The block diagram of the high-frequency electronics ballast as is shown in Fig. 15-5b consists of a diode rectifier bridge discussed in Chapter 3 and a dc-to–high-frequency-ac inverter. The inversion of dc to high-frequency ac can be obtained in one of the several ways discussed in the previous chapters: for example, a class-E resonant converter discussed in Chapter 7 can be used to produce sinusoidal lamp voltage and current; another possibility is to use a switch-mode converter, for example a half-bridge topology, as

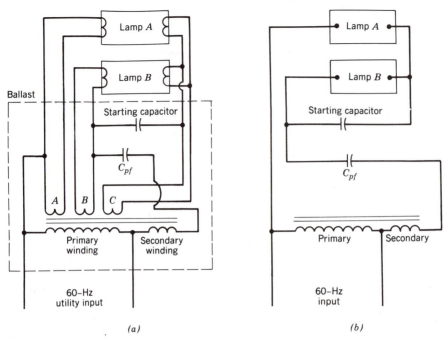

FIGURE 15-4: Conventional 60-Hz rapid-start fluorescent lamp: (*a*) circuit schematic; (*b*) simplified schematic.

discussed in Chapter 8, but without the isolation transformer and the output rectifying stage. An EMI filter is used before the rectifier bridge to suppress the conducted EMI. As in most power electronics equipment, the current drawn by the ballast from the utility system will contain significant harmonics and hence the electronic ballast will operate at a poor power factor. The problem of harmonics can be remedied efficiently by an input current waveshaping circuit described in Chapter 17.

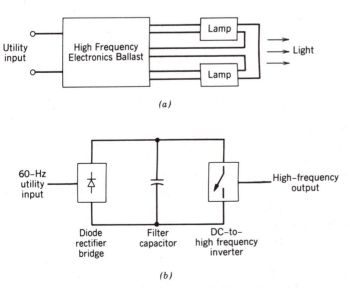

FIGURE 15-5: High-frequency fluorescent lighting system: (*a*) system block diagram; (*b*) ballast block diagram.

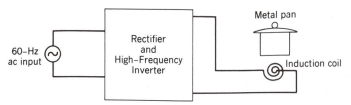

FIGURE 15-6: Induction cooking.

Because a large electromagnetic ballast associated with a standard 60-Hz fluorescent system is not required, the electronic ballasts in general are more energy efficient compared to the standard ballasts. A dimming control can be incorporated in the 60-Hz as well as the high-frequency lighting systems to compensate for the daylight coming in through the windows. In addition, a dimming control can lead to significant energy savings in the following manner: the lumen capacity of a lamp diminishes with time. Therefore, the new lamps are selected to have a lumen capacity that is approximately 30% higher than the nominal requirement. With a dimming control, new lamps can be operated at a reduced power to deliver the nominal requirement, thus resulting in energy savings during the period while the lamps have a high lumen capacity.

15-2-3　Induction Cooking

In a standard electric or gas cooking range, a significant amount of heat escapes to the surroundings, thus resulting in poor thermal efficiency. This can be avoided by means of induction cooking, which is shown in Fig. 15-6 in a block diagram form. The 60-Hz ac input is converted to a high-frequency ac in a range of 25 to 40 kHz, which is supplied to an induction coil. This induces circulating currents in the metal pan on top of the induction coil, thus directly heating the pan. Similar circuits as discussed in connection with the high-frequency electronics ballasts for fluorescent lights can be used to convert 60-Hz ac input into high-frequency ac.

15-3　INDUSTRIAL APPLICATIONS

Industrial applications such as induction heating and welding are discussed here in terms of the converter circuits discussed in the previous chapters.

15-3-1　Induction Heating

In induction heating, the heat in the electrically conducting workpiece is produced by circulating currents caused by electromagnetic induction. Induction heating is clean, quick, and efficient. It allows a defined section of the workpiece to be heated accurately. The magnitude of the induced currents in workpiece decreases exponentially with the distance x from the surface as given by the following equation

$$I(x) = I_0 e^{-x/\delta} \tag{15-2}$$

where I_0 is the current at the surface and δ is the penetration depth at which the current is reduced to I_0 times a factor $1/e$ ($\simeq 0.368$). The penetration depth is inversely proportional to the square-root of frequency f and proportional to the square-root of the workpiece resistivity ρ

$$\delta = k\sqrt{\frac{\rho}{f}} \qquad (15\text{-}3)$$

where k is a constant. Therefore, the induction frequency is selected based on the application. A low frequency such as the utility frequency may be used for induction melting of large workpieces. High frequencies of up to a few hundred kilohertz are used for forging, soldering, hardening, and annealing.

The circulating currents are caused in the workpiece by the currents in the induction coil. The induction coil is inductive and the induction load can be represented by an equivalent resistance in series with the coil inductance or by an equivalent parallel resistance as shown in Figs. 15-7a and 15-7b, respectively. A resonant capacitor is used to supply a sinusoidal current to the induction coil and to compensate for the poor power factor due to the coil inductance. This leads to the following two basic circuit configurations:

1. Voltage-source, series-resonant inverters as shown in Fig. 15-7a.

2. Current-source, parallel-resonant inverters as shown in Fig. 15-7b.

The voltage-source series-resonant inverter configuration of Fig. 15-7a is similar to the series-loaded resonant (SLR) converters discussed in Chapter 7. The inverter input is a dc voltage and the output is a square-wave voltage at the desired frequency. If the operating frequency is chosen to be near the resonant frequency, then the

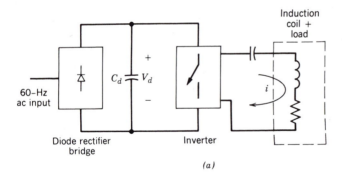

(a)

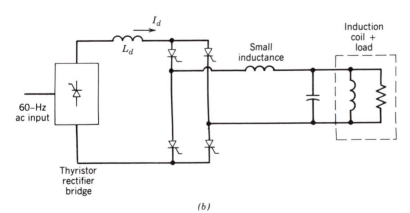

(b)

FIGURE 15-7: Induction heating: (*a*) voltage-source series-resonant induction heating; (*b*) current-source parallel-resonant induction heating.

current i will be essentially sinusoidal due to the impedance characteristic shown in Fig. 7-7 of Chapter 7 for a series-resonant circuit. Up to a few tens of kilohertz, it is possible to use thyristors as switches in the inverter. This will require that the operating frequency be below the resonant frequency so that the circuit impedance is capacitive and the current through the thyristors is naturally commutated. The power to the load can be controlled by controlling the inverter frequency.

Current-source, parallel-resonant inverters of Fig. 15-7b for induction heating were discussed in Chapter 7 in connection with resonant converters.

15-3-2 Electric Welding

In electric arc welding, the melting energy is provided by establishing an arc between two electrodes, one of which is the metallic workpiece being welded.

The voltage-current characteristic of the welder depends on the type of welding process employed. Typical rated voltage and current are 50 V and 500 A dc, respectively. It is desirable to have a very low ripple in the current once an arc is established. In all welding applications, the output needs to be electrically isolated from the utility input. This electrical isolation is provided by either a 60-Hz power transformer or a high-frequency transformer.

In welders with a 60-Hz power transformer, the input ac voltage is first stepped down to a suitably low voltage. Then, it is converted to a controlled dc by means of the one of the three schemes shown in Fig. 15-8 in a block diagram form. In Fig. 15-8a, a full-bridge thyristor rectifier is used. A large inductor is needed at the input to limit the current ripple. The other alternative as shown in Fig. 15-8b is to use a diode rectifier bridge that produces an uncontrolled dc. This uncontrolled dc voltage is controlled by means of a transistor series regulator. The transistor operates in its active region and acts as an adjustable resistor in order to regulate the welder's output. In the scheme of Fig. 15-8c, a switch-mode, step-down dc–dc converter is used to control the output voltage and current. All of these schemes suffer from the weight, size, and the losses in the 60-Hz power transformer. The energy efficiency is particu-

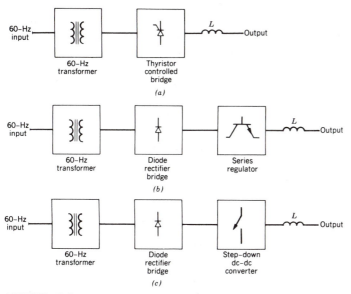

FIGURE 15-8: Welders with a 60-Hz transformer: (a) controlled thyristor bridge; (b) series regulator; (c) step-down dc–dc converter.

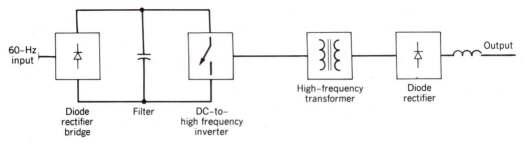

FIGURE 15-9: Switch-mode welder.

larly low in the scheme of Fig. 15-8*b* with a series regulator, where a substantial power loss takes place in the transistor operating in its active region.

The block diagram of a switch-mode welder is shown in Fig. 15-9, where the electrical isolation is provided by a high-frequency transformer. The various blocks shown in Fig. 15-9 are very similar to those used for the switching dc power supplies discussed in Chapter 8. One of the resonant concepts discussed in Chapter 7 may be used to invert dc into a high-frequency ac. A small inductance is needed at the output to limit the output current ripple at high frequencies. The efficiency of such a welder is in the 85 to 90% range, in addition to a much smaller weight and size compared with the welders employing a 60-Hz power transformer.

15-3-3   Integral Half-Cycle Controllers

In industrial applications requiring resistive heating or melting where the thermal time-constants of the process are much longer than the 60-Hz time period, it

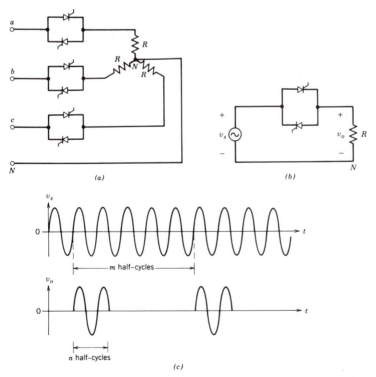

FIGURE 15-10: Integral half-cycle controllers: (*a*) three-phase circuit; (*b*) per-phase circuit; (*c*) waveforms.

is possible to employ an integral half-cycle control. This is shown in Fig. 15-10a for a resistive Y-connected load supplied through three triacs or back-to-back connected thyristors. If the neutral wire is accessible, this circuit can be analyzed on a per-phase basis as shown in Fig. 15-10b. The waveforms are drawn in Fig. 15-10c. By controlling the ratio n/m, keeping m constant, the average power supplied to the load is controlled.

15-4 SUMMARY

Some of the residential and industrial applications of power electronics are discussed. The associated power quality issues of the injected harmonic currents and the power factor of operation are discussed in Chapter 17.

PROBLEMS

15-1. In Fig. 15-2 for a single-speed heatpump, assume that each on and off period is 10 minutes long, that is, there are three cycles per hour. When the compressor is turned on, its output increases exponentially, reaching 99% of its maximum capacity at the end of the 10-minute on-interval. Once the compressor is turned off, the heating (or cooling) decays with a much smaller time constant and can be assumed to be instantaneous.

    (a) If the rated electrical power is drawn throughout the on-interval, calculate the loss in efficiency due to the exponential rise in the compressor output.

    (b) A load-proportional capacity-modulated heatpump is used to eliminate the above on/off cycling. The efficiency of the solid-state controller is 96% and the motor efficiency is lower by 1 percentage point because of reduced speed, reduced load operation, and inverter harmonics. Assume that the compressor efficiency remains unchanged.

Compare the system efficiency with the single-speed compressor system of part a, if the motor efficiency in part a can be assumed to be 85%.

15-5 REFERENCES

Space Heating and Air Conditioning

1. N. MOHAN AND J. W. RAMSEY, *Comparative Study of Adjustable-Speed Drives for Heat Pumps*, EPRI Final Report, EPRI EM-4704, Project 2033-4, Aug. 1986, Palo Alto, CA.

High-Frequency Fluorescent Lighting

2. E. E. HAMMER and T. K. McGOWAN, "Characteristics of Various F40 Fluorescent Systems at 60 Hz and High Frequency," *IEEE/IAS Transactions*, Vol. IA-21, No. 1, pp. 11–16, Jan./Feb 1985.
3. Illuminating Engineering Society (IES), *Lighting Handbook*, 1981 Reference Volume.

Induction Heating and Electric Welding

4. SIEMENS and JOHN WILEY & SONS, *Electrical Engineering Handbook*, John Wiley & Sons, New York, 1985.

16

Electric Utility Applications

16-1 INTRODUCTION

Power electronic systems that have unique electric utility applications such as high-voltage dc transmission, static var control, and the interconnection of renewable energy sources and energy storage systems to the utility grid are discussed in this chapter.

16-2 HIGH-VOLTAGE DC (HVDC) TRANSMISSION

Electrical plants generate power in the form of ac voltages and currents. This power is transmitted to the load centers on three-phase, ac transmission lines. However, under certain circumstances, it becomes desirable to transmit this power over dc transmission lines. This alternative becomes economically attractive where a large amount of power is to be transmitted over a long distance from a remote generating plant to the load center. This breakeven distance for hvdc overhead transmission lines usually lies somewhere in a range of 300 to 400 miles and is much smaller for underwater cables. In addition, there are many other factors such as the improved transient stability and the dynamic damping of the electrical system oscillations that may influence the selection of dc transmission in preference to the ac transmission. It is possible to interconnect two ac systems, which are at two different frequencies or which are not synchronized, by means of an hvdc transmission line.

Figure 16-1 shows a typical one-line diagram of an hvdc transmission system for interconnecting two ac systems (where each ac system may include its own generation and load) by an hvdc transmission line. Power flow over the transmission line can be reversed. If we assume the power flow to be from system A to B, the system A voltage, in a 69 to 230 kV range, is transformed up to the transmission level and then rectified by means of the converter terminal A and transmitted over the hvdc transmission line. At the receiving end, the dc power is inverted by means of the converter terminal B and the voltage is transformed down to match the ac voltage of system B. The power

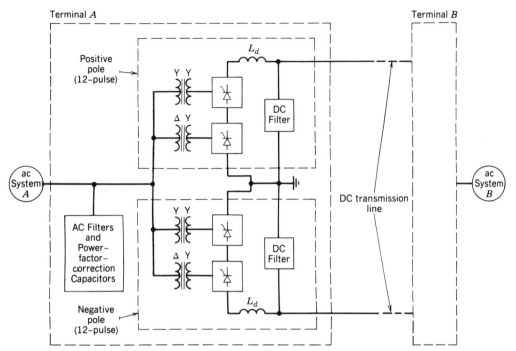

FIGURE 16-1: A typical HVDC transmission system.

received over the hvdc transmission line is then transmitted over ac transmission and distribution lines to wherever it is needed in system B.

Each converter terminal in Fig. 16-1 consists of a positive pole and a negative pole. Each pole consists of two 6-pulse, line-frequency bridge converters connected through a Y-Y and a Δ-Y transformer to yield a 12-pulse converter arrangement. On the ac-side of the converter, the filters are required to reduce the current harmonics generated by the converters from entering the ac system. Moreover, the power-factor-correction capacitors are included along with the ac filter banks to supply the lagging reactive power (or the inductive vars) required by the converter in the rectifier as well as in the inverter mode of operation. On the dc side of the converter, the ripple in the dc voltage is prevented from causing excessive ripple in the dc transmission line current by means of smoothing inductors L_d and the dc side filter banks, as shown in Fig. 16-1.

16-2-1   Twelve-Pulse Line-Frequency Converters

The 6-pulse line-frequency controlled converters were discussed in detail in Chapter 4. Because of high power levels associated with the hvdc transmission application, it is important to reduce the current harmonics generated on the ac side and the voltage ripple produced on the dc side of the converter. This is accomplished by means of a 12-pulse converter operation, which requires two 6-pulse converters connected through a Y-Y and a Δ-Y transformer, as is shown in Fig. 16-2. The two 6-pulse converters are connected in series on the dc side and in parallel on the ac side. The series connection of two 6-pulse converters on the dc side is important to meet the high voltage requirement of an hvdc system.

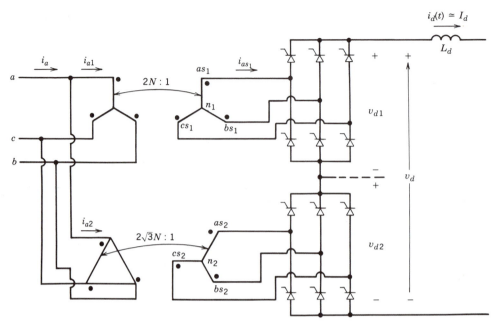

FIGURE 16-2: Twelve-pulse converter arrangement.

In Fig. 16-2, $V_{as_1n_1}$ leads $V_{as_2n_2}$ by 30°. The voltage and current waveforms can be drawn by assuming the current I_d on the dc side of the converter to be a pure dc in the presence of the large smoothing inductor L_d shown in Fig. 16-2. Initially, for simplicity, we will assume that the per-phase ac-side commutating inductance L_s is negligible, thus resulting in rectangular current pulses. In practice, however, substantial commutating inductances are present as a result of the transformer leakage inductances. The effects of these commutating inductances on the 12-pulse waveform are discussed later on.

With the foregoing assumptions of $L_s = 0$ and $i_d(t) \simeq I_d$, and recognizing that $V_{as_1n_1}$ leads $V_{as_2n_2}$ by 30°, we can draw the current waveforms as in Fig. 16-3a. Each 6-pulse converter operates at the same delay angle α. The waveform of the total per-phase current $i_a = i_{a1} + i_{a2}$ clearly shows that it contains fewer harmonics than either i_{a1} or i_{a2} drawn by the 6-pulse converters. In terms of their Fourier components

$$i_{a1} = \frac{2\sqrt{3}}{2N\pi}I_d\left(\cos\theta - \frac{1}{5}\cos 5\theta + \frac{1}{7}\cos 7\theta - \frac{1}{11}\cos 11\theta + \frac{1}{13}\cos 13\theta \ldots\right) \quad (16\text{-}1)$$

and

$$i_{a2} = \frac{2\sqrt{3}}{2N\pi}I_d\left(\cos\theta + \frac{1}{5}\cos 5\theta - \frac{1}{7}\cos 7\theta - \frac{1}{11}\cos 11\theta + \frac{1}{13}\cos 13\theta \ldots\right) \quad (16\text{-}2)$$

where $\theta = \omega t$ and the transformer turns ratio N is indicated in Fig. 16-2. Therefore, the combined current drawn is

$$i_a = i_{a1} + i_{a2} = \frac{2\sqrt{3}}{N\pi}I_d\left(\cos\theta - \frac{1}{11}\cos 11\theta + \frac{1}{13}\cos 13\theta \ldots\right) \quad (16\text{-}3)$$

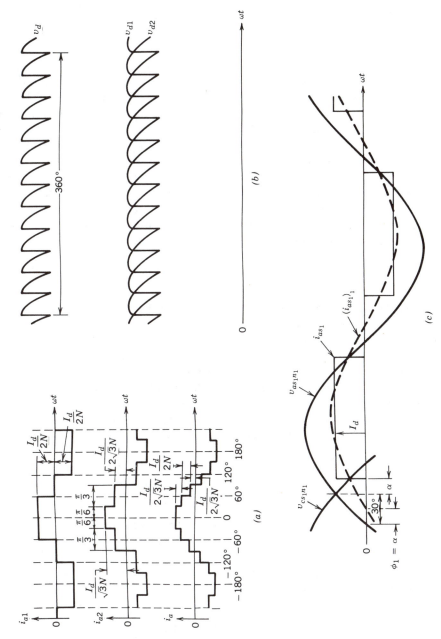

FIGURE 16-3: Idealized waveforms assuming $L_s = 0$.

This Fourier analysis shows that the combined line current has harmonics of the order

$$h = 12k \pm 1 \quad \text{(where } k = \text{an integer)} \tag{16-4}$$

resulting in a 12-pulse operation, as compared with a 6-pulse operation where the ac current harmonics are of the order $6k \pm 1$ (where $k = $ an integer). The harmonic current amplitudes in Eq. 16-3 for a 12-pulse converter are inversely proportional to their harmonic order and the lowest order harmonics are the eleventh and the thirteenth. The currents on the ac side of the two 6-pulse converters add, confirming that the two converters are effectively in parallel on the ac side.

On the dc side, the voltage waveforms v_{d1} and v_{d2} for the two 6-pulse converters are shown in Fig. 16-3b. These two voltage waveforms are shifted by 30° with respect to each other. Since the two 6-pulse converters are connected in series on the dc side, the total dc voltage $v_d = v_{d1} + v_{d2}$ has 12 ripple pulses per fundamental frequence ac cycle. This results in the voltage harmonics of the order h in v_d, where

$$h = 12k \quad (k = \text{an integer}) \tag{16-5}$$

and the twelfth harmonic is the lowest order harmonic. Magnitudes of the dc-side voltage harmonics vary significantly with the delay angle α.

In practice, L_s is substantial because of the leakage inductance of the transformers. The presence of L_s does not change the order of characteristic harmonics produced either on the ac side or on the dc side, provided that the two 6-pulse converters operate under identical conditions. However, the harmonic magnitudes depend significantly on L_s, delay angle α, and the dc current I_d. The effect of L_s on the ac current waveform and harmonics was discussed in Chapter 4.

Based on the derivation in Chapter 4, the average dc voltage can be written as

$$V_{d1} = V_{d2} = \frac{V_d}{2} = \frac{3\sqrt{2}}{\pi} V_{LL} \cos \alpha - \frac{3\omega L_s}{\pi} I_d \tag{16-6}$$

where V_{LL} is the line-to-line rms voltage applied to each of the 6-pulse converters and L_s is per-phase leakage inductance of each of the transformers, referred to their converter side.

As we explained in Chapter 4, α greater than 90° corresponds to an inverter mode of operation with a transfer of power from the dc to the ac side of the converter.

16-2-2   Reactive Power Drawn by Converters

As was alluded to earlier, the line-frequency, line-voltage-commutated converters operate at a lagging power factor and, hence, draw reactive power from the ac system. Even though the ac-side currents associated with the converter contain harmonics in addition to their fundamental frequency components, the harmonic currents are "absorbed" by the ac-side filters, whose design must be based on the magnitude of the generated harmonic current magnitudes, as we will discuss later on. Therefore, only the fundamental frequency components of the ac currents are considered for the real power transfer and the reactive power drawn. It is necessary to consider only one of the two 6-pulse converters, since the real and the reactive power for the 12-pulse converter arrangement making up a pole are twice the per-converter values.

16-2-2-1 RECTIFIER MODE OF OPERATION

With the initial assumption that $L_s = 0$ in Fig. 16-2, Fig. 16-3c shows the phase-to-neutral-voltage $v_{as_1 n_1}$ and the current i_{as_1} (corresponding to converter 1 in Fig. 16-3) with $i_d(t) \simeq I_d$ at a delay angle α. The fundamental frequency current component $(i_{as_1})_1$ shown by the dotted curve lags behind the phase voltage $v_{as_1 n_1}$ by the displacement power factor angle ϕ_1 where

$$\phi_1 = \alpha \tag{16-7}$$

Therefore, the three-phase reactive power (lagging) required by the 6-pulse converter because of the fundamental frequency reactive current components, which lag their respective phase voltages by 90°, equals

$$Q_1 = \sqrt{3} \, V_{LL}(I_{as_1})_1 \sin \alpha \tag{16-8}$$

where V_{LL} is the line-to-line voltage on the ac side of the converter.

From the Fourier analysis of i_{as_1} in Fig. 16-3c, the rms value of its fundamental frequency component is

$$(I_{as_1})_1 = \frac{\sqrt{6}}{\pi} I_d \simeq 0.78 I_d \tag{16-9}$$

Therefore, from Eqs. 16-8 and 16-9

$$Q_1 = \sqrt{3} \, V_{LL}\left(\frac{\sqrt{6}}{\pi} I_d\right) \sin \alpha = 1.35 V_{LL} I_d \sin \alpha \tag{16-10}$$

The real power transfer through each of the 6-pulse converters can be calculated from Eq. 16-6 with $L_s = 0$ as

$$P_{d1} = V_{d1} I_d = 1.35 V_{LL} I_d \cos \alpha \tag{16-11}$$

For a desired power transfer P_{d1}, the reactive power demand Q_1 should be minimized as much as possible. Similarly, I_d should be kept as small as possible to minimize $I^2 R$ losses on the dc transmission line. To minimize I_d and Q_1, noting that V_{LL} is essentially constant in Eqs. 16-10 and 16-11, we should choose a small value for the delay α in the rectifier mode of operation. For practical reasons, the minimum value is α is chosen in a range of 10 to 20 degrees.

16-2-2-2 INVERTER MODE OF OPERATION

In the inverter mode, the dc voltage of the converter acts like a counter-emf in a dc motor. Therefore, it is convenient to define the dc voltage polarity as shown in Fig. 16-4a, so that the dc voltage is positive when written specifically for the inverter mode of operation. In Chapter 4, the extinction angle γ for the inverter was defined in terms of α and u as

$$\gamma = 180° - (\alpha + u) \tag{16-12}$$

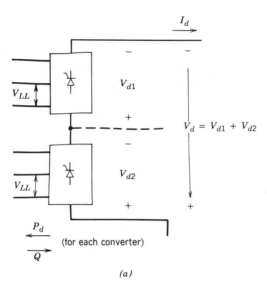

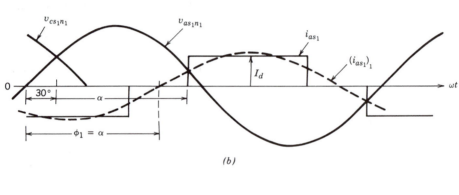

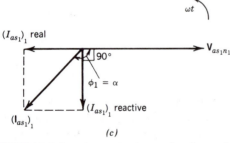

FIGURE 16-4: Inverter mode of operation (assuming $L_s = 0$).

where α is the delay angle and u is the commutation or the overlap angle. The inverter voltages in Fig. 16-4 can be obtained as (see Problems 16-7)

$$V_{d1} = V_{d2} = \frac{V_d}{2} = 1.35 V_{LL} \cos \gamma - \frac{3\omega L_s}{\pi} I_d \qquad (16\text{-}13)$$

Again with the assumption that $L_s = 0$ for simplicity, Fig. 16-4b shows the idealized waveforms for $v_{as_1 n_1}$ and i_{as_1} at an $\alpha > 90°$, corresponding to the inverter mode of operation. The fundamental frequency component $(i_{as_1})_1$ of the phase current is shown by the dotted curve. In the phasor diagram of Fig. 16-4c, the fundamental frequency reactive current component lags behind the phase-to-neutral voltage, indicating that even in the inverter mode, where the direction of power flow through the converter has reversed, the converter requires reactive power (lagging) from the ac system.

With $L_s = 0$, u equals zero in Eq. 16-12 and $\gamma = 180° - \alpha$. Therefore, the expressions for per-converter Q_1 and P_{d1} in Eqs. 16-10 and 16-11 can be obtained specifically for the inverter mode in terms of γ as

$$Q_1 = 1.35 V_{LL} I_d \sin \gamma \qquad (16\text{-}14)$$

and

$$P_{d1} = 1.35 V_{LL} I_d \cos \gamma \qquad (16\text{-}15)$$

where the directions of the reactive power (lagging) and the real power are as shown in Fig. 16-4a.

In Eqs. 16-14 and 16-15, γ should be as small as possible for a given power transfer level to minimize I^2R losses in the transmission line due to I_d and to minimize the reactive power demand by the converter. As we discussed in Chapter 4, the minimum value that γ is allowed to attain is called the minimum extinction angle γ_{min} that is based on allowing sufficient turn-off time to the thyristors.

In a 12-pulse converter arrangement, the reactive power requirement is the sum of the reactive powers required by each of the two 6-pulse converters. The ac-side filter banks and the power-factor-correction capacitors partially provide the reactive power demand of the converters, as discussed in Section 16-2-4.

16-2-3 Control of hvdc Converters

It is possible to discuss the control of converters in an hvdc system on a per-pole basis, since both the positive and the negative poles are operated under identical conditions. Figure 16-5a shows the positive pole, for example, consisting of the 12-pulse converters A and B. Terminal A is assumed to be operating as a rectifier, and its dc voltage is defined as V_{dA}. Terminal B is assumed to be operating as an inverter, and its dc voltage V_{dB} is shown with a polarity that is specific to the inverter mode of operation, so that V_{dB} has a positive value.

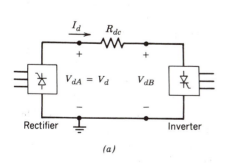

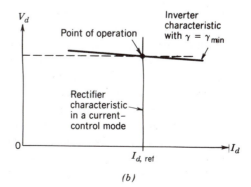

FIGURE 16-5: Control of HVDC system.

In steady state in Fig. 16-5a

$$I_d = \frac{V_{dA} - V_{dB}}{R_{dc}} \tag{16-16}$$

where R_{dc} is the dc resistance of the positive transmission line conductor. In practice, R_{dc} is small and I_d results as a consequence of a small difference between two very large voltages in Eq. 16-16. Therefore, one converter is assigned to control the voltage on the transmission line and the other to control I_d. Since the inverter should operate at a constant $\gamma = \gamma_{min}$, it is natural to choose the inverter (converter B in Fig. 16-5a) to control V_d. Then, I_d and, hence, the power level is controlled by the rectifier (converter A in Fig. 16-5a).

Figure 16-5b shows the rectifier and the inverter control characteristics in the V_d-I_d plane, where V_d is chosen to be the voltage at the rectifier, that is, $V_d = V_{dA}$. At the constant extinction angle $\gamma = \gamma_{min}$, the inverter produces a voltage V_d in Fig. 16-5a, which is given as

$$V_d = 2 \times \left[1.35 V_{LL} \cos \gamma_{min} - \frac{3\omega L_s}{\pi} I_d \right] + R_{dc} \cdot I_d$$

$$= 2 \times 1.35 V_{LL} \cos \gamma_{min} - \left(\frac{6\omega L_s}{\pi} - R_{dc} \right) I_d \tag{16-17}$$

Assuming the quantity within the bracket in Eq. 16-17 to be positive, the constant extinction angle operation of the inverter results in a V_d-I_d characteristic as shown in Fig. 16-5b.

The rectifier can be controlled to maintain I_d equal to its commanded or reference value $I_{d, ref}$. The actual current I_d is measured, and the error $(I_d - I_{d, ref})$, if positive, increases the rectifier delay angle α; if the error is negative, α is decreased. A high-gain current controller results in a nearly vertical rectifier characteristic in Fig. 16-5b at $I_{d, ref}$. The intersection of the two characteristics in Fig. 16-5b establishes the transmission line voltage V_d and the current I_d.

The foregoing discussion shows how the power flow $P_d = V_d I_d$ from terminal A to terminal B can be controlled in Fig. 16-5a by controlling I_d, while maintaining the transmission line voltage as high as possible to minimize $I_d^2 R_{dc}$ power loss in the transmission line. This type of control also results in a small value of α in the rectifier and a small $\gamma = \gamma_{min}$ in the inverter, thus minimizing the reactive power demand by both the rectifier and the inverter. In practice, the transformers at both the terminals consist of tap changers, which can control the ac voltage V_{LL} supplied to the converters in a small range, thus providing an additional degree of control.

The control characteristics shown in Fig. 16-5b can be extended for negative values of V_d so that the power flow can be controlled smoothly in magnitude as well as in direction. A detailed discussion can be found in Reference 1. This capability to be able to reverse the power flow is useful if the two ac systems interconnected by the dc transmission line have loads that vary differently with seasons or the time of day. The same may be the case if one of the ac systems contains hydro generation whose output depends on seasons. Another application of this control capability is to modulate the power flow on the dc line to damp out the ac system oscillations.

16-2-4   Harmonic Filters and Power-Factor-Correction Capacitors

16-2-4-1   DC-SIDE HARMONIC FILTERS

To minimize the inductively coupled harmonic interference produced in the telephone system and other types of control/communication channels in parallel with the hvdc transmission lines, it is important to minimize the magnitudes of the current harmonics on the dc transmission line. The voltage harmonics are of the order $12k$, where k is an integer. The magnitudes of the harmonic voltages depend on α, L_s, and I_d for a given ac system voltages. Under a balanced 12-pulse operating condition, the 12-pulse converter can be represented by an equivalent circuit as shown in Fig. 16-6a, where the harmonic voltages are connected in series with the dc voltage V_d.

A large smoothing inductor L_d of the order of several hundred millihenries (mH) is used in combination with a high-pass filter, as shown in Fig. 16-6a, in order to limit the flow of harmonic currents on the transmission line. The impedance of the high-pass filter in Fig. 16-6a is plotted in Fig. 16-6b, where the filter is designed specifically to provide a low impedance at the dominant twelfth harmonic frequency.

16-2-4-2   AC-SIDE HARMONIC FILTERS AND POWER-FACTOR-CORRECTION CAPACITORS

In a 12-pulse converter, the ac currents consist of the characteristic harmonics of the order $12k \pm 1$ (k = an integer), as given by Eq. 16-4. The harmonic currents can be represented by means of an equivalent circuit, as shown in Fig. 16-7a. It is desirable to prevent these harmonic currents from entering the ac network, where they cause power losses and may also cause interference with the other electronic communication equipment. For this purpose, the per-phase filters shown in Fig. 16-7a are commonly used. The series-tuned filters are used for the two lower order harmonics: the eleventh and the thirteenth. A high-pass filter as shown in Fig. 16-7a is used to eliminate the rest of the higher order harmonics. The combined impedance of all the harmonic filters is plotted in Fig. 16-7b.

Note that the filter design very much depends on the ac system impedance at the harmonic frequencies in order to provide adequate filtering and to avoid certain resonance conditions. The system impedance depends on the system configuration based on loads, generation pattern, and the transmission lines in service. Therefore, the filter design must anticipate the changes to occur in the foreseeable future (e.g., an additional interconnection), which may alter the ac system impedance.

The harmonic filters also provide a large percentage of the reactive power required by the converters in the rectifier and the inverter mode. In the ac-side filters discussed earlier (both series-tuned and the high-pass), the capacitive impedance dominates at the 60-Hz frequency over the inductive elements connected in series with the capacitor. Therefore, the effective shunt capacitance offered per-phase by the ac-filter bank at the fundamental or line frequency can be approximated as

$$C_f \approx C_{11} + C_{13} + C_{\text{hp}} \qquad (16\text{-}18)$$

At the 60-Hz fundamental system frequency, the per-phase reactive power (vars) supplied by the filter bank equals

$$Q_f \approx 377 C_f V_s^2 \qquad (16\text{-}19)$$

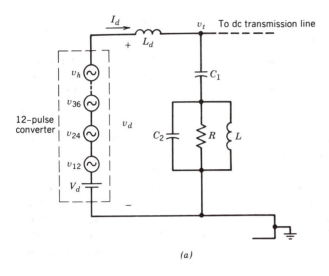

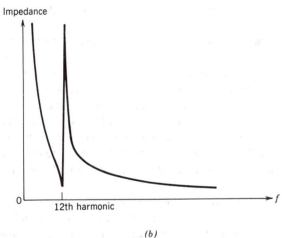

FIGURE 16-6: Filter for dc-side voltage harmonics: (a) dc-side
equivalent circuit; (b) high-pass filter impedance
versus frequency.

where, V_s is the rms phase voltage applied across the filters. Thus, the ac filters play an important role in meeting the reactive power demand of the converters, in addition to filtering the current harmonics.

As was discussed in Section 16-2-3, which deals with the control of hvdc systems, the power flow on the dc line is controlled by adjusting I_d in Fig. 16-5. Therefore, the reactive power demand by the converters increases with the increasing power transfer level, as is suggested by Eq. 16-10. The filter capacitors are chosen such that the reactive power supplied by them does not exceed the reactive power demand of the converters at the minimum power level of the hvdc system operation. The reason is that if the reactive vars supplied by the filters exceed the converter demand, the problem of the system overvoltage, which tends to occur at the light load conditions in the first place will increase. Therefore, to compensate for the higher reactive power demand by the converters at higher power transfer levels, additional power-factor-correction capacitors C_{pf} are switched in, as shown in Fig. 16-7a.

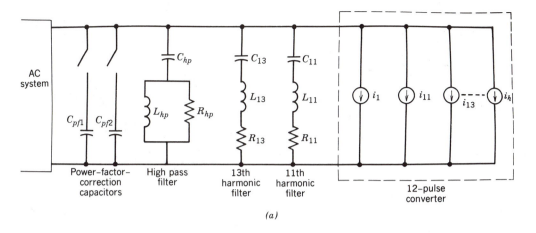

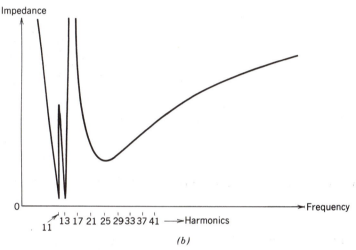

FIGURE 16-7: AC side filters and power-factor-correction capacitors: (*a*) per-phase equivalent circuit; (*b*) combined per-phase filter impedance versus frequency.

16-3 STATIC VAR CONTROL (SVC)

In an electric utility network, it is desirable to regulate the voltage within a narrow range of its nominal value. Most utilities attempt to maintain the voltage deviations in a $+5\%$, -10% range around their nominal values. Moreover, it is desirable to have a balanced load on all three phases to eliminate negative- and zero-sequence currents that can have undesirable consequences such as additional heating in electrical equipment, torque pulsations in generators and turbines, and so on. The load on the power system fluctuates and can result in voltages outside of their acceptable limits. In view of the fact that the internal impedance of the ac system seen by the load is mainly inductive (since the transmission and distribution lines, transformers, generators, etc., have mainly inductive impedance at the line frequency of 60 Hz), it is the reactive power change in the load that has the most adverse effect on the voltage regulation.

Consider a simple per-phase system equivalent circuit shown in Fig. 16-8*a* by means of the ac system Thévenin equivalent, where the internal impedance of the ac system is assumed to be purely inductive. Figure 16-8*b* shows the phasor diagram for a

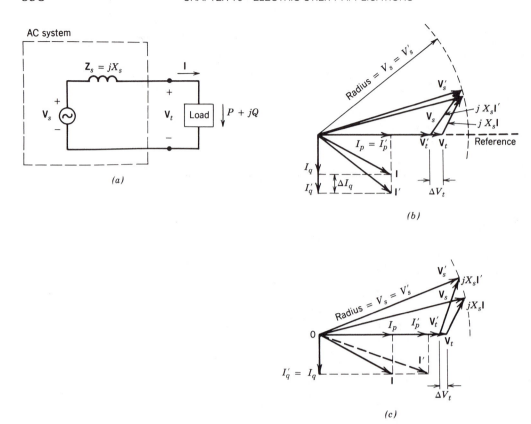

FIGURE 16-8: Effect of I_p and I_q on V_t: (a) equivalent circuit; (b) change in I_q; (c) change in I_p.

lagging power factor load $P + jQ$ with a current $\mathbf{I} = I_p + jI_q$, which lags the terminal voltage $\mathbf{V}_t$. The terminal voltage magnitude is assumed to be at its nominal value. An increase ΔQ in the lagging vars drawn by the load causes the reactive current component to increase to $I_q + \Delta I_q$, while I_p is assumed to be unchanged. The phasor diagram for the increased Q is indicated by "primed" quantities in Fig. 16-8b where the terminal voltage is again chosen to be the reference phasor for simplicity; the magnitude of the internal system voltage V_s remains the same as before. The phasor diagram of Fig. 16-8b shows a drop in the terminal voltage by ΔV_t caused by an increase in the lagging reactive power drawn by the load. In this case, even if I_p remains constant, the real power P will decrease because of the reduction in V_t. For comparison purposes, Fig. 16-8c shows the phasor diagram where the percentage change in I_p is the same as the percentage change in I_q in Fig. 16-8b, while I_q is assumed to be unchanged. Figure 16-8c shows that the voltage change ΔV_t is small due to a change in I_p.

Most utility systems utilize power-factor-correction capacitor banks, which are switched in and out by means of mechanical contactors, to compensate for the slow changes in the reactive power of the load in order to keep the overall load power factor as close to unity as possible. The reasons for the power factor correction are twofold: (1) it facilitates regulation of the system voltage within a range of $+5\%$, -10% around its nominal value and, (2) a close-to-unity power factor of the load-capacitor combination results in the lowest magnitude of the current drawn for a given real power demand. This in turn reduces the I^2R losses in the various equipment within the ac system. Moreover, the capacity of the equipment, which is rated in terms of current-handling capability, is more effectively utilized.

In this section, however, the objective is to discuss static var controllers, which by means of a power electronic interface can provide a quick control over the reactive power. These static var controllers are used to prevent annoying voltage flickers caused by industrial loads such as arc furnaces, which cause very rapid changes in the reactive power and also introduce a fluctuating load unbalance between the three phases. Another use for the static var controllers is to provide a dynamic voltage regulation to enhance the stability of the interconnection between two ac systems.

There are primarily three types of static var controllers, as listed here:

1.   Thyristor-controlled inductors (TCI),

2.   Thyristor-switched capacitors (TSC) and,

3.   Switching converters with a minimum energy storage elements.

A hybrid arrangement of a thyristor-controlled inductor with a thyristor-switched capacitor minimizes the no-load losses.

16-3-1   Thyristor-Controlled Inductors (TCI)

Thyristor-controlled inductors act as variable inductors where the inductive vars supplied can be varied very quickly. The system may require either inductive or capacitive vars, depending on the system conditions. This requirement can be met by paralleling TCI with a capacitor bank.

The basic principle of a thyristor-controlled inductor can be understood by considering the per-phase circuit of Fig. 16-9a where an inductor L is connected to the ac source through a bidirectional switch, consisting of two back-to-back connected thyristors. If the resistive component of the inductor is assumed to be negligibly small, the current through the inductor in steady state can be obtained as a function of the thyristor delay angle α. An equal value of α is used for both thyristors.

As a base case, Fig. 16-9b shows the current waveform where the thyristor gate pulses are always present corresponding to $\alpha = 0$, as if thyristors were replaced by diodes. With $\omega = 2\pi f$, this results in a sinusoidal current i_L whose rms value equals

$$I_L = I_{L1} = \frac{V_s}{\omega L}(\omega = 2\pi f)$$ (16-20)

where the inductor current consists solely of the fundamental frequency component without any harmonics. Since i_L lags behind V_s by $90°$ as shown in Fig. 16-9b, the delay angle α in a range of 0 to $90°$ has no control over i_L and its rms value remains the same as given by Eq. 16-20.

If α is increased beyond $90°$, i_L can be controlled as is shown in Figs. 16-9c and 16-9d corresponding to $\alpha = 120°$ and $135°$, respectively. Clearly as α is increased, I_{L1} is reduced, thus, allowing a control over the effective value of inductance connected to the utility voltage since

$$L_{\text{eff}} = \frac{V_s}{\omega I_{L1}}$$ (16-21)

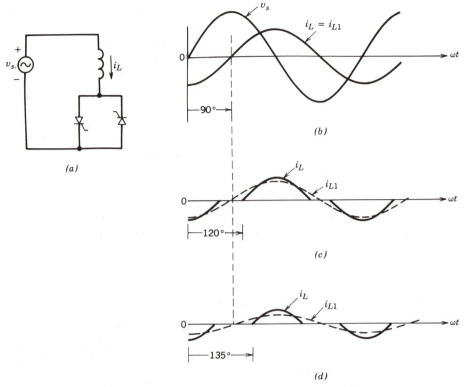

FIGURE 16-9: TCI — basic principle: (a) per-phase TCI; (b) $0 \le \alpha \le 90°$; (c) $\alpha = 120°$; (d) $\alpha = 135°$.

where by Fourier analysis (see Problem 16-8)

$$I_{L_1} = \frac{V_s}{\pi \omega L}(2\pi - 2\alpha + \sin 2\alpha) \qquad \frac{\pi}{2} \le \alpha \le \pi \qquad (16\text{-}22)$$

Therefore, the lagging reactive power drawn by the per-phase thyristor-controlled inductor (TCI) at the fundamental frequency is

$$Q_I = V_s I_{L1} = \frac{V_s^2}{\omega L_{\text{eff}}} \qquad (16\text{-}23)$$

The inductor current is not a pure sine wave at $\alpha > 90°$, as can be seen by the waveforms in Fig. 16-9c and 16-9d. A Fourier analysis of the inductor current waveform shows that i_L consists of odd harmonics h of the order 3, 5, 7, 9, 11, 13, etc., whose amplitudes as a ratio of I_{L1} depend on α. To prevent the third- and the multiples of third-order harmonics, it is a common practice to connect three-phase TCI in a Δ so that these harmonics circulate through the inductors and do not enter the ac system. The capacitor, in parallel with the TCI to meet the system var requirements, filters out high-frequency harmonics. Similar to the discussion in Section 16-2-4-2, the fifth and the seventh harmonics are filtered out by series-tuned filters. These series-tuned filters also provide the capacitive vars as described by Eqs. 16-18 and 16-19.

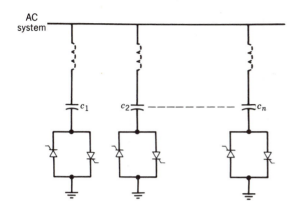

AC
system

c_1 c_2 —————— c_n

FIGURE 16-10: TSC arrangement.

16-3-2   Thyristor-Switched-Capacitors (TSC)

Figure 16-10 shows the basic arrangement where several (three or four) capacitors can be connected to the supply voltage through a bidirectional switch consisting of back-to-back connected thyristors. Unlike the phase control used in TCI to vary the effective value of the inductor, TSC employs integral half-cycle control where the capacitor is either fully in or out of the circuit.

The capacitor bank can be switched out by blocking the gate pulses to both the thyristors. The current flow stops at the instant of its zero crossing, which also corresponds to the capacitor voltage equal to the maximum ac system voltage. The polarity of the capacitor voltage depends on the instant when the thyristor gate pulses are blocked. At switch-on, the thyristor must be gated at the proper instant of maximum ac voltage to avoid large overcurrents. Moreover, inductors, shown dotted in Fig. 16-10, are used to limit overcurrents at switch-on. By using a large number of thyristor-switched small capacitor banks, it is possible to vary the reactive power Q_c in small but still in discrete steps.

16-3-3   Instantaneous var Control Using Switching Converters with Minimum Energy Storage

The static var control schemes discussed in the previous sections consist of large energy storage inductors and capacitors to meet the var demand. Moreover, they cannot provide an instantaneous var control because of their inherent time delays.

In Chapter 6, dealing with the switch-mode converters (inverters and rectifiers), it was shown that their ac current can be controlled by operating them in a current-control mode. Such a converter is shown in Fig. 16-11. The ac current of such switch-mode converters can be quickly controlled in magnitude as well as in their phase relationship (leading or lagging) to the ac voltages. Since the average power drawn or supplied by these converters is desired to be zero, a dc source at the dc input of the converter is not necessary. Only a small capacitor with a minimum energy storage is sufficient, and its dc voltage is maintained by the switch-mode converter, which transfers sufficient real power from the ac system to compensate for its own losses and to maintain a constant dc voltage across the capacitor, in addition to controlling the vars.

The discussion of instantaneous reactive power can be found in Reference 7, where the reactive power calculator calculates the instantaneous reference currents i_a^*, i_b^*, and i_c^*, which the switch-mode converter of Fig. 16-11 supplies under a current-

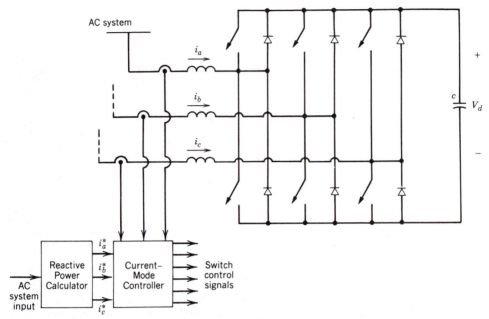

FIGURE 16-11: Instantaneous var controller.

mode control. Note that the resonant converter concepts discussed in Chapter 7 can be used to reduce switching losses in the converter.

16-4   INTERCONNECTION OF RENEWABLE ENERGY SOURCES AND ENERGY STORAGE SYSTEMS TO THE UTILITY GRID

A power electronic interface is needed to connect renewable energy sources such as photovoltaic, wind, and small hydro to the utility system. The same is true for the interconnection of energy storage systems for utility load-leveling (also called load-peak shaving), such as batteries, fuel-cells, and superconductive energy storage inductors. Some of these systems are briefly discussed here.

16-4-1   Photovoltaic Array Interconnection

A large number of solar cells, connected in series and parallel, make up the photovoltaic or solar arrays. These cells produce a dc voltage when they are exposed to sunlight. Figure 16-12 shows the $i-v$ characteristics of such a cell for various insolation (sunlight intensity) levels and temperatures. Figure 16-12 shows that the cell characteristic at a given insolation and temperature basically consists of two segments: (1) the constant voltage segment and (2) the constant current segment. The current is limited as the cell is short-circuited. The maximum power condition occurs at the knee of the characteristic where the two segments meet. It is desirable to operate at the maximum power point. Ideally, a pure dc current should be drawn from the solar array, though the reduction in delivered power is not very large even in the presence of a fair amount of ripple current. As an example, a 5% peak ripple current results in a power reduction of less than 1%.

To ensure that the array keeps on operating at the maximum power point, a perturb-and-adjust method (also called the "dithering" technique) is used where at regular intervals (once every few seconds) the amount of current drawn is perturbed

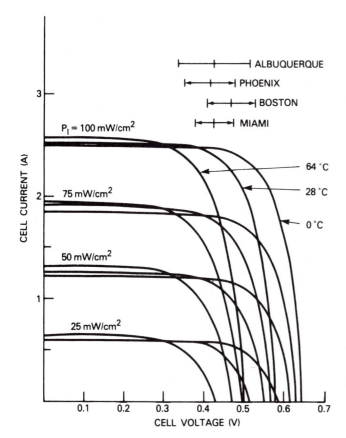

FIGURE 16-12: *I-V* characteristics of solar cells. (*Source:* reference 9, DOE report.)

and the resulting power output is observed. If an increased current results in a higher power, it is further increased until power output begins to decline. On the other hand, if an increase in current results in less power than before, then the current is decreased until the power output stops increasing and begins to go down.

The solar array is interconnected to the utility grid through an electrical isolation. The current supplied to the utility grid should be sinusoidal at nearly a unity power factor. The guidelines for the harmonic currents injected into the utility system and the total harmonic distortion THD are discussed in Chapter 17.

16-4-1-1 SINGLE-PHASE INTERCONNECTION

It is possible to use a line-frequency, phase-controlled converter of the type discussed in Chapter 4, where the converter always operates in an inverter mode and the electrical isolation is provided by a 60-Hz transformer. However, ac-side filters and the reactive power compensation would be needed, since the output current will contain harmonics and will be at a lagging power factor. Alternatively, a switch-mode pulse-width-modulated converter of the type discussed in Chapter 6 can be used where the electrical isolation is again provided by a 60-Hz transformer. However, since the current output is controlled to be in phase with the utility voltage, it may be economically feasible to utilize a high-frequency transformer to provide electrical isolation.

A circuit diagram of an interface that utilizes a high-frequency transformer is shown in Fig. 16-13. The dc voltage input is inverted to produce a high-frequency ac across the primary of the high-frequency transformer. Its secondary voltage is rectified

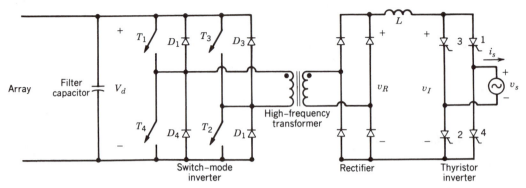

FIGURE 16-13: High-frequency photovoltaic interface.

and the resulting dc is interfaced with the line voltage through a line-frequency line-voltage-commutated thyristor inverter of the type discussed in Chapter 4. Since the line current is required to be sinusoidal and in phase with the line voltage, the line voltage waveform is measured to establish the reference waveform for the sinusoidal line-current i_s^*, whose amplitude is determined by the maximum power controller using a "dithering" scheme discussed earlier. The current i_s^* multiplied by the transformer turns-ratio acts as the reference current at the switch-mode inverter output. The inverter can be controlled to deliver the reference current, by means of current-regulated control as described in Chapter 6. One way to implement this control is described in References 8 and 9. The line-frequency thyristor-converter in Fig. 16-13 can be operated at a very small value of extinction-angle γ since the current through it is controlled to be very small near the zero-crossing of the ac system voltage.

16-4-1-2   THREE-PHASE INTERCONNECTION

At a power level above a few kilowatts, it is preferable to use a three-phase interconnection. Sinusoidal ac currents at a unity power factor can be delivered by using a switch-mode dc-to-ac inverter of the type discussed in Chapter 6 under a current-mode control. A 60-Hz three-phase transformer would be required to provide electrical isolation.

16-4-2   Wind and Small Hydro Interconnection

In the case of wind, the power available varies with the cube of the wind velocity. For small hydro, the power available depends on the pressure head and the flow. For both wind and small hydro, to extract the maximum amount of power, it is desirable to let the turbine speed vary over a wide range to an optimum value dependent on the operating conditions. This would not be possible if a synchronous generator were directly connected to the utility (60-Hz) system that dictated a constant speed (synchronous speed). The induction generators connected to the utility system would allow the speed to vary only in a very narrow speed range. Therefore, to allow the generator-turbine speed to vary to optimize efficiency of power generation, the three-phase generator output is rectified into dc and then interfaced with the three-phase utility source by means of a switch-mode converter of the type discussed in Chapter 6. A block diagram is shown in Fig. 16-14 where a 60-Hz isolation transformer is included.

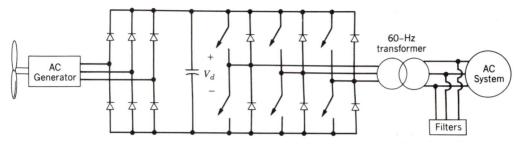

FIGURE 16-14: Interconnection of wind/hydro generator.

Because of the medium power levels (a few tens of kilowatts and higher) usually associated with the wind and small hydro generators, a three-phase utility interconnection is preferable.

16-4-3 Interconnection of Energy Storage Systems for Utility Load Leveling

Fuel-cells, batteries, and superconductive energy storage inductors are some of the means under consideration to reduce utility load peaks. As a brief explanation, it is desirable to operate the most efficient utility generators (such as nuclear and the newer high-efficiency coal-fired plants) at their rated capacity at all times. However, the load on the utility system does not remain constant with the time of day and also fluctuates based on the weather conditions. To meet the peak load, either oil- or gas-fired generators, also called the peaking plants, have to be used, which are expensive to operate because of the high cost of fuel. An alternative to this is to store the electrical energy generated by the efficient generating plants during low-load conditions and to supply it back to the utility grid under peak-load conditions, thus, reducing or eliminating the need for peaking plants powered by gas or oil. The electric energy can be stored in batteries or in the form of a magnetic field in a superconductive inductor. Another option is to use the electric energy during low-load conditions to produce oxygen by electrolysis, which can then be used in fuel cells to provide electrical output.

Both batteries and fuel-cells produce a dc voltage. To interconnect them to the utility system, a scheme similar to that used for single-phase or three-phase interconnection of photovoltaic arrays can be used.

The most economical way to interconnect large superconductive inductors would be to use 12-pulse line-commutated converters, as shown in Fig. 16-15. Under the control of the delay angle, the converter operation can be continuously varied from a full-rectifier (charge) mode to the full-inverter (discharge) mode, while the current flows continuously in the same direction.

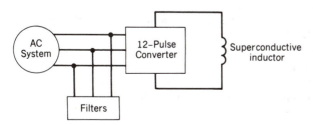

FIGURE 16-15: Superconductive energy storage inductor interconnection.

16-5 SUMMARY

In this chapter, high-power electric utility applications of power electronics are discussed. These include hvdc transmission, static var control, interconnection of renewable energy sources, and the energy storage systems for the utility load leveling.

PROBLEMS

16-1. Verify the current waveforms in Fig. 16-3a and the expressions given by Eqs. 16-1 through 16-3.

16-2. With a constant input ac voltage V_{LL} and a constant dc current I_d, plot the locus in the $P-Q$ plane as the delay angle α of the converters in Fig. 16-2 is varied. Repeat this for a family of I_d values.

16-3. A dc transmission link interconnects two 230-kV ac systems. It has four bridges at each terminal (two per pole) with each pole rated at ± 250 kV, 1000 A. The parameters for each pole of the DC link are given in the following table.

	Rectifier	Inverter
Actual open-circuit voltage ratio, for line–line voltages on primary and secondary of converter transformers. Secondary voltage divided by primary voltage.	0.468	0.435
Number of three-phase converter bridges in series on the dc side.	2	2
Converter transformer leakage reactance per bridge in ohms, referred to secondary side.	16.28	14.27
DC line resistance per pole = 15.35 Ω		
Minimum extinction angle of inverter = 18°		

In this system, $\gamma_I = \gamma_{\min} = 18°$. At the rectifier terminal, the voltage is as close to 250 kV as possible.

Calculate all the currents, voltages, real and reactive powers, and angles at each end of the dc link.

16-4. Repeat Problem 16-3 if each converter transformer is equipped with a tap-changer. Now, it is possible to operate the rectifier at a firing angle as close to 18° as possible, while the inverter operates as close to the minimum extinction angle of 18° (but $\gamma_I \geq 18°$) as possible. The tap information is given below where the nominal line–line, primary voltage for each converter transformer is 230 kV (rms):

Maximum value of the converter transformer tap ratio in per unit: 1.15 at the rectifier terminal and 1.10 at the inverter terminal.

Minimum value of the converter transformer tap ratio in per unit: 0.95 at the rectifier and 0.90 at the inverter terminal.

Converter transformer tap step in per unit: 0.0125 at both terminals.

16-5. The nominal line-to-line voltage at a bus in a three-phase ac system is 230 kV (rms), when it is supplying a three-phase inductive load of $P + jQ = 1500\,MW + j750\,MVAR$.

The per-phase ac system impedance Z_s seen by the bus can be approximated to be purely inductive with $Z_s = j5.0\,\Omega$.

(a) Calculate the percentage change in the bus voltage magnitude for a 10% increase in P.

(b) Calculate the percentage change in the bus voltage magnitude for a 10% increase in Q.

16-6. A hybrid arrangement of a thyristor-controlled inductor and a thyristor-switched capacitors is connected at the ac bus in Problem 16-5. The TCI can draw a maximum of 50 MVARs per-phase, where as the TSC consists of four-capacitor banks, each with a per-phase rating of 50 MVARs. Holding the ac-bus voltage to its nominal value for a 10% increase in Q in Problem 16-5b, calculate the number of capacitor banks that should be switched in, the delay angle α at which the TCI should operate, and the per-phase effective inductance of the TCI.

16-7. Derive Eq. 16-13.

16-8. Derive Eq. 16-22.

16-6 REFERENCES

HVDC Transmission

1. E. W. KIMBARK, *Direct Current Transmission*, Vol. I, Wiley-Interscience, New York, 1971.
2. C. ADAMSON and N. G. HINGORANI, *High Voltage Direct Current Transmission*, Garraway, London, 1960, available from University Microfilms, Ann Arbor, MI.

Static Var Control — TCI and TSC

3. L. GYUGYI and W. P. MATTY, "Static VAR Generator with Minimum No Load Losses for Transmission Line Compensation," *Proceedings of the 1979 American Power Conference*.
4. T. J. E. MILLER, (Ed.), *Reactive Power Control in Electric Systems*, Wiley-Interscience, New York, 1982.
5. L. GYUGYI and E. R. TAYLOR, "Characteristic of Static, Thyristor-Controlled Shunt Compensators for Power Transmission System Applications," *IEEE Transactions on Power Apparatus and Systems*, Vol. PAS-99, No. 5, pp. 1795–1804, Sept./Oct. 1980.

Static Var Control — Minimum Energy Storage

6. Y. SUMI, et al., "New Static Var Control using Force-Commutated Inverters," *IEEE Transactions on Power Apparatus and Systems*, Vol. PAS-100, No. 9, pp. 4216–4224, Sept. 1981.
7. H. AKAGI, Y. KANAZAWA and A. NABAE, "Instantaneous Reactive Power Compensators Comprising Switching Devices without Energy Storage Components," *IEEE Transactions on Industry Applications*, Vol. IA-20, No. 3, pp. 625–630, May/June 1984.

Photovoltaic Array Interconnection

8. R. L. STEIGERWALD, A. FERRARO, and F. G. TURNBULL, "Application of Power Transistors to Residential and Intermediate Rating Photovoltaic Array Power Conditioners," *IEEE Transactions on Industry Applications*, Vol. IA-19, No. 2, pp. 254–267, March/April 1983.
9. R. L. STEIGERWALD, A. FERRARO, and R. E. TOMPKINS, "Final Report—Investigation of a Family of Power Conditioners Integrated into Utility Grid—Residential Power Level," DOE Contract DE-AC02-80ET29310, Sandia National Lab., Rep. SAND81-7031, 1981.

10. K. Tsukamoto and K. Tanaka, "Photovoltaic Power System Interconnected with Utility," *1986 Proceedings of the American Power Conference*, pp. 276–281.

Battery Storage and Fuel Cells

11. EPRI Report, "AC/DC Power Converters for Batteries and Fuel Cells," EPRI EM-2031, Project 841-1, Final Report, Sept. 1981.

Superconductive Energy Storage

12. H. A. Peterson, N. Mohan, and R. W. Bloom, "Superconductive Energy Storage Inductor-Converter Units for Power Systems," *IEEE Transactions on Power Apparatus and Systems*, Vol. 94, No. 4, pp. 1337–1348, July/Aug. 1975.

17

Optimizing the Utility Interface with Power Electronic Systems

---

17-1 INTRODUCTION

In Chapter 9, we discussed various powerline disturbances and how power electronic converters can perform as power conditioners and uninterruptible power supplies to prevent these powerline disturbances from disrupting the operation of critical loads such as computers used for controlling important processes, medical equipment, and the like. However, as discussed in the previous chapters, all power electronic converters (including those used to protect critical loads) can add to the inherent powerline disturbances by distorting the utility waveform due to harmonic currents injected into the utility grid and by producing electromagnetic interference. To illustrate the problems due to current harmonics i_h in the input current i_s of a power electronic load, consider the simple block diagram of Fig. 17-1. Due to the finite (non-zero) internal impedance of the utility source which is simply represented by L_s in Fig. 17-1, the voltage waveform at the point of common coupling to the other loads will become distorted, which may cause them to malfunction. In addition to the voltage waveform distortion, some other problems due to the harmonic currents are as follows: additional heating and possibly overvoltages (due to resonance conditions) in the utility's distribution and transmission equipment, errors in metering and malfunction of utility relays, interference with communication and control signals, and so on. In addition to these problems, phase-controlled converters discussed in Chapter 4 cause notches in the utility voltage waveform and many draw power at a very low displacement power factor which results in a very poor power factor of operation.

The foregoing discussion shows that the proliferation of power electronic systems and loads has the potential for significant negative impact on the utilities themselves, as well as on their customers. One approach to minimize this impact is to filter the harmonic currents and the electromagnetic interference produced by the power electronic loads. A better alternative, in spite of a small increase in the initial cost, may be to design the power electronic equipment such that the harmonic currents and

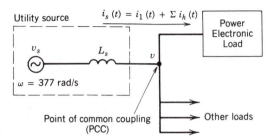

FIGURE 17-1: Utility interface.

the EMI are prevented or minimized from being generated in the first place. Both, the concerns about the utility interface and the design of power electronic equipment to minimize these concerns are discussed in this chapter.

17-2 GENERATION OF CURRENT HARMONICS

In most power electronic equipment, such as switch-mode dc power supplies, uninterruptible power supplies (UPS), and ac and dc motor drives, ac-to-dc converters are used as the interface with the utility voltage source. Commonly, a line-frequency diode rectifier bridge as shown in Fig. 17-2 is used to convert line frequency ac into dc. The rectifier output is a dc voltage whose average magnitude V_d is uncontrolled. A large filter capacitor is used at the rectifier output to reduce the ripple in the dc voltage v_d. The dc voltage v_d and the dc current i_d are unipolar and unidirectional, respectively. Therefore, the power flow is always from the utility ac input to the dc side. These line-frequency rectifiers with a filter capacitor at the dc side were discussed in detail in Chapter 3.

A class of power electronic systems utilizes line-frequency thyristor-controlled ac-to-dc converters as the utility interface. In these converters, which were discussed in detail in Chapter 4, the average dc output voltage V_d is controllable in magnitude and polarity, but the dc current i_d remains unidirectional. Because of the reversible polarity of the dc voltage, the power flow through these converters is reversible. As was pointed out in Chapter 4, the trend is to use these converters only at very high power levels such as in high-voltage dc transmission systems. Because of the very high power levels, the techniques to filter the current harmonics and to improve the power factor of operation are quite different in these converters, as discussed in Chapter 16, than those for the line-frequency diode rectifiers. In the general discussion presented in this chapter, therefore, only the diode rectifiers, where the dc voltage V_d remains essentially constant, are considered.

The diode rectifiers are used to interface with both the single-phase and the three-phase utility voltages. Typical ac current waveforms with minimal filtering were shown in Chapter 3. Typical harmonics in a single-phase input current waveform are

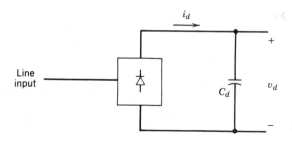

FIGURE 17-2: Diode rectifier bridge.

TABLE 17-1
Typical Harmonics in a Single-phase Input Current Waveform with No Line Filtering

h	3	5	7	9	11	13	15	17
$\left(\dfrac{I_h}{I_1}\right)\%$	73.2	36.6	8.1	5.7	4.1	2.9	0.8	0.4

listed in Table 17-1, where the harmonic currents I_h are expressed as a ratio of the fundamental current I_1. As is shown by Table 17-1, such current waveforms consist of large harmonic magnitudes. Therefore, for a finite internal per-phase source impedance L_s, the voltage distortion at the point of common coupling in Fig. 17-1 can be substantial. The higher the internal source inductance L_s, the greater would be the voltage distortion.

17-3 CURRENT HARMONICS AND POWER FACTOR

As we discussed in Chapter 3, the power factor PF at which an equipment operates is the product of the current ratio I_1/I_s and the displacement power factor DPF:

$$PF = \frac{\text{Power}}{\text{Volt-ampere}} = \frac{I_1}{I_s} \cdot DPF \qquad (17\text{-}1)$$

In Eq. 17-1, the displacement power factor equals the cosine of the angle ϕ_1 by which the fundamental frequency component in the current waveform is displaced with respect to the input voltage waveform. The current ratio I_1/I_s in Eq. 17-1 is the ratio of the rms value of the fundamental frequency current component to the rms value of the total current. The power factor indicates how effectively the equipment draws power from the utility; at a low power factor of operation for a given voltage and power level, the current drawn by the equipment will be large, thus requiring increased volt-ampere ratings of the utility equipment such as transformers, transmission lines, and generators. The importance of the high power factor has been recognized by residential and office equipment manufacturers for their own benefit to maximize the power available from a wall outlet. For example from a 120-V, 15-A electrical circuit in a building, the maximum power available is 1.8 kW, provided the power factor is unity. The maximum power that can be drawn without exceeding the 15-A limit decreases with decreasing power factor. The foregoing arguments indicate the responsibility and desirability on the part of the equipment manufacturers and users to design power electronic equipment with a high power factor of operation. This requires that the displacement power factor DPF should be high in Eq. 17-1. Moreover, the current harmonics should be low to yield a high current ratio I_1/I_s in Eq. 17-1.

17-4 HARMONIC STANDARDS AND RECOMMENDED PRACTICES

In view of the proliferation of the power electronic equipment connected to the utility system, various national and international agencies have been considering limits on harmonic current injection to maintain good power quality. As a consequence, various standards and guidelines have been established that specify limits on the magnitudes of harmonic currents and harmonic voltage distortion at various harmonic frequencies. Some of these are as follows:

1. EN 50 006, "The Limitation of Disturbances in Electricity Supply Networks caused by Domestic and Similar Appliances Equipped with Electronic Devices," European Standard prepared by Comité Européen de Normalisation Electrotechnique, CENELEC.

2. IEC Norm 555-3, prepared by the International Electrical Commission.

3. West German Standards VDE 0838 for household appliances, VDE 0160 for converters, and VDE 0712 for fluorescent lamp ballasts.

4. *IEEE Guide for Harmonic Control and Reactive Compensation of Static Power Converters*, ANSI/IEEE Std. 519-1981, which is expected to be revised.

CENELEC, IEC, and VDE standards specify the limits on the voltages (as a percentage of the nominal voltage) at various harmonic frequencies of the utility frequency, when the equipment-generated harmonic currents are injected into a network whose impedances are specified.

In the revised IEEE-519, which will contain recommended practices and requirements for harmonic control in electric power systems, the present proposal is to specify requirements on the user as well as on the utility. Table 17-2 lists the limits on the harmonic currents that a user of power electronic equipment and other nonlinear loads is allowed to inject into the utility system. Table 17-3 lists the quality of voltage that the utility must furnish the user. A utility will be able to furnish the voltage as listed in Table 17-3, provided that the harmonic currents injected by the users on a distribution feeder are limited in accordance with Table 17-2. Tables 17-2 and 17-3 are very broad in their scope and apply to wide voltage and power ranges. They are primarily intended for three-phase systems but can also be used as a guide to limit distortion in single-phase systems.

The principal justification for the harmonic limits specified in Table 17-2 is explained below. The voltage distortion at the point of common coupling (PCC) in Fig. 17-1 depends on the internal impedance of the ac source and the magnitudes of

TABLE 17-2

Harmonic Current Distortion (I_h/I_1) in %: Harmonic current limits for nonlinear load connected to a public utility at the point of common coupling (PCC) with other loads at voltages of 2.4 to 69 kV

I_{sc}/I_1	Odd Harmonic Order h					Total Harmonic Distortion
	$h < 11$	$11 < h < 17$	$17 < h < 23$	$23 < h < 35$	$35 < h$	
< 20	4.0	2.0	1.5	0.6	0.3	5.0
20–50	7.0	3.5	2.5	1.0	0.5	8.0
50–100	10.0	4.5	4.0	1.5	0.7	12.0
100–1000	12.0	5.5	5.0	2.0	1.0	15.0
> 1000	15.0	7.0	6.0	2.5	1.4	20.0

NOTES: 1. I_{sc} is the maximum short-circuit current at PCC.

2. I_1 is the maximum fundamental frequency load current at PCC.

3. Even harmonics are limited to 25% of the odd harmonic limits above.

Source: Reference 1.

TABLE 17-3
*Harmonic Voltage Limits (V_h/V_1) in % for Power Producers
(Public Utilities or Cogenerators)*

	2.3–69 kV	69–138 kV	> 138 KV
Maximum for individual harmonic	3.0	1.5	1.0
Total harmonic distortion (THD)	5.0	2.5	1.5

NOTE: This table lists the quality of the voltage that the power producer is required to furnish a user. It is based on the voltage level at which the user is supplied.

Source: Reference 1.

the injected current harmonics. In practice, the internal impedance of the source is highly inductive and therefore is represented by L_s in Fig. 17-1. At a harmonic h of the line frequency ω, the rms harmonic voltage at PCC is

$$V_h = (h\omega L_s)I_h \qquad (17\text{-}2)$$

where I_h is the h harmonic current injected into the ac source.

The internal inductance L_s in Eq. 17-2 is often specified in terms of the short-circuit-current I_{sc} at the point of common coupling. On a per-phase basis, I_{sc} will be the per-phase rms current supplied by the ac source to the fault, if all three phases are shorted to ground at PCC:

$$I_{sc} = \frac{V_s}{\omega L_s} \qquad (17\text{-}3)$$

where V_s is the rms value of the per-phase internal voltage of the ac source, which is assumed to be sinusoidal. A large I_{sc} represents a large capacity of the ac system at PCC. From Eqs. 17-2 and 17-3, the harmonic voltage can be expressed as a ratio of the nominal system voltage V_s in percentage

$$\%V_h = \frac{V_h}{V_s} \times 100 = h\frac{I_h}{I_{sc}} \times 100 \qquad (17\text{-}4)$$

If I_1 is the line-frequency component of the current drawn by the power electronic load, then dividing both I_h and I_{sc} in Eq. 17-4 by I_1 results in

$$\%V_h = h\frac{(I_h/I_1)}{(I_{sc}/I_1)} \times 100 \qquad (17\text{-}5)$$

In Eq. 17-5, I_{sc}/I_1 represents the capacity of the utility system with respect to the fundamental-frequency volt-amperes of the load. Equation 17-5 shows that for an acceptable harmonic voltage distortion in percentage, the harmonic current ratio I_h/I_1 can be higher (although not in a linear proportion) for a higher I_{sc}/I_1 ratio in Table 17-2. Moreover, because the internal impedance of the ac system is mostly inductive, the harmonic voltage distortion in Eq. 17-5 is proportional to the harmonic order h. Therefore, the maximum allowable harmonic current ratio (I_h/I_1) decreases (although not linearly) with increasing value of h, as shown in Table 17-2. As in

Chapter 3, the total harmonic distortion (THD) in the input current is defined as

$$THD = \frac{\sqrt{\sum_{h=2}^{\infty} I_h^2}}{I_1} \tag{17-6}$$

The total current harmonic distortion allowed in Table 17-2 increases with I_{sc}/I_1. It should be noted that there are other factors such as higher losses at high frequencies which also contribute to the allowable limits in Table 17-2.

The total harmonic distortion in the voltage can be calculated in a manner similar to that given by Eq. 17-6. Table 17-3 specifies the individual harmonics and the THD limits on the voltage that the utility supplies to the user at the point of common coupling.

17-5 NEED FOR IMPROVED UTILITY INTERFACE

Because of the large harmonic content as indicated by Table 17-1, typical diode rectifiers used for interfacing power electronic equipment with the utility system may exceed the limits on individual current harmonics and THD specified in Table 17-2. In addition to the effect on the power-line quality, the poor waveform of the input current also affects the power electronic equipment itself in the following ways:

- The power available from the wall outlet is reduced to approximately two-thirds.
- DC-side filter capacitor in Fig. 17-2 is severely stressed due to large peak pulse currents.
- The losses in the diodes of the rectifier bridge are higher due to a current-dependent forward voltage drop across the diodes.
- The components in EMI filter used at the input to the rectifier bridge must be designed for higher peak pulse currents.
- If a line-frequency transformer is used at the input, it must be highly overrated.

In view of these drawbacks, some of the alternatives for improving the input current waveforms are discussed, along with their relative advantages and disadvantages.

17-6 IMPROVED SINGLE-PHASE UTILITY INTERFACE

Various options for improving the single-phase utility interface of power electronic equipment are discussed.

17-6-1 Passive Circuits

Inductors and capacitors can be used in conjunction with the diode rectifier bridge to improve the waveform of the current drawn from the utility grid. The simplest approach is to add an inductor on the ac side of the rectifier bridge in Fig. 17-2. This added inductor results in a higher effective value of the ac-side inductance L_s, which improves the power factor and reduces harmonics as shown by means of Figs. 3-6 and 3-7 in Chapter 3. The impact of adding an inductor can be summarized

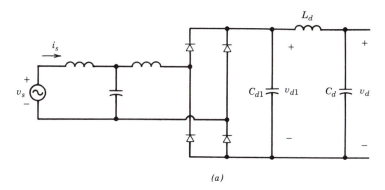

(a)

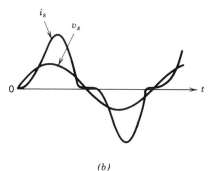

(b)

FIGURE 17-3: Passive filters to improve i_s waveform: (a) passive filter arrangement; (b) current waveform.

as follows:

- Because of an improved current waveform, the power factor is improved from very poor to somewhat acceptable.
- The output voltage V_d is dependent on the output load and is substantially ($\sim 10\%$) lower compared with the no-inductance case.
- Inductance and C_d together in Fig. 17-2 form a low-pass filter and, therefore, the peak-to-peak ripple in the rectified output voltage v_d is less.
- The overall energy efficiency remains essentially the same; there are additional losses in the inductor, but the conduction losses in the diodes are lower.

It is possible to further improve the input current waveform (Fig. 17-3b) by using a circuit arrangement, as is shown in Fig. 17-3a. In Fig. 17-3a, C_{d1} directly across the rectifier bridge is small relative to C_d. This allows a larger ripple in v_{d1} but results in an improved waveform of i_s. The ripple in v_{d1} is filtered out by the low-pass filter consisting of L_d and C_d. Obvious disadvantages of such an arrangement are cost, size, losses, and the significant dependence of the average dc voltage V_d on the power drawn by the load.

17-6-2 Active Shaping of the Input Line Current

By using a power electronic converter for current shaping, as is shown in the diagram of Fig. 17-4a, it is possible to shape the input current drawn by the rectifier bridge to be sinusoidal and in phase with the input voltage. The choice of the power

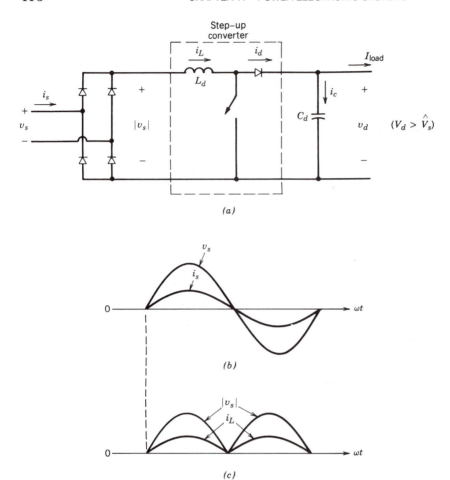

FIGURE 17-4: Active harmonic filtering: (*a*) step-up converter for current shaping; (*b*) line waveforms; (*c*) $|v_s|$ and i_L.

electronic converter is based on the following considerations:

- In general, electrical isolation between the utility input and the output of the power electronic system is either not needed (for example in ac- and dc-motor drives), or it can be provided in the second converter stage as in the switch-mode dc power supplies.

- In most applications it is acceptable, and in many cases it is desirable to stabilize the dc voltage V_d slightly in excess of the peak of the maximum of the ac input voltage.

- The input current drawn should ideally be at a unity power factor so that the power electronic interface emulates a resistor supplied by the utility source. This also implies that the power flow is always unidirectional, from the utility source to the power electronic equipment.

- The cost, power losses, and size of the current shaping circuit should be as small as possible.

Based on these considerations, a line-frequency transformer isolation is ruled out. Also, it is acceptable to have $V_d > \hat{V}_s$, where $\hat{V}_s$ is the peak of the ac input voltage.

Therefore, the obvious choice for the current shaping circuit is a step-up dc–dc converter, similar to that discussed in Chapter 5. This converter is shown in Fig. 17-4a, where C_d is used to minimize the ripple in v_d and to meet the energy storage requirement of the power electronic system. As in Chapter 3, a dc current I_{load} represents the power supplied to the rest of the system (the high-frequency component in the output current is effectively filtered out by C_d). For simplicity, the internal inductance L_s of the utility source is not included in Fig. 17-4a.

The basic principle of operation is straightforward. At the utility input, the current i_s is desired to be sinusoidal and in phase with v_s, as is shown in Fig. 17-4b. Therefore, at the full-bridge rectifier output in Fig. 17-4a, i_L and $|v_s|$ have the same waveform as shown in Fig. 17-4c. In practice, the power losses in the rectifier bridge and the step-up dc–dc converter are fairly small. These are neglected in the following theoretical analysis. From the waveforms of Fig. 17-4b where $\hat{V}_s = \sqrt{2}\,V_s$ and $\hat{I}_s = \sqrt{2}\,I_s$, the input power $p_{\text{in}}(t)$ from the ac source is

$$p_{\text{in}}(t) = \hat{V}_s|\sin \omega t| \cdot \hat{I}_s|\sin \omega t| = V_sI_s - V_sI_s \cos 2\omega t \qquad (17\text{-}7)$$

Because of a fairly large capacitance C_d, the voltage v_d can be initially assumed to be dc, that is, $v_d(t) = V_d$. Therefore, the output power is

$$p_d(t) = V_d i_d(t) \qquad (17\text{-}8)$$

where in Fig. 17-4a

$$i_d(t) = I_{\text{load}} + i_c(t) \qquad (17\text{-}9)$$

If the step-up converter in Fig. 17-4a is idealized and can be assumed to be operating at a switching frequency approaching infinity, then required L_d would be negligibly small. This allows the assumption in Fig. 17-4a that $p_{\text{in}}(t) = p_d(t)$ on an instantaneous basis. Therefore, from Eqs. 17-7 through 17-9,

$$i_d(t) = I_{\text{load}} + i_c(t) = \frac{V_sI_s}{V_d} - \frac{V_sI_s}{V_d} \cos 2\omega t \qquad (17\text{-}10)$$

where the average value of i_d is

$$I_d = I_{\text{load}} = \frac{V_sI_s}{V_d} \qquad (17\text{-}11)$$

and the current through the capacitor is

$$i_c(t) = -\frac{V_sI_s}{V_d} \cos 2\omega t = -I_d \cos 2\omega t \qquad (17\text{-}12)$$

Even though this analysis is carried out by assuming the voltage across the capacitor to be ripple-free dc, the ripple in v_d can be estimated from Eq. 17-12 as

$$v_{d,\text{ripple}}(t) \approx \frac{1}{C_d} \int i_c \cdot dt \approx -\frac{I_d}{2\omega C_d} \sin 2\omega t \qquad (17\text{-}13)$$

which can be kept low by selecting a suitably large value of C_d. A series-tuned L-C

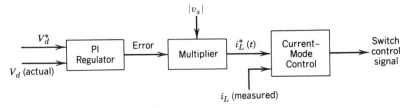

FIGURE 17-5: Control block diagram.

filter, tuned for twice the ac frequency according to Eq. 17-12 may be put in parallel with C_d to minimize the ripple in the dc voltage. It should be noted that the switching-frequency components of currents in i_d and the high-frequency components in the load current will also flow through C_d.

Because the input current to the step-up converter is to be shaped, the step-up converter is operated in a current-regulated mode, as discussed in Chapter 6 in connection with dc-to-ac inverters. The feedback control is shown in a block diagram form in Fig. 17-5, where i_L^* is the reference or the desired value of the current i_L in Fig. 17-4a. i_L^* has the same waveform as $|v_s|$. The amplitude of i_L^* should be such as to maintain the output voltage at a desired or reference level V_d^*, in spite of the variation in load and the fluctuation in the line voltage from its nominal value. The waveform of i_L^* in Fig. 17-5 is obtained by measuring $|v_s|$ in Fig. 17-4a by means of a resistive potential divider and multiplying it with the amplified error between the reference value V_d^* and the actual measured value of V_d. The actual current i_L is sensed, usually by measuring the voltage across a small resistor inserted in the return path of i_L. The status of the switch in the step-up converter is controlled by comparing the actual current i_L with i_L^*.

Once i_L^* and i_L in Fig. 17-5 are available, there are various ways to implement the current-mode control of the step-up converter. Some of these were discussed in connection with the current-mode control of switch-mode dc power supplies in Chapter 8. Four such control modes are discussed as follows where f_s is the switching frequency and I_{rip} is the peak-to-peak ripple in i_L during one time period of the switching frequency. Only the constant frequency control is described in some detail.

1. *Constant frequency control.* Here, the switching frequency f_s is kept constant. When i_L reaches i_L^*, the switch in the step-up converter is turned off. The switch is turned on by a clock at a fixed frequency f_s, which results in i_L as shown in Fig. 17-6a. As discussed in Chapter 8, a slope compensation ramp must be used; otherwise i_L will be irregular at switch duty ratios in excess of 0.5. Normalized I_{rip} is plotted in Fig. 17-6b.

2. *Constant tolerance-band control.* Here, the current i_L is controlled such that the peak-to-peak ripple I_{rip} in i_L remains constant. With a preselected value of I_{rip}, i_L is forced to be within the tolerance band $(i_L^* + I_{rip}/2)$ and $(i_L^* - I_{rip}/2)$ by controlling the switch status.

3. *Variable tolerance-band control.* Here, the peak-to-peak ripple current I_{rip} is increased in proportion to the instantaneous value of $|v_s|$. Otherwise, this approach is similar to the constant tolerance-band control.

4. *Discontinuous current control.* In this scheme, the switch is turned off when i_L reaches $2i_L^*$. The switch is kept off until i_L reaches zero, at which instant the switch is turned back on. This can be considered as a special case of a variable tolerance-band control.

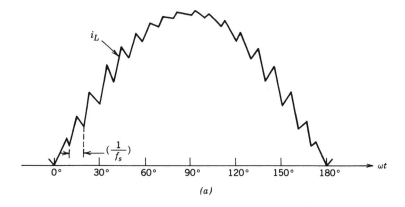

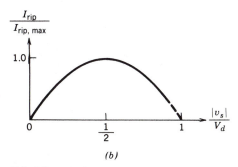

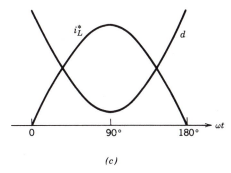

FIGURE 17-6: Constant-frequency control.

During a switching-frequency time period, the output voltage is assumed to be constant as V_d and the input voltage to the step-up converter is assumed to be constant at that instant of time; I_{rip} is the peak-to-peak ripple current during one time period of the switching frequency. The following equations can be written from Fig. 17-4a during the on interval t_{on} and the off interval t_{off} of the switch

$$t_{on} = \frac{L_d I_{rip}}{|v_s|} \tag{17-14}$$

and

$$t_{off} = \frac{L_d I_{rip}}{V_d - |v_s|} \tag{17-15}$$

where the switching frequency f_s is given as

$$f_s = \frac{1}{t_{on} + t_{off}} = \frac{(V_d - |v_s|)|v_s|}{L_d I_{rip} V_d} \tag{17-16}$$

In a *constant frequency control scheme*, f_s in Eq. 17-16 is constant and hence

$$I_{rip} = \frac{(V_d - |v_s|)|v_s|}{f_s L_d V_d} \tag{17-17}$$

Figure 17-6b shows the plot of the normalized I_{rip} as a function of $|v_s|/V_d$, noting that in a step-up converter $|v_s|/V_d$ must be less than or equal to 1. The maximum ripple current is given as

$$I_{\text{rip,max}} = \frac{V_d}{4f_sL_d} \qquad \text{when} \quad |v_s| = \frac{V_d}{2} \tag{17-18}$$

In the active current shaping circuit using a step-up dc–dc converter, the following additional observations are made:

- The output voltage v_d across the capacitor C_d contains a 120-Hz ripple at twice the line frequency. The feedback control circuit used to control V_d at a desired value cannot compensate this voltage ripple without distorting the input line current.

- If the switching frequency ripple in i_L is kept to a small amplitude, then a laminated iron core inductor can possibly be used that will be smaller in size due to its higher saturation flux density compared to the high-frequency ferrite materials.

- A higher switching frequency allows a lower value of L_d and an increased ease of filtering high-frequency ripple. However, the switching frequency is chosen as a compromise between the foregoing advantages and the increased switching losses.

- V_d much larger than 10% beyond the peak input ac voltage $\hat{V}_s$ will cause efficiency to decline.

- To limit the in-rush current at start-up, a current limiting resistor in series with L_d can be used. Subsequent to the initial transient, the resistor is bypassed by a contactor or a thyristor in parallel with the current limiting resistor.

- The step-up converter topology is well suited for the input current shaping because when the switch is off, the input current directly (through the diode) feeds the output stage. In a constant-frequency current control, as an example, the switch duty ratio d is as shown in Fig. 17-6c as a function of ωt, recognizing from Chapter 5 that in a step-up converter with an input voltage $|v_s|$ and an output voltage V_d, $(|v_s|/V_d) = (1 - d)$. Therefore

$$d = 1 - \frac{|v_s|}{V_d} \tag{17-19}$$

Figure 17-6c shows that d is smallest at the peak of i_L^*. Thus, the large values of i_L flow through the switch only during a small fraction of the switching time-period.

- A small filter capacitor must be used across the output of the diode rectifier bridge to prevent the ripple in i_L from entering the utility system. An EMI filter at the input is still required as in a conventional circuit without the active current shaping.

In addition to an almost sinusoidal input waveform at nearly a unity power factor, the other advantages of an active input current shaping can be summarized as follows:

- The dc voltage V_d can be stabilized to a nearly constant value for large variations in the line voltage. With V_d equal to 1.1 times the peak of the nominal input voltage, for example, this circuit will continue to draw sinusoidal current for line overvoltages of up to 10%. By selecting proper current ratings of the components, this current-shaping circuit can easily handle large undervoltages at the utility input.

- Since V_d is stabilized to a nearly constant value, the volt-ampere ratings of the semiconductor devices in the converter fed from V_d are significantly reduced.

- Because of the absence of large peaks in the input current, the size of the EMI filter components is smaller.

- For the equal ripple in v_d, only one-third to one-half the capacitance C_d is needed compared with the conventional circuit, thus, resulting in a reduced size.

- The energy efficiency from v_s to V_d of such a circuit is typically 96% compared with the efficiency of 99% in the conventional arrangement without active current shaping.

At present, the cost, slightly higher power losses, and the complexity of active current shaping have prevented their widespread usage. This may change in the future because of increased device integration leading to lower semiconductor cost, a strict enforcement of harmonic standards, and some of the advantages mentioned above. Another factor in favor of the active line-current shaping is as follows: in power supplies to computers, a sinusoidal line current is important to avoid the added kilovolt-ampere (kVA) rating and, hence, the increased cost of uninterruptible power supplies and standby diesel generators, which often supply computer systems. In such applications, the current-shaping techniques described above are being applied. Therefore, ICs and other components suitable for these applications will become available, which will lower the cost of development and the components of the active current-shaping circuit.

17-6-4 Interface for a Bidirectional Power Flow

In certain applications, for example, in motor drives with regenerative braking, the power flow through the utility interface converter reverses during the regenerative braking while the kinetic energy associated with the inertias of the motor and load is recovered and fed back to the utility system. One approach used in the past is to employ two back-to-back connected line-frequency thyristor converters, as is shown in Fig. 17-7. During the normal mode, converter 1 acts as a rectifier and the power flows from the ac input to the dc side. During regenerative braking, the gate pulses to the thyristors of converter 1 are blocked and converter 2 operates in an inverter mode

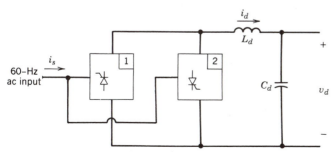

FIGURE 17-7: Back-to-back connected converters for bidirectional power flow.

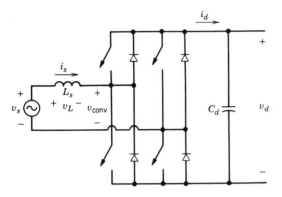

FIGURE 17-8: Switch-mode converter for the utility interface.

where the polarity of v_d remains the same but the direction of i_d is reversed. Each of these converters is similar to those discussed in detail in Chapter 4. There are several drawbacks associated with this approach: (1) the input current i_s has a distorted waveform and the power factor is low, (2) the dc voltage V_d is limited in the inverter mode because of the minimum extinction angle requirement of converter 2 while it operates in an inverter mode, and (3) there is a possibility of commutation failure in the inverter mode due to ac line disturbances.

It is possible to overcome these limitations by using a switch-mode converter, as shown in Fig. 17-8. This converter is identical to the four-quadrant switch-mode inverters discussed in Chapter 6 where the inverter mode, in which the power flows from the dc to the ac side was discussed in detail. The rectifier mode was only briefly discussed in Section 6-7.

Rectifier being the dominant mode of operation, i_s is defined with a direction, as shown in Fig. 17-8. An inductance L_s (which augments the internal inductance of the utility source) is included to reduce the ripple in i_s at a finite switching frequency. In the circuit of Fig. 17-8,

$$v_s = v_{conv} + v_L \tag{17-20}$$

where

$$v_L = L_s \frac{di_s}{dt} \tag{17-21}$$

Assuming v_s to be sinusoidal, the fundamental frequency components of v_{conv} and i_s in a Fig. 17-8 can be expressed as phasors $\mathbf{V}_{conv1}$ and $\mathbf{I}_{s1}$, respectively. Choosing $\mathbf{V}_s$ arbitrarily as the reference phasor $\mathbf{V}_s = V_s e^{j0°}$, at the line frequency $\omega = 2\pi f$

$$\mathbf{V}_s = \mathbf{V}_{conv1} + \mathbf{V}_{L1} \tag{17-22}$$

where

$$\mathbf{V}_{L1} = j\omega L_s \mathbf{I}_{s1} \tag{17-23}$$

A phasor diagram corresponding to Eqs. 17-22 and 17-23 is shown in Fig. 17-9a, where $\mathbf{I}_{s1}$ lags $\mathbf{V}_s$ by an arbitrary phase angle θ. The real power P *supplied* by the ac

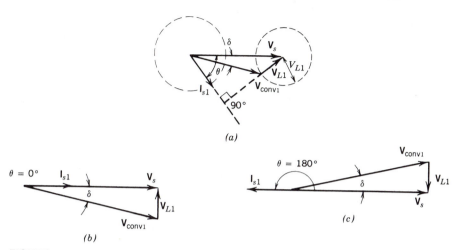

FIGURE 17-9: Rectification and inversion: (a) general phasor diagram; (b) rectification at unity power factor; (c) inversion at unity power factor.

source to the converter is

$$P = V_s I_{s1} \cos \theta = \frac{V_s^2}{\omega L_s} \left(\frac{V_{conv1}}{V_s} \sin \delta \right) \qquad (17\text{-}24)$$

since in Fig. 17-9a, $V_{L1} \cos \theta = \omega L_s I_{s1} \cos \theta = V_{conv1} \sin \delta$.

In the phasor diagram of Fig. 17-9a, the reactive power Q *supplied* by the ac source is positive. It can be expressed as

$$Q = V_s I_{s1} \sin \theta = \frac{V_s^2}{\omega L_s} \left(1 - \frac{V_{conv1}}{V_s} \cos \delta \right) \qquad (17\text{-}25)$$

since in Fig. 17-9a, $V_s - \omega L_s I_{s1} \sin \theta = V_{conv1} \cos \delta$. Note that Q is the sum of the reactive power absorbed by the converter and the reactive power consumed by the inductance L_s. However, at very high switching frequencies, L_s can be made to be quite small; thus, Q can be approximated as the reactive power absorbed by the converter.

The important equations are summarized below:

$$P = \frac{V_s^2}{\omega L_s} \left(\frac{V_{conv1}}{V_s} \sin \delta \right) \qquad [\text{Eq. 17-24, repeated}]$$

$$Q = \frac{V_s^2}{\omega L_s} \left(1 - \frac{V_{conv1}}{V_s} \cos \delta \right) \qquad [\text{Eq. 17-25, repeated}]$$

and

$$\mathbf{I}_{s1} = \frac{(\mathbf{V}_s - \mathbf{V}_{conv1})}{(j\omega L_s)} \qquad (17\text{-}26)$$

From these equations it is clear that for a given line voltage v_s and the chosen inductance L_s, desired values of P and Q can be obtained by controlling the magnitude and the phase of v_{conv1}. Figure 17-9a shows how $\mathbf{V}_{conv1}$ can be varied, keeping the magnitude of $\mathbf{I}_{s1}$ constant. The two circles in Fig. 17-9a are traced by the loci of $\mathbf{I}_{s1}$ and $\mathbf{V}_{conv1}$ phasors.

In the general analysis discussed earlier, two cases are of special interest: rectification and inversion at a unity power factor. These two are shown by phasor diagrams in Figs. 17-9b and 17-9c. In both cases

$$V_{conv1} = \left[V_s^2 + \left(\omega L_s I_{s1} \right)^2 \right]^{1/2} \qquad (17\text{-}27)$$

If a high switching frequency is used, only a small inductance L_s is needed. Therefore, from Eq. 17-27

$$V_{conv1} \approx V_s \qquad (17\text{-}28)$$

For the desirable magnitude and the direction of power flow as well as Q, the magnitude V_{conv1} and the phase angle δ with respect to the line voltage must be controlled. In the circuit of Fig. 17-8, V_d is established by charging the capacitor C_d through the switch-mode converter. V_d should be of a sufficiently large magnitude so that v_{conv1} at the ac side of the converter is produced by a pulse-width-modulation that corresponds to PWM in a linear region (i.e., $m_a \leq 1.0$ as discussed in Chapter 6). This is necessary to limit ripple in the input current i_s. Therefore from Eq. 6-19 of Chapter 6 and Eq. 17-28, V_d must be greater than the peak of the input ac voltage, that is

$$V_d > \sqrt{2}\, V_s \qquad (17\text{-}29)$$

The control circuit to regulate V_d in Fig. 17-8 at its reference value V_d^* and to achieve a unity power factor of operation is shown in Fig. 17-10. The amplified error between V_d and V_d^* is multiplied with the signal proportional to the input voltage v_s waveform to produce the reference current signal i_s^*. A current-mode control such as a tolerance-band control or a fixed-frequency control as discussed in Chapter 6 can be used to deliver i_{s1} equal to i_s^* and in phase or $180°$ out of phase with the line voltage v_s. The magnitude and direction of power flow are automatically controlled by regulating V_d at its desired value. It is possible to obtain a phase shift between v_s and i_{s1}, and hence a finite reactive power flow by introducing the corresponding phase shift in the signal proportional to v_s in the control circuit of Fig. 17-10. The steady-state waveforms in the circuit of Fig. 17-8 for a unity power factor rectifier operation are shown in Fig. 17-11.

As we discussed in the previous section, in a single-phase circuit with i_s nearly sinusoidal and V_d essentially dc, the current i_d in Fig. 17-8 consists of a ripple at twice

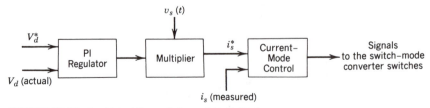

FIGURE 17-10: Control of the switch-mode interface.

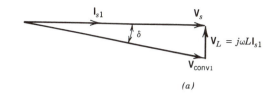

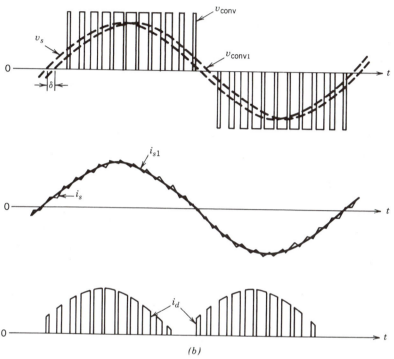

FIGURE 17-11: Waveforms in the circuit of Fig. 17-8 at unity power factor of operation: (*a*) phasor diagram; (*b*) circuit waveforms.

the line frequency. This ripple in i_d results in a ripple in the dc voltage across C_d, which can be minimized by means of an L-C filter (series tuned at twice the line frequency) in parallel with C_d. This is done in high power applications such as electric locomotives.

17-7 IMPROVED THREE-PHASE UTILITY INTERFACE

Three-phase diode rectifier bridges with a filter capacitor on the dc side were discussed in Chapter 3. The input current in these rectifiers is also highly distorted. One of the ways to improve the input current waveform is to include a dc-side inductor L_d between the rectifier and the filter capacitor. Improvement in the input power factor as a function of L_d was shown in Fig. 3-14 in Chapter 3.

In a three-phase case, active input line current shaping can be achieved by three separate transformer-isolated current-shaping circuits. At least two of the current-shaping circuits must have their outputs electrically isolated from their inputs, since the outputs of the three current-shaping circuits feed the same output capacitor. The

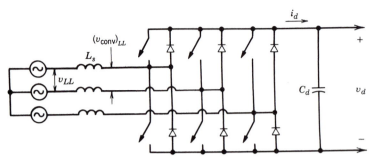

FIGURE 17-12: Three-phase, switch-mode converter.

requirement of electrical isolation can be met by using high-frequency transformer-iso-lated dc–dc converters discussed in Chapter 8 in the current-shaping circuit. Because the three-phase input is generally used for higher power level equipment, an alterna-tive would be to use 60-Hz isolation transformers at the input.

Since in many applications the electrical isolation between the utility input and the output is not needed, the requirement of transformer isolation will cause unneces-sary losses in the transformers and will be more expensive. In such cases, therefore, a better alternative is to use a four-quadrant switch-mode inverter as shown in Fig. 17-12. This converter is capable of supplying nearly sinusoidal input current at a unity power factor; in addition, the power flow through such a converter is reversible. This three-phase switch-mode dc-to-ac inverter was considered in detail in Chapter 6. Its operation in the rectifier mode was briefly discussed in Section 6-7 of Chapter 6 on a per-phase basis.

The block diagram for controlling such a converter is the same as Fig. 17-10 for the single-phase case in Section 17-6-4 where the dc voltage V_d is regulated to its reference value V_d^*. For the converter to be capable of controlling the input current waveforms to be sinusoidal, V_d^* should be appropriately chosen. If a high switching frequency is used, the ac-side inductances L_s in Fig. 17-12 can be minimized. Therefore, the voltage drops across L_s are small and, the rms voltages

$$\left(V_{\text{conv}}\right)_{LL} \approx V_{LL} \tag{17-30}$$

If the converter in Fig. 17-12 is to be pulse-width-modulated in a linear range with $m_a \le 1.0$ to control the input currents to be sinusoidal, then from Eq. 6-57

$$V_d > 1.634 \, V_{LL} \tag{17-31}$$

An important difference between the dc current i_d in the three-phase converters as compared with single-phase converters is that it consists of a dc component I_d and the high switching frequency component (the ripple at twice the line frequency does not exist as in the single-phase case). As discussed in Chapter 6

$$I_d = \frac{3V_s I_s}{V_d} \cos \phi_1 \tag{17-32}$$

where V_s and I_s are the sinusoidal per-phase line quantities and ϕ_1 is the angle by which the phase current lags the phase voltage. For rectification at a unity power

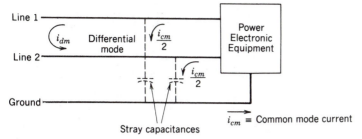

FIGURE 17-13: Conducted interference.

factor, $\phi_1 = 0$ and

$$I_d = \frac{3V_sI_s}{V_d} \qquad (17\text{-}33)$$

Since only the high switching frequency current flows through C_d, only a small capacitance is needed.

17-8   ELECTROMAGNETIC INTERFERENCE (EMI)

Because of rapid changes in voltages and currents within a switching converter, power electronic equipment is a source of electromagnetic interference with other equipment as well as with its own proper operation. EMI is transmitted in two forms: radiated and conducted. The switching converters supplied by the power lines generate conducted noise into the power lines that is usually several orders of magnitude higher than the radiated noise into free space. Metal cabinets used for housing power converters reduce the radiated component of the EMI.

Conducted noise as shown in Fig. 17-13 consists of two categories commonly known as the differential mode and the common mode. The differential mode noise is a current or a voltage measured between the lines of the source, that is, a line-to-line voltage or a line current i_{dm} in Fig. 17-13. The common mode noise is a voltage or current measured between the power lines and ground, such as i_{cm} in Fig. 17-13. Both differential mode and common mode noise are present in general on both the input lines and the output lines. Any filter design has to take into account both of these modes of noise.

17-8-1 Generation of EMI

Switching waveforms such as that shown in Fig. 17-14, for example, are inherent in all switching converters. Because of short rise and fall times, these waveforms contain significant energy levels at harmonic frequencies in the radio frequency (RF) region, several orders above the fundamental frequency.

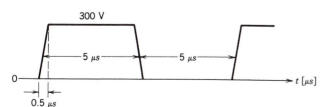

FIGURE 17-14: Switching waveform.

The transmission of the differential-mode noise is through the input line to the utility system and through the dc-side network to the load on the power converter. Moreover, conduction paths through stray capacitances between components and due to magnetic coupling between circuits must also be considered.

The transmission of the common-mode noise is entirely through "parasitic" or stray capacitors and stray electric and magnetic fields. These stray capacitances exist between various system components and between components and ground. For safety reasons, most power electronic equipment have a grounded cabinet. The noise appearing on the ground line contributes significantly to the electromagnetic interference.

17-8-2 EMI Standards

There are various CISPR, IEC, VDE, FCC, and the Military Standards that specify the maximum limit on the conducted EMI. Figure 17-15 shows the FCC and the VDE standards for the radio frequency equipment used in industrial, commercial, and residential applications. Two separate limits are specified, where the limit A is for equipment used in commercial, industrial, and office environment and the limit B is for residential equipment. To compare against these limits, the conducted noise is measured by means of a specified impedance network called LISN (Line Impedance Stabilization Network). Standards for the radiated EMI are also specified by the various agencies.

17-8-3 Reduction of EMI

As is discussed in Reference 12, the most cost-effective way of dealing with EMI is to prevent the EMI from being generated at its source, which can significantly reduce radiated and conducted interference before the application of filters, shielding,

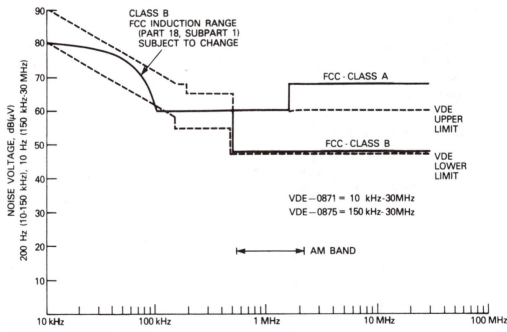

FIGURE 17-15: FCC and VDE standards for conducted EMI.

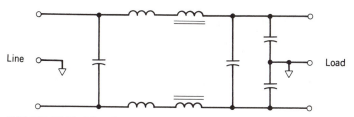

FIGURE 17-16: Filter for conducted EMI.

and the like. Another advantage to this approach is that a system that is not generating a high level of EMI will not be susceptible to its own noise and therefore will be more reliable.

From the point of view of EMI reduction, a properly designed snubber is quite effective, since it reduces both the dv/dt and the di/dt of the circuit. The snubber must be connected directly on the component being snubbed with as short leads as possible. Another approach to reducing EMI generation is to use the resonant converter concepts discussed in Chapter 7. In addition to these, the magnitudes of the coupling fields should be reduced by proper mechanical layout, wiring, and shielding.

To reduce magnetic fields, it is important to minimize the net area enclosed by a current loop. All current loops with switching transients should be made to have as small an area as possible. All current-carrying conductors should be run in close proximity to the return wire, such as by copper strips discussed in Chapter 26. A twisted pair of wires will reduce the generated external field to a minimum.

To reduce stray capacitances, the area of exposed metal at the switching potential should be minimized and kept as far from ground as possible by proper mechanical design.

In addition to these steps to minimize the generation of EMI, EMI filters such as that shown in Fig. 17-16 are used to meet the conducted EMI limits.

Generally, the radiated noise is effectively shielded by the metal cabinets used for housing the power electronic equipment. Additional steps may be necessary if the power electronic equipment is operating near sensitive communication or medical equipment.

17-9 SUMMARY

1. Power electronic equipment are a source of current harmonics and electromagnetic interference (EMI). Appropriate steps must be taken to prevent these from degrading power quality.

2. Harmonic standards and recommended practices have been established by various agencies to limit the harmonics injected by power electronic equipment.

3. In power electronic equipment with diode bridge rectifiers, input inductance reduces the input current harmonics.

4. In circuits with a single-phase utility input, the input line current can be actively shaped to be sinusoidal.

5. In single-phase and three-phase circuits, if bidirectional power flow is needed, the switch-mode converters such as those discussed in Chapter 6 can be used to interface with the utility system. These produce sinusoidal utility currents at a unity power factor.

6. EMI-generating mechanisms, EMI standards, and EMI-reduction techniques are discussed.

TABLE P17-1

h	3	5	7	9	11	13	15	17	19	21	23
$\left(\dfrac{I_h}{I_1}\right)\%$	34.0	5.3	1.8	1.8	1.6	1.2	0.9	0.8	0.8	0.4	0.4

PROBLEMS

17-1. A residential load of 240-V, 5-kW load-proportional, capacity-modulated heatpump is supplied through a single-phase, 25-kVA secondary distribution transformer with a leakage reactance of 4%. The input current harmonic components are given in Table P17-1 as a percentage of the fundamental frequency current component. The displacement power factor is approximately 1.

    (a) Calculate the short-circuit capacity of the system at the point of coupling, assuming the impedance of the rest of the system other than the secondary distribution transformer to be negligible.

    (b) Are the harmonics and THD within the limits specified in Table 17-2?

    (c) Calculate the power factor of the load.

17-2. In a single-phase, 240-V, 60-Hz, 2-kW diode-rectifier interface, the displacement power factor is essentially 1.0 and the input current harmonics are listed in Table 17-1 as a percentage of the fundamental frequency component. Calculate the rms value of the ripple current through the filter capacitor C_d due to the input interface. Neglect all losses.

17-3. In a single-phase interface for a bidirectional power flow as shown in Fig. 17-8, $V_s = 240$ V (rms) at 60 Hz and $L_s = 2.5$ mH. Neglect all losses and assume that the switch-mode converter is pulse-width-modulated in its linear range with $m_a \leq 1.0$. The converter is controlled such that i_s is either in phase or out of phase with v_s. Calculate the minimum value of V_d if the power flow through the converter is 2 kW (a) from the ac to the dc side, and (b) from the dc to the ac side.

17-10 REFERENCES

1. C. K. DUFFEY and R. P. STRATFORD, "Update of Harmonic Standard IEEE-519 IEEE Recommended Practices and Requirements for Harmonic Control in Electric Power Systems," IEEE/IAS Petroleum and Chemical Industry Conference (PCIC), Dallas, TX, Sept. 12–14, 1988.

2. EN 50006, "The Limitation of Disturbances in Electric Supply Networks caused by Domestic and Similar Appliances Equipped with Electronic Devices," European Standards prepared by Comité Européen de Normalisation Electrotechnique, CENELEC.

3. IEC Norm 555-3 prepared by the International Electrical Commission.

4. VDE Standards 0838 for Household Appliances and 0712 for Fluorescent Lamp Ballasts, West Germany.

5. *IEEE Guide for Harmonic Control and Reactive Compensation of Static Power Converters*, IEEE Project No. 519/05, July 1979; to be updated in 1989.

6. C. P. HENZE and N. MOHAN, "A Digitally Controlled AC to DC Power Conditioner that Draws Sinusoidal Input Current," *1986 IEEE Power Electronics Specialist Conference*, pp. 531–540.

7. M. HERFURTH, "TDA 4814—Integrated Circuit for Sinusoidal Line Current Consumption," Siemens Components publication, 1987.

8. M. HERFURTH, "Active Harmonic Filtering for Line Rectifiers of Higher Power Output," Siemens Components publication, 1986.

9. N. MOHAN, T. UNDELAND, and R. J. FERRARO, "Sinusoidal Line Current Rectification with a 100 kHz B-SIT Step-up Converter," *1984 IEEE Power Electronics Specialists Conference*, pp. 92–98.

10. VDE Standards 0875/6.77 for radio interference suppression of electrical appliances and systems.

11. VDE Standards 0871/6.78 for radio interference suppression of radio frequency equipment, for industrial, scientific and medical (ISM) and similar purposes.

12. N. MOHAN, "Techniques for Energy Conservation in AC Motor Drive Systems," EPRI Final Report EM-2037, September 1981.

13. L. M. SCHNEIDER, "Take the Guesswork out of Emission Filter Design," *EMC Technology*, pp. 23–32, April–June 1984.

P A R T

6

Semiconductor Devices and Converter Design

———
———
———

18

Basic Semiconductor Physics

_____
_____
_____

18-1 INTRODUCTION

In the previous chapters, it has been assumed that the semiconductor power devices had nearly ideal characteristics. These properties included

1. Large breakdown voltages.

2. Low on-state voltages and resistances.

3. Fast turn-on and turn-off.

4. Large power-dissipation capability.

In spite of significant progress in the development of power devices, there are none available that simultaneously have all of these properties. In all device types, there is a trade-off between breakdown voltages and on-state losses. In bipolar (minority carrier) devices, there is also a trade-off between on-state losses and switching speeds.

Such trade-offs mean that there is not one device type that can be used for all applications. The requirements of the specific application must be matched to the capabilities of the available devices. This often requires clever and innovative design approaches. For example, several devices may have to be combined in parallel or series connections in order to control larger amounts of power.

In this environment, it is important for the user to have a firm qualitative understanding of the physics of power devices and of how they are fabricated and packaged. A superficial knowledge of low-power solid-state devices is insufficient for the effective use of high-power devices. Indeed a naïve extrapolation of low-power device knowledge to the high-power regime could lead to the damage or even destruction of the high-power device.

In the next several chapters, the operation, fabrication, and packaging of high-power devices will be explored. In this chapter, the fundamental properties of semiconductors will be reviewed.

18-2   CONDUCTION PROCESSES IN SEMICONDUCTORS

18-2-1   Metals, Insulators, and Semiconductors

Electrical current will flow in a material if there are charge carriers (usually electrons) in the material that are free to move in response to an applied electric field. The number of free carriers in various materials varies over an extraordinarily wide range. In metals such as copper or silver, the free-electron density is on the order of 10^{23} cm^{-3} whereas in insulators such as quartz or aluminum oxide the free-electron density is less than 10^3 cm^{-3}. This difference in free-electron densities is the reason why the electrical conductivity in metals can be on the order of 10^6 mhos-cm^{-1} whereas it is on the order of 10^{-15} mhos-cm^{-1} or smaller in a good insulator. A material such as silicon or gallium arsenide, which has a free-carrier density intermediate between that of an insulator and a metal (10^8 to 10^{19} cm^{-3}), is termed a semiconductor.

In a metal or an insulator, the free-carrier density is a constant of the material and cannot be changed to any significant degree. However, in a semiconductor the free-carrier density can be changed by orders of magnitude either by introduction of impurities into the material or by the application of electric fields to appropriate semiconductor structures. This ability to manipulate the free-carrier density by large amounts is what makes the semiconductor such a unique and useful material for electrical applications.

18-2-2   Electrons and Holes

A single crystal of a semiconductor such as silicon, which has four valance electrons, is composed of a regular array or lattice of silicon atoms. Each silicon atom is bonded to four nearest neighbors, as illustrated in Fig. 18-1, by covalent bonds composed of electrons shared between the two adjacent atoms. At temperatures above absolute zero, some of these bonds are broken by energy carried by the silicon atom due to its random thermal motion about its equilibrium position. This process, known

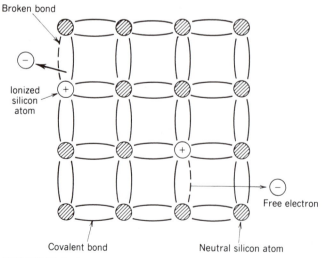

FIGURE 18-1: A silicon lattice showing thermal ionization and the creation of free electrons.

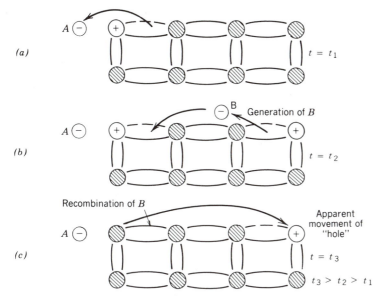

FIGURE 18-2: Hole movement in semiconductors.

as thermal ionization, creates a free hole and leaves behind a fixed positive charge on the nucleus of the silicon atom where the bond was broken as is shown in Fig. 18-1.

At some later time such as t_2 indicated in Fig. 18-2*b* another free electron may be attracted to the positive charge and become trapped in the bond which was broken at an earlier time t_1, as indicated in Fig. 18-2*a*, thus becoming bound. However, the silicon atom where this free electron originated now has a positive charge, and the end result of the transaction is the movement of the positive charge, as shown in Fig. 18-2*c* at time t_3. This moving positive charge is termed a hole because it originates from an empty bond normally occupied by an electron.

The thermal ionization mechanism generates an equal number of electrons and holes. The thermal equilibrium density of electrons and holes, n_i, in a pure (intrinsic) semiconductor is given by

$$n_i^2 \approx C \exp\left(\frac{-qE_g}{kT}\right) \tag{18-1}$$

where E_g is the energy gap of the semiconductor (1.1 eV for silicon), q is the magnitude of the electron charge, k is Boltzmann's constant, T is the temperature in degrees Kelvin, and C is a constant of proportionality. At room temperature (300°K), $n_i \approx 10^{10}$ cm^{-3} in silicon.

18-2-3 Doped Semiconductors

The thermal equilibrium density of electrons and holes can be changed by adding appropriate impurity atoms to the semiconductor. In the case of silicon, the appropriate impurities are elements from column III of the periodic table, such as boron, or from column V, such as phosphorous.

Elements such as boron have only three electrons (valance electrons) available for bonding to other atoms in a crystal and thus when boron is introduced into a silicon crystal, it needs an additional electron to bond to the four neighboring silicon

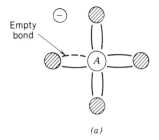

(a)

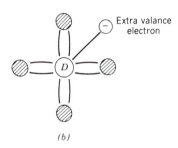

(b)

FIGURE 18-3: Doping by acceptors (a) and donors (b) to create p-type material and n-type material: (a) p-type silicon; (b) n-type silicon.

atoms, as shown in Fig. 18-3a. The boron will very quickly acquire or accept the needed electron from the silicon lattice by capturing a free electron. This immobilizes a free electron and leaves a hole free to move through the crystal. The result is that the silicon now has more free holes, now termed majority carriers, than free electrons, now termed minority carriers. The silicon is said to be doped p-type with an acceptor impurity.

Column V elements such as phosphorous have five valance electrons but only four are needed for bonding in a silicon lattice. Such atoms are easily thermally ionized when placed in a silicon crystal and the fifth electron becomes free, as is illustrated in Fig. 18-3b. The resulting positive charge on the donor impurity (so named because it donates the fifth electron to the silicon lattice) represents a trapped or bound hole. Electrons are now the majority carriers and holes are the minority carriers. The silicon is said to be doped n-type.

The impurity levels commonly used in semiconductor devices (10^{19} cm^{-3} or less) are orders of magnitude smaller than the density (about 10^{23} cm^{-3}) of semiconductor atoms. Thus, the presence of impurities in a semiconductor will not affect the rate at which covalent bonds are broken by thermal ionization and subsequently refilled by free electrons (electron–hole recombination). This means that the product of the thermal equilibrium electron density, now termed n_0, and the thermal equilibrium hole density, now termed p_0, must still equal n_i^2 as is shown below:

$$p_0 n_0 = n_i^2 \tag{18-2}$$

even though p_0 and n_0 are no longer equal. This relationship, Eq. 18-2, is sometimes called the law of mass action or the principle of detailed balance.

A doped (extrinsic) semiconductor is electrically neutral even though n_0 is no longer equal to p_0. The positive charge per unit volume in the extrinsic material is the sum of the hole density p_0 and the ionized donor density N_d, whereas the negative-charge density is the sum of the electron density n_0 and the ionized acceptor density N_a. The space charge neutrality condition in the general case where both donors and acceptors are assumed to be present in the material thus becomes

$$p_0 + N_d = n_0 + N_a \qquad (18\text{-}3)$$

Equations 18-2 and 18-3 can be solved simultaneously to find p_0 and n_0 separately. In a p-type material, the simultaneous solution of Eqs. 18-2 and 18-3, assuming $N_a \gg n_i$, yields the approximate result

$$n_0 \approx n_i^2/N_a \qquad \text{and} \qquad p_0 \approx N_a \qquad (18\text{-}4)$$

Similar expressions can be developed for strongly n-type material $N_d \gg n_i$. In either type of material, the minority carrier density is proportional to the square of the intrinsic carrier density (see Eq. 18-4) and thus is strongly temperature dependent.

18-2-4 Recombination

Fixed numbers of free electrons and holes require that mechanisms exist for the disappearance or recombination of them at the same rate as they are generated in thermal equilibrium. These mechanisms include direct recombination of electrons and holes (capture of a free electron in an empty covalent bond) and the trapping of carriers by impurities or imperfections in the crystal. In our largely qualitative examination of device physics, a simple rate equation describing the approximate time behavior of the excess carrier density (δn, the free-carrier density in excess of p_0 and n_0) is sufficient for our purposes. Space charge neutrality forces the excess hole density δp to equal δn.

This rate equation is given by

$$\frac{d(\delta n)}{dt} = -\frac{\delta n}{\tau} \qquad (18\text{-}5)$$

assuming that there is no generation of excess carriers during the time interval when this equation is to be applied. If $\delta n > 0$ at $t = 0$, Eq. 18-5 predicts that $\delta n(t)$ decays exponentially with time for $t > 0$. The characteristic decay time or time constant, τ, is termed the excess carrier lifetime, an important characteristic of minority carrier devices.

In most situations it is convenient to consider the lifetime as a constant of the material. However, in two situations encountered in power semiconductor devices, the lifetime varies with device operating conditions. First, the excess carrier lifetime will increase somewhat as the internal temperature of the devices increases. This will lead to a lengthening of the switching times of some devices (the minority carrier or bipolar devices such as BJTs, thyristors, and GTOs). In simplistic terms, the minority carriers are more energetic at higher temperatures and thus are somewhat less likely to be captured by a recombination center. The details of why the lifetime increases with temperature are beyond the scope of this discussion.

Second, at large excess carrier densities, δn, the carrier lifetime becomes dependent on the value of δn. As excess carrier densities approach a value n_b approximately equal to 10^{17} cm^{-3} and larger, another recombination process, Auger recombination, becomes important and causes the lifetime to decrease as δn increases. At these large carrier densities, the lifetime is given by

$$\tau = \frac{\tau_0}{1 + \dfrac{(\delta n)^2}{n_b^2}} \tag{18-6}$$

where τ_0 is the lifetime for $\delta n \ll n_b$. This decrease in excess-carrier lifetime will increase the on-state losses of some power devices at high current levels and thus will be a limiting factor in operation of these devices.

The value of the excess-carrier lifetime has important effects on the characteristics of minority-carrier (also called bipolar) power devices. Larger values of the lifetime minimize the on-state losses but also tend to slow down the switching transition from on to off and vice versa. Hence, the device manufacturer strives for fairly precise and reproducible control of the lifetime during the fabrication process. Two commonly used methods of lifetime control are the use of gold doping and the use of electron irradiation. Gold is an impurity in silicon devices that acts as a recombination center. The higher the gold doping density, the shorter the lifetimes will be. When electron irradiation is used, high-energy (a few million electron volts of kinetic energy) electrons penetrate deeply (the depth of penetration is a function of the energy) into a semiconductor before they collide with the crystalline lattice. When a collision occurs, imperfections in the crystalline lattice are created that act as recombination centers. The impinging dose of high-energy electrons is easily controlled, so the final density of recombination centers and thus the lifetime is under good control. In recent years this method has become the preferred method of lifetime control because it can be applied during the final stages of fabrication as a final "tuning" or "tweaking" of the device characteristics that depend on the value of the lifetime.

18-2-5   Drift and Diffusion

The flow of current in a semiconductor is the sum of the net flow of holes in the direction of the current and the net flow of electrons in the opposite direction. The free carriers can move via two mechanisms, drift and diffusion.

When an electric field is impressed across a semiconductor, the free holes are accelerated by the field and acquire a velocity component parallel to the field while electrons acquire a velocity component antiparallel to the field, as shown in Fig. 18-4. This velocity is termed the drift velocity and is proportional to the strength of the electric field. The drift component of current is given by

$$J_{\text{drift}} = q\mu_n nE + q\mu_p pE \tag{18-7}$$

where E is the applied electric field, μ_n is the electron mobility, μ_p is the hole mobility, and q is the charge on an electron. At room temperature in moderately doped silicon (less than 10^{15} cm^{-3}), $\mu_n \approx 1500$ cm^2/V-s and $\mu_p \approx 500$ cm^2/V-s. The carrier mobilities decrease with increasing temperature T (approximately T^{-2}).

If there is a variation in the spatial density of the free carriers such as is illustrated in Fig. 18-5, then there will be a movement of carriers from regions of higher concentration to regions of lower concentration. This movement is termed diffusion and is due to the random thermal velocity that each free carrier has. Such a spatial variation in carrier density could be obtained by a variety of methods including a variation in doping density. The movement of carriers by diffusion will

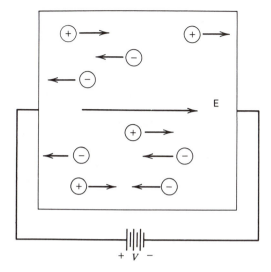

FIGURE 18-4: The drift of electrons and holes under the influence of an applied electric field.

produce a component of current density which, in one dimension, is given by

$$J_{\text{diff}} = J_n + J_p = qD_n\frac{dn}{dx} - qD_p\frac{dp}{dx} \qquad (18\text{-}8)$$

where D_n is the electron diffusion constant and D_p is the hole diffusion constant. The diffusion constants and mobilities are related by the Einstein relation, which is given by

$$\frac{D_p}{\mu_p} = \frac{D_n}{\mu_n} = \frac{kT}{q} \qquad (18\text{-}9)$$

At room temperature, $kT/q = 0.026$ eV. In a particular situation, current flow will usually be either predominantly by drift or by diffusion. In the general case, current flow by both mechanisms may have to be considered simultaneously.

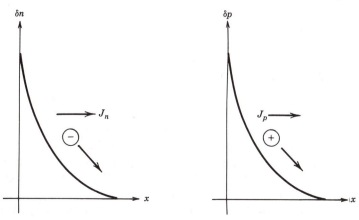

FIGURE 18-5: Carrier movement and current flow by diffusion.

Metallurgical junction

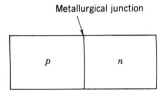

FIGURE 18-6: A *pn* junction.

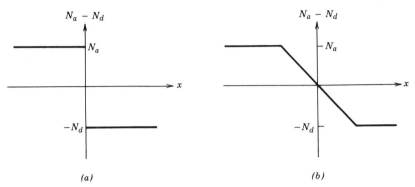

(a)   *(b)*

FIGURE 18-7: Impurity density versus position for (*a*) an abrupt (step) junction, and
(*b*) a linearly graded junction.

18-3 *PN* JUNCTIONS

A *pn* junction is formed when an *n*-type region in a silicon crystal is adjacent to or abuts a *p*-type region in the same crystal, as illustrated in Fig. 18-6. Such a junction can be formed by diffusing acceptor impurities into an *n*-type silicon crystal, for example. The opposite sequence (diffusing donors into *p*-type silicon) can also be used.

The junction is often characterized by how the doping changes from *n*-type to *p*-type as the junction is crossed. A so-called step or abrupt junction is shown in Fig. 18-7*a*. A more gradual change in doping density is the linearly graded junction shown in Fig. 18-7*b*. The junction is also characterized by the relative doping densities on each side of the junction. If the acceptor density on the *p*-type side is very large compared to the donor density on the *n*-type side, the junction is sometimes termed a p^+n junction. If the donor density is not much larger than n_i in the previous example, the junction might be termed a p^+n^- junction. Other variations on this theme are possible and are found in power semiconductor devices.

18-3-1 Potential Barrier at Thermal Equilibrium

Some of the majority carriers on either side of the junction will diffuse across it to the opposite side, where they are in the minority. This will create a space charge layer on either side of the junction, as is illustrated in Fig. 18-8, because the diffusing carriers will leave behind ionized impurities that are immobile and that are now not screened by enough free carriers for electrical neutrality. The resulting space charge density ρ, whose spatial variation for a step junction is shown in Fig. 18-9*a*, gives rise to an electric field. The electric field spatial variation that arises from a step junction is plotted in Fig. 18-9*b* and the potential barrier for a hole associated with this field is plotted in Fig. 18-9*c*.

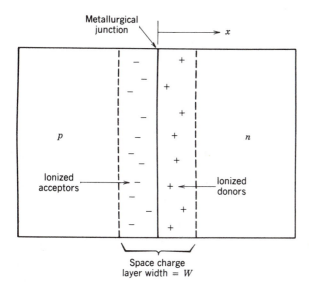

Metallurgical junction

x

p

Ionized acceptors

n

Ionized donors

Space charge layer width $= W$

FIGURE 18-8: A *pn* junction with a space charge or depletion layer shown.

The electric field increases in strength as more ionized impurities are exposed because of the diffusing carriers. The field, however, tends to retard the diffusion process because it acts to push the electrons back to the *n*-type side and holes back to the *p*-type side. An equilibrium is reached when the carrier flux caused by diffusion is counterbalanced by the carrier flux due to the electric field (drift). In equilibrium the hole flux and electron flux separately sum to zero, rather than merely the total current being zero. Otherwise, there would be a buildup of electrons and holes on one side of the junction.

The height of the potential barrier that supports the retarding electric field is often called the contact potential and is given by

$$\Phi_c = \frac{kT}{q} \ln \left[\frac{N_a N_d}{n_i^2} \right] \tag{18-10}$$

where N_a and N_d are the assumed spatially constant doping densities, as is shown in Fig. 18-7a. This contact potential cannot be measured directly with a voltmeter because equal and opposite contact potentials are developed when any measuring probes are attached to the *n* and *p* sides of the junction. At room temperature, Φ_c is less than E_g. For example, in a silicon *pn* junction at room temperature with $N_a = N_d = 10^{16}$ cm^{-3}, $\Phi_c = 0.72$ eV.

18-3-2 Forward and Reverse Bias

When an external voltage is applied between the *p* and *n* regions, as shown in Fig. 18-10a, it appears entirely across the space charge region because of the large resistance of the depletion layer compared to the rest of the material. If the applied potential is positive on the *p*-side, it opposes the contact potential and reduces the height of the potential barrier, and the junction is said to be forward biased.

When the applied voltage makes the *n*-side more positive, as is indicated in Fig. 18-11a, the barrier height is increased and the junction is said to be reverse biased. The width of the space charge region (or depletion region as it is also termed because

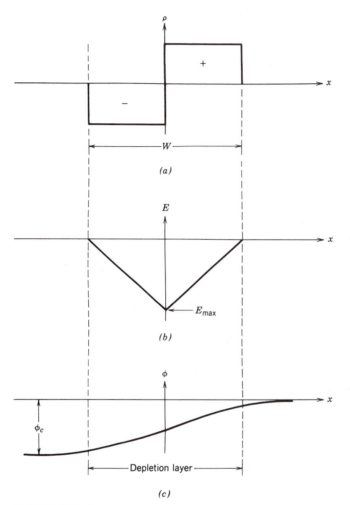

FIGURE 18-9: Spatial variation of (a) the space charge density ρ, (b) the electric field E, and (c) the potential ϕ across the space charge layer of a step pn junction. The electric field and potential are shown as negative because the applied voltages are positive when the p-region is biased positive with respect to the n-region (see Figure 18-10).

of the absence of free carriers) must grow or shrink as the height of the potential barrier either grows or shrinks. This occurs because a change in the potential requires a change in the magnitude of the total amount of charge on each side of the junction. Since the charge density equals the impurity density (a constant), a change in the total charge can occur only if there is a change in the dimensions of the depletion region. This variation in the dimensions of the depletion region with applied voltage has important consequences for the design of power semiconductor devices.

The behavior of the step junction depletion region with applied voltage is summarized later. The width of the step junction depletion region is given by

$$W(V) = W_0\sqrt{1 - (V/\Phi_c)} \qquad (18\text{-}11)$$

where W_0 is the depletion layer width at zero bias and V is the applied diode voltage

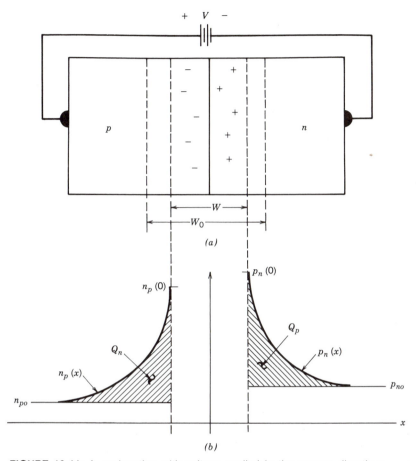

FIGURE 18-11: A *pn* junction with voltage applied in the reverse direction: (a) reversed biased *pn* junction; (b) minority carrier densities versus position in reverse biased *pn* junction.

(negative for reverse bias). W_0 is given by

$$W_0 = \sqrt{\frac{2\varepsilon\Phi_c(N_a + N_d)}{qN_aN_d}} \tag{18-12}$$

where ε is the dielectric constant of the semiconductor ($\varepsilon = 11.7\,\varepsilon_0$ for silicon where $\varepsilon_0 = 8.85 \times 10^{-14}$ F/cm). The electric field is a maximum at the metallurgical junction and is given by

$$E_{max} = \frac{2\Phi_c}{W_0}\sqrt{1 - (V/\Phi_c)} \tag{18-13}$$

Other doping profiles will produce slightly different functional dependencies of W and E_{max} with applied voltage and doping levels, but qualitatively their behavior is the same as the step junction. Thus, we will use the step junction as our model for discussing how the properties of a *pn* junction affect the operation and performance of power semiconductor devices.

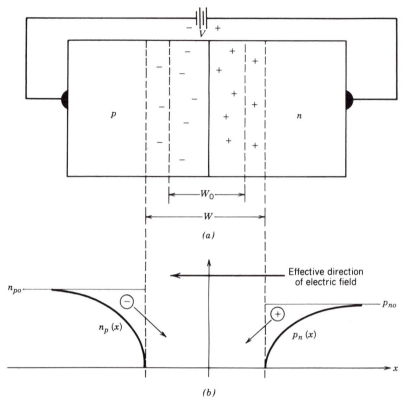

FIGURE 18-11: A *pn* junction with voltage applied in the reverse direction: (*a*) reversed biased *pn* junction; (*b*) minority carrier densities versus position in reverse biased *pn* junction.

18-4 CHARGE CONTROL DESCRIPTION OF *PN* JUNCTION OPERATION

A reverse bias voltage increases the potential barrier, which in turn makes the probability of any carrier diffusing across the junction vanishingly small. The minority carrier densities become nearly zero at the edge of the depletion region, as is shown in Fig. 18-11*b*. The small gradients in the minority carrier densities will cause a small flux of diffusing minority carriers toward the depletion region, as is indicated in the figure. When these diffusing carriers reach the depletion layer, the large electric fields in the space charge layer will immediately sweep them across the layer into the electrical neutral region on the other side of the junction. Electrons will be swept to the *n*-type side and holes to the *p*-type side. This will constitute a small leakage current termed the reverse saturation current I_s, which is diagrammed in the *i-v* curve shown in Fig. 18-12. This current is independent of the reverse voltage. There will also be a contribution to the leakage current by the electrons and holes created in the space charge layer by thermal ionization processes.

A forward bias voltage lowers the potential barrier and upsets the equilibrium between drift and diffusion in favor of diffusion. As is shown in Fig. 18-10*b*, this results in an enormous increase in the minority carrier densities (electrons and holes) in the electrically neutral regions on both sides of the junction immediately adjacent to the depletion regions. This increase, sometimes termed carrier injection, is an excess

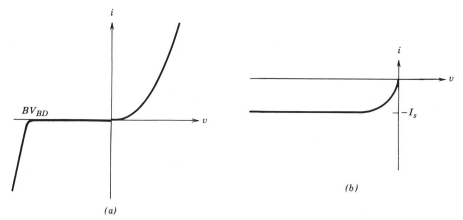

FIGURE 18-12: *I-V* characteristic of a *pn* junction in both forward and reverse bias (*a*). The reverse bias portion is redrawn in (*b*) because the reverse saturation current is too small to be seen on the same linear scales as the forward current.

carrier density, that is, above the thermal equilibrium values, and it has a significant impact on the diode's characteristics, especially the switching behavior.

The injected minority carriers eventually recombine with the majority carriers as they diffuse farther into the electrically neutral drift regions shown schematically in the one-dimensional diode model illustrated in Fig. 18-10*b*. This leads to an exponential decrease in the excess minority carrier density with distance, which is illustrated in the figure. The characteristic decay length is termed the minority carrier diffusion length, which for electrons in *p*-type material is given by

$$L_n = \sqrt{D_n \tau_n} \tag{18-14}$$

and for holes in *n*-type material is given by

$$L_p = \sqrt{D_p \tau_p} \tag{18-15}$$

In these equations, D_n and D_p are the electron and hole diffusion constants and τ_n and τ_p are the corresponding minority carrier lifetimes, which can range from nanoseconds to tens of microseconds depending on how the diode is fabricated.

The minority carrier density at the edge of the depletion region, $p_n(0)$ in Fig. 18-10*b*, is exponentially dependent on the forward bias voltage and is given by

$$p_n(0) = \frac{n_i^2}{N_D} e^{(qV/kT)} \tag{18-16}$$

A similar expression exists for $n_p(0)$. At zero bias, Eq. 18-16 reduces to the correct thermal equilibrium value. As a result of this voltage dependence, the minority carrier densities will vary by orders of magnitude as the voltage is changed by relatively small amounts. This will result in large gradients in the minority carrier densities in the regions adjacent to the depletion layer (within a few diffusion lengths) and consequently large diffusion currents (see Eq. 18-8). In fact, in these regions the diffusion component of the current is the dominant component of the total current (the drift component is negligible in this region because the large injected carrier densities short out the electric field needed for any substantial drift current).

In the steady-state (dc) forward bias situation, the excess carrier distributions shown in Fig. 18-10b neither grow nor decay in time. For this to happen, the carriers that are lost per unit time from the distribution via recombination must be replaced. They are replaced by new carriers that are injected across the junction by the forward bias current density J ($J = I/A$ where I is the terminal current and A is the cross-sectional area of the diode). This leads to the simple relation

$$J = \frac{Q_n}{\tau_n} + \frac{Q_p}{\tau_p} \tag{18-17}$$

where Q_n and Q_p are the areas under the excess minority carrier distributions shown in Fig. 18-10b. Since the excess carrier distributions depend exponentially on position (decaying with distance away from the junction), they can easily be integrated to find Q_n and Q_p. The resulting expressions are

$$Q_p = q\left[p_n(0) - \frac{n_i^2}{N_d} \right] L_p \tag{18-18}$$

and

$$Q_n = q\left[n_p(0) - \frac{n_i^2}{N_a} \right] L_n \tag{18-19}$$

It should be noted in Eqs. 18-18 and 18-19 that the excess minority carrier density at the edge of the depletion region [$p_n(0) - n_i^2/N_d$ in Eq. 18-18 and $n_p(0) - n_i^2/N_a$ in Eq. 18-19] is used rather than the total minority carrier density [$p_n(0)$ in Eq. 18-18 or $n_p(0)$ in Eq. 18-19]. This of course is because only the excess minority carriers have any spatial density gradient that can lead to diffusion currents. Putting the expressions for Q_p and Q_n into Eq. 18-17 using Eq. 18-16 yields

$$J = qn_i^2\left[\frac{L_n}{N_a\tau_n} + \frac{L_p}{N_d\tau_p} \right]\left[e^{qV/(kT)} - 1 \right] \tag{18-20}$$

Equation 18-20 is plotted in Fig. 18-12 and is the i-v characteristic of the pn junction diode for both forward and reverse bias. The term $qn_i^2[\{ L_n/(N_a\tau_n) \} + \{ L_p/(N_d\tau_n) \}]$ is readily identified as the reverse saturation current I_s shown in Fig. 18-12b. This current has an extreme temperature sensitivity because of its dependence on n_i^2 (see Eq. 18-1). This temperature sensitivity must be considered in all diode applications. In the forward direction the current depends exponentially on the forward voltage. The large increase in current in Fig. 18-12a at the reverse bias voltage BV_{BD} is caused by impact ionization and is not predicted by Eq. 18-20.

18-5   IMPACT IONIZATION

If a free electron with sufficient kinetic energy strikes a silicon atom, it can break a covalent bond and liberate an electron from the bond. If the kinetic energy is gained from an applied electric field, such as reverse voltages applied across a space charge layer, the liberation of the electron from the bond is termed impact ionization. This process is important because the newly liberated electron can gain enough energy from the applied field to break a covalent bond when it strikes a silicon atom thus liberating an additional electron. This process can cascade (avalanche) very quickly in a chain reaction-like manner producing a large number of free electrons and thus a large current. If this occurs, the junction is said to be in avalanche breakdown and the power dissipation (the product of the large applied voltage and the large avalanche current) will quickly destroy the device unless the applied voltage is reduced below the value needed to sustain avalanche breakdown very quickly.

An approximately constant value of electric field, E_{BD}, is required to cause appreciable impact ionization according to experimental observations. This value can be estimated from a very simple model, which highlights the important mechanism of impact ionization. The amount of kinetic energy required to break a bond is the energy gap E_g, assuming that all of the kinetic energy of the incident free electron is transferred into breaking the bond and that both the incident and liberated electrons have little kinetic energy after the collision. If it is now assumed that these electrons start essentially at rest and are accelerated by the electric field until their next collisions and that the time between collisions is t_c, then the value of E_{BD} for impact ionization is

$$E_{BD} = \sqrt{\frac{2E_g m}{q t_c^2}} \tag{18-21}$$

In this equation m is the mass of the electron (about 10^{-27} g) and q is the electron charge. The average time between collisions of electrons with the lattice is on the order of 10^{-12} to 10^{-14} s. Taking an intermediate value of 10^{-13} s as typical for silicon and using it in Eq. 18-13 predicts a value of about 300,000 V/cm for E_{BD}. This estimate is surprisingly close to the experimental value of 200,000 V/cm determined from avalanche breakdown measurements in power devices.

A more correct picture of the impact ionization process would have to take into account that not all electrons will lose their entire kinetic energy at each collision. Also, electrons will have a wide range of kinetic energies obtained from the thermal energy in the lattice. This considerably complicates the calculation of the electric field needed for appreciable ionization, although it has been done. However, the more correct theory yields a value for E_{BD} that is close to the value estimated in the preceding paragraph. Thus it is concluded that the simple picture of impact ionization given is adequate for the qualitative study of voltage breakdown in power semiconductor devices.

18-6 SUMMARY

In this chapter the basic properties of semiconductors were reviewed. The important concepts are listed.

1. Current in a semiconductor is carried by both electrons and holes.

2. Electrons and holes move by both drift and diffusion.

3. Intentional doping of the semiconductor with impurities will cause the density of holes and electrons to be vastly different.

4. The density of minority carriers increases exponentially with temperature.

5. A *pn* junction can be formed by doping one region *n*-type and the adjacent region *p*-type.

6. A potential barrier is set up across a *pn* junction in thermal equilibrium that balances out the drift and diffusion of carriers across the junction so that no net current flows.

7. In reverse bias a depletion region forms on both sides of the *pn* junction and only a small current can flow by drift.

8. In forward bias large numbers of electrons and holes are injected across the *pn* junction and large currents flow by diffusion with small applied voltages.

9. Large numbers of excess electron–hole pairs are created by impact ionization if the electric field in the semiconductor exceeds a critical value.

PROBLEMS

18-1. The intrinsic temperature T_i of a semiconductor device is that temperature at which n_i equals the doping density. What is T_i of a silicon *pn* junction that has 10^{18} cm^{-3} acceptors on the *p*-type side and 10^{14} cm^{-3} donors on the *n*-type side?

18-2. What are the resistivities of the *p*-region and *n*-region of the *pn* junction described in Problem 18-1?

18-3. Estimate p_0 and n_0 in a silicon sample where both donor and acceptor impurities are simultaneously present and $N_d - N_a = 10^{13}$ cm^{-3}.

18-4. What change in temperature ΔT doubles the minority carrier density in the *n*-type side of the *pn* junction described in Problem 18-1 compared with the room temperature value?

18-5. Show that a decade increase in the forward current of a *pn* junction is accompanied by an increase in the forward voltage of about 60 mV.

18-6. Consider a step silicon *pn* junction with 10^{14} cm^{-3} donors on the *n*-type side and 10^{15} cm^{-3} acceptors on the *p*-type side.
   (a) Find the width of the depletion layer on each side of the junction.
   (b) Sketch and dimension the electric field distribution versus position through the depletion layer.
   (c) Estimate the contact potential Φ_c.
   (d) Using a parallel plate capacitor formalism, estimate the capacitance per unit area of the junction at 0 V and at -50 V.
   (e) Estimate the current flowing through this junction if the forward voltage is 0.7 V, the excess carrier lifetime is 1 μs, and the cross-sectional area is 100 μm square.

18-7. A bar of *n*-type silicon with 10^{14} cm^{-3} donors is 200 μm long and 1 mm $\times$ 1 mm square. What is its resistance at room temperature and at 200°C?

18-7 REFERENCES

1. Ben G. Streetman, *Solid State Electronic Devices*, 2nd edition, Prentice-Hall, Englewood Cliffs, NJ, 1980, Chapters 1–6.
2. Sorab K. Ghandhi, *Semiconductor Power Devices*, John Wiley & Sons, New York, 1977, Chapters 1–3.
3. A. S. Grove, *Physics and Technology of Semiconductor Devices*, John Wiley & Sons, New York, 1967, Chapters 4–6.
4. Adel S. Sedra and Kenneth C. Smith, *Microelectronics Circuits*, 2nd Ed., Holt, Rinehart, and Winston, New York, 1987, Chapter 4.

19

Power Diodes

19-1 INTRODUCTION

Power semiconductor devices, even diodes, are more complicated in structure and operational characteristics than their low-power counterparts with which most of us have some degree of familiarity. The added complexity arises from the modifications made to the simpler low-power devices to make them suitable for high-power applications. These modifications are essentially generic in nature, that is, the same basic modifications are made to all low-power semiconductor devices in order to scale up their respective power capabilities. Thus, if the modifications can be understood in the context of one specific type of device, then it will be much easier to see the effects of these modifications in the other types of power devices.

The study of power semiconductor devices begins with the diode, both *pn* junction and Schottky barrier devices, because they are the simplest of all semiconductor devices. The modifications for high-power operation will thus be most easily considered first in these devices. Additionally, the diode or *pn* junction is the basic building block of all other power semiconductor devices. A comprehension of the other power devices will be more easily obtained if first the characteristics of the diode are clearly understood.

19-2 BASIC STRUCTURE AND *I-V* CHARACTERISTICS

The ideal *pn* junction diode geometry was discussed in Chapter 18. The practical realization of the diode for power applications is shown in Fig. 19-1. It consists of a heavily doped *n*-type substrate on top of which is grown a lightly doped n^- epitaxial layer of specified thickness. Finally the *pn* junction is formed by diffusing in a heavily doped *p*-type region that forms the anode of the diode. Typical layer thicknesses and doping levels are shown in Fig. 19-1. The cross-sectional area A of the diode will vary according to the amount of total current the device is designed to carry. For diodes that can carry several thousand amperes, the area can be several

451

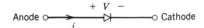

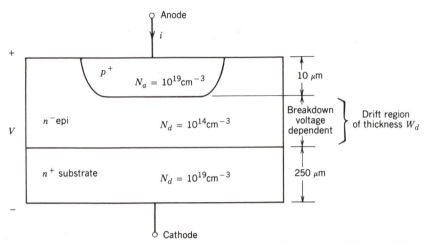

FIGURE 19-1: A cross-sectional view of a *pn* junction diode intended for power applications. The circuit symbol is also shown with anode and cathode designations. The cross-sectional view of a high power diode is shown in Figures 19-6 and 19-7.

square centimeters (wafers with diameters as large as four inches are used in the production of power devices and for the largest diodes, only one is made from the wafer). The circuit symbol for the diode is shown in Fig. 19-1 and is the same as that used for low-power signal level diodes.

The n^- layer in Fig. 19-1, which is often termed the drift region, is the prime structural feature not found in low-power diodes. Its function is to absorb the depletion layer of the reverse biased p^+n^- junction. This layer can be quite wide at large reverse voltages. The drift region establishes what the reverse breakdown voltage will be. This relatively long lightly doped region would appear to add significant ohmic resistance to the diode when it is forward biased, a situation that would apparently lead to unacceptably large power dissipation in the diode when it is conducting current. However, other mechanisms to be described shortly greatly reduce this apparent problem.

The i-v characteristic of the diode is shown in Fig. 19-2. In the forward direction it appears the same as that discussed in Chapter 18 except that the current grows linearly with the forward bias voltage rather than exponentially. The large currents in a power diode create ohmic drops that mask the exponential i-v characteristic. The voltage drop in the lightly doped drift region accounts for part of this ohmic resistance. The nature of the on-state voltage drop will be discussed in a later section of this chapter.

In reverse bias only a small leakage current, which is independent of the reverse voltage, flows until the reverse breakdown voltage BV_{BD} is reached. When breakdown is reached, the voltage appears to remain essentially constant while the current increases dramatically, being limited only by the external circuit. The combination of a large voltage at breakdown and a large current leads to excessive power dissipation that can quickly destroy the device. Operation of the diode in breakdown must therefore be avoided.

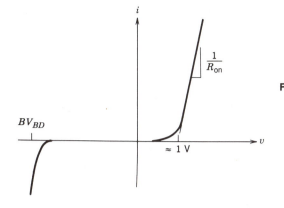

FIGURE 19-2: I-V characteristic of a *pn* junction diode. The reverse bias portion of the characteristic shows avalanche breakdown at BV_{BD}. The exponential i-v relationship in forward bias expected from signal level diode characteristics is masked by the ohmic resistance R_{on} in power diodes.

19-3 BREAKDOWN VOLTAGE CONSIDERATIONS

19-3-1 Breakdown Voltage Estimates

The voltage breakdown illustrated in the i-v curve shown in Fig. 19-2 occurs when the reverse bias voltage impressed across the diode by the circuit in which it is embedded attempts to exceed the breakdown value BV_{BD}. Since the reverse voltage is dropped entirely across the depletion region, the larger the voltage, the larger the electric field in the region and the closer it approaches the value E_{BD}, where substantial impact ionization begins. The value of voltage that causes the electric field to reach the critical value depends on the doping profile (step, linearly graded, diffused, etc.) of the junction and on the magnitudes of the doping densities. The breakdown behavior of the step junction is sufficiently representative of all *pn* junctions that a detailed examination of it will yield the basic features of breakdown that are needed for a qualitative study of power semiconductor devices without the quantitative details of more complicated doping profiles obscuring the essential characteristics.

For the step junction, the relation between the maximum electric field in the depletion region and the applied voltage is given by Eq. 18-13. Setting the maximum field equal to E_{BD} in this equation, assuming the reverse bias voltage is large compared with the contact potential, and solving Eqs. 18-12 and 18-13 for the breakdown voltage yields for the breakdown voltage of a step junction

$$BV_{BD} \approx \frac{\varepsilon E_{BD}^2}{2qN_d} \tag{19-1}$$

In this equation it is assumed that the doping density N_d on the n^- side is much less than that on the p^+ side. In this situation the depletion layer is contained almost entirely on the more lightly doped side. Putting in the numerical values of E_{BD} (2×10^5 V/cm) and the dielectric constant of silicon (1.05×10^{-12} farads/cm) yields

$$BV_{BD} \approx \frac{1.3 \times 10^{17}}{N_d} \tag{19-2}$$

where the doping density is in number per cubic centimeter. The corresponding

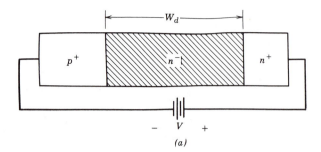

(a)

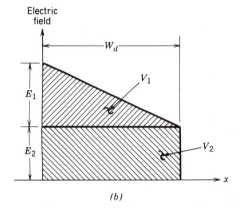

(b)

FIGURE 19-3: Punch-through in a reverse biased power diode: (*a*) reverse biased diode with the depletion layer extending completely across the drift region, sometimes called the punch-through condition; (*b*) electric field profile of punch-through condition in a reverse biased diode.

depletion layer width (in cm) of the step junction at breakdown is (using Eqs. 18-11 and 18-12 and N_d as given by Eq. 19-2)

$$W_d \geq W(BV_{\mathrm{BD}}) \approx \frac{2BV_{\mathrm{BD}}}{E_{\mathrm{BD}}} = 1 \times 10^{-5}BV_{\mathrm{BD}} \qquad (19\text{-}3)$$

Recall from Fig. 19-1 that W_d is the drift region thickness.

Two basic facts are evident from Eqs. 19-2 and 19-3. First, large breakdown voltages require lightly doped junctions, at least on one side. Second, the electrically neutral drift layer in the diode must be fairly long in high-voltage devices to accommodate the long depletion layers. For example, a breakdown voltage of 1000 V requires a doping level of approximately 10^{14} cm^{-3} or less and a minimum drift region thickness W_d of about 100 μm to accommodate the depletion region. These requirements are satisfied by the lightly doped drift region shown in the cross-sectional diagram of the power diode illustrated in Fig. 19-1.

In some situations, it is possible to have shorter drift region lengths and still have the diode block large reverse voltages. Consider the situation shown in Fig. 19-3, where the depletion region has extended all the way across the drift region and is in contact with the n^+ layer. When this occurs (which is commonly termed punch-through), further increases in reverse voltage will not cause the depletion region to widen any further because the large doping density in the n^+ layer effectively blocks further growth of the depletion layer. Instead the electric field profile begins to flatten out as is shown in Fig. 19-3b, becoming less triangular and more rectangular.

As shown in the figure, the electric field profile can be considered to be composed of a triangular-shaped component with a peak electric field value of E_1 at the junction and a rectangular-shaped component of constant electric field value E_2.

The triangular-shaped component is due to the ionized donors in the drift region and hence E_1 is given by

$$E_1 = \frac{qN_dW_d}{\varepsilon} \qquad (19\text{-}4)$$

and the area under the triangular component represents a voltage V_1, which is given by

$$V_1 = \frac{qN_dW_d^2}{2\varepsilon} \qquad (19\text{-}5)$$

The area under the rectangular component is a voltage given by

$$V_2 = E_2W_d \qquad (19\text{-}6)$$

When the junction having this punch-through profile breaks down, the following conditions exist:

$$E_1 + E_2 = E_{BD} \qquad (19\text{-}7)$$

and

$$BV_{BD} = V_1 + V_2 \qquad (19\text{-}8)$$

Substituting Eq. 19-5 for V_1 into Eq. 19-8 and simultaneously substituting $V_2 = [E_{BD} - E_1]W_d$ with E_1 being given by Eq. 19-4 yields

$$BV_{BD} = E_{BD}W_d - \frac{qN_dW_d^2}{2\varepsilon} \qquad (19\text{-}9)$$

If the doping density in the drift region is made much smaller than the values permitted by Eq. 19-2, then V_1 will be much less than V_2 (the electrical field profile will be essentially constant and independent of position) so that

$$BV_{BD} \approx E_{BD}W_d \quad \text{or} \quad W_d \approx \frac{BV_{BD}}{E_{BD}} \qquad (19\text{-}10)$$

which is about one-half the value of W_d given by Eq. 19-3 for the same value of breakdown voltage. The low doping density means that the ohmic resistivity of the punch-through drift region is much larger than the nonpunch-through drift region resistivity.

The higher resistivity of the drift region of the punch-through structure has no significant effect on the operation of the diode because the conductivity modulation that occurs during on-state operation (a subject to be discussed shortly) shorts out the drift region. As we will discuss in detail in later sections of this chapter, the shorter drift region length of the punch-through diode permits this diode to have lower on-state voltages compared with the conventional nonpunch-through diode, assuming the same lifetime in each diode and the same breakdown voltage. However, the punch-through structure cannot be used in majority carrier devices because there is no conductivity modulation in the on-state operation of these devices. Hence, the large

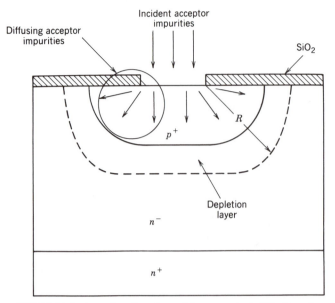

FIGURE 19-4: A *pn* junction formed by a masked diffusion of impurities into the substrate. The lateral diffusion of impurities gives rise to a curvature of the *pn* junction boundary and thus to the depletion layer. The smaller the radius of curvature *R*, the lower the breakdown voltage becomes.

resistance of the punch-through drift region will not be shorted out and there will be large amounts of on-state power dissipation.

19-3-2 Depletion Layer Boundary Control

The plane parallel junctions that have been implicitly assumed thus far in these discussions are an unrealistic idealization in practice. Junctions in actual devices are formed via masked diffusions of impurities (such as acceptors shown in Fig. 19-4) that will inevitably cause the resulting *pn* junction to have some degree of curvature, as shown in Fig. 19-4. The amount of curvature, which is specified by the radius of curvature (indicated in Fig. 19-4), will depend on the size of the diffusion mask, the length of the diffusion time, and the magnitude of the diffusion temperature. The curvature is caused by the fact that the impurities diffuse as fast laterally as they do vertically into the substrate. In such junctions, the plane parallel description becomes inaccurate when the radius of curvature becomes comparable to the depletion layer width. In these circumstances, the electric field in the depletion layer becomes spatially nonuniform and has its largest magnitude where the radius of curvature is the smallest. This will lead to a smaller breakdown voltage compared to a plane parallel junction of similar doping.

The obvious step in combating this potential reduction in breakdown voltage is to keep the radius of curvature as large as possible. Modeling studies with a cylindrical *pn* junction of radius *R* indicate that the radius must be six or more times larger than the depletion layer thickness, at breakdown, of a comparable plane parallel junction to keep the breakdown voltage of the cylindrical junction within 90% of the plane parallel junction breakdown voltage. In high voltage diodes where $BV_{\rm BD}$

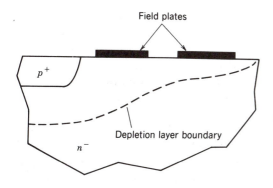

FIGURE 19-5: Use of field plates in a *pn* junction diode to control the depletion layer boundary curvature in order to keep breakdown voltages from being reduced.

is 1000 V and larger, the required radius of curvature R, using the estimate of 100 μm for depletion layer thickness, at breakdown, of the plane parallel junction given earlier, would be 600 μm. The realization of such large values would require deep diffusions (depths comparable to R) into the substrate, which would be impractical because of impossibly long diffusion times.

Thus, the radius of curvature of the depletion layer boundary must be controlled by other means. One method is via the use of electrically floating field plates that are illustrated in Fig. 19-5. The field plates act as an equipotential surface, and by their proper placement, as is indicated in the figure, they can redirect the electric field lines and prevent the depletion layer from having too small a radius of curvature. The price to be paid for this approach is that the field plates require a considerable amount of silicon real estate.

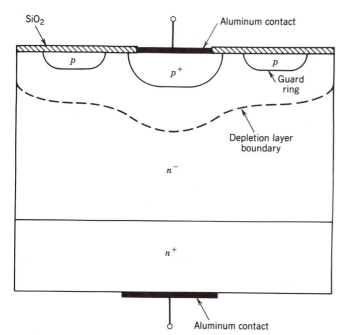

FIGURE 19-6: A *pn* junction diode with both an n^- type drift region and guard rings for improved breakdown voltage capabilities. The guard rings help to prevent the depletion layer from having too small a radius of curvature.

Another method of controlling the depletion layer is the use of guard rings, illustrated for a simple *pn* junction in Fig. 19-6. The *p*-type guard rings are allowed to float electrically. The depletion layers of the guard rings merge with the growing depletion layer of the reverse-biased *pn* junction, which prevents the radius of curvature from getting too small. Since the guard rings are electrically floating, they do not acquire the full reverse bias voltage and, thus, breakdown will not occur across their depletion layers even though their radii of curvature may be somewhat small, as Fig. 19-6 implies.

In some devices the metallurgical junction extends to the surface of the silicon and the high field depletion layer intersects the semiconductor–air boundary, as illustrated in Fig. 19-7a. This situation will also cause curvature of the depletion layer boundary even if the metallurgical junction itself is a plane parallel structure. The fringing electric fields at or near the surface may cause premature breakdown or interact with surface impurities that will ultimately degrade the performance of the device. Experience indicates that this causes a 20% to 30% reduction in E_{BD} at the

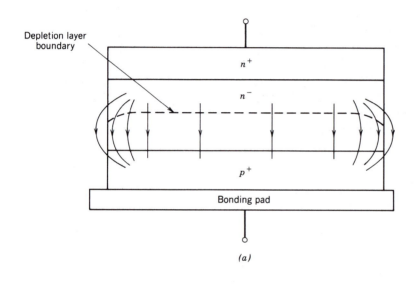

(a)

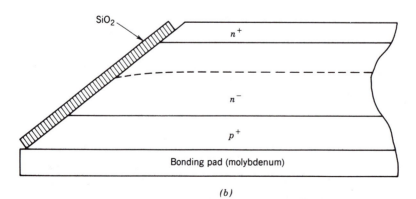

(b)

FIGURE 19-7: (a) Depletion layer intersection with a semiconductor surface and attendant field crowding, and (b) the use of topological contouring (beveling in this example) to minimize field crowding due to depletion layer curvature.

surface compared to the bulk. Thus, it is often necessary to shape the topological contours of the device to minimize the surface electric fields. An example of beveling is shown in Fig. 19-7b. Additionally coating the sample surfaces with appropriate materials such as silicon dioxide or some other insulator will aid in controlling the electric fields at the surface.

19-4   ON-STATE LOSSES

19-4-1 Conductivity Modulation

Nearly all of the power dissipated in a diode occurs when it is in the on-state (forward bias). At high switching frequencies a significant amount of dissipation can occur during the switching transient from one state to the other, a subject that will be considered later. In low-power applications, the forward bias voltage is the junction voltage and is often considered approximately constant, since the voltage depends logarithmically on the current. In a silicon diode this constant voltage is 0.7 to 1.0 V and the on-state dissipation would then be $P = 0.7 \, I$, where I is the current through the diode.

In power diodes, this estimate would be satisfactory only at low current levels. At large current levels, it would severely underestimate the total dissipation because the dissipation in the drift region of the power diode is ignored. It is these losses that limit the diode's ultimate power capability. However, in estimating the power dissipated in the drift region, care must be exercised because the effective value of resistance of this region in the on-state is much less than the apparent ohmic value calculated on the basis of the geometric size and the thermal equilibrium carrier densities. In the on-state there is a substantial reduction in the resistance of the drift region because of the large amount of excess carrier injection into the drift region. This conductivity modulation, as it is sometimes termed, substantially reduces the power dissipation over what would be estimated on the basis of the thermal equilibrium conductivity of the drift region.

Consider the one-dimensional power diode model shown in Fig. 19-8. In the on-state the forward biased pn junction injects holes into the n-type drift region. At low injection levels, $\delta p \ll n_{no}$, the thermal equilibrium electrons n_{no} easily neutralize the space charge of the holes. But at high injection levels, $\delta p > n_{no}$, the hole space charge is large enough to attract electrons from the n^+ region into the drift region. This leads to the injection of electrons across the $n^+ n^-$ interface into the drift region with densities $\delta n = \delta p$. These injected electrons and holes diffuse into the drift region toward each other, recombining as they diffuse. This is the origin of so-called double injection.

If the diffusion length L, where $L = \sqrt{D\tau}$ (see Eqs. 18-14 and 18-15) is greater than the drift region length W_d, then the spatial distribution of the excess carriers will be fairly flat, as shown in Fig. 19-8 and equal to an average value n_a. Since the doping density n^- of the drift region is quite small, typically 10^{14} cm^{-3}, n_a will be typically much greater than n^-, a condition termed high-level injection. The conductivity of the drift region will thus be greatly enhanced over its ohmic or low injection level.

Under these conditions the current in the drift region can be written approximately as

$$I_F \approx \frac{q[\mu_n + \mu_p]n_a A V_d}{W_d} \qquad (19\text{-}11)$$

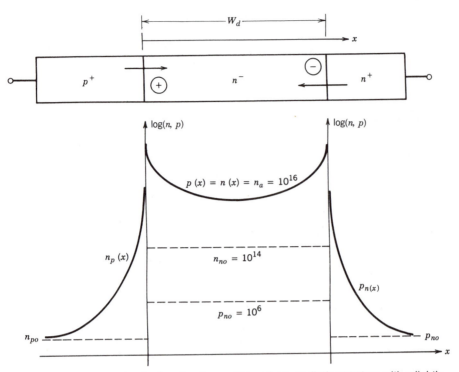

FIGURE 19-8: Carrier distributions in a forward biased power diode structure with a lightly doped drift region. Note how the excess carriers are injected into the drift region from both ends. The average value of the excess carriers, n_a is large compared with n_{no}, the majority carrier density in the drift region.

where V_d/W_d is the average electric field in the conductivity modulated region, A is the cross-sectional area of the diode, and V_d is the approximate voltage drop across the drift region. Now the excess carriers stored in the drift region during the on-state can be related to the current I_F using a stored charge formulation (Q_F being the stored charge in the drift region) so that

$$I_F \approx \frac{qAW_d n_a}{\tau} \approx \frac{Q_F}{\tau} \tag{19-12}$$

Setting Eq. 19-11 equal to Eq. 19-12 yields for V_d

$$V_d \approx \frac{W_d^2}{[\mu_n + \mu_p]\tau} \tag{19-13}$$

Note that the voltage given by this equation is not the total forward bias diode voltage. The total voltage $V = V_j + V_d$, where V_j is the voltage across the pn junction and can be estimated using an equation similar to Eq. 18-20.

The voltage V_d cannot be reduced to arbitrarily small values. Two mechanisms become active which cause V_d to increase with increasing current density. First as the excess carrier density, n_a, gets large enough, on the order of $10^{17} \, cm^{-3}$, the lifetime τ begins to decrease because of Auger recombination as discussed in Ch. 18 (see Eq. 18-6). The lifetime reduction caused by this mechanism will cause the voltage drop across the drift region to increase substantially at large current densities where the carrier density is corresponding large. Secondly, at about the same excess carrier density, the carrier mobilities begin to decrease with increasing excess carrier density, becoming inversely proportional to n_a as is shown in Eq. 19-14a

$$\mu_n + \mu_p = \frac{\mu_o}{\left[1 + \dfrac{n_a}{n_b}\right]} \tag{19-14a}$$

where μ_o is the low injection level value of $\mu_n + \mu_p$. This decrease occurs because the carrier densities are large enough that the free carriers collide with each other almost as often as they do with the crystalline lattice. These additional collisions reduce the mobilities just as do collisions with the lattice. This carrier–carrier scattering represents another mechanism that causes V_d to increase at large current densities.

Insertion of the density-dependent lifetime given by Eq. 18-6 into Eq. 19-12 yields

$$J_F = \frac{qW_d n_a}{\tau_o}\left(1 + \frac{n_a^2}{n_b^2}\right) \tag{19-14b}$$

Insertion of Eq. 19-14a for the density-dependent mobilities into Eq. 19-11 and solving for n_a as a function of V_d, J, and other parameters and using this resulting expression for n_a in Eq. 19-14b yields, after a few manipulations,

$$\left[V_d - \frac{J_F W_d}{q\mu_o n_b}\right]^2\left[V_d - \frac{W_d^2}{\mu_o \tau_o}\right] = \frac{J_F^2 W_d^4}{q^2\mu_o^3 n_b^2 \tau_o} \tag{19-15}$$

As n_a approaches n_b, V_d will become larger than $W_d^2/\mu_o\tau_o$ so that Eq. 19-15 becomes approximately

$$V_d \approx \frac{J_F W_d}{q\mu_o n_b} + \sqrt[3]{\frac{J_F^2 W_d^4}{q^2\mu_o^3 n_b^2 \tau_o}} \tag{19-16}$$

Thus at large currents where J_F is large, the drift region voltage has an ohmic-like dependence upon the current density J_F. Thus it is not surprising that the total on-state diode voltage drop is approximately given by

$$V \approx V_j + R_{on}I \tag{19-17}$$

as is indicated in Fig. 19-2.

19-4-2 Impact on On-state Losses

This simplified analysis summarizes several important features about the on-state losses of not only diodes but other minority carrier-based devices such as bipolar junction transistors and thyristors. First, if the lifetime can be made large enough so that the diffusion length L is comparable with the drift region length W_d, then the voltage drop across the drift region can be made quite small and approximately independent of the current. This means that the power dissipation in bipolar devices will be much less than for majority carrier devices such as MOSFETs or JFETs carrying about the same current density. Second, the price of the reduced on-state losses is a large amount of stored charge, which will compromise the switching times, a subject to be examined in the next section. Third, the larger the breakdown voltage of the device, the larger the voltage drop V_d across the drift region will be since this voltage is proportional to square of the length W_d, and this length must be larger for larger breakdown voltages.

A natural question to ask at this point is how much lower are the on-state losses in a bipolar device compared with a majority carrier device. One way to address the question is to estimate the current density that can flow in the drift region of both types of devices as a function of the breakdown voltage rating of the device and the maximum desired voltage drop V_d across the drift region. The drift region is chosen for comparison because in both types of devices, this is where most of the on-state losses will occur. The current density for a minority carrier device can be developed by using Eq. 19-11 and expressing the drift region length W_d in terms of the breakdown voltage BV_{BD} using Eq. 19-3. Using this with $\mu_n + \mu_p \approx 900$ cm^2/V-s (the approximate value at excess carrier densities of 10^{17} cm^{-3}) yields

$$J(\text{minority}) \approx 1.4 \times 10^6 \frac{V_d}{BV_{BD}} \qquad (19\text{-}18)$$

In the drift region of a majority carrier device, the current density, $J(\text{majority})$ can be written as

$$J(\text{majority}) \approx \frac{q\mu_n N_d V_d}{W_d} \qquad (19\text{-}19)$$

where the drift region is assumed to be n-type in order to take advantage of the higher mobility of electrons. Using Eq. 19-2 to express N_d in terms of BV_{BD} and Eq. 19-3 to express W_d in terms of BV_{BD}, and setting $\mu_n \approx 1500$ cm^2/V-s, the value of μ_n in silicon at doping densities of 10^{14} cm^{-3}, yields

$$J(\text{majority}) \approx 3.1 \times 10^6 \frac{V_d}{BV_{BD}^2} \qquad (19\text{-}20)$$

These equations clearly show that as breakdown voltages increase, bipolar devices and majority carrier devices suffer reductions in their current-carrying capabilities. However the reduction in bipolar devices is less severe compared to majority carrier devices and hence bipolar devices are generally the device of choice at larger blocking voltages (several hundred volts and larger).

19-5   SWITCHING CHARACTERISTICS

19-5-1 Observed Switching Waveform

A power diode requires a finite time to switch from the blocking state (reverse bias) to the on-state (forward bias) and vice versa. The user must be concerned not only with the time required for the transitions, but also with how the diode current and voltage vary during the transitions. Both the transition times and the shapes of the waveforms are affected by the intrinsic properties of the diode and by the circuit in which the diode is embedded.

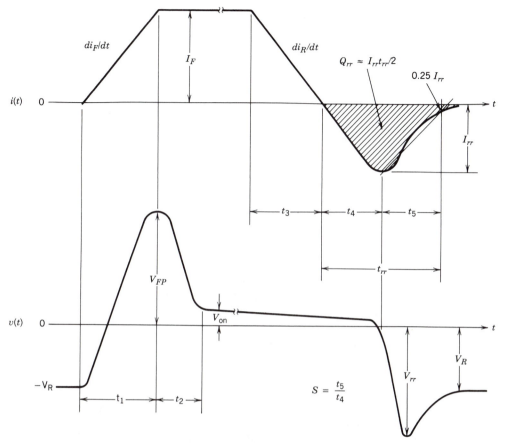

FIGURE 19-9: Voltage and current waveforms for a power diode driven by currents with a specified rate of rise during turn-on and a specified rate of fall during turn-off.

The switching properties of a diode are often given on specification sheets for diode currents with a specified time rate of change, di/dt, as is shown in Fig. 19-9. The reason for this selection is that power diodes are very often used in circuits containing inductances that control the rate of change of the current, or the diodes are used as free-wheeling diodes where the turn-off of a solid-state device controls di/dt. The resulting diode voltage and current versus time are shown in Fig. 19-9. The features of particular interest in these waveforms are the voltage overshoot during turn-on and the sharpness of the fall of the reverse current during the turn-off phase. The overshoot of the voltage during turn-on is not observed with signal level diodes.

19-5-2 Turn-on Transient

The turn-on portion of the diode waveforms in Fig. 19-9 is encompassed by the times labeled t_1 and t_2. During these intervals two physical processes occur in sequence. First the space charge stored in the depletion region (located mainly in the drift region) because of the large reverse bias voltage is removed (discharged) by the growth of the forward current. When the depletion layer is discharged to its thermal equilibrium level, the metallurgical junction becomes forward biased and the injection of excess carriers across the junction into the drift region commences at time t_1, thus, marking the start of the second phase and the end of the first. During the second

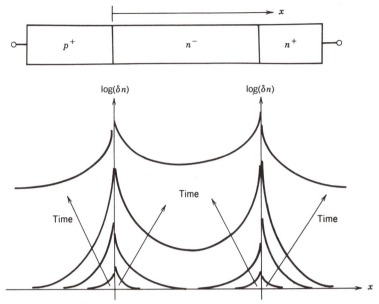

FIGURE 19-10: Growth of the excess carrier distribution during the turn-on of a power *pn* junction diode. Note that the carriers are injected into the drift region from both ends.

phase, the excess carrier distribution in the drift region grows toward the steady-state value that can be supported by the forward diode current I_F. The approximate growth of the excess carrier distribution in time is diagrammed in Fig. 19-10. Note that excess carriers are injected into the drift region from both ends with holes being injected from the p^+n^- junction and electrons from the n^+n^- junction.

A simple interpretation of this sequence of events would lead one to expect that the diode voltage would rise smoothly and monotonically from its initial large negative value to a steady-state forward bias value of about 1.0 V. Only one distinct time interval would be observed, which would be the discharge of the space charge layer, which is analogous to the discharge of a capacitance. Indeed, the depletion layer under reverse bias conditions is often modeled as a capacitor (space charge capacitance) whose value (per square centimeter using a parallel plate capacitor formalism) is given by

$$C_{sc}(V) \approx \frac{\varepsilon}{W(V)} \tag{19-21}$$

where the depletion layer width $W(V)$ is given by Eq. 18-11. The interval duration should scale with I_F and inversely with di/dt.

However, this interpretation does not account for the voltage overshoot shown in the waveform in Fig. 19-9. The shortcoming of the interpretation is that it fails to consider the effect of the ohmic resistance of the drift region and the inductance of the silicon wafer and of the bonding wires attached to it. As the forward current grows in time, there is an increasingly large voltage drop across the drift region, since there is no conductivity modulation of the region until the space charge layer is discharged to its thermal equilibrium value. The inductance also adds a significant voltage drop if large values of di/dt are applied. The combined effect of these two factors is an overshoot that can be as large as several tens of volts, which is large enough to seriously affect the operation of some power electronic circuits.

The growth of the diode voltage slows and eventually turns over as the drift region becomes shorted out by the large amount of carrier injection into it. In addition, the inductive contribution ends when the diode current stabilizes at I_F. The interval during which the voltage falls from the peak overshoot value to the steady-state forward value marks the completion of the transient growth of the excess charge distribution in the drift region.

The duration of the space charge layer discharge and the growth of the excess carrier distribution in the drift region is governed by both the intrinsic properties of the diode and by the external circuit in which the diode is embedded. A large value of di/dt will minimize the time needed to discharge the space charge layer. However, a large value of I_F and of carrier lifetime in the drift region will lengthen the time needed for the excess carrier portion of the transient to be completed. Typical values for these switching times in high voltage diodes are in the hundreds of nanoseconds for t_1 and in the microsecond range for t_2. Devices with faster turn-on times are available, but their improved performance in this respect is achieved only by reducing the lifetime, as was explained previously. Thus, there is an inherent trade-off between shorter turn-on transients and higher on-state losses.

19-5-3 Turn-off Transient

The turn-off portion of the switching waveform is encompassed by the times labeled t_3, t_4, and t_5 and is essentially the inverse of the turn-on process. First the excess carriers stored in the drift region must be removed before the metallurgical junctions can become reverse biased. The decay of the excess carrier distribution is approximately the inverse of the growth illustrated in Fig. 19-10. Once the carriers are removed by the combined action of recombination and sweepout by negative diode currents, the depletion layer acquires a substantial amount of space charge from the reverse bias voltage and expands into the drift region from both ends (junctions).

As long as there are excess carriers at the ends of the drift region, the p^+n^- and n^+n^- junctions must be forward biased. Thus, the diode voltage will be little changed from its on-state value except for a small decrease due to ohmic drops caused by the reverse current. But after the current goes negative and carrier sweepout has proceeded for a sufficient time (t_4) to reduce the excess carrier density at one or both of the junctions to zero, the junction or junctions become reverse biased. At this point the diode voltage goes negative and rapidly acquires substantial negative values as the depletion regions from the two junctions expand into the drift region toward each other. At this time the negative diode current demanded by the stray inductance of the external circuit cannot be supported by the excess carrier distribution because too few carriers remain. The diode current ceases its growth in the negative direction and quickly falls, becoming zero after a time t_5. The reverse current has its maximum reverse value, I_{rr}, at the end of the t_4 interval.

19-5-4 Reverse Recovery

The time interval $t_{rr} = t_4 + t_5$ shown in Fig. 19-9 is often termed the reverse recovery time. Its characteristics are important in almost all power electronic circuits where diodes are used. Diode specification sheets often give detailed plots of t_{rr}, reverse recovery charge Q_{rr}, and "snappiness" factor S (all defined in Fig. 19-9) as functions of the time rate of change of the reverse current, di_R/dt. These quantities are all interrelated to each other and to other diode parameters such as breakdown voltage and on-state voltage drop across the drift region.

A useful quantitative description of these relationships can be obtained from the following considerations. From Fig. 19-9, we note that I_{rr} can be written as

$$I_{rr} = \frac{di_R}{dt} t_4 = \frac{di_R}{dt} \frac{t_{rr}}{(S+1)} \tag{19-22}$$

since $t_4 = t_{rr} - t_5 = t_{rr}/(S+1)$. From Fig. 19-9, $Q_{rr} \approx I_{rr}t_{rr}/2$ so that

$$Q_{rr} = \frac{di_R}{dt} \frac{t_{rr}^2}{2(S+1)} \tag{19-23}$$

Solving Eq. 19-23 for the reverse recovery time yields

$$t_{rr} = \sqrt{\frac{2Q_{rr}(1+S)}{di_R/dt}} \tag{19-24}$$

Using Eq. 19-24 in Eq. 19-22 yields

$$I_{rr} = \sqrt{\frac{2Q_{rr}(di_R/dt)}{(S+1)}} \tag{19-25}$$

The charge Q_{rr} represents the portion of the total charge Q_F (the charge stored in the diode during forward bias), which is swept out by the reverse current and not lost to internal recombination. Most of Q_F is stored in the drift region (especially in higher voltage diodes) and is given by $Q_F = \tau I_F$ (see Eq. 18-17 and 19-12). Since Q_{rr} must be less than Q_F (because of internal recombination in the diode), Eqs. 19-24 and 19-25 can be rewritten as

$$t_{rr} < \sqrt{\frac{2\tau I_F}{di_R/dt}} \tag{19-26}$$

and

$$I_{rr} < \sqrt{2\tau I_F \frac{di_R}{dt}} \tag{19-27}$$

In these last two equations we have made use of the observation that $S < 1$ in most diodes.

It was explained earlier that the diffusion length $L = \sqrt{D\tau}$ must be at least as large as the drift region length W_d to have small voltage drops across the drift region. This yields an expression for the lifetime using Eq. 18-9, which is

$$\tau = \frac{W_d^2}{(kT/q)\{\mu_n + \mu_p\}} \tag{19-28}$$

In this equation, the width of the drift region W_d must be at least as large as the depletion layer width at the diode breakdown voltage (because the drift region must contain the depletion layer). Setting W_d equal to $W(BV_{BD})$ using Eq. 19-3 and using the result in Eq. 19-28 along with using $\mu_n + \mu_p \approx 900$ cm^2/V-s yields

$$\tau \approx 4 \times 10^{-12} BV_{BD}^2 \tag{19-29}$$

Insertion of this expression for the carrier lifetime into Eqs. 19-26 and 19-27 gives

$$t_{rr} \approx 2.8 \times 10^{-6} BV_{BD} \sqrt{\frac{I_F}{di_R/dt}} \tag{19-30}$$

and

$$I_{rr} \approx 2.8 \times 10^{-6} BV_{BD} \sqrt{I_F \, di_R/dt} \tag{19-31}$$

In Eqs. 19-29 to 19-31 the times are in seconds, the currents in amperes, the voltages in volts, and the time derivative of current in amperes per second.

These last three equations are approximate estimates in several respects. First, they are based only on an approximate analysis of one type of diode, the abrupt junction. Second, it is assumed that the drift region width is the minimum allowable width given by Eq. 19-3. A larger value of W_d will make the estimates larger by the same amount. Third, they are based on the approximation that $Q_{rr} = Q_F$, a result that is most accurate for large values of di_R/dt (short values of t_{rr}), which minimizes the excess carriers lost to recombination in the diode. Thus, numerical estimates made with these equations may not be precise, but they do indicate the general trends. Most important, they summarize the trade-offs that must be made in the design of high-voltage pn junction diodes between low on-state losses (small V_d), faster switching times (small carrier lifetime τ and short values of t_{rr}), and larger breakdown voltages BV_{BD}.

19-6 SCHOTTKY DIODES

19-6-1 Structure and *I-V* Characteristics

A Schottky diode is formed by placing a thin film of metal in direct contact with a semiconductor. The metal film is usually deposited on an n-type semiconductor as is shown in Fig. 19-11, although appropriate metal films on p-type material could also be used. In Fig. 19-11 the metal film is the positive electrode and the semiconductor is the cathode.

Such a structure has a rectifying *i-v* characteristic very similar to that of a *pn* junction diode. However, the on-state voltage is significantly lower, typically 0.3 to 0.4 V, than that of a silicon diode. Thus, the Schottky diode may be preferable for use in some power applications such as those discussed in Chapter 8. In the reverse direction, the Schottky diode has a reverse leakage current that is larger than that of a comparable silicon *pn* junction diode. The breakdown voltage of a Schottky diode at present cannot be made reliably larger than 75 to 100 V.

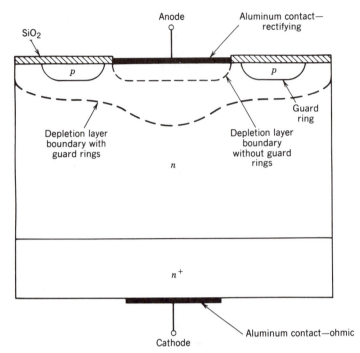

FIGURE 19-11: Cross-sectional view of a Schottky diode. A guard ring structure is also shown for improving the breakdown voltage capability of the diode.

19-6-2 Principle of Operation

The key to the operation of the Schottky diode is the fact that electrons in different materials have different absolute potential energies compared with electrons at rest in free space (the potential energies are lower in materials, indicating that the electrons are bound in the solid). Consider Fig. 19-11 where an n-type semiconductor is in contact with a metal whose electrons have a lower absolute potential energy than the electrons in the semiconductor. There is a flow of electrons in both directions across the metal–semiconductor interface when contact is first made. However, the flux of electrons from the semiconductor into the metal will be much larger because of the higher absolute potential energy of the electrons in the semiconductor. As a consequence, the metal will become negatively charged and the semiconductor will acquire a positive charge by forming a depletion region adjacent to the interface. The overall picture is quite similar to that shown in Figs. 18-8 and 18-9 except that the metal replaces the p-type side and the negative space charge comes from free electrons. The electrostatic potential barrier that accompanies the space charge region will grow in magnitude and oppose the continued flow of electrons from the semiconductor into the metal.

Eventually the potential barrier gets large enough so that the flux of electrons from the semiconductor into the metal is just equal to the flux from the metal to the semiconductor. At this point thermal equilibrium is established and there is no net current flow across the interface. Note that in establishing this equilibrium, no minority carriers, holes in this situation, were involved. Only majority carriers, electrons in this example, took part. This is the key difference between a Schottky diode and a pn junction diode. For this reason, Schottky diodes are termed majority carrier devices and *pn* junctions are labeled minority carrier devices or bipolar devices, since they use both electrons and holes in their basic operation.

When a voltage is applied to the structure of Fig. 19-11 that biases the metal positive with respect to the semiconductor, it opposes the built-in potential and makes it easier for current to flow. Biasing the metal negative with respect to the semiconductor increases the potential barrier to majority carrier current flow. Thus, the metal–semiconductor interface has rectifying characteristics similar to those of a *pn* junction. The major difference is that at any given forward current, the voltage across the Schottky diode is smaller than that across a *pn* junction. The difference amounts to roughly 0.3 V. The reason for the smaller voltage drop across the Schottky diode is that the reverse saturation current of a Schottky diode is significantly larger than that of a *pn* junction diode of the same cross-sectional area. The details of why the Schottky diode reverse current is larger is beyond the scope of this discussion. Suffice it to say that the lower forward voltage of the Schottky means that it is less lossy than a conducting *pn* junction diode.

19-6-3 Ohmic Contacts

The metal-semiconductor structure can form ohmic contacts (for example the cathode contact in Fig. 19-11) to semiconductor materials of all types. Such ohmic contacts are used in all types of semiconductor power devices. By an ohmic contact, we mean a contact that has no rectifying characteristics. The slope of the *i-v* characteristic of the contact is extremely steep (low resistance) and the same regardless of the voltage and current polarity.

The possibility of ohmic contacts utilizing a Schottky geometry arises because not all metals have electrons with lower absolute potential energies than electrons in a semiconductor. If such a metal is brought into contact with a semiconductor, an electric field and hence a potential barrier is set up across the interface that opposes the movement of electrons from the metal to the semiconductor. The source of the barrier is the accumulation of electrons in the semiconductor in the vicinity of the interface. This accumulation is extremely large and greatly increases the conductivity of the interface. The enhancement of the conductivity is so great that it obscures the nonlinearities (rectification) of the junction and makes the voltage drop across the interface proportional to the current through it. Naturally such a junction is termed an ohmic contact.

Even the cathode structure of Fig. 19-11 can be made into an ohmic contact if the doping in the *n*-region is made very heavy. In this circumstance the depletion region that is set up is extremely narrow (examine Eqs. 19-2 and 19-3) and the electric field that is set up is very large, approaching impact ionization values. Under these circumstances, electrons move very easily across the interface under the influence of small applied voltages. The mechanism that leads to this is tunneling, a quantum-mechanical effect that is beyond the scope of this discussion. This is why the cathode end of the Schottky diode in Fig. 19-11 is an ohmic contact and not a rectifying contact.

19-6-4 Breakdown Voltage

As we mentioned previously, the breakdown voltage of a Schottky diode cannot reliably be made much larger than 100 volts. The reason for this is twofold. First, the basic geometry of a Schottky diode leads to depletion layers, which have an extremely small radius of curvature at the edges of the contact metal, as illustrated in Fig. 19-11.

This leads to electric field crowding and low breakdown voltages, as explained earlier in this chapter. The use of field plates can alleviate this problem to some degree.

Another contributing factor is the lower breakdown field strength of the silicon at the surface compared with the bulk interior. There are a variety of imperfections in the silicon surface, some of which are process related and others more fundamental in nature, all of which lower the breakdown electric field strength. Since the geometry of the Schottky diode places the depletion layer (where large electric fields occur) right at the silicon surface, the reduced breakdown voltages compared to *pn* junction diodes is not surprising. Improvements in device processing and fabrication have reduced the surface imperfections and contaminants, which in turn have led to improvements in the breakdown voltage to the present-day levels.

A third contribution to the smaller breakdown voltages in Schottky diodes is the lack of any stored minority carriers that can short out the ohmic resistance of the drift region that must support the depletion region in reverse bias. This basic trade-off was discussed in Section 19-4-2 in conjunction with on-state losses, but bears repeating here because keeping on-state losses in a Schottky diode within set limits means using doping densities much larger than say 10^{14} cm^{-3} and so the achievable breakdown voltage is circumscribed. In comparison, the on-state losses and breakdown voltage of a *pn* junction diode can be designed much more independently of each other.

19-6-5 Switching Characteristics

A Schottky diode turns on and off faster than a comparable *pn* junction diode. The basic reason is that Schottky diodes are majority carrier devices and have no stored minority carriers that must be injected into the device during turn-on and pulled out during turn-off.

The lack of any significant stored charge changes the shape of observed switching waveforms in important ways. During turn-off, there will be no reverse current associated with removal of stored charge. However reverse current, associated with the growth of the depletion layer charge in reverse bias, will flow. This current may be comparable to the reverse current observed during switching of a *pn* junction because the space charge capacitance of a Schottky diode is larger (by as much as a factor of five) than in a comparable *pn* junction. The reason is that the depletion layer in a Schottky diode is thinner than that of a *pn* junction because of the heavier doping used in the *n*-region of the Schottky to keep the ohmic losses under control.

Schottky diodes have much less voltage overshoot during device turn-on than comparable *pn* junction diodes. The basic reason is that the ohmic resistance of the drift regions in a Schottky diode must be made much less than that of a *pn* junction diode in order to carry the same forward current because there is no excess carrier injection to short out high-resistivity drift regions. Some voltage overshoot associated with parasitic inductance will be observed if di/dt is large.

19-7   DIODE SNUBBERS

19-7-1 Capacitive Snubber

Snubbers are needed in diode circuits to minimize overvoltages. These overvoltages occur in circuits like the step-down converter shown in Fig. 19-12*a* because of the stray or leakage inductance in series with the diode and the snap-off of the diode reverse recovery current at the turn-on of the switch *T*. The design of the snubber circuit that will protect the diode will be based on this step-down converter circuit

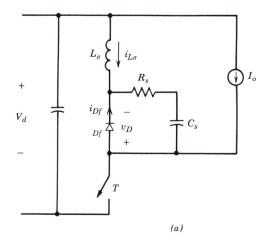

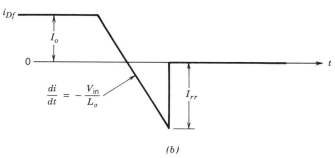

FIGURE 19-12: (*a*) A step-down converter circuit with stray inductance and a snubber circuit for the free-wheeling diode. (*b*) The diode reverse-recovery current.

where L_σ is the stray inductance. It is shown later that for the purposes of snubber analysis, this circuit is an equivalent circuit for almost any converter where diodes are used. An R_s-C_s snubber is commonly used across the diode for overvoltage protection, as is shown in Fig. 19-12*a*. To simplify the analysis, the diode reverse recovery current is assumed to snap off instantaneously, as is shown in Fig. 19-12*b*. The load is inductive, and it is assumed that the load current I_o is constant during the switching transient.

Although the capacitive snubber ($R_s = 0$) is not used in practice (for reasons that will become apparent), it provides an easily analyzed starting point that illustrates the basic concepts. In obtaining an equivalent circuit, the switch in Fig. 19-12*a* is assumed to be ideal, which results in a worst case analysis of this circuit. Treating the instant of diode snap-off at the peak reverse recovery current I_{rr} as $t = 0$, the initial inductor current in the equivalent circuit of Fig. 19-13 is I_{rr} and the initial snubber capacitor voltage is zero. To establish a baseline circuit, the snubber resistance R_s is assumed to be zero, as is shown in Fig. 19-13*b*. The capacitor voltage (which is the negative of the diode voltage in this baseline circuit) can be obtained from Eq. 7-4 as

$$v_{Cs} = V_d - V_d \cos(\omega_0 t) + I_{rr}\sqrt{\frac{L_\sigma}{C_s}} \sin(\omega_0 t) \qquad (19\text{-}32)$$

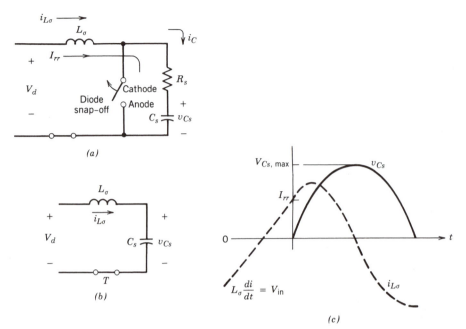

FIGURE 19-13: (a) Equivalent circuit of the step-down converter at the instant of diode reverse-recovery current snap-off, and (b) the simplification that results when the snubber resistance is zero. (c) The voltage and current waveforms for $R_s = 0$ and $C_s = C_{\text{base}}$.

where

$$\omega_0 = \frac{1}{\sqrt{L_\sigma C_s}} \tag{19-33}$$

Introducing a baseline capacitance C_{base} given by

$$C_{\text{base}} = L_\sigma \left[\frac{I_{rr}}{V_d}\right]^2 \tag{19-34}$$

it is possible to express Eq. 19-32 as

$$v_{Cs} = V_d \left[1 - \cos(\omega_0 t) + \sqrt{\frac{C_{\text{base}}}{C_s}} \sin(\omega_0 t)\right] \tag{19-35}$$

Either by a time derivative or a phasor approach, the maximum value of v_{Cs} in Eq. 19-35 can be calculated as

$$V_{Cs,\max} = V_d \left[1 + \sqrt{1 + \frac{C_{\text{base}}}{C_s}}\right] \tag{19-36}$$

The waveforms for v_{Cs} and inductor current $i_{L\sigma}$ are shown in Fig. 19-13c for $C_s = C_{\text{base}}$. In this case the maximum reverse diode voltage is the same as $V_{Cs,\max}$

calculated from Eq. 19-36. For small values of C_s, the maximum diode voltage becomes excessive.

19-7-2 Effect of Adding a Snubber Resistance

When the diode snubber resistance R_s is included, the equivalent circuit for the snubber becomes as shown in Fig. 19-13a. In analyzing this circuit, the instant of diode snap-off is treated as $t = 0$, and the initial inductor current is I_{rr} and the initial capacitor voltage is zero. The circuit waveforms for $t > 0$ can be found by analytical calculations or computer simulation and are shown in Fig. 19-14 for $C_s = C_{base}$. In these waveforms, the oscillations are damped out by R_s and the maximum diode voltage depends on the values of R_s and C_s used. For a selected value of C_s, the maximum diode voltage varies with R_s. As an example, for $C_s = C_{base}$, the normalized maximum diode voltage is plotted in Fig. 19-15 as a function of R_s/R_{base}, where the baseline resistance $R_{base} = V_d/I_{rr}$. It can be seen that for a given C_s, there is an optimum value of $R_s = R_{opt}$ which minimizes V_{max}.

The foregoing analysis has been extended to various values of C_s. This allows the optimum resistance and corresponding V_{max} to be plotted as a function of C_s, as is shown in Fig. 19-16, where all quantities are normalized.

The energy loss in the resistor R_s at diode turn-off is given by Reference 1 as

$$W_R = \frac{1}{2}L_\sigma I_{rr}^2 + \frac{1}{2}C_s V_d^2 \qquad (19\text{-}37)$$

W_R, normalized with respect to $(L_\sigma I_{rr}^2/2)$, the peak energy stored in the leakage inductance as the diode snaps off, is also plotted in Fig. 19-16. At the end of the current oscillations, the energy stored in C_s is equal to

$$W_{Cs} = \frac{1}{2}C_s V_d^2 \qquad (19\text{-}38)$$

which is dissipated in R_s at the next turn-on of the diode assuming instantaneous turn-on of the diode. The total energy dissipation in the diode and its snubber

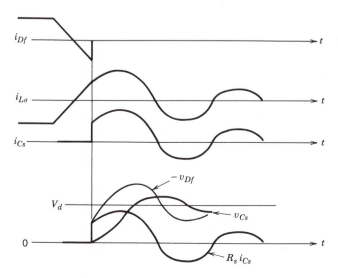

FIGURE 19-14: The current and voltage waveforms after the diode snaps off at $t = 0$.

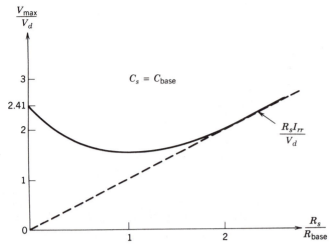

FIGURE 19-15: Maximum overvoltage across the diode as a function of snubber resistance for a fixed value of snubber capacitance C_s.

resistance is given by

$$W_{\text{tot}} = W_R + W_{Cs} = \frac{1}{2}L_\sigma I_{rr}^2 + C_s V_d^2 = \frac{1}{2}L_\sigma I_{rr}^2\left(1 + 2\frac{C_s}{C_{\text{base}}}\right) \quad (19\text{-}39)$$

and the normalized W_{tot} is also plotted in Fig. 19-16 as a function of the normalized C_s. It can be seen from Fig. 19-16 that the maximum voltage decreases only slightly by increasing C_s beyond C_{base}. However, the total energy dissipation increases linearly with C_s; therefore, a snubber capacitor with C_s in a range close to C_{base} would normally be used. Once C_s has been selected, $R_s = R_{\text{opt}}$ can be obtained directly from Fig. 19-16.

In this analysis, it is assumed that the reverse recovery current of the diode snaps off instantaneously. In practice, the diode reverse recovery current can be assumed to

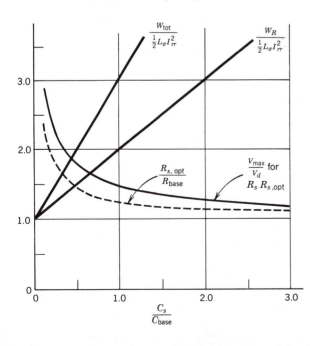

FIGURE 19-16: Snubber energy loss and the maximum diode voltage for the optimum value of snubber resistance R_s as a function of the snubber capacitance C_s.

decay exponentially. This can be accommodated in the equivalent circuit of Fig. 19-13a by adding a time-varying current source. The analysis can be carried out by computer simulation, and the results show that the snubber design remains essentially the same as before.

19-7-3 Implementation

The circuit of Fig. 19-13a has been used for the detailed analysis of the diode snubber. In the following paragraphs we consider how to derive similar equivalent circuits for diode snubber analysis in some commonly used converters. In the flyback converter of Fig. 19-17a operating in an incomplete demagnetization mode, when the switch is in its off-state, the diode is conducting. When the switch is turned on, the secondary side of the circuit can be represented by the circuit of Fig. 19-17b, where the current in the leakage inductance and the diode decreases. Considering the instant at which the diode current I_{rr} snaps back as the time origin, the equivalent circuit of Fig. 19-17c is obtained identical to Fig. 19-13a (that was obtained for the step-down converter circuit of Fig. 19-12a).

Another example is that of a center-tapped secondary in a push–pull, half-bridge, or a full-bridge dc–dc converter circuit shown in Fig. 19-18a. If we assume that the converter is operating in a continuous current conduction mode, when all primary switches are OFF, one-half of the output current flows through each diode. When a converter switch turns on such that a positive voltage is applied across the primary, the secondary side of the circuit can be represented by Fig. 19-18b, where the current through D_1 will increase and through D_2 will decrease. When we consider the instant at which the peak reverse recovery current I_{rr} snaps off in D_2 as the initial instant, the equivalent circuit of Fig. 19-18c, which is similar to that of Fig. 19-17c, results. There should be a snubber across each diode in this circuit.

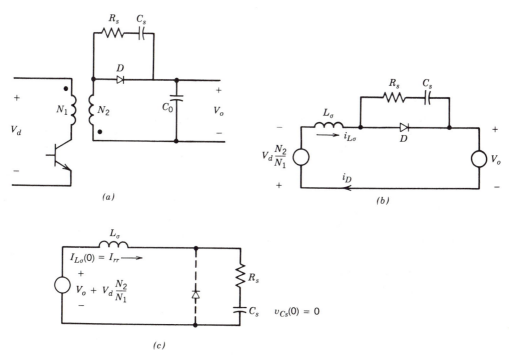

FIGURE 19-17: (a) Flyback converter circuit operating in an incomplete demagnetization mode. (b) Equivalent circuit on the secondary side, and (c) the simplified equivalent circuit after the snap-off of the diode current. L_σ is the transformer leakage inductance.

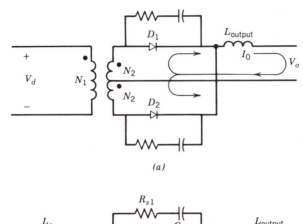

(a)

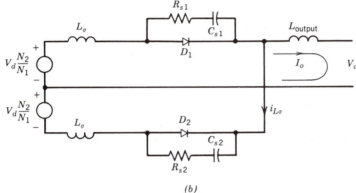

(b)

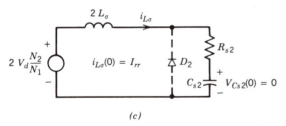

(c)

FIGURE 19-18: (a) Full-wave rectifier using a transformer with a center tapped secondary. (b) Equivalent circuit on the secondary side, and (c) the equivalent circuit at the instant of the snap-off of diode D_2.

In the foregoing examples, converters consisted of an isolation transformer that introduced substantial leakage inductance. Therefore, diode snubbers were needed. However, the step-down converter circuit of Fig. 19-12a, which was used only to present the useful equivalent circuit for analyzing diode snubbers, may not actually require a diode snubber when BJTs and MOSFETs are used (where di/dt through the diode at turn-off can be controlled and hence I_{rr} minimized) as the controllable switch and L_σ is minimized by proper circuit layout, as is discussed in Chapters 20 and 21. The same comments also apply to half- and full-bridge switch-mode converters using BJTs and MOSFETs.

In single-phase line-frequency diode converters discussed in Chapter 3, if the diode bridge rectifier feeds a dc capacitor and operates in a discontinuous mode, then if filter inductors are used, they should be placed on the ac side, as is shown in Fig. 19-19a. Such a configuration will provide overvoltage protection of the diodes against

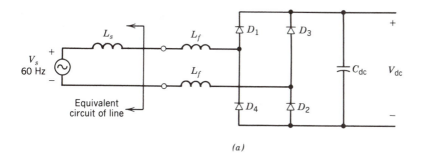

(a)

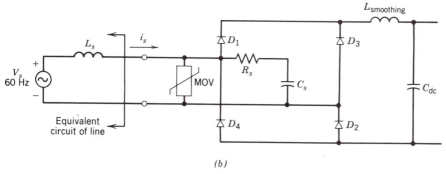

(b)

FIGURE 19-19: *(a)* Single-phase line-frequency diode rectifier. *(b)* Single-phase diode converter with an R-C snubber to protect against overvoltages due to unknown inductive reactances on the ac side of the rectifier. A MOV (metal-oxide varistor) is also shown for further overvoltage protection.

incoming line voltage transients and against voltages induced by the di/dt at reverse recovery. There is no need for diode snubbers since, for example, when the reverse recovery current of diode D_1 snaps off, then D_4 provides a path for the inductive current and the reverse voltage across D_1 is clamped to the dc capacitor voltage V_{dc}.

In the case of continuous conduction in a single-phase full-bridge rectifier, the filter inductor is placed on the dc side, as is shown in Fig. 19-19b. In practice, there is a finite inductance on the ac side, which is normally not known. For a worst case analysis, the ac-side reactance $X_s(=\omega L_s)$ can be assumed to be 5%, which implies that

$$X_s = \omega L_s = 0.05\frac{V_s}{I_{s1}} \tag{19-40}$$

where V_s is the rms line voltage and I_{s1} is the rms value of the fundamental frequency component of the current at full-load. In the circuit of Fig. 19-19b, one R-C snubber may be used to protect all the diodes. In Fig. 19-19b, the waveforms due to the diode reverse recovery snap-off are much faster than the variations in the 60-Hz line voltage input v_s. Therefore, the value of v_s at the instant of diode snap-off can be treated as a constant dc, again allowing the use of the equivalent circuit of Fig. 19-13a. The detailed analysis of snubbers for such a circuit is given for a thyristor rectifier in the thyristor chapter where the diode bridge can be treated as a special case. The analysis for single-phase rectifiers also applies to three-phase diode bridge rectifiers.

It should be noted that in the converters connected to the line, the diode snubbers should also provide overvoltage protection against incoming line voltage transients. In fact, this consideration may supersede the snubber design based on the reverse recovery snap-off. Often Metal Oxide Varistors (MOVs) are used for this transient overvoltage protection.

19-8   SUMMARY

This chapter has explored the characteristics of *pn* junction diodes intended for power applications. The characteristics of power Schottky diodes were also briefly discussed. Protective circuits or snubbers for diodes were also investigated. The important points are listed as follows.

1. Power diodes are constructed with a vertically oriented structure that includes a n^- drift region to support large blocking voltages.

2. The breakdown voltage is approximately inversely proportional to the doping density of the drift region and required minimum length of the drift region scales with the desired breakdown voltage.

3. Achievement of large breakdown voltages requires special depletion layer boundary shaping techniques.

4. Conductivity modulation of the drift region in the on-state keeps the losses in the diode to manageable levels even for large on-state currents.

5. Low on-state losses requires long carrier lifetimes in the diode drift region.

6. Minority carrier devices have lower on-state losses than majority carrier devices such as MOSFETs at high blocking voltage ratings.

7. During the turn-on transient the forward voltage in a diode may have a substantial overshoot, on the order of tens of volts.

8. Short turn-off times require short carrier lifetimes, so a trade-off between switching times and on-state losses must be made by the device designer.

9. During turn-off, fast reverse recovery may lead to large voltage spikes because of stray inductance.

10. The problems with the reverse recovery transient are most severe in diodes with large blocking voltage ratings.

11. Schottky diodes turn on and off faster than *pn* junction diodes and have no substantial reverse recovery transient.

12. Schottky diodes have lower on-state losses than *pn* junction diodes but also have low breakdown voltage ratings, rarely exceeding 100 V.

13. Diodes require snubber circuits to prevent damaging overvoltages that might otherwise result from stray inductance in series with the diode.

14. A simple R-C circuit in parallel with the diode with appropriately chosen values of R and C will act as an efficient snubber for the diode.

15. There are optimum values for the snubber capacitance and resistance that will provide the minimum overvoltage across the device being protected by the snubber.

PROBLEMS

19-1. The silicon diode shown in Fig. 19-1 is to have a breakdown voltage of 2500 V. Estimate what the doping density of the drift region should be and what the minimum width of the drift region should be. The diode is a nonpunch-through device.

19-2. A silicon diode similar to that shown in Fig. 19-1 has a drift region doping density of 5×10^{13} cm^{-3} donors and a drift region width of 50 μ. What is the breakdown voltage?

19-3. The diode in Problem 19-2 has a cross-sectional area of 2 cm² and carrier lifetime τ_0 of 2 μs. Approximately sketch and dimension the on-state voltage including the junction drop versus forward current. Consider currents as large as 3000 A.

19-4. The diode of Problem 19-3 is forward biased by a 1000-A current that rises at the rate of 250 A/μs.

   (a) Assuming that no carrier injection takes place until the current reaches its steady-state value, sketch and dimension the forward voltage for $0 < t < 4$ μs.

   (b) Now assume that carrier injection commences at $t = 0$ and that this causes the drift region to drop linearly in time from its ohmic value at $t = 0$ to its on-state value at $t = 4$ μs. Sketch and dimension the forward voltage for $0 < t < 4$ μs.

19-5. A silicon diode with a breakdown voltage of 2000 V that is conducting a forward current of 2000 A is turned off with a constant $di_R/dt = 250$ A/μs. Roughly estimate the time required for the diode to turn-off.

19-6. Derive the equation (Eq. 19-36) for the maximum overvoltage across a purely capacitive snubber.

19-7. Consider the flyback converter circuit shown in Fig. 19-17. The input voltage is 100 V, as is the dc output voltage. The transformer has a 1 : 1 turns ratio and a leakage inductance of 10 μH. The transistor, which can be considered as an ideal switch, is driven by a square wave with a 50% duty cycle. The snubber resistance is zero. The diode has a reverse recovery time t_{rr} of 0.3 μs.

   (a) Draw an equivalent circuit suitable for snubber design calculations.

   (b) Find the value of snubber capacitance C_s that will limit the peak overvoltage to 2.5 times the dc output voltage.

19-8. Repeat Problem 19-7 with a resistance R_s included in the snubber circuit. Find both the value of snubber capacitance and optimum value of snubber resistance.

19-9. Estimate the power dissipated in the snubber resistance found in Problem 19-8 if the square wave switching frequency is 20 kHz.

19-9 REFERENCES

1. W. McMurray, "Optimum Snubbers For Power Semiconductors," IEEE IAS 1971 Annual Meeting.

2. *SCR Manual*, 6th edition, General Electric Company, Syracuse, NY, 1979.

3. M. H. Rashid, *Power Electronics: Circuits, Devices, and Applications*, Prentice-Hall, Englewood Cliffs, NJ, 1988, Chapter 15.

4. Ben G. Streetman, *Solid State Electronic Devices*, 2nd edition, Prentice-Hall, Englewood Cliffs, NJ, 1980, Chapter 5.

5. Sorab K. Ghandhi, *Semiconductor Power Devices*, John Wiley & Sons, New York, 1977, Chapters 2–3.

6. B. M. Bird and K. G. King, *An Introduction to Power Electronics*, John Wiley & Sons, New York, 1983, Chapter 6.

7. M. S. Adler and V. A. K. Temple, "Analysis and Design of High-Power Rectifiers," *Semiconductor Devices for Power Conditioning*, Roland Sittig and P. Roggwiller, Editors, Plenum Press, New York, 1982.

8. R. J. Grover, "Epi and Schottky Diodes," *Semiconductor Devices for Power Conditioning*, Roland Sittig and P. Roggwiller (Eds.), Plenum Press, New York, 1982.

9. B. W. Williams, *Power Electronics, Devices, Drivers, and Applications*, John Wiley & Sons, New York, 1987, Chapters 1–4.

20

BJTs with Drive and Snubber Circuits

20-1 INTRODUCTION

The need for a large blocking voltage in the off-state and a high current carrying capability in the on-state means that a power bipolar junction transistor (BJT) must have a substantially different structure than its logic level counterpart. The modified structure leads to significant differences in the i-v characteristics and switching behavior between the two types of devices. These differences also mean that the circuits that drive the transistor on and off are different from the simple circuits used in low-level logic circuits. Often special protective measures must be included in the circuits in which the power transistor is embedded. In this chapter these and other topics will be explored for both power BJTs and monolithic Darlington-connected devices. The drive and snubber circuits discussed with BJTs as examples are also applicable to the other controllable switches discussed in later chapters.

20-2 VERTICAL POWER TRANSISTOR STRUCTURES

A power transistor has a vertically oriented four-layer structure of alternating p-type and n-type doping such as the npn transistor shown in Fig. 20-1. The transistor has three terminals, as is indicated in the figure, and they are respectively labeled collector, base, and emitter. In most power applications, the base is the input terminal, the collector is the output terminal, and the emitter is common between input and output (the so-called common emitter configuration). The circuit symbol for the BJT is shown in the same figure. A pnp transistor, whose circuit symbol is also shown in Fig. 20-1, would have the opposite type of doping in each of the layers shown in the figure. NPN transistors are much more widely used than pnp transistors as power switches.

480

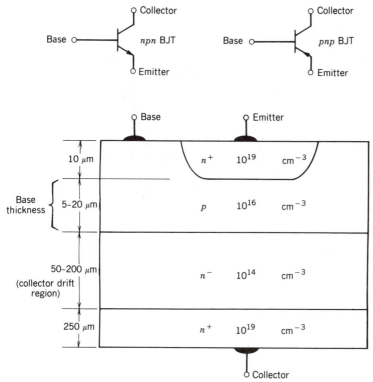

FIGURE 20-1: Vertical cross section of a typical *npn* power bipolar junction transistor. The circuit symbol for the transistor is also shown.

The vertical structure is preferred for power transistors because it maximizes the cross-sectional area through which the current in the device is flowing. This minimizes the on-state resistance and thus the power dissipation in the transistor. In addition, having a large cross-sectional area minimizes the thermal resistance of the transistor, thus also helping to keep power dissipation problems under control.

The doping levels in each of the layers and the thicknesses of the layers have a significant effect on the characteristics of the device. The doping in the emitter layer is quite large (typically 10^{19} cm^{-3}), whereas the base doping is moderate (10^{16} cm^{-3}). The n^- region that forms the collector half of the C-B (collector-base) junction is usually termed the collector drift region and has a light (10^{14} cm^{-3}) doping level. The n^+ region that terminates the drift region has a doping level similar to that found in the emitter. This region serves as the collector contact to the outside world. The thickness of the drift region determines the breakdown voltage of the transistor and thus can range from tens to hundreds of microns in extent. The base thickness is made as small as possible in order to have good amplification capabilities, as will be explained in later sections. However, if the base thickness is too small, the breakdown voltage capability of the transistor is compromised, as will be explained. Thus, base thicknesses in power devices are a compromise between these two competing considerations and are typically several microns to a few tens of microns in thickness, compared with the small fraction of a micron in thickness for logic level transistors.

Practical power transistors have their emitters and bases interleaved as narrow fingers, as is shown in Fig. 20-2. The purpose of this arrangement is principally to

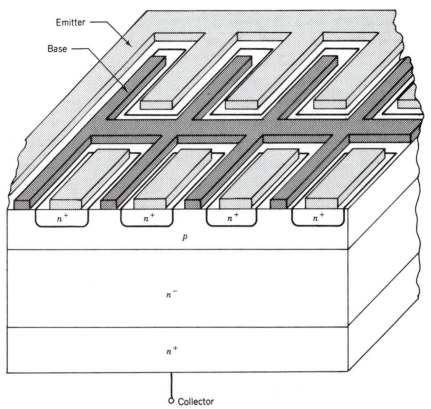

FIGURE 20-2: Vertical cross section of a multiple emitter *npn* transistor. Some of the layers such as the emitter metallization are only partially shown (i.e., are cut away) to provide better clarity.

reduce the effects of current crowding, a phenomenon that can lead to second breakdown and possible device failure. These topics will be considered later in this chapter. This multiple emitter layout also reduces the parasitic ohmic resistance in the base current path, which helps to reduce power dissipation in the transistor.

The relatively thick base found in power transistor structures causes the current gain, $\beta = I_C/I_B$, to be rather small, typically 5 to 10. This is undesirably small for some applications and, hence, monolithic designs for Darlington-connected BJT pairs shown in Fig. 20-3 have been developed. The current gain of a Darlington pair is given by

$$\beta = \beta_M \beta_D + \beta_M + \beta_D \tag{20-1}$$

so that even though each individual transistor has a small beta, the effective beta of the pair can still be quite large. The vertical cross section of a monolithic Darlington is shown in Fig. 20-4. A discrete diode D_1 is added as shown in Fig. 20-3 to speed up the turn-off time of the main transistor, as will be explained shortly. The discrete diode D_2 also shown in Fig. 20-3 is added for half- and full-bridge circuit applications.

20-3   *I-V* CHARACTERISTICS

The output characteristics (i_C versus v_{CE}) of a typical *npn* power transistor are shown in Fig. 20-5. The various curves are distinguished from each other by the value

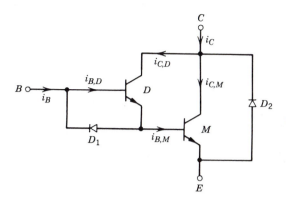

FIGURE 20-3: Power transistors in a Darlington configuration in order to obtain a larger effective current gain beta. The discrete diodes are added to aid turn-off (D_1) and for full-bridge applications (D_2).

of the base current. The characteristics of monolithic Darlingtons are quite similar to those shown in the figure. Several features of the characteristics should be noted. First, there is a maximum collector–emitter voltage that can be sustained across the transistor when it is carrying substantial collector current. This voltage is usually labeled BV_{SUS}. In the limit of zero base current, the maximum voltage between collector and emitter that can be sustained increases somewhat to a value labeled BV_{CEO}, the collector–emitter breakdown voltage when the base is open circuited. This latter voltage is often used as the measure of the transistor's voltage standoff capability because usually the only time the transistor will see large voltages is when the base current is zero and the BJT is in cutoff. The voltage BV_{CBO} is the collector–base breakdown voltage when the emitter is open circuited. The fact that this voltage is larger than BV_{CEO} is used to advantage in so-called open-emitter transistor turn-off circuits.

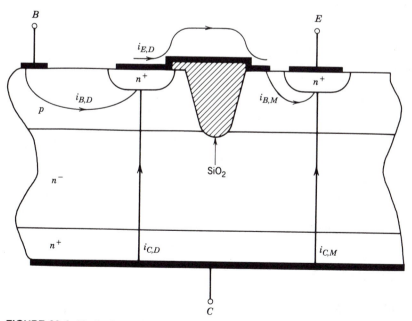

FIGURE 20-4: Vertical cross section of a pair of monolithic Darlington-connected bipolar transistors. The silicon dioxide protrusion through the upper *p*-layer (the base region of both transistors) electrically isolates the two bases from each other.

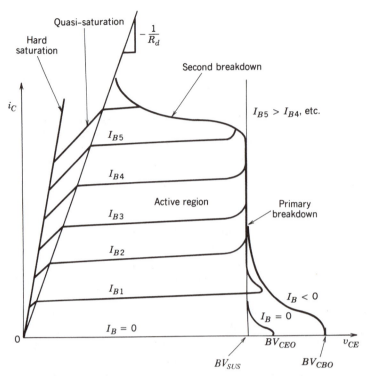

FIGURE 20-5: Current-voltage characteristics of a *npn* power bipolar junction transistor showing second breakdown and quasi-saturation.

The region labeled primary breakdown is due to conventional avalanche breakdown of the collector–base junction and the attendant large flow of current. This region of the characteristics is to be avoided because of the large power dissipation that clearly accompanies such breakdown. The region labeled second breakdown must also be avoided because large power dissipation also accompanies it, particularly at localized sites within the semiconductor. The origin of second breakdown is different from that of avalanche breakdown and will be considered in detail later in this chapter. BJT failure is often associated with second breakdown.

The major observable difference between the *i-v* characteristics of a power transistor and those of a logic level transistor is the region labeled quasi-saturation on the power transistor characteristics of Fig. 20-5. As we will explain in detail in later sections of this chapter, quasi-saturation is a consequence of the lightly doped collector drift region found in the power transistor. Logic level transistors do not have this drift region and so do not exhibit quasi-saturation. Otherwise all of the major features of the power transistor characteristic are also found on those of logic level devices.

20-4　PHYSICS OF BJT OPERATION

20-4-1　Basic Gain Mechanism and Beta

An understanding of how the BJT provides current (power) amplification is most easily obtained by considering the simplified one-dimensional transistor structure shown in Fig. 20-6*a*. In this model, which is essentially the structure of a logic level

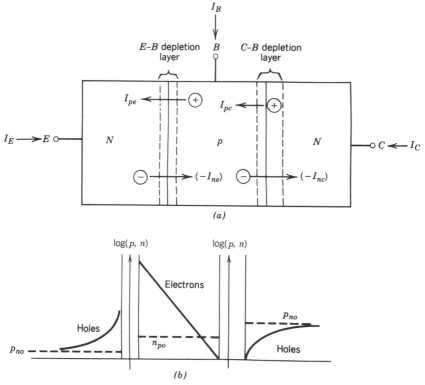

FIGURE 20-6: (a) Simplified model of a BJT, and (b) the stored charge distributions that exist in the BJT biased in the normal active region. The internal current components that flow in the active region are also shown on the model.

transistor, there is no lightly doped collector drift region. It is further assumed that the transistor is in the active region. In the active mode of BJT operation, the drift region does not play a major role, and retaining it would needlessly complicate the discussion. The effect of the drift region will be considered in detail when breakdown voltage, on-state losses, and switching times are considered since the presence of the drift region has a significant effect on these items, but not on the active mode of operation.

In the active region, the B-E (base–emitter) junction is forward biased and C-B junction is reverse biased. Electrons are injected into the base from the emitter and holes are injected from the base into the emitter. This produces the minority carrier distributions shown in Fig. 20-6b. These distributions have large density gradients, especially in the base region, which support significant diffusion currents. In fact, the total current flowing across the B-E junction will be almost entirely diffusion current, the same as for the pn junction diode described in Chapter 18. Unlike the forward-biased pn junction diode of Chapter 18, the base current entering the p-type side of the B-E junction from the B-E junction bias source will not equal to the current (emitter current) that leaves the n-type side. The currents will be unequal because the transistor structure provides an alternative besides the base terminal for electrons injected into the base from the emitter to exit the region.

Electrons injected into the base from the emitter are most likely to exit the base via the collector rather than the base terminal for three reasons. First, the thickness of the base region is made quite small compared with the electron diffusion length $L_{nb} = (D_{nb}\tau_{nb})^{1/2}$ in the base so they are unlikely to recombine there. Second, the area of the collector is made much larger than that of the emitter or the base contact, as is shown in Fig. 20-1, so that the electrons diffusing away from the emitter are much more likely to encounter the collector than anything else because of the short distance between emitter and collector. Third, the density of electrons at the collector–base junction is essentially zero as is shown in Fig. 20-6b because the high electric fields in the reverse biased C-B junction sweep all the diffusing electrons at the edge of the space charge region across the junction and into the collector region. The large density of injected electrons at the B-E junction and essentially zero excess electrons at the collector side of the base means that a very large gradient of electrons exist in the base, as shown in Fig. 20-6b. This density gradient carries most of the injected electrons and very few of them exit the base region via the base lead.

This means that the base current will be much less than the emitter current, and the collector current will almost equal the emitter current. A small base current causes the flow of a much larger current between the collector and emitter and thus a substantial amount of gain is obtained between the input base current and the output collector current. This is the basic gain mechanism of the bipolar junction transistor. The gain is quantitatively characterized by the ratio of the collector current i_C to the base current i_B (beta or β of the transistor).

The features of the transistor structure that lead to large values of beta can be more clearly understood by considering the currents that flow internally in the transistor. These currents can be conveniently divided into the four components shown in Fig. 20-6a. The current I_{pe} is the diffusion current that originates from holes injected from the base into the emitter to sustain the hole distribution in the emitter. Similarly, I_{ne} is the diffusion current that originates from electron injection from the emitter into the base in order to support the electron distribution in the base. As is explained in the preceding paragraphs, these electrons then diffuse across the base, and those that survive recombination arrive at the C-B depletion layer. These excess electrons, along with a much smaller number that are thermally generated in the depletion layer, are swept across the depletion layer by the large electric fields into the collector layer. This flow of electrons is the I_{nc} current. The current I_{pc} arises from holes that are thermally generated in the C-B depletion layer and then are swept into the base layer by the large electric fields in the depletion layer. It is much smaller than the other current components because the hole density in the C-B space charge layer is much less than the carrier densities in the other regions. Hence, I_{pc} will be neglected in all further discussions.

The terminal currents of the transistor, I_C and I_B, can be expressed in terms of these internal currents. The collector current is given by

$$I_C = I_{nc} \tag{20-2}$$

and the base current is given by

$$I_B = -I_C - I_E = -I_{nc} + I_{ne} + I_{pe} \tag{20-3}$$

Beta can be expressed in terms of the internal currents as

$$\frac{I_B}{I_C} = \frac{1}{\beta} = \frac{I_{ne} - I_{nc}}{I_{nc}} + \frac{I_{pe}}{I_{nc}} \tag{20-4}$$

In order for beta to be large, the numerators of the two terms in Eq. 20-4 must be small compared with I_{nc}. The I_{p_e} term can be minimized by doping the emitter very heavily so that the stored hole distribution there is made small (see Eq. 18-18). The term $(I_{ne} - I_{nc})$ represents the difference between the electrons injected into the base at the B-E junctions and those swept across the C-B junction into the collector. This difference is caused by the recombination of some of the injected electrons in the base region, and it is minimized by having a large electron lifetime in the base region and by making the base thickness small (fractions of a micron in logic level BJTs) compared with the electron diffusion length (see Eq. 18-14).

In summary, there are three prime requirements for large values of beta in a bipolar junction transistor. These are (1) heavy doping of the emitter, (2) long minority carrier lifetimes in the base, and (3) short base thicknesses. These factors will conflict with other characteristics desired for the transistor and, hence, a trade-off will be required between large gain and other parameters such as fast switching times. The consequence of these trade-offs is that the base thickness in a power transistor is larger than in a logic level transistor and the beta of power transistors is typically 5 to 20.

A feature of BJTs including power transistors not predicted by the foregoing discussion is the fall-off of the current gain at collector current values larger than some value which is characteristic for a specific type of transistor. This fall-off, which is illustrated in Fig. 20-7, commences at currents of less than 1 A in logic level transistors and at current levels of 100 A in some power devices. Several different mechanisms operative in the transistor simultaneously contribute to this fall-off in beta, of which two of the most significant are conductivity modulation in the base and emitter current crowding. Conductivity modulation of the base is essentially the same as the conductivity modulation of the diode drift region discussed in the previous chapter. In the case of the BJT base layer, conductivity modulation occurs when the minority carrier density in the base becomes comparable to the majority carrier doping density. For example, in the BJT diagram of Fig. 20-1, high-level injection in the base would occur when the excess electron density reaches about 10^{16} cm^{-3}. When the electron density gets this large, excess holes of the same density must also be injected into the base, and the only way for this to occur is for the base current to supply them. This represents an increase in the base current without a similar increase in the collector current and, hence, a fall in the value of beta. At the larger values of collector current, beta is approximately inversely proportional to the collector current.

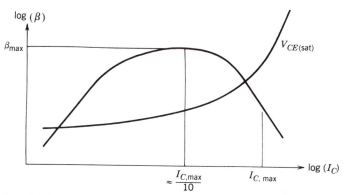

FIGURE 20-7: Variation in the BJT current gain β and $V_{CE(sat)}$ as a function of the dc collector current showing the fall-off of beta and increase in $V_{CE(sat)}$ at large collector currents.

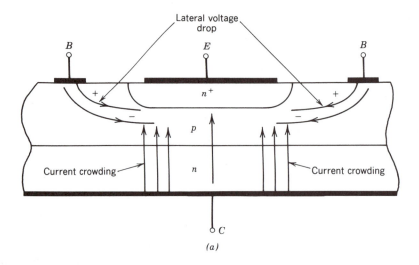

(a)

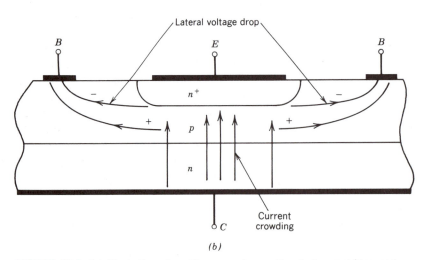

(b)

FIGURE 20-8: (a) Illustration of emitter current crowding in forward bias and
(b) reverse bias (turn-off transient) caused by lateral voltage drops
induced by large base currents.

Emitter current crowding is another mechanism that causes a decrease in beta. Consider the simplified BJT cross section shown in Fig. 20-8a, where both the base current and collector current flow paths are shown, assuming the BJT is in the active region. Because of the device geometry, there is a lateral ohmic voltage drop in the base region, as indicated in the figure, which is caused by the lateral base current. This lateral ohmic voltage drop subtracts from the externally applied B-E voltage and this means that the voltage drop across the B-E junction is larger at the emitter periphery near the base contact than it is in the center of the emitter area. This in turn causes a larger current density to flow across the junction at the emitter edge near the base terminal compared with the current density in the center of the emitter area. The current crowding clearly will mean that the onset of high-level injection and the attendant reduction in beta will occur at lower total currents than if the current density were uniformly spread over the entire emitter area. As we mentioned earlier, modern power BJTs have their emitters separated into many narrow rectangular areas, as is shown in Fig. 20-2, to minimize current crowding.

20-4-2 Quasi-saturation

To understand the phenomenon of quasi-saturation, the one-dimensional model of the BJT is now generalized to include a collector drift region, as is shown in Fig. 20-9. As in the previous section, it is assumed that the transistor is initially in the active region and now the base current is allowed to increase. As the collector current

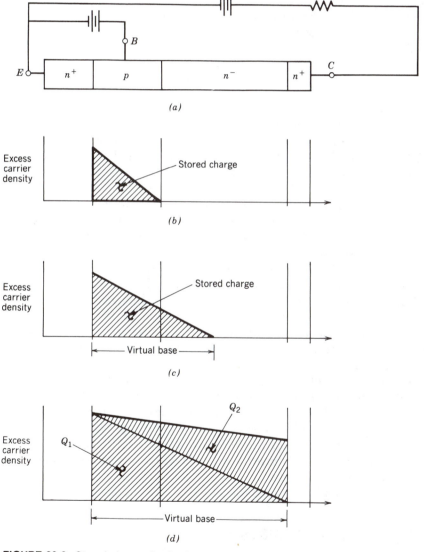

FIGURE 20-9: Stored charge distribution in the base and collector drift regions of a power BJT: (a) power transistor cross section; (b) active; (c) quasi-saturation; (d) hard saturation. Q_1 is the amount of stored charge that puts the BJT at the edge of hard saturation and Q_2 drives the transistor deeper into hard saturation.

rises in response to the base current, the C-E voltage drops because of the increased voltage drop across the collector load. However, there is a simultaneous increase in the voltage drop in the drift region as a result of its ohmic resistance because of the increase in i_C. This means that the reverse bias across the actual C-B junction, the n^-p junction, is getting smaller and at some point the junction will become forward biased.

When this occurs, injection of holes from the base into the collector drift region commences. At the same time, space charge neutrality requires that electrons also be injected into the drift region in about the same numbers as the holes. These electrons are conveniently obtained from the very large number of electrons being supplied to the C-B junction via injection from the emitter and subsequent diffusion across the base. As this excess carrier build-up in the drift region begins to occur, the quasi-saturation region of the i-v characteristic is entered. If the ohmic resistance of the drift region is R_d, then the boundary between the quasi-saturation region and the active region in Fig. 20-5 is given by

$$i_C = \frac{v_{CE}}{R_d} \tag{20-5}$$

In quasi-saturation, double injection is occurring in the drift region in a manner similar to that in the drift region of the forward biased power diode. However, the stored charge accumulates in the drift region from only one side of the drift region, the C-B junction side (or pn^- side) as is diagrammed in Fig. 20-9. In the transistor, electron injection across the n^-n^+ junction is much less noticeable because there is a much more plentiful supply of electrons at the pn^- junction (due to electrons injected from the emitter, as discussed in the previous paragraph) compared to the situation in the pn junction diode where there is no such supply of electrons. As the injected carriers increase, the drift region is gradually shorted out and the voltage across the drift region drops even though the collector current is large. It is also apparent from Fig. 20-9c that as the hole injection from the base across the C-B junction commences, the thickness of the effective or virtual base is increasing. This means that the effective value of beta decreases and, hence, the collector current magnitude that a given base current can support must also decrease, as diagrammed in the i-v characteristics of Fig. 20-5. In quasi-saturation, the drift region is not completely shorted out by high-level injection; hence, the power dissipation in the BJT is larger than when hard saturation is entered.

Hard saturation is obtained when the excess carrier density reaches the other side (n^+ side) of the drift region, as is diagrammed in Fig. 20-9d. This requires a minimum amount of stored charge Q_1, which is indicated on the figure. In this situation, the effective base thickness is approximately the sum of the normal base thickness plus the length of the drift region. Any additional stored charge, such as Q_2, illustrated also on Fig. 20-9d, will drive the transistor deeper into hard saturation. The voltage drop across the drift region is small, roughly given by Eq. 19-13, and the on-state power dissipation is minimized compared to quasi-saturation.

20-5   SWITCHING CHARACTERISTICS

20-5-1   BJT Turn-on

From the basic description of how the transistor works given in the previous section, we know that to switch the transistor from the off-state to the on-state, charge must be supplied to the transistor so that stored charge distributions similar to those

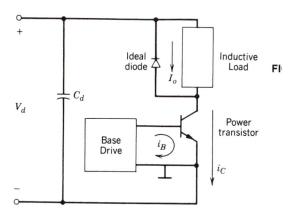

FIGURE 20-10: Inductively loaded BJT switching circuit with a free-wheeling diode clamp. The L/R time constant in the inductive load is large compared with the switching frequency so that it approximates a constant current source I_o. Note that the power BJT is a four-terminal device having two emitter leads — one for the large collector current and the other for the base current.

shown in Fig. 20-9 are established and maintained in the transistor. The characteristics of the transistor and of the external circuit in which the device is embedded interact to determine just how fast the stored charge can be injected and, thus, how fast the device can turn on. To make this interaction as clear as possible, we shall assume that the BJT is embedded in the diode-clamped circuit shown in Fig. 20-10 which arises in the converters discussed in Chapters 5 and 6.

The external circuit determines the collector current that can flow in the on-state. This value of collector current together with the carrier lifetimes in the transistor, particularly in the collector drift region, determines what minimum amount of stored charge must be maintained in the BJT in order that it be on. The current gain of the transistor then establishes what minimum base current must be provided to the device in order to establish and maintain this stored charge distribution. Forward bias base currents in excess of this minimum amount will build up the stored charge distributions faster and thus shorten the switching times from the off-state to the on-state. However, such a base current overdrive will build up the stored charge to values larger than that needed to just maintain hard saturation.

The approximate manner in which the stored charge distribution grows during turn-on is shown in Fig. 20-11 for a forward bias base current applied at $t = 0$. The input voltage that drives the base current, the resulting collector current, and other transistor voltages and currents of interest as functions of time are shown in Fig. 20-12. For an initial time period called the turn-on delay time $t_{d(on)}$, there is no build-up of stored charge because the negative charge on the B-E space charge capacitance must be discharged and the junction forward biased so that carrier injection can commence. During this interval only base current flows and only the base-emitter voltage changes.

After the $t_{d(on)}$ interval, the B-E junction is forward biased and the growth of the stored charge proceeds as diagrammed in Fig. 20-11 and the collector current rises quickly, reaching its on-state value in a time t_{ri}, the current rise time. The voltage v_{CE} is unchanged during this interval because of the diode clamp so that the transistor is still in the active region. After the t_{ri} interval, the collector–emitter voltage falls quickly since the diode no longer can act as a clamp (no forward bias current through the diode). After a short interval labeled t_{fv1}, quasi-saturation is entered as carrier injection into the drift region begins from the C-B junction. During quasi-saturation, the rate of the collector voltage fall slows because of the reduction in beta that accompanies transistor operation in quasi-saturation. Hard saturation commences when the excess carriers have completely swept across the drift region, which occurs after the time interval t_{fv2} indicated on the switching waveforms. The shaded area

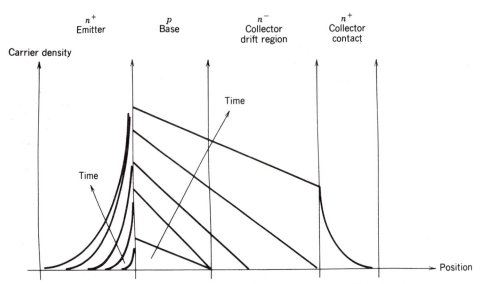

FIGURE 20-11: Growth of the stored charge distribution in a power BJT during the turn-on transient.

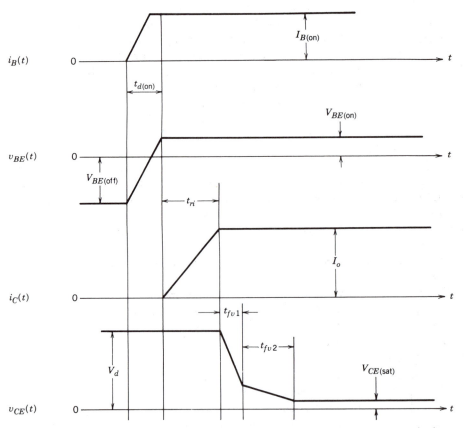

FIGURE 20-12: Power BJT current and voltage waveforms as the transistor turns on in the clamped inductive load circuit of Fig. 20-10.

labeled Q_1 in Fig. 20-9 represents the stored charge that just puts the transistor at the edge of hard saturation. The area labeled Q_2 in Fig. 20-9 represents the excess stored charge that puts the transistor deeper into hard saturation and, in a sense, represents the overdrive of the transistor.

20-5-2   Transistor Turn-off

Turn-off of the transistor involves removing all of the stored charge in the transistor. This could be accomplished by merely reducing the base current to zero and relying on the internal recombination processes in the transistor to remove the charge. However, this would take far too long for practical applications, so the base current is driven negative to speed up the charge removal by carrier sweep-out processes. The process is initiated at $t = 0$, when the base current is either abruptly (step function) or more gradually (ramped with a controlled di_B/dt) changed to negative bias value, as is indicated in Fig. 20-13. The other transistor voltages and currents of interest are shown in the same figure.

For a time interval labeled the storage time t_s in Fig. 20-13, the collector current remains at its on-state value while the excess stored charge Q_2 (refer to Fig. 20-9) is removed. After the t_s interval, quasi-saturation is entered and the voltage begins to rise with a rather shallow slope. When the stored charge distribution is reduced to zero at the C-B end of the drift region after a time interval t_{rv1}, the transistor enters the

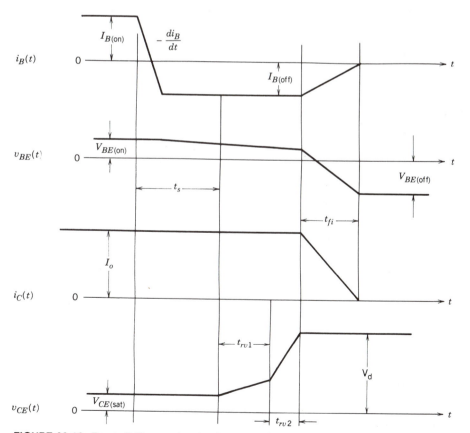

FIGURE 20-13: Power BJT current and voltage waveforms as the transistor turns off in the clamped inductively loaded circuit of Fig. 20-10.

active region. The increased beta of the transistor causes v_{CE} to complete its growth to the power supply voltage with a much steeper slope as the still constant collector current charges up the space charge capacitance of the C-B junction. The growth of v_{CE} ends after the t_{rv2} interval shown in Fig. 20-13, and the collector current begins to fall as current is commutated into the diode clamp. After a time interval t_{fi}, the rest of the stored charge is removed from the transistor and the collector current becomes zero. The BJT now enters cutoff and the B-E space charge capacitance acquires a negative charge as v_{BE} goes negative.

The waveforms shown in Fig. 20-13 are predicated on the base current making a controlled transition from positive to negative values. If a large negative base current with a fast transition is made at $t = 0$, as is shown in Fig. 20-14, there will be significant changes in the collector current response. The t_s interval would be shortened, as would the two v_{CE} time intervals because of the larger negative base current at earlier times in the transient. Significantly more of the stored charge in the base region would be removed compared to the ramped i_B transient, as is shown in the stored charge distributions plotted in Fig. 20-15 for this situation. However, the amount of charge removed from the drift region would not be increased by the same proportion. Most of the drift region charge is removed by the collector current and not by the increased base current. The smaller amount of stored charge left in the base

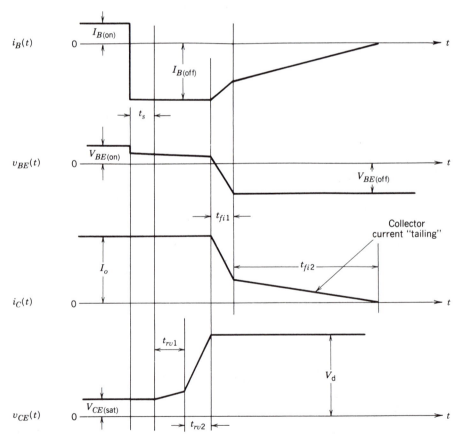

FIGURE 20-14: Power BJT current and voltage waveforms as the transistor turns off in the clamped inductively loaded circuit of Fig. 20-10 with a large reverse base current. Note the long "tail" current that leads to excessive power dissipation.

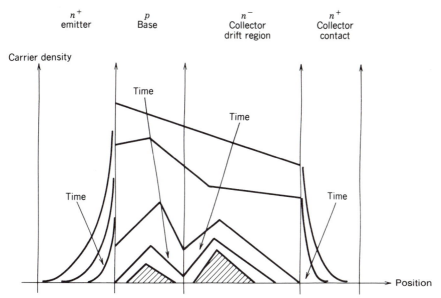

FIGURE 20-15: Decay of the stored charge distributions in a power BJT during a turn-off transient by an abrupt step function reverse base current $I_{B(off)}$. The shaded area represents stored charge remaining in the transistor after the B-E junction cuts off.

means that a shorter time would be required to remove sufficient stored charge in the base so that the B-E emitter junction could become reverse biased with the emitter current going to zero.

If this happens, there will still be stored charge left in the drift region (shown shaded in Fig. 21-15) that must be removed before the collector current can become zero and the BJT can enter cutoff. The only ways for this remaining charge to be removed is via internal recombination and by the negative base current that is flowing. The carrier sweep-out rate in this circumstance will be slow compared to the previous situation because the collector current must equal the negative base current and not beta times larger, as was the previous case. This loss of current gain produces the long tail in the collector current waveform shown in Fig. 20-14 during the time interval labeled t_{fi2}. This long "tailing" time is undesirable because it can lead to increased switching losses.

20-5-3   Switching of Monolithic Darlingtons

The turn-on transient behavior of a monolithic Darlington (MD) embedded in the circuit of Fig. 20-10 will have the same qualitative features as those just described for the single BJT. However, there are two important quantitative differences. First, the main transistor cannot go into hard saturation because the on-state voltage of the driver transistor keeps the voltage across the C-B terminals of the main transistor large enough so it stays in quasi-saturation. This means that the on-state power dissipation of MDs will be larger than those of an otherwise comparable single power BJT. Second, the overall switching time to the on-state will be faster for the MD because the main transistor will be driven by a larger base current than a comparable single BJT. The base current to the main BJT is β_D (beta of the drive BJT) times larger

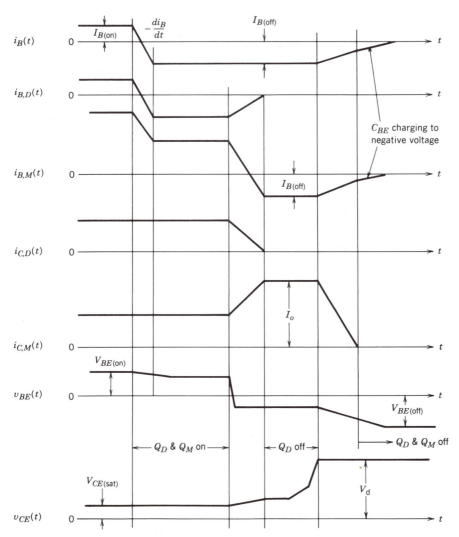

FIGURE 20-16: Current and voltage waveforms in a power Darlington during turn-off in the clamped inductive load circuit of Fig. 20-10.

than what base current would be provided to a single BJT in the same circuit shown in Fig. 20-10.

The most significant differences show up during the turn-off transient, as the waveforms in Fig. 20-16 illustrate. Once the driver transistor turns off, the base current of the main transistor goes negative and its collector current increases, since it must now carry that portion of the load current that the driver BJT had been carrying. The negative base current to the main transistor soon depletes enough stored charge out of the base and collector drift regions that the transistor goes active and completes the traverse of the switching locus to cutoff. As in the case of the conventional power transistor, a controlled rate of change of negative base current should be considered in preference to step function changes. Regardless of which type of turn-off base current drive is selected, the overall turn-off time of the MD will be somewhat longer than that of a conventional power BJT in the same circuit.

If the diode D_1 shown in the circuit of Fig. 20-3 was not present, the turn-off time would be much longer. This would occur because in the circuit of the MD, once the driver transistor cuts off, negative base current for the main transistor would not be able to flow. The only mechanism that would remain active for the removal of the stored charge in the main transistor would be internal recombination, which would take far longer than the removal of the charge by carrier sweepout via the negative base current and collector current.

20-6 BREAKDOWN VOLTAGES

When the BJT is in the blocking state, the C-B junction must withstand the applied voltage. A BJT cannot block the opposite polarity voltage because the B-E junction has a much lower breakdown voltage than the C-B junction because of the very heavy emitter doping used to increase the beta. Typical E-B breakdown voltages are 5 to 20 V.

In designing a transistor to withstand a specified voltage, the doping in the drift region on the collector side of the C-B junction is made much smaller than the base doping. This is done so that the depletion region will be predominantly on the collector side, where there is room for it. As in the case of high-voltage diodes, a lightly doped drift region (the collector drift region) is carefully designed (Eqs. 19-2 and 19-3 apply approximately to this situation) to accommodate the width of depletion layer at the maximum applied voltage without being overly long, which would lead to increased on-state losses. The base thickness must be kept small so that respectable values of beta can be realized. This means that no significant encroachment of the base by the C-B depletion layer can be tolerated. Some encroachment is unavoidable and decreases the effective base thickness and causes an apparent increase in beta. This effect is known as base thickness modulation and shows up as the finite slope in the active region portion of the i_C-v_{CE} curves, such as is shown in Fig. 20-5. If the transistor base has significant encroachment by the C-B depletion layer, then it must be made thicker, which has the undesirable effect of lowering the beta.

However, the principal reason for strictly limiting the C-B depletion layer encroachment into the base is to avoid reach-through. This occurs when the depletion layer from the C-B junction stretches completely through the base layer to E-B junction. If this happens, the enormous number of electrons in the emitter (or holes for a *pnp* BJT) will be drawn from the emitter into the base by the large electric fields in the depletion layer. This will lead to large current flows and to a breakdown-like behavior and attendant large power dissipation. To avoid reach-through, the base thickness must be large enough to accommodate the expected depletion layer encroachment and the doping level in the base must be large enough to keep the encroachment small. The thickness of the base in a power BJT is thus a compromise between being small for large beta and being large to minimize reach-through problems. The compromise leads to larger base thicknesses than are found in logic level transistors and consequently smaller betas, with values of 5 to 10 being typical.

As we indicated in Section 20-3, in the common emitter configuration the breakdown voltage BV_{CEO} is smaller than BV_{CBO}. There is a semi-empirical relationship between these two parameters, which is given by

$$BV_{CEO} = \frac{BV_{CBO}}{\beta^{1/n}} \tag{20-6}$$

where $n = 4$ for npn transistors and $n = 6$ for pnp transistors. The consequence of this relationship is that transistors with high breakdown voltages will have small values of beta. For high-voltage npn transistors where beta is between 10 and 20, the value of BV_{CEO} will be about one-half of BV_{CBO}.

The lowering of BV_{CEO} compared with BV_{CBO} is the result of excess carrier injection into the base from the emitter (note that in the common emitter configuration, the B-E junction is forward biased even with $I_B = 0$ due to the reverse bias current of the C-B junction). These excess carriers effectively increase the reverse bias current (termed I_{CEO} in the common emitter configuration) of the C-B junction over the reverse bias current I_{CBO} of the same junction when the emitter is open. Qualitatively the larger value of I_{CEO} compared with I_{CBO} means that more carriers are crossing the C-B depletion region at any given value of voltage. Consequently, the rate of impact ionization must be larger in the common emitter mode compared with the emitter open mode. A larger rate of impact ionization means that the breakdown voltage will be lower.

20-7   SECOND BREAKDOWN

Bipolar junction transistors and to some degree other types of minority carrier devices have a potential failure mode, usually termed second breakdown. It appears on the output characteristics of the BJT as a precipitous drop in the collector–emitter voltage at large collector currents. As the collector voltage drops, there is often a significant increase in the collector current and a substantial increase in the power dissipation. What makes this situation particularly dangerous for the BJT is that the dissipation is not uniformly spread over the entire volume of the device, but is concentrated in highly localized regions where the local temperature may grow very quickly to unacceptably high values. If this situation is not terminated in a very short time, device destruction results. When devices that have been so destroyed are analyzed, they often show dramatic evidence of the localized power dissipation and attendant heating in the form of melted and then recrystallized silicon.

Second breakdown does not originate from impact ionization and an attendant avalanche breakdown of a pn junction. This is clear from the fact that a drop in voltage accompanies second breakdown, whereas no such drop is observed in avalanche breakdown.

Several intrinsic aspects of the transistor combine to give the BJT its susceptibility to second breakdown. First there is the general propensity of minority carrier devices to thermal runaway when the voltage across them is held approximately constant as the device temperature increases. Minority carrier devices have a negative temperature coefficient of resistivity [the resistivity drops as the temperature increases because the minority carrier densities are proportional to the intrinsic carrier density n_i, which increases exponentially with temperature, see Eq. 18-20]. This means the power dissipation will increase as the resistance drops as long as the voltage remains constant. If the rate of increase in power dissipation with temperature is greater than linear with temperature (the rate of heat removal is linear with temperature, i.e., characterized by a thermal resistance), then an unstable situation will result when the power dissipated exceeds the rate at which heat energy can be removed (a function of the thermal resistance). The situation becomes a classic case of positive feedback in which the power dissipation leads to an increase in temperature, which leads to further increases in power dissipation, and so on until device destruction results. It is often and quite appropriately termed thermal runaway.

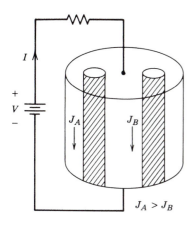

FIGURE 20-17: Semiconductor device with regions of current density nonuniformities that could lead to the formation of current filaments and possible second breakdown.

This potential for thermal runaway is made much more dangerous if the current density in the device is nonuniform across the device cross section. Current filaments where the current density is substantially larger than in surrounding areas may occur and localized thermal runaway becomes likely. Consider the situation illustrated in Fig. 20-17, where the current density J_A in region A is assumed to be greater than the current density J_B in region B. The power dissipation density will be greater in A than in B, which will lead to an increase in the temperature T_A compared to T_B. This in turn will lead to further increases in J_A compared with J_B and the temperature T_A will increase further. If the local temperature T_A exceeds the intrinsic temperature T_i [the temperature at which the intrinsic carrier density n_i equals the majority carrier doping density], then thermal runaway will be in progress in that local region and will lead to intense localized heating and device failure if not terminated very quickly.

The formation of the current filaments and subsequent localized thermal runaway requires only a nonuniformity in the current density and enough localized power dissipation to cause a substantial rise in the temperature of the filament. Indeed, the increase in carrier density in the current filament may often cause a drop in the external voltage across the device if the external resistance in series with the device is appreciable, and yet device destruction may still occur. When the shorting effect of the filament is strong enough to cause this voltage drop, the device is said to be in second breakdown.

The key to avoiding second breakdown would then seem to be (1) keeping the total power dissipation under control and, more important, (2) avoiding any current density nonuniformities, especially during turn-on and turn-off, when the instantaneous power dissipation is largest. However, as we have already seen, the basic construction of the transistor leads to current constrictions via mechanisms such as emitter current crowding. While current crowding can be postponed until specific current levels are reached, once these levels are exceeded, the current constriction can be severe enough to lead to the formation of a current filament and to possible localized thermal runaway.

When device turn-off is initiated, the flow of negative base current induces a lateral voltage drop of the opposite polarity to that described in Fig. 20-8a. This causes a crowding of the emitter current toward the center of the emitter, as is shown in Fig. 20-8b and once again the conditions are favorable for thermal runaway. If, however, the width of the emitter is made smaller, then the lateral voltage drop will be smaller (less resistance for the lateral flowing base current to develop a voltage drop across). This means that the severity of current crowding will be less and the attendant possibility for second breakdown will be smaller. For this reason, power BJTs are constructed with many narrow emitter fingers, as is shown in Fig. 20-2, in parallel rather than a few very large cross-sectional area emitters. Other measures to reduce the possibility of second breakdown include the use of a controlled rate of change of base current during turn-off, the use of protective circuitry such as snubbers and free-wheeling diodes, and the positioning of the switching trajectory within the safe operating area (SOA) boundaries, a topic to be discussed shortly.

20-8   ON-STATE LOSSES

Except at high switching frequencies, nearly all the power dissipated in the switch-mode operation of a bipolar junction transistor occurs when the transistor is in the on-state, usually hard saturation. In this circumstance the power dissipation is given by (ignoring base current losses)

$$P_{on} = I_C V_{CE(sat)} \qquad (20\text{-}7)$$

The collector–emitter saturation voltage $V_{CE(sat)}$ increases with increasing collector current.

Several internal voltage drops in a power transistor contribute to $V_{CE(sat)}$. These voltage drops and some of their origins are indicated schematically on the vertical cross section of a power transistor, as shown in Fig. 20-18. Adding these together yields

$$V_{CE(sat)} = V_{BE(on)} - V_{BC(sat)} + V_d + I_C(R_e + R_c) \qquad (20\text{-}8)$$

The voltages $V_{BE(on)}$ and $V_{BC(sat)}$ are the voltages appearing across the forward biased B-E and B-C collector junctions, respectively. These voltages differ from each other by 0.1 to 0.2 V because the two junctions are significantly different from each other. The C-B junction is much larger in area than the B-E junction and the doping levels are much lower across the B-C junction compared with B-E junction. This voltage difference is relatively independent of collector current.

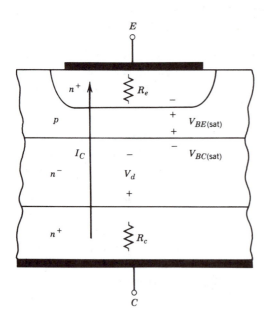

FIGURE 20-18: Vertical cross-section of a power BJT showing the origins of the components of the on-state collector-emitter voltage $V_{CE(sat)}$.

The resistances R_e and R_c represent the ohmic resistance of the heavily doped emitter and collector regions, respectively. At low to moderate collector currents, the voltage drops across these resistances are negligible. But at larger currents, these drops become important and add to the increase in $V_{CE(\text{sat})}$ with collector current.

The major contribution to the increase in $V_{CE(\text{sat})}$ with collector current is the voltage drop V_d across the collector drift region. This voltage can be made reasonably small because of conductivity modulation and relatively independent of collector current, as was described in Chapter 19 for the pn junction power diode, which has an analogous drift region. However, at larger collector currents, the excess carrier densities in the drift region approach large enough values that the carrier lifetime begins to decrease (Auger recombination) as well as the mobility (carrier-carrier scattering). When this occurs, V_d begins to increase significantly, thus increasing $V_{CE(\text{sat})}$. As was true with the diode, the magnitude of V_d is dependent on the excess carrier lifetime (see Eq. 19-13), whose value is a compromise between large values that minimize V_d and shorter values that minimize switching times.

The increase in V_d with collector current will be most severe and commence at lower values of collector current in high-voltage BJTs because of the long drift region that these transistors must have to hold off large C-E voltages in the off-state. This aspect of the transistor behavior is analogous to the behavior of high-voltage pn junction diodes. Based on these observation, we would expect the current capabilities of the BJT to be similar to those of diodes with the same voltage rating.

However, BJTs have a significantly lower current density capability versus breakdown voltage than this optimistic estimate. Mechanisms such as emitter current crowding and conductivity modulation of the base, which lower the value of beta, commence at current densities lower than current densities that diodes can handle. The decrease in beta with increasing collector current means that the base current must be increased at a greater rate than the collector current to maintain the device in hard saturation or, at least, near it so that the voltage drop across the transistor is not too large. The transistor cannot be allowed to enter very far into quasi-saturation at larger collector currents because v_{CE} (and, hence, the power dissipation) increases very rapidly as is shown in Fig. 20-7. Since there is a practical limit to how much base current the user is willing to put into the transistor, there is a practical upper limit to the collector current and, hence, to the current density. Approximately speaking, the transistor is usable to collector currents about 10 times larger than the value at which the current gain peaks and begins to fall with increasing collector current (see Fig. 20-7).

Emitter current crowding and other mechanisms that reduce beta at large collector currents are so significant that the transistor manufacturer designs the transistor structure with a specific value of collector current in mind at which the beta begins to decrease. Once this current level has been determined, the device designer then adjusts the carrier lifetime in the drift region so that the voltage drop across it is kept at the desired levels. In this design approach, the same basic trade-offs between breakdown voltage, on-state losses, and switching speeds described for the power diode are still valid; the only basic change is that they occur at lower current densities.

20-9 SAFE OPERATING AREAS

Safe operating areas or SOAs, a concept described in earlier chapters of this book, are a very convenient and compact method of summarizing maximum values of current and voltage to which the bipolar junction transistor should be subjected. Two

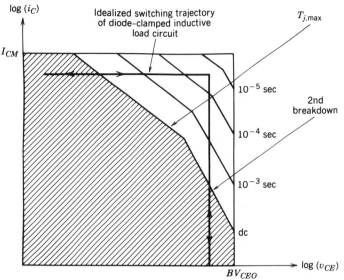

FIGURE 20-19: Forward bias safe operating area FBSOA of a power BJT. The dc FBSOA is shown as shaded and the expansion of the area for pulsed operation of the BJT is shown with shorter switching times leading to a larger FBSOA.

separate SOAs are used in conjunction with BJTs and both are commonly given on specification sheets. The so-called forward bias safe operating area or FBSOA is shown in Fig. 20-19 and the reverse bias safe operating area is shown in Fig. 20-20. The terms forward bias and reverse bias refer to whether the base current bias source forward biases the B-E junction or reverse biases it (which would be appropriate for turning off the BJT).

Several different physical mechanisms are active in determining the boundaries of the FBSOA shown in Fig. 20-19. The current I_{CM} is the maximum collector current even as a pulse that should be applied to the transistor. Exceeding this current may

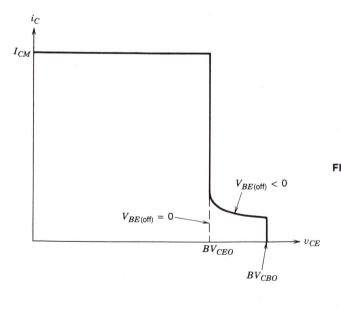

FIGURE 20-20: Reverse bias safe operating area (RBSOA) of a power BJT. Reverse bias refers to the base current being in the opposite direction to the normal on-state direction.

cause bonding wires or metallizations on the wafer to vaporize or otherwise fail. The thermal limit is a power dissipation limit set by the thermal resistance of the transistor and the maximum allowable junction temperature. The second breakdown boundary represents the maximum permissible combinations of voltage and current without getting into the regions of the i_C-v_{CE} plane where second breakdown may occur. The final portion of the boundary of the FBSOA is breakdown voltage limit BV_{CEO}.

If the transistor is operated as a switch, then the boundaries of the FBSOA expand as is indicated in the figure. Crudely speaking, the expansion of the SOA occurs for switch-mode operation because the silicon wafer and its packaging have a thermal capacitance and, hence, an ability to absorb a finite amount of energy without the junction temperature rising to excessive levels. If the transistor turns on in a few microseconds or less, the amount of energy that is absorbed is too small to cause any appreciable rise in the junction temperature and, as a result, the FBSOA is essentially square, being limited only by I_{CM} and BV_{CEO}. The safe operating area is particular useful when the switching trajectory, such as the one for the circuit of Fig. 20-10, is plotted on it as in Fig. 20-19 because such a construction makes it immediately clear if the circuit operation puts the transistor outside of its specification range.

In a similar fashion, the RBSOA shown in Fig. 20-20 is constructed. The area encompassed by the RBSOA, which is a pulsed safe operating area, is somewhat larger than the FBSOA because of the extension of the area to higher voltages than BV_{CEO}, up to BV_{CBO}, at low collector currents. The operation of the transistor up to the higher voltage is possible because the combination of low collector current and reverse base current has made the beta small so that the breakdown voltage rises toward BV_{CBO} as Eq. 20-6 predicts.

20-10 DESIGN OF DRIVE CIRCUITS FOR BJTS

20-10-1 Design Considerations

The design of drive circuits for power transistors is considerably more complicated than for logic level devices for several reasons. First, the low beta of power transistors means that their base currents will be large, sometimes tens of amperes, and consequently logic circuits cannot directly drive power transistors. An intermediate gain stage made up of transistors of moderate power and current capability must be used to provide the large base current needed to drive the high-power device. This leads to significant power dissipation in the drive circuit, which has to be considered as well as the dissipation in the main power transistor. Second, it is mandatory that a negative base current be used in turning off the power BJT because otherwise the turn-off time will be too long and will lead to far too much power dissipation during the turn-off interval. Third, it is sometimes desirable to put several power BJTs in parallel to increase the total current capability of the composite switch, and the drive circuit must ensure that all the parallel BJTs turn on and off simultaneously.

The design of drive circuits for BJTs operating in converters switching inductive load currents is especially demanding. In designing such circuits it is very important to pay attention to circuit layout. Otherwise overvoltages (discussed in Section 20-11) and unwanted oscillations across the BJT may occur.

Consider the simple step-down converter circuit shown in Fig. 20-10 where the drive circuit is shown as a block. A single bar ground (which is the BJT emitter) is used as the voltage reference point for the base drive circuit. The power circuit may be safety grounded (connected to the utility system ground). However if the dc power is supplied from a rectifier connected to the ac line without a transformer, the dc terminals will be alternately connected to the ac line terminals depending on which of the rectifier devices are conducting. This makes safety grounding impossible.

The auxiliary power supplies needed for the base drive circuits must be referred to the emitter reference potential of the power transistor and must be supplied through an isolation transformer. In larger converters, the signal electronics, which include the voltage regulators and pulse width modulation circuits, are grounded to the safety ground for noise reduction and safety considerations. In this situation a transformer or an optocoupler must be used for isolation between the base drive circuit and the logic control input. This will be discussed in more detail shortly.

20-10-2   Base Drive Circuits

A very simple base drive circuit suitable for converters with a single-switch topology is shown in Fig. 20-21a. At turn-on, the *pnp* driver transistor is turned on by saturating one of the internal transistors in the comparator (type LM311 for example). This provides a base current for the main power BJT, which can be calculated by noting in the circuit of Fig. 20-21a

$$V_{BB} = V_{CE(\text{sat})}(T_B) + R_1 I_1 + V_{BE(\text{on})} \tag{20-9}$$

and

$$I_{B(\text{on})} = I_1 - \frac{V_{BE(\text{on})}}{R_2} \tag{20-10}$$

For the specified maximum collector current I_C that the application demands of the transistor, the necessary base current $I_{B(\text{on})}$ and corresponding $V_{BE(\text{on})}$ can be found from the power transistor data sheets. Similarly, the $V_{CE(\text{sat})}$ for the *pnp* transistor in the base drive circuit can be obtained from its data sheets. In selecting R_1, R_2, and V_{BB}, it should be recognized that a small R_2 will allow a faster turn-off but will also cause the power dissipation in the drive circuit to be large. The approximate turn-off waveforms are shown in Fig. 20-21b, where V_{BE} is shown as larger during the on-state compared with the storage interval.

A step-by-step design procedure is shown below.

1. Based on the turn-off speed required, the negative base current $I_{B,\text{storage}}$ during the storage time is estimated. From this, R_2 in Fig. 20-21a can be calculated as

$$R_2 = \frac{V_{BE,\text{storage}}}{I_{B,\text{storage}}} \tag{20-11}$$

2. Since we know the required on-state base current $I_{B(\text{on})}$ and the corresponding $V_{BE(\text{on})}$ and R_2 (from the previous step), I_1 becomes

$$I_1 = I_{B(\text{on})} + \frac{V_{BE(\text{on})}}{R_2} \tag{20-12}$$

3. Two unknowns remain in Eq. 20-9, V_{BB} and R_1. The on-state losses in the drive circuit are approximately equal to $V_{BB}I_1$, which suggests that V_{BB} should be

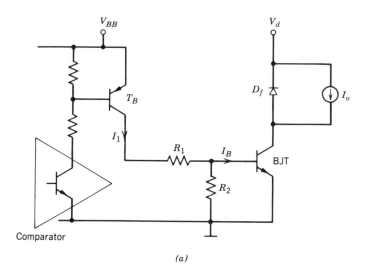

(a)

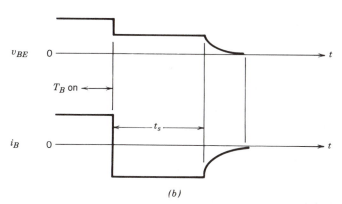

(b)

FIGURE 20-21: (a) Simple base current drive circuit for a power BJT, and (b) the associated current and voltage waveforms at turn-off.

small. On the other hand, to reduce the influence of variations in $V_{BE(\text{on})}$, V_{BB} should be large. In practice, a V_{BB} of about 8 V is optimum. With $V_{BB} = 8$ V, R_1 can then be estimated using Eq. 20-9.

This base drive circuit should not be used in pulse-width-modulated bridge converter circuits for reasons that will be discussed shortly.

A fast turn-off can be provided by the base drive circuit shown in Fig. 20-22, where both a positive and negative voltage supply with respect to the emitter are used. During the turn-on interval of the BJT, the output transistor of the comparator is off, thus turning the transistor T_{B+} on. The on-state base current is

$$I_{B(\text{on})} = \frac{V_{BB+} - V_{CE(\text{sat})}(T_{B+}) - V_{BE(\text{on})}}{R_B} \qquad (20\text{-}13)$$

Arguments similar to step 3 of the previous drive circuit design apply in the selection of V_{BB+} and R_B. The optional capacitor C_{on}, shown as dotted, acts as a speed-up

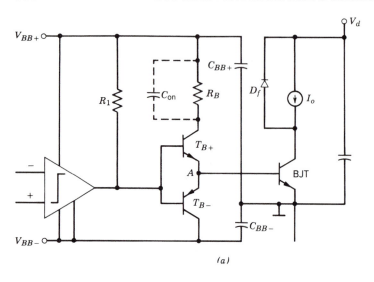

(a)

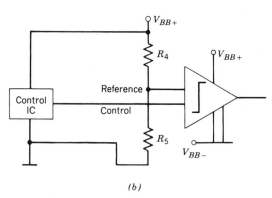

(b)

FIGURE 20-22: (a) Base current drive circuit with both positive and negative voltages with respect to the BJT emitter for faster turn-off of the power device. (b) The BJT can be controlled by logic level circuits.

capacitor by providing a large transient base current to the power transistor at the instant of turn-on to speed up the turn-on sequence.

For turning the BJT off, the internal output transistor of the comparator turns on, thus turning the *pnp* transistor T_{B-} on (and automatically turns the *npn* transistor T_{B+} off). For a fast turn-off, no external resistance is used in series with T_{B-}. The magnitude of the negative voltage V_{BB-} must be less than the B-E breakdown voltage of the BJT, which is given on the data sheets and is normally in the 5 to 7 V range. The switching waveforms will be similar to those described in Section 20-5 if the BJT is used in a similar circuit. If the BJT has a tendency for collector current tailing due to a too rapid turn-off of the B-E junction compared to the C-B junction, as described in Section 20-5, then a resistor or, if necessary, an inductor can be added in the turn-off base drive between points A and the emitter of T_{B-} in Fig. 20-22a.

In applications where the base drive required is small, the totem-pole output of a buffer IC, such as the UC3707, can be used as the driver circuit in Fig. 20-22a. If the control signal is supplied by a logic circuit connected between V_{BB+} and the emitter of the BJT, then the reference input to the comparator should be at the

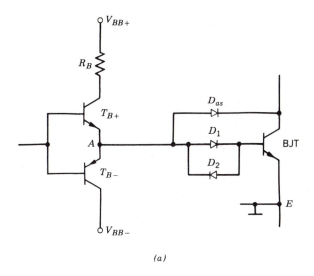

(a)

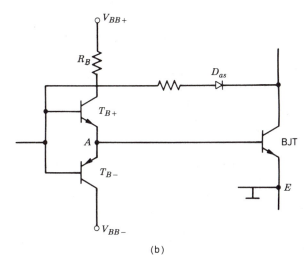

(b)

FIGURE 20-23: (a) Base drive circuit with anti-saturation diodes to minimize the storage time of the BJT and thus the turn-off time. The modifications in (b) permit the antisaturation diode to have a lower current rating compared to the situation in (a).

mid-potential between V_{BB+} and the BJT emitter terminal, as is shown in Fig. 20-22b, where $R_4 = R_5$.

The modifications shown in Fig. 20-23a further enhance the turn-off performance of the drive circuit of Fig. 20-22a. An antisaturation diode D_{as} is added to keep the BJT voltage v_{CE} slightly above its saturation value $V_{CE(sat)}$. This can be seen in Fig. 20-23a, where

$$V_{AE} = V_{BE(on)} + V_{D1} = V_{CE(on)} + V_{Das} \qquad (20\text{-}14)$$

and, therefore,

$$V_{CE(on)} = V_{BE(on)} \qquad (20\text{-}15)$$

since $V_{D1} = V_{Das}$. Since $V_{BE(on)}$ is in general larger than $V_{CE(sat)}$, the presence of the antisaturation diode keeps the transistor slightly out of saturation, thus reducing the storage time at the expense of increased on-state losses in the BJT. Therefore, the antisaturation diode should be used only if the BJT is to be used in a high switching frequency application. If still faster turn-off switching is needed, the on-state voltage $V_{CE(on)}$ can be adjusted by putting one or more diodes in series with D_1.

In the circuit of Fig. 20-23a, the diode D_2 is needed to provide a path for the negative base current. D_{as} should be a fast recovery diode with a reserve recovery time smaller than the storage time of the BJT. Moreover, its reverse voltage rating must be similar to the voltage rating of the power transistor.

An improved version of the circuit of Fig. 20-23a is shown in 20-23b, where the power loss in the positive portion of the base drive circuit is reduced compared to the original circuit. Here the antisaturation diode adjusts the base current of the drive transistor turns T_{B+} such that T_{B+} operates in the active mode and the current drawn from V_{BB+} now is equal only to the actual I_B needed to keep the BJT in quasi-saturation. Moreover, the current rating required of D_{as} is reduced. A small resistance in series with the antisaturation diode can significantly help reduce oscillations at turn-on. Since T_{B+} operates in the active region, it must be mounted on a small heatsink.

20-10-3   Overcurrent Protection

In some applications the potential may exist for currents that exceed the transistor's capabilities. If the BJT is not somehow protected against these overcurrents, it may be destroyed. The BJT cannot be protected against the overcurrents by fuses because they cannot act fast enough. Overcurrents can be detected by measuring the transistor current and comparing it against a limit. At currents above this limit, the power transistor is turned off by a protection network in the base drive circuit.

A cheaper and normally better way of providing overcurrent protection is to monitor the instantaneous collector–emitter on-state voltage. Figure 20-24a shows a simple circuit to provide overcurrent protection based on this principle. The voltage during the on-state at point C will be one forward bias diode drop above $V_{CE(sat)}$. This voltage signal is one of the inputs to the overcurrent protection block that requires the control signal as another input. When the transistor is supposed to be on, if the voltage at point C remains above some predetermined threshold longer than a pre-set duration, the overcurrent is detected and the protection block causes the base drive to turn the BJT off. Depending on the design philosophy, the overall system may be shut down after such an overcurrent detection and may have to be manually reset. The overcurrent detection network can be combined with the antisaturation network, as shown in the subcircuit of Fig. 20-24b.

In the worst case the output of a step-down converter circuit may be accidently shorted, as is shown in Fig. 20-25a. The instantaneous short circuit current through the BJT can be estimated from the static i-v characteristics shown in Fig. 20-25b where v_{CE} equals V_d under the short circuit conditions. If the instantaneous short circuit current is to be limited to a safe value, for example, twice the continuous current rating of the transistor, then the corresponding base current $I_{B, max}$ can be obtained from Fig. 20-25b. If the base current provided by the drive circuit remains less than $I_{B, max}$, the instantaneous short circuit current would also be limited to the required safe value. The overcurrent protection circuit must act within a few microseconds to turn off the BJT; otherwise it will be destroyed.

20-10-4   Circuit Layout Considerations

Besides overcurrent protection there are other practical considerations in the design of base drive circuits. The circuit shown in Fig. 20-26a serves as the focus of

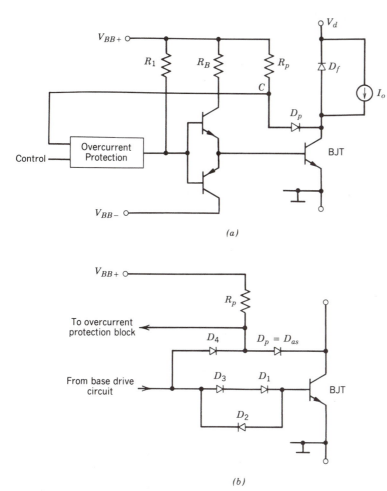

FIGURE 20-24: (*a*) Overcurrent protection by measuring the instantaneous on-state collector-emitter voltage of the power transistor. (*b*) This protection circuit can be used with an antisaturation network.

these considerations. First the length of the conductor that connects the base drive circuit to the emitter of the power transistor should be as short as possible to minimize the stray inductance illustrated in Fig. 20-26*b*. Otherwise the turn-off will be slowed down and unwanted oscillations may occur. Consider a positive base current i_B that turns the BJT on which in turn causes the collector current i_C to increase rapidly. The stray inductance illustrated in Fig. 20-26*b* will induce a voltage that will tend to reduce the base current. If this then causes a reduction in the collector current, there will be a subsequent negative di_C/dt and a voltage induced that will cause an increase in i_B. This then represents the start of unwanted oscillations. In minimizing the stray inductance, all power semiconductor devices, including BJTs, should be treated as four terminal devices having two control terminals and two power terminals (as is illustrated for the BJT in Fig. 20-26*c*). To facilitate the reduction of this stray inductance in high-power transistor modules, manufacturers provide a separate emitter terminal for the connection of the drive circuit.

Stray inductance must also be minimized in the circuit loop of Fig. 20-10 consisting of the power transistor, the free-wheeling diode, and the dc link capacitor

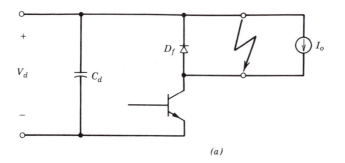

(a)

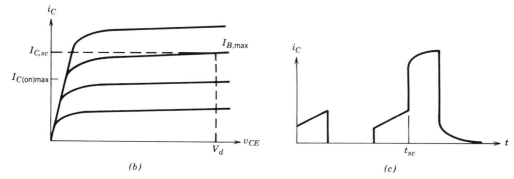

(b)   (c)

FIGURE 20-25: (*a*) Step-down converter circuit with an accidental short circuit. (*b*) The short circuit current can be estimated from the transistor *i-v* characteristics. (*c*) The current waveform at the onset of the short.

C_d. At turn-off of the BJT, large overvoltages may develop across it due to large values of di_C/dt if the stray inductance is not minimized. Even with careful layout, it may be necessary to further reduce the overvoltages at turn-off by means of snubber circuits such as discussed in Section 20-11.

In many designs, the basic drive circuit may be on a printed circuit board at some distance away from the power transistor, which is mounted on a heat sink. A twisted pair of wires or even a shielded cable where the shield is connected to the emitter terminal should be used to minimize the stray inductance and the inductive pick-up of noise in the base drive circuit. A small filter capacitor C_f and damping resistor R_d can be added across the base and emitter terminals as shown in Fig. 20-26*a* to avoid oscillations and the problem of retriggering at the turn-off of the BJT.

The output stages of base drive circuits such as shown in Fig. 20-22*a* should be put close together in a corner of the base drive printed circuit board close to the terminals connecting it to the power transistor. This includes the components R_B, T_{B+}, T_{B-}, C_{BB+}, C_{BB-}, and D_{as}. This will minimize the noise generated in the base drive circuit as well as minimizing the leakage inductances in both the positive and negative base current loops so that the transistor can switch as rapidly as possible.

If more than one base drive circuit is put on the same printed circuit board, they must be put on separately dedicated areas of the board with a minimum distance of at least 1 cm between the areas. This is especially important on double-sided or multiple-layer circuit boards. There must never be an intermixing of the printed wires of the different isolated base drive circuits on any area of the card.

In the rack of printed circuits of a converter, the logic and control electronic circuits should be put on one side and the base drive circuit on the other side. A

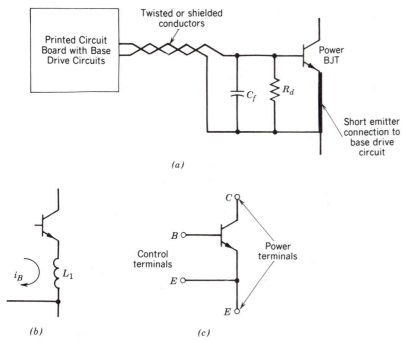

FIGURE 20-26: (*a*) Circuit layout and the interconnection considerations in connecting base drive circuits to power BJTs in order to minimize stray inductance (*b*) and other potential problems. Some BJTs have an extra emitter connection as is shown in (*c*) to help minimize such potential problems.

dummy aluminum "card" used as a shield may be necessary on each side of the base drive circuit card in the rack for further noise reduction.

20-10-5   Electrical Isolation of Base Drive Circuits

Very often, there is a need for electrical isolation between the logic level control signals and the BJT base drive circuits. This is illustrated in Fig. 20-27 for the case of a single phase ac supply where one of the power terminals is a grounded neutral wire. Now the positive dc bus is close to the ground potential during the negative half-cycle of v_s and the negative dc bus is near ground potential during the positive half-cycle of v_s. Under these conditions the emitter terminals of both BJTs must be treated as "hot" with respect to power neutral (safety ground). The logic level control signals are with respect to logic ground, which is at the same potential as the power neutral, since the logic circuits are connected to the neutral by means of a safety ground wire. The basic ways to provide electrical isolation are either by fiber optics, optocouplers, or a transformer.

The optocoupler may be used, as is shown in Fig. 20-28, where isolated positive and negative dc voltage supplies are needed with respect to the emitter terminal of the BJT. The optocoupler consists of a light-emitting diode, the output transistor, and a built-in Schmitt trigger. The optocoupler output can be used as the control input to the comparator in the base drive circuit of Fig. 20-22. The capacitance between the light-emitting diode and the base of the receiving transistor within the optocoupler

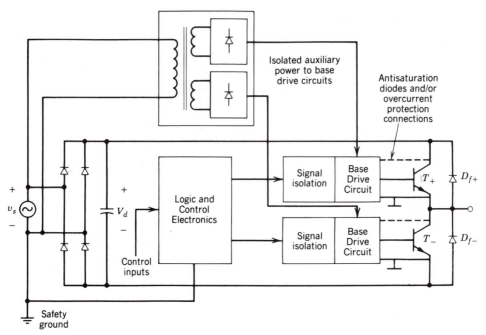

FIGURE 20-27: Power BJT base drive system showing the need for electrical isolation between the base drive circuitry and the logic level control circuitry.

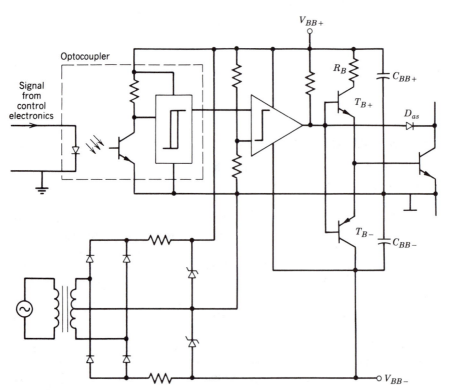

FIGURE 20-28: Optocoupler isolation of base drive circuits.

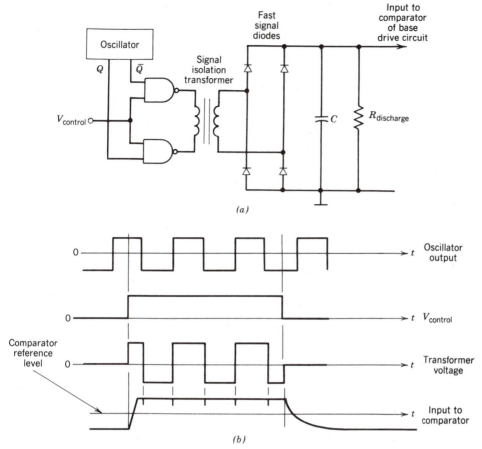

FIGURE 20-29: Base drive circuit using a transformer to couple in the control signals.

should be as small as possible to avoid retriggering at both turn-on and turn-off of the power transistor due to the jump in the potential between the power transistor emitter reference point and the ground of the control electronics. To reduce this problem, optocouplers with electrical shields between the LED and the receiver transistor should be used. As an alternative, fiber optic cables can be used to completely eliminate this problem and to provide very high electrical isolation and creepage distance. When using fiber optic cables, the LED is kept on the printed circuit board of the control electronics and the optical fiber transmits the signal to the receiver transistor that is put on the base drive printed circuit board.

Instead of using an optocoupler, we can modulate the control signal by a high frequency (for example, 1 MHz) oscillator output before applying it to the primary of a high-frequency signal transformer, as is shown in Fig. 20-29a. Since a high frequency transformer can be made quite small, it is easy to avoid stray capacitances between the input and the output windings, and the transformer will be inexpensive. The transformer secondary output is rectified and filtered and then applied to the comparator and the rest of the base drive circuit, which is similar to that in Fig. 20-28. The waveforms for this transformer isolated drive circuit are shown in Fig. 20-29b. After the control voltage goes to zero, the comparator input voltage decays with a time constant equal to CR discharge.

20-10-6   Blanking Times in Bridge Circuits

In bridge circuits, where two power transistors are connected in series in one converter leg, it is important to provide a blanking time so that the turn-on control input to one transistor is delayed with respect to the turn-off control input of the other transistor in the inverter leg. This blanking time should be chosen conservatively to be greater than the worst case maximum storage time of the transistors being used to avoid cross conduction. Under normal operation, such a conservatively chosen blanking time will cause a dead time equal to the blanking time minus the actual storage time to occur in which both the transistors in the inverter leg are off. This dead time introduces an unwanted nonlinearity in the converter transfer characteristic, as we discussed in Chapter 6. This dead time can be minimized by the use of antisaturation diodes that reduce the storage time as well as the variation in storage times of the BJTs.

This blanking time in the control inputs can be introduced by means of the circuit shown in Fig. 20-30a, where the control signal is common to both BJTs of the converter leg. When the control signal is high, the upper transistor T_+ should be on and vice versa. The polarized R-C network and the Schmitt trigger introduce a significant time delay in the turn-on of the BJT and almost no time delay in the turn-off of the transistor. The difference of these two time delays is the blanking time needed. The waveforms are shown in Fig. 20-30b, where the bridge control input goes low and a significant time delay occurs in the control signal to turn-on the bottom transistor T_- and almost no time delay occurs in turning off the upper transistor T_+. The blanking time and the dead time are also shown in Fig. 20-30b.

As we mentioned previously, the simple base drive circuit of Fig. 20-21 cannot be used in a pulse width modulated bridge configuration where one of the freewheeling diodes may conduct when the parallel-connected transistor turns on. The waveforms for such a situation are shown in Fig. 20-31. Assuming I_o to be positive, the freewheeling diode D_{F-} will conduct when T_+ is off and when T_- is controlled to be on. Since T_{B2} is conducting during this interval, there will be a base current i_{B1} for T_-. This will make T_- conduct in its reverse direction, thus acting as a transistor in the reverse active region with the roles of collector and emitter being interchanged. In addition, due to the on-state voltage of D_{F-}, T_- will be provided with a base current i_{B2} supplied through R_2 and flowing through the base-collector diode of T_-. Assuming a D_{F-} forward voltage drop of 2 V and 0.7 V across the forward biased base–collector junction of T_-, then i_{B2} equals $1.3/R_2$.

During the blanking time, both T_{B1} and T_{B2} are off. However, this does not prevent the reverse conduction of T_-, since i_{B2} will flow as long as D_{F-} is conducting. Instead of providing a negative base current, R_2 is a source of positive base current in this situation. The reverse current of T_- will now decay slowly to a steady-state value dictated by i_{B2}. Turning T_{B1} and thus T_+ on will cause a large forward current to flow through the barely saturated T_- in addition to the reverse recovery current of D_{F-}. This current will normally destroy T_- and make it a short circuit and in turn T_+ will be destroyed. Base drive circuits for bridge configurations must therefore be similar to that shown in Fig. 20-22, where the negative base current is supplied from a negative auxiliary power supply. A further improvement is to use an antisaturation diode. Then there will be no base current in the power transistor when its parallel freewheeling diode is conducting, even if the transistor is controlled to be on.

It may seem odd that we describe in detail what should *not* be done. However, in some monolithic Darlingtons, the R_2 resistor of Fig. 20-21 is provided by overlap-

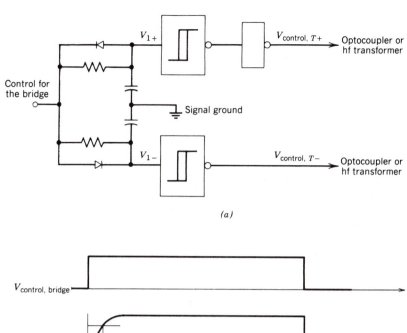

(a)

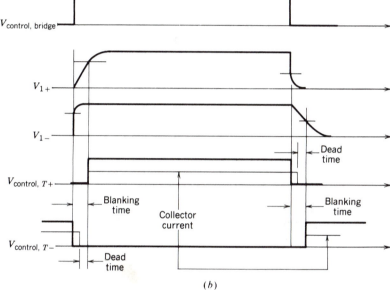

(b)

FIGURE 20-30: *(a)* Circuit for providing blanking times to the base drives of BJTs in a bridge configuration so as to avoid cross-conduction of the BJTs. The dead times are the result of the BJT storage times, which are shown on *(b)* the collector current waveforms.

ping the emitter metallization of the main transistor onto the base region of the main transistor (see Fig. 20-4). Such MDs will conduct reverse currents during the blanking time even with the base connected to V_{BB-}. Hence, for the reasons explained in the preceding paragraphs, MDs constructed in this manner *cannot* be used in bridge configurations when one transistor is turned on while its freewheeling diode is conducting.

20-10-7 Drive Circuits for Bridge Configurations in Switch-Mode DC Power Supplies

In low-power, high-switching-frequency applications where the variation in the duty ratio is limited such as in switch-mode dc power supplies, both the control signal

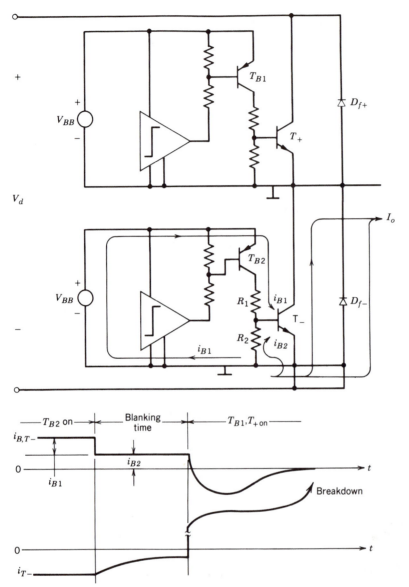

FIGURE 20-31: Reverse conduction of T_- at the turn-on of T_+, which causes destructive breakdown when the drive circuit of Fig. 20-21 is used.

and the base current can be supplied through an isolation transformer. To avoid saturating the transformer, such designs are fairly complicated and are not considered here. Descriptions of such drive circuits can be found in the application notes of power transistor manufacturers.

Another base drive for high-switching-frequency applications where the variations in the duty cycle are limited is shown in Fig. 20-32. This is an isolated base drive in which the base current is made to be proportional to the collector current. Here the need for an auxiliary dc power supply with respect to the emitter terminal is avoided. The transformer is a combination of a flyback converter transformer and a current transformer. When the drive transistor T_1 is on, the BJT is off and vice versa. When T_1 is conducting and the BJT is off, the transformer core is magnetized to the limit of

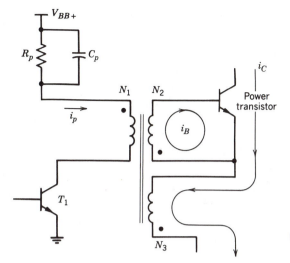

FIGURE 20-32: Proportional "flyback" base drive.

saturation with $i_p = V_{BB+}/R_p$. Because of the stored energy in the slightly gapped transformer core, turning T_1 off forces a current to flow in the secondary winding as in a flyback converter, thus resulting in a positive base current to the BJT. This causes the BJT to start conducting, and its base current is mainly provided by the transformer action between windings 2 and 3, causing $i_B = N_3 i_C/N_2$. During the off interval of T_1, the voltage across the capacitor C_p discharges to zero because of the resistance R_p. Therefore, when T_1 is turned on in order to turn the BJT off, a voltage essentially equal to V_{BB+} is applied across winding 1, causing i_p to be large.

During the turn-off of the BJT, its base current is given as

$$i_B = \frac{N_3 i_C}{N_2} - \frac{N_1 i_p}{N_2} \tag{20-16}$$

The drive circuit must be designed so that this base current during turn-off is negative and of adequate magnitude and duration.

20-10-8 Open-Emitter or Cascode Switching

An attractive alternative to conventional base-emitter drive circuits for switching the power BJT is the so-called open-emitter or cascode switching circuit shown in Fig. 20-33a. The switch in series with the BJT is chosen to be a MOSFET since it can switch very fast, is easy to control, and provides a very low on-state resistance during conduction since the breakdown voltage required for this application is very low, on the order of a few tens of volts. To turn the BJT on, the MOSFET is switched on, which causes the BJT base current to flow, thus turning the transistor on. When conducting, the main current flow is through the BJT and the MOSFET. To turn the BJT off, the MOSFET is quickly turned off, which causes the collector current to flow out of the base terminal through the capacitor, thus making the negative base current equal to the collector current. This negative base current quickly turns off the BJT and the problem of premature cutoff of the B-E junction discussed in Section 20-5 that can occur in conventional base drive circuits and lead to collector current tailing is avoided. The potential of the base terminal is clamped by the zener diode and therefore the MOSFET breakdown voltage is limited.

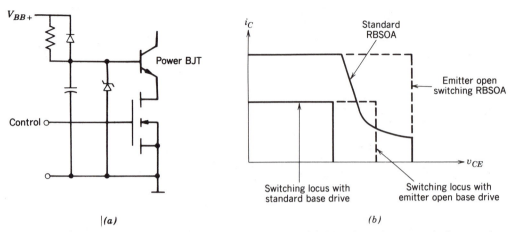

FIGURE 20-33: The emitter-open or cascode switching circuit (*a*) that takes advantage of a larger safe operating area (*b*).

The safe operating area under emitter-open turn-off, which is shown in Fig. 20-33*b*, is much greater than the conventional RBSOA because the v_{CE} limit is BV_{CBO}, which can be as much as twice as large as BV_{CEO}. The larger voltage limit is a consequence of the BJT being used during turn-off as a simple diode composed of the C-B junction. The degradation of the breakdown voltage to BV_{CEO} because of emitter current flow that was discussed in Section 20-6 is absent in this instance because the external circuit does not permit any emitter current to flow in the off-state. Thus, the BJT used in this circuit is chosen on the basis of its BV_{CBO} rating and not its BV_{CEO} rating. This provides for a combined switch that may have lower on-state conduction losses compared with a conventional BJT switch, since the conventional switch would have a larger breakdown rating (the conventional BJT switch would be chosen on the basis of BV_{CEO} exceeding the off-state applied voltage) and, consequently, a much larger on-state voltage drop $V_{CE(\text{sat})}$.

The emitter-open switching circuit can be modified to provide the base current proportional to the collector current by means of a two-winding transformer. This eliminates the need for a dc voltage supply referenced to the BJT emitter. The simple version of the open-emitter drive circuit shown in Fig. 20-33*a* cannot be used in bridge configurations for the same reasons as illustrated in Fig. 20-31.

20-10-9   Parallel Connections of BJTs

Because of high quality control and high volume production, modern transistors with identical part numbers from the same manufacturer can easily be paralleled by directly connecting all their terminals together as shown in Fig. 20-34, provided their physical layout is symmetrical. Equal current sharing is critical when transistors are operating close to their current ratings. The transistor that carries the highest current will have the highest junction temperature because the voltage across each of them is the same. The $V_{BE(\text{on})}$ has a positive temperature coefficient at large I_C, as shown in Fig. 20-34*b*, and the current gain h_{FE} has a negative temperature coefficient at large I_C, as shown in Fig. 20-34*c*. These factors combined have a stabilizing effect on the on-state current sharing by the BJTs.

During turn-on transitions, the transistor starts to conduct only when v_{BE} passes a threshold voltage of about 0.7 V, as described in Section 20-4. Therefore, all parallel transistors will have the same turn-on delay time because of equal values of v_{BE} forced by the parallel connection. The current rise time is small and the forward bias SOA is

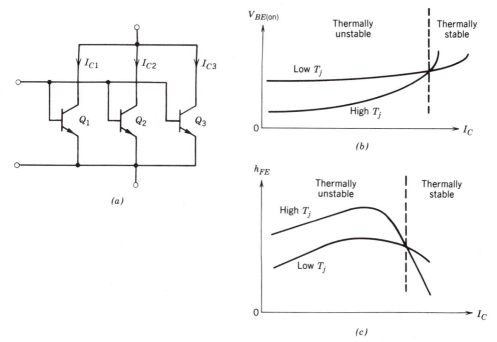

FIGURE 20-34: (a) Parallel connection of BJTs, (b) the variation in $V_{BE(on)}$ with temperature and (c) the variation in beta with temperature that makes the parallel operation practical.

large. Therefore, a minor mismatch in current sharing during turn-on can easily be tolerated.

The turn-off transition in paralleled BJTs is more critical because of large storage times and variations in storage times, which may also be large. At first it appears that the slowest transistor will end up with all the current prior to the voltage rise time. This would cause its destruction because of the limited RBSOA. However, due to charge in the base of the slowest transistor, v_{BE} across it is positive during its storage time. Since v_{BE} is the same for all transistors because of tightly connected terminals, the currents in the faster devices will decrease only slightly toward the end of the storage time of the slowest transistor. This causes the current share of the slowest BJT not to be as high during turn-off switching as anticipated. The current sharing during the storage time can be further improved by the use of antisaturation diodes to minimize the storage times and the spread of storage times, as explained earlier. Some current derating may still be needed for a safe design, for example, five transistors may be paralleled where only four would be needed based on ideal current sharing.

20-10-10   Base Drive for Darlingtons

Figure 20-3 shows a Darlington connection of two discrete BJTs. The turn-off switching waveforms are shown in Fig. 20-16. During its on-state, most of the current is carried by the main transistor M and the rest of the current is carried by the drive transistor D, as is shown in Fig. 20-16. When the Darlington pair turns off, the base current $i_{B,D}$ of the driver transistor goes negative. During the storage time of the transistor D, the base current $i_{B,M}$ of the main transistor decreases only slightly due to i_{CD} still flowing. When the drive transistor turns off, the base current $i_{B,M}$ ($= i_B$) becomes negative and flows through the diode, which is needed to provide a path for

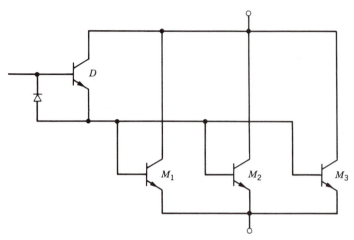

FIGURE 20-35: Discrete Darlington with several paralleled main transistors driven from a single driver BJT.

the negative base current of the main transistor. At the end of the storage time of the main transistor, the voltage v_{CE} across the Darlington connection rises and eventually the current turns off.

Since the driver transistor turns off at zero voltage, its reverse bias safe operating area can be fairly small, allowing its current gain to be high (see Eq. 20-6). Therefore, for the same overall current gain, the current gain of the main transistor can be decreased, possibly improving its reverse bias safe operating area and thus that of the Darlington connection.

The drive circuits for a Darlington connection with two discrete BJTs are identical to those described for single BJTs. For high-power Darlington modules, hybrid ICs are available from the power transistor manufacturers.

Note that, from the load standpoint, the main and driver transistors are in parallel. The total on-state voltage drop across the Darlington connection is given as

$$V_{CE(\text{sat})} = V_{BE(\text{on})M} + V_{CE(\text{sat})D} \qquad (20\text{-}17)$$

In a Darlington connection, several main transistors can be paralleled and driven by a single driver transistor, as shown in Fig. 20-35. The mismatch in the current sharing is minimized because the drive transistor prevents the main transistors from operating in hard saturation, thus improving the current sharing similar to the role of antisaturation diodes in paralleled BJTs discussed earlier.

In a Darlington connection, it may be preferable to use a MOSFET as a driver transistor for ease of control.

20-11 SNUBBER CIRCUITS FOR BJTs AND DARLINGTONS

20-11-1 Need for Snubber Circuits

Snubber circuits are used to protect the transistors by improving their switching trajectory. There are three basic types of snubbers:

1. Turn-off snubbers.

2. Turn-on snubbers.

3. Overvoltage snubbers.

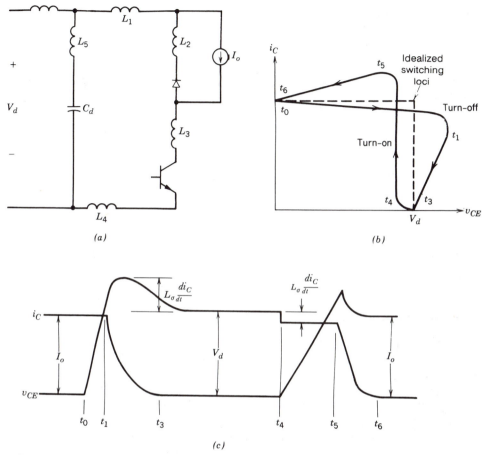

FIGURE 20-36: (a) Step-down converter circuit with stray inductance shown explicitly with (b) associated switching trajectory and (c) the current and voltage waveforms during turn-on and turn-off.

To explain the need for these snubbers, a step-down converter without any snubbers is shown in Fig. 20-36a, where the stray inductances in the various parts of the circuit are shown explicitly. Initially, the transistor is conducting and $i_C = I_o$. During the turn-off switching in Fig. 20-36c, at $t = t_0$, the transistor voltage begins to rise but the currents in the various parts of the circuit remain the same until t_1, when the freewheeling diode begins to conduct. Then the transistor current begins to decrease, and the rate at which it decreases is dictated by the transistor properties and its base drive. The transistor voltage can be expressed as

$$v_{CE} = V_d - L_\sigma \frac{di_C}{dt} \tag{20-18}$$

where $L_\sigma = L_1 + L_2 + \dots$. The presence of stray inductances results in an overvoltage since di_C/dt is negative. At t_3, at the end of the current fall time, the voltage drops to V_d and stays at that value.

During the turn-on transition, the transistor current begins to rise at t_4 at a rate dictated by the transistor properties and the base drive circuit. Equation 20-18 is still valid but due to a positive di_C/dt, the transistor voltage v_{CE} is slightly less than V_d. Because of the reverse recovery current of the freewheeling diode, i_C exceeds I_o. The freewheeling diode recovers at t_5 and the voltage across the BJT decreases to zero at t_6 at a rate dictated by the device properties.

These switching waveforms can be represented by switching loci as shown in Fig. 20-36b. The dotted lines represent idealized switching loci both for turn-on and turn-off, assuming zero stray inductances and no reverse recovery current through the diode. They show that the transistor experiences high stresses at turn-on and turn-off when both its voltage and current are high simultaneously, thus causing a high instantaneous power dissipation. Moreover, the stray inductances result in overvoltage beyond V_d and the diode reverse recovery current causes overcurrent beyond I_o. If necessary, snubber circuits are used to reduce these stresses.

An important assumption that simplifies the snubber analysis is that the transistor current changes linearly in time with a constant di/dt, which is dictated by the transistor and its base drive circuit. Therefore di/dt, which may be different at turn-on and turn-off, is assumed not to be affected by the addition of the snubber circuit. This assumption provides the basis for a simple design procedure for a laboratory prototype. The final design may be somewhat different depending upon what is revealed by laboratory measurements on the prototype circuit.

20-11-2   Turn-off Snubber

To avoid problems at turn-off, the goal of a turn-off snubber is to provide a zero voltage across the transistor while the current turns off. This can be approached by connecting a R-C-D network across the BJT, as is shown in Fig. 20-37a, where the stray inductances are ignored initially for ease of explanation. Prior to turn-off, the transistor current is I_o and the transistor voltage is essentially zero. At turn-off in the presence of this snubber, the transistor current i_C decreases with a constant di/dt and $(I_o - i_C)$ flows into the capacitor through the snubber diode D_s. Therefore, for a current fall time of t_{fi}, the capacitor current can be written as

$$i_{Cs} = I_o t/t_{fi} \qquad 0 < t < t_{fi} \tag{20-19}$$

where i_{Cs} is zero prior to turn-off at $t = 0$. The capacitor voltage, which is the same as the voltage across the transistor when D_s is conducting, can be written as

$$v_{Cs} = v_{CE} = \frac{1}{C_s} \int_0^t i_{Cs} \, dt = \frac{I_o t^2}{2C_s t_{fi}} \tag{20-20}$$

which is valid during the current fall time so long as the capacitor voltage is less than or equal to V_d. The equivalent circuit that represents this condition is shown in Fig. 20-37b.

The voltage and current waveforms are shown in Fig. 20-37c for three values of the snubber capacitance C_s. For a small value of capacitance, the capacitor voltage reaches V_d before the current fall time is over. At that time, the freewheeling diode D_f turns on and clamps the capacitor and the transistor to V_d. i_{Cs} drops to zero because of dv_{Cs}/dt being equal to zero.

The next set of waveforms in Fig. 20-37c are drawn for a value of $C_s = C_{s1}$ that causes the capacitor voltage to reach V_d exactly at the current fall time t_{fi}. C_{s1} can be

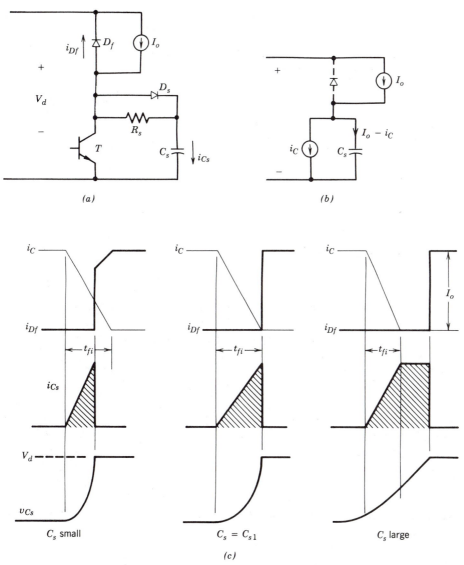

FIGURE 20-37: (a) Turn-off snubber circuit, (b) its equivalent circuit during the transient, and (c) current and voltage waveforms during the turn-off transient. The shaded area represents the charge put on the snubber capacitance during turn-off that will be dissipated in the BJT at the next turn-on.

calculated by substituting $t = t_{fi}$ and $v_{Cs} = V_d$ in Eq. 20-20 and is given as

$$C_{s1} = \frac{I_o t_{fi}}{2V_d} \tag{20-21}$$

For a large snubber capacitance with $C_s > C_{s1}$, the waveforms in Fig. 20-37c show that the transistor voltage rises slowly and takes longer than t_{fi} to reach V_d. Beyond t_{fi}, the capacitor current equals I_o and the capacitor and the transistor voltages rise linearly to V_d. The turn-off switching loci with the three values of C_s used in Fig. 20-37c are shown in Fig. 20-38.

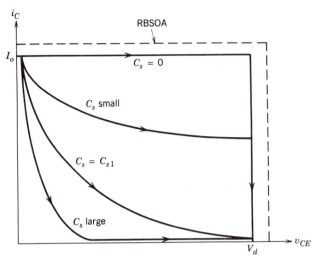

FIGURE 20-38: Switching trajectory during turn-off with various values of snubber capacitance C_s.

To optimize the snubber design it is necessary to consider the transistor turn-on in the presence of the turn-off snubber. To understand the transistor behavior at turn-on, initially it is assumed that the resistor is essentially zero, that is, a pure capacitor without R_s and D_s is used as the turn-off snubber, as is shown in Fig. 20-39a. The presence of C_s causes the turn-on current to increase beyond I_o and the freewheeling diode reverse recovery current. The shaded area in Fig. 20-39a represents the charge on the capacitor that is discharged into the transistor. This charge is equal to the area of one of the shaded areas in Fig. 20-37c depending on the value of C_s used. In the absence of the snubber capacitor C_s, the transistor voltage would have fallen almost instantaneously (since the voltage fall time is usually quite small), as is shown by the dotted line in Fig. 20-39a, and, hence, the energy dissipated in the transistor during turn-on would have been small. The presence of C_s lengthens the voltage fall time so that additional energy is dissipated in the transistor. The additional energy dissipated in the transistor during the capacitor discharge time can be expressed as

$$\Delta W = \int_{t_{ri}+t_{rr}}^{t_2} i_C v_{CE} \, dt = \int_{t_{ri}+t_{rr}}^{t_2} i_{Cs} v_{CE} \, dt + \int_{t_{ri}+t_{rr}}^{t_2} I_o v_{CE} \, dt \qquad (20\text{-}22)$$

The first term in the right-hand side equals the energy stored in the capacitor, which is dissipated in the transistor at turn-on. However, there is additional energy dissipation in the transistor, as is expressed by the second term in Eq. 20-22, which may be larger than the first term. This energy dissipation is due to the lengthening of the voltage fall time brought about by the presence of C_s.

The transistor turn-on waveforms in the presence of the snubber resistance R_s is shown in Fig. 20-39b. Here, unlike the pure capacitively snubbed transistor, the voltage can be assumed to fall almost instantaneously and, therefore, no additional energy dissipation due to the snubber occurs in the transistor at turn-on. The capacitor energy, which is dissipated in the snubber resistor, is given by

$$W_R = \frac{C_s V_d^2}{2} \qquad (20\text{-}23)$$

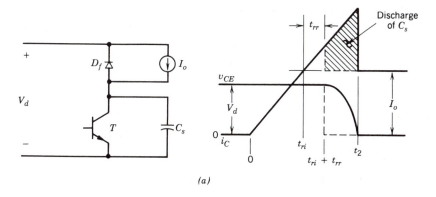

(a)

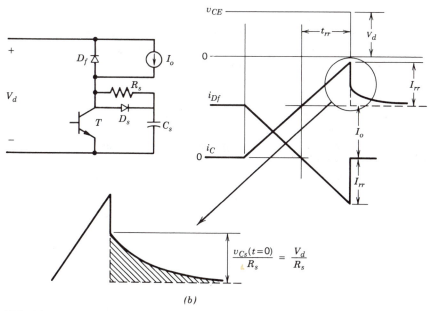

(b)

FIGURE 20-39: Effect of the turn-off snubber capacitance C_s on the (a) turn-on transient without a snubber resistance R_s and (b) with the resistance.

In Fig. 20-39b, the snubber resistance should be chosen so that the peak current through it is less than the reverse recovery current I_{rr} of the freewheeling diode, that is,

$$\frac{V_d}{R_s} < I_{rr} \tag{20-24}$$

The circuit designer usually attempts to limit I_{rr} to $0.2I_o$ or less, so that Eq. 20-24 becomes approximately

$$\frac{V_d}{R_s} = 0.2I_o \tag{20-25}$$

Based on these assumptions, the comparisons of Fig. 20-39a and 20-39b indicate that including the resistance R_s has the following beneficial effects during the transistor turn-on:

1. All the capacitor energy is dissipated in the resistor, which is easier to cool than the transistor.

2. No additional energy dissipation occurs in the transistor due to the turn-off snubber.

3. The peak current that the transistor must conduct is not increased due to the turn-off snubber.

As an aid in choosing the appropriate value of C_s, the energy dissipated in the transistor during turn-off and the energy dissipated in the snubber resistance R_s during turn-on are plotted as functions of C_s in Fig. 20-40. Based on the previous assumptions, these plots are independent of R_s, and there is no additional energy dissipation in the transistor during turn-on because of the presence of the turn-off snubber. C_s should be chosen based on (1) keeping the turn-off switching locus within the reverse bias safe operating area, (2) reducing the transistor losses based on its cooling considerations, and (3) keeping the sum (shown as the dotted line in Fig. 20-40) of transistor turn-off energy dissipation and snubber resistance energy dissipation low.

Having made initial selections of R_s based on Eq. 20-25 and C_s based on the design trade-offs just discussed, the designer must ensure that the capacitor has sufficient time to discharge down to a low voltage, say $0.1 V_d$, during the minimum on-state time of the transistor so that the turn-off snubber will be effective at the next turn-off interval. During the on-state of the transistor, the capacitor discharges with a time constant $\tau_c = R_s C_s$ and

$$v_{Cs} = V_d e^{-t/\tau_c} \tag{20-26}$$

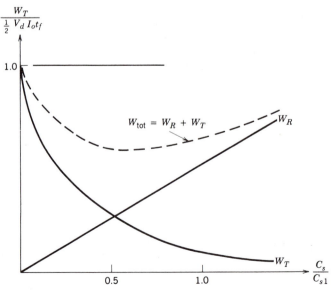

FIGURE 20-40: Turn-off energy dissipation in the BJT and the snubber resistance as a function of the snubber capacitance C_s.

and, therefore, discharging v_{Cs} down to $0.1\ V_d$ requires a time interval of $2.3\ \tau_c$ and, thus,

$$t_{\text{on-state}} > 2.3 R_s C_s \qquad (20\text{-}27)$$

As an example, if we choose $C_s = C_{s1}$ (given in Eq. 20-21) and R_s using Eq. 20-25, then the minimum on-state time of the transistor must be six times the transistor current fall time t_{fi}.

20-11-3 Overvoltage Snubber

In describing the turn-off snubber, the stray inductances were neglected and, hence, there was no overvoltage. The overvoltages at turn-off due to stray inductances such as shown in Fig. 20-36a can be minimized by means of the overvoltage snubber circuit shown in Fig. 20-41a, assuming it is possible to lump all the stray inductances together as indicated in Eq. 20-18. The operation of the overvoltage snubber can be described as follows.

Initially the transistor is conducting and the voltage $v_{C_{ov}}$ across the overvoltage snubber capacitor equals V_d. At turn-off assuming the BJT current fall time to be small, the current through L_σ is essentially I_o when the transistor current decreases to

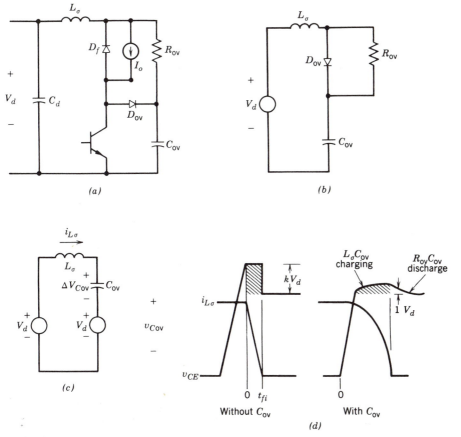

FIGURE 20-41: (a) Overvoltage snubber and (b,c) its equivalent circuit during transistor turn-off. (d) The collector-emitter voltage with and without the snubber.

zero and the output current then freewheels through the freewheeling diode D_f. At this stage the equivalent circuit is as shown in Fig. 20-41b, where the D_f, I_o combination appears as a short circuit and the transistor is an open circuit. Now the energy stored in the stray inductance gets transferred to the overvoltage capacitor through the diode D_{ov} and the overvoltage ΔV_{CE} across the transistor (noting that, in this state, the capacitor C_{ov} and the transistor have the same voltage) can be obtained by replacing the precharged capacitor with its equivalent circuit as shown in Fig. 20-41c. By using energy considerations and noting that $\Delta V_{Cov} = \Delta V_{CE}$, we obtain

$$\frac{C_{ov}\,\Delta V_{CE,\,max}^2}{2} = \frac{L_\sigma I_o^2}{2} \tag{20-28}$$

This equation shows that a large value of C_{ov} will minimize the overvoltage $\Delta V_{CE,\,max}$. Once the current through L_σ has decreased to zero, it can reverse its direction due to the resistor R_{ov}, and the overvoltage on the capacitor decreases to V_d through the resistor R_{ov}. The capacitor discharge time constant $R_{ov}C_{ov}$ should be small enough so that the capacitor voltage has decayed approximately to V_d prior to the next turn-off of the transistor.

To aid in the estimation of the proper value of C_{ov}, the circuit waveforms with and without the overvoltage snubber are shown in Fig. 20-41d. The observed overvoltage of kV_d without the overvoltage snubber is used to estimate L_σ as

$$kV_d = \frac{L_\sigma I_o}{t_{fi}} \tag{20-29}$$

If an overvoltage, for example $\Delta V_{CE,\,max} = 0.1\,V_d$, is acceptable, then using this in Eq. 20-28 and substituting for L_σ from Eq. 20-29 yields

$$C_{ov} = \frac{100kI_o t_{fi}}{V_d} \qquad (\Delta V_{CE,\,max} = 0.1V_d) \tag{20-30}$$

In terms of C_{s1} given by Eq. 20-21, C_{ov} from Eq. 20-30 can be rewritten as

$$C_{ov} = 200kC_{s1} \tag{20-31}$$

which shows that substantially large capacitance is needed for overvoltage protection compared with the values used in the turn-off snubber, which are on the order of C_{s1}. It can be shown that even with a large value of C_{ov}, the energy dissipated in R_{ov} is of the same order as the energy dissipated in the resistor of the turn-off snubber.

Both the turn-off and overvoltage protection snubbers can be used simultaneously.

20-11-4   Turn-on Snubber

Because of the large FBSOA of transistors, turn-on snubbers are used only to reduce turn-on switching losses at high switching frequencies. Turn-on snubbers work by reducing the voltage across the switch as the current builds up. A turn-on snubber can be in series with the transistor as in Fig. 20-42a or in series with the freewheeling diode as in Fig. 20-42b. In both circuits the turn-on and turn-off switching waveforms across the transistor and freewheeling diode are identical. The reduction in the voltage

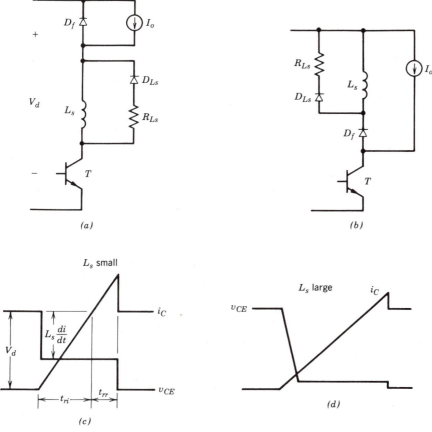

FIGURE 20-42: Turn-on snubber circuit (a) in series with the BJT or (b) in series with the free-wheeling diode. (c) The transistor voltage and current waveforms for small values of L_s and (d) for large values of L_s.

across the transistor during turn-on is due to the voltage drop across L_s. This reduction is given by

$$\Delta V_{CE} = \frac{L_s I_o}{t_{ri}} \tag{20-32}$$

where t_{ri} is the current rise time, as is shown in Fig. 20-42c for small values of L_s. For such small values, di/dt is dictated only by the transistor and its base drive circuit and is assumed to be the same as without the turn-on snubber. Therefore, the diode peak reverse recovery current is also the same as without the turn-on snubber.

If it is important to reduce the diode peak reverse recovery current, this can be achieved with a large value of L_s, as shown by the waveforms in Fig. 20-42d. Here the di/dt of the current is controlled by L_s and the voltage across the transistor is almost zero during the current rise time.

During the on-state of the transistor, L_s conducts I_o. When the transistor turns off, the energy stored in the snubber inductor, $L_s I_o^2/2$, will be dissipated in the snubber resistor R_{Ls}. The snubber time constant is $\tau_L = L_s/R_{Ls}$. In selecting R_{Ls} the following two factors must be considered. First, during transistor turn-off, this turn-on

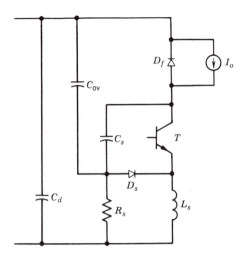

FIGURE 20-43: A modified circuit with an overvoltage snubber, a turn-on snubber, and a turn-off snubber; the Undeland-snubber for step-down converters.

snubber will generate an overvoltage across the transistor given by

$$\Delta V_{CE,\,max} = R_{Ls}I_o \qquad (20\text{-}33)$$

Second, during the off-state the inductor current must decay to a low value, for example, $0.1I_o$, so that the snubber can be effective during the next turn-on. Therefore, the minimum interval for the off-state of the BJT should be

$$t_{off\text{-}state} > 2.3\frac{L_s}{R_{Ls}} \qquad (20\text{-}34)$$

Thus, a large inductance will result in lower turn-on voltages and lower turn-on losses. But it will cause overvoltages during turn-off, lengthen the minimum required off-state interval, and result in higher losses in the snubber. Therefore, L_s and R_s must be selected based on the foregoing design trade-offs following a procedure similar to that described for the turn-off snubber. Since the turn-on snubber inductance must carry the load current, which makes this snubber expensive, it is seldom used.

It is possible to use all three snubbers simultaneously or in any other combination. A snubber configuration with a reduced component count is shown in Fig. 20-43.

20-11-5   Snubbers for Bridge Configurations

As we discussed in Chapter 6, in pulse-width-modulated switch-mode converters with half- or full-bridge configurations such as in motor drives and uninterruptible power supplies, the load current can be treated as a constant I_o over the switching cycle. In Fig. 20-44a, I_o is shown to be going into the converter leg, although it could be in the opposite direction as well. The turn-off snubber shown in Fig. 20-44a, which was shown to be effective in the step-down converter circuit, should *not* be used without including a turn-on snubber. With the direction of I_o as indicated, the diode D_{f+} is conducting during the off-state of T_- and the voltage across C_{s+} is zero. When T_- is turned on, there will be a capacitive discharge current through T_- at the recovery of D_{f+}, as shown by the current loop in Fig. 20-44a. This will result in additional turn-on losses in T_-, as was explained in the discussion of the step-down converter circuit of Fig. 20-39a with a turn-off snubber consisting of only C_s. T_+ will experience an identical problem at turn-on.

An R-C snubber of the type shown in Fig. 20-44b will also suffer from the same drawbacks as the snubber of Fig. 20-44a, if the same degree of improvement in the

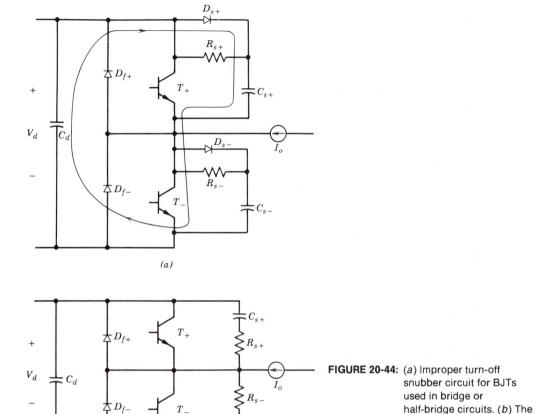

FIGURE 20-44: (a) Improper turn-off snubber circuit for BJTs used in bridge or half-bridge circuits. (b) The turn-off snubber circuit is satisfactory if the goal is to reduce *dv/dt* at turn-off to minimize EMI problems.

turn-off performance is required. However, the R-C snubber of the type in Fig. 20-44*b* with small capacitance values is currently used in bridge configurations if the primary goal is to reduce *dv/dt* to reduce EMI problems.

The turn-on snubber of the type discussed previously, which will protect both the transistor and the freewheeling diode, can be used in the bridge configuration, as is shown in Fig. 20-45*a*. Here the R-C-D turn-off snubber can also be used. The reason why can be seen by looking at the current path when T_- is turned on and D_{f+} recovers. This path includes the turn-on snubber inductance, thus reducing the problems compared to using such a turn-off snubber without a turn-on snubber. The two turn-off snubbers of Fig. 20-45*a* can be combined into one turn-off snubber, as is shown in Fig. 20-45*b*, which will protect the two transistors and the freewheeling diodes at turn-on in the same manner as the separate snubbers of Fig. 20-45*a*.

In the circuits of Fig. 20-45, it is easy to implement an overvoltage protection by connecting a capacitor C_{ov} as shown, and R_{Ls} also serves as R_{ov} of the overvoltage protection capacitor. This overvoltage snubber protects both the upper and lower transistors and the freewheeling diodes. Also, the turn-off snubber capacitors for both the transistors are combined into a single capacitor, which will halve the losses at turn-on compared with those in the circuit of Fig. 20-45*a*. All the snubber losses in the circuit of Fig. 20-45*b* occur in only one resistor, which can be replaced by a dc–dc converter for loss recovery. The circuit of Fig. 20-45*b* is analyzed in Reference 11.

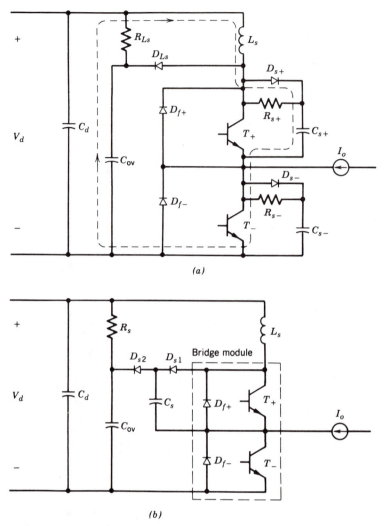

FIGURE 20-45: (*a*) Bridge circuit with both turn-on and turn-off snubbers.
(*b*) A modified arrangement; the Undeland-snubber for
bridge configurations.

20-12 SUMMARY

This chapter has explored the structure and operating characteristics of bipolar junction transistors intended for power switch-mode applications. Several types of generic drive circuits for the transistor and snubber circuits to protect the BJT were examined. The important conclusions are listed:

1. The power BJT has a vertically oriented structure with a highly interdigitated base–emitter structure and a lightly doped collector drift region.

2. The drift region determines the blocking voltage rating of the BJT and also causes the so-called quasi-saturation region of the *i-v* characteristics.

3. The BJT is a normally-off device that is turned on by the application of a sufficiently large base current to cause injection of large numbers of minority carriers into the base from the emitter region. The subsequent diffusion of these carriers across the base to the collector forms the collector current.

4. Power BJTs have low current gain, especially at larger breakdown voltage ratings. This has led to the development of monolithic Darlington transistors, which have larger current gains.

5. Lateral current flow in the base is the basic limiting factor in BJT performance. It causes lateral voltage drops, which lead to emitter current crowding, which in turn causes decreases in current gain. If the current crowding is excessive, second breakdown and device destruction will occur.

6. Heavy conductivity modulation of the drift region in order to minimize on-state losses requires large carrier lifetimes. But this leads to long turn-off times, so a trade-off must be made in the design of the BJT between lower on-state losses or shorter switching times.

7. Turn-off of some types of BJTs should be done with a controlled rate of change of negative base current in order to avoid isolating excessive stored charge in the BJT, which would result in excessively long turn-off times and large power dissipation.

8. Drive circuits for the BJT require careful design to meet the demanding requirements for large positive and negative base currents for short switching times along with the requirements for electrical isolation and interfacing with logic level control circuits.

9. The safe operating areas of the BJT are limited by second breakdown. The RBSOA is normally the limiting factor.

10. BJTs with limited safe operating areas may require that their switching trajectory be controlled with snubber circuits during both turn-on and turn-off.

PROBLEMS

20-1. Plot BV_{CEO} as a function of beta (β) for $5 < \beta < 100$ for identical *npn* and *pnp* silicon transistors. Assume that both BJTs have the same value of BV_{CBO}.

20-2. When emitter-open switching is used to turn off a power BJT, the BJT is less susceptible to second breakdown compared with the normal turn-off situation, where negative base current flows while emitter current is still flowing. Qualitatively explain, with the aid of diagrams, why this is true.

20-3. Consider the step-down converter circuit of Fig. 20-39a with a purely capacitive snubber having $C_s = C_{s1}$. Assume $I_o = 25$ A, an input voltage $V_d = 200$ V, and a reverse recovery time $t_{rr} = 0.2$ μs for the free-wheeling diode.
(a) Calculate the reduction in the turn-off losses in the BJT due to the use of the snubber capacitor. Assume a switching frequency of 20 kHz and a current fall time t_{fi} of 0.4 μs.
(b) Calculate the increase in BJT losses during turn-on due to C_s. Assume that during the turn-on $di_C/dt = 50$ A/μs.

20-4. Repeat Problem 20-3 but now use a polarized R_s-C_s snubber such as is shown in Fig. 20-37a.

20-5. Consider the step-down converter circuit of Fig. 20-39a without any snubber circuit. The free-wheeling diode is ideal and the power transistor has the following parameters: $\beta = 10$, $V_{CE(on)} = 2$ V, $R_{\theta j - a} = 1°$C/W, $T_{j,max} = 150$C, $t_{ri} = t_{fi} = 200$ ns, $t_{fv1} = t_{fv2} = t_{rv1} = t_{rv2} = 50$ ns, and $t_{d(on)} = t_{d(off)} = 100$ ns. The BJT is driven by a square wave (50% duty cycle) of variable frequency. Assume $I_o = 40$ A and $V_d = 100$ V.
(a) Sketch and dimension the average power dissipated in the transistor versus the switching frequency.
(b) Estimate the maximum permissible switching frequency.

20-6. The transistor in the circuit of Problem 20-5 is driven by a 25-kHz square wave (50% duty cycle). The switching times increase by 40% as the junction temperature increases by 100°C (from 25°C to 125°C). If the circuit is operating in an ambient temperature of

$50°C$, estimate the allowable range of values of the thermal resistance $R_{\theta j - a}$ that will keep the junction temperature T_j less than $110°C$.

20-7. A transistor similar to that shown in Fig. 20-1 has an effective emitter area A of 1 cm^2 and a base width of 3 μs. At approximately what value of collector current does the beta of the device begin to drop as the current is increased? *Hint*: Recall from the discussion in Section 20-4-1 that beta begins to fall when high-level injection conditions are obtained in the base. Also recall that $I_C \approx -I_{ne} = qD_n A \, dn_b(x)/dx$.

20-8. Consider a bipolar junction transistor and a *pn* junction diode each having the same cross-sectional area and same drift region length (and thus the same carrier lifetimes and blocking voltage capabilities). Which device can carry the larger forward current and why?

20-9. A BJT similar to that shown in Fig. 20-1 has been designed by a novice device designer. The voltage rating of the device is supposed to be 1000 V. The drift region doping is as indicated in the figure and the drift region length is 100 μm. But the base doping density is 10^{15} cm^{-3} and the base width is 3 μm. What is the actual voltage rating of this device? Assume all junctions are step junctions.

20-13 REFERENCES

1. Sorab K. Ghandhi, *Semiconductor Power Devices*, John Wiley & Sons, New York, 1987, Chapter 4.

2. P. L. Hower, "Bipolar Transistors," *Semiconductor Devices for Power Conditioning*, Roland Sittig and P. Roggwiller (Eds.), Plenum Press, New York, 1982.

3. Adolf Blicher, *Field Effect and Bipolar Power Transistor Physics*, Academic Press, New York, 1981, Chapters 6–10.

4. A. S. Grove, *Physics and Technology of Semiconductor Devices*, John Wiley & Sons, New York, 1967, Chapter 7.

5. Adel S. Sedra and Kenneth C. Smith, *Microelectronics Circuits*, 2nd Ed., Holt, Rinehart, and Winston, New York, 1987, Chapter 5.

6. M. H. Rashid, *Power Electronics: Circuits, Devices, and Applications*, Prentice-Hall, Englewood Cliffs, NJ, 1988, Chapter 15.

7. *Power Transistor in Its Environment*, Thompson-CSF, Semiconductor Division, 1978.

8. B. W. Williams, *Power Electronics, Devices, Drivers, and Applications*, John Wiley & Sons, New York, 1987, Chapters 3, 4, 7, 9.

9. B. Jayant Baliga and Dan Y. Chen, Editors, *Power Transistors, Device Design and Applications*, IEEE Press, Institute of Electrical and Electronic Engineers, New York, 1984, Part I, *Power Bipolar Transistors*, pp. 19–122.

10. Michael S. Adler, King W. Owyang, B. Jayant Baliga, and Richard A. Kokosa, "The Evolution of Power Device Technology," *IEEE Transactions on Electron Devices*, Vol. ED-31, No. 11, pp. 1570–1591, Nov. 1984.

11. T. M. Undeland, F. Jenset, A. Steinbakk, T. Rogne, and M. Hernes, "A Snubber Configuration for Both Power Transistor and GTO PWM Inverters," *Proc. of 1984 Power Electronics Specialists' Conference*, pp. 42–53.

21

Power MOSFETs

21-1 INTRODUCTION

MOSFETs (metal-oxide-semiconductor field effect transistors) with appreciable on-state current-carrying capability and off-state blocking voltage capability and, thus, potential for power electronic applications have been available since the early 1980's. They have become as widely used as power BJTs and in fact are replacing BJTs in many applications, especially those where high switching speeds are important. MOSFETs operate on different physical mechanisms than BJTs, and a clear understanding of these differences is essential for the effective utilization of both BJTs and MOSFETs. This chapter considers the basic physical mechanisms that govern the operation of MOSFETs, the factors that establish the current and voltage limits of the MOSFET and possible failure modes if these limits are exceeded, and finally various circuits for driving the MOSFET and for protecting it against overvoltages and overcurrents.

21-2 BASIC STRUCTURE

A power MOSFET has the vertically oriented four-layer structure of alternating p-type and n-type doping shown in Fig. 21-1a for a single cell of the many paralleled cells of a complete device. The $n^+pn^-n^+$ structure is termed an enhancement mode n-channel MOSFET (for reasons that will become apparent shortly). A structure with the opposite doping profile can also be fabricated and is termed a p-channel MOSFET. The doping in the two n^+ end layers, labeled source and drain in Fig 21-1, is approximately the same in both layers and is quite large, typically 10^{19} cm^{-3}. The p-type middle layer is usually termed the body and is the region where the channel (to be discussed in the next section) is established between source and drain and is typically doped at 10^{16} cm^{-3}. The n^- layer is the drain drift region and is typically doped at 10^{14} to 10^{15} cm^{-3}. This drift region determines the breakdown voltage of the

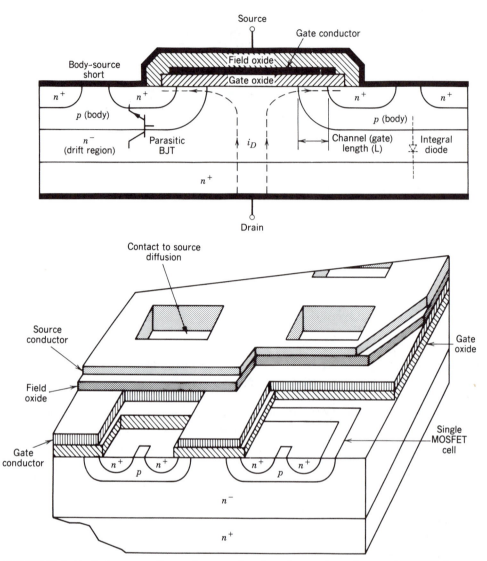

FIGURE 21-1: Vertical cross section and perspective view of an *n*-channel power MOSFET. Some of the layers in the perspective view have been cut away to enhance the clarity of the drawing.

At first glance, it would appear that there is no way that current can flow between the drain and source terminals of the device because one of the *pn* junctions (either the body–source junction or the drain–body junction) will be reverse biased by either polarity of applied voltage between the drain and source. There can be no injection of minority carriers into the body region via the gate terminal because the gate is isolated from the body by a layer of silicon dioxide [usually termed the gate oxide and typically about 1000 Å (angstroms) thick], which is a very good insulator and, hence, there is no bipolar junction transistor operation. However, an application of a voltage which biases the gate positive with respect to the source will convert the silicon surface beneath the gate oxide into an *n*-type layer or channel, thus connecting the source to the drain and allowing the flow of appreciable currents. The thickness of the gate oxide, the width of the gate (as diagrammed in Fig. 21-1) and the number of gate/source regions connected electrically in parallel are important in determining how much current will flow for a given gate-to-source voltage. The mechanisms that lead to the creation of the channel will be discussed later in this chapter.

The structure shown in Fig. 21-1 is usually termed VDMOS, meaning vertical diffused MOSFET. The name crudely describes the fabrication sequence of the device. The starting substrate is usually the n^+ drain onto which the n^- drift region of specified thickness is grown epitaxially. Then the p-type body region is diffused into the wafer from the source side of the wafer, followed by the n^+ source diffusion. These two diffusions are masked diffusions, meaning that portions of the wafer are protected by silicon dioxide so that the dopants cannot reach the wafer where the SiO_2 has been left. The remaining steps involve the deposition of the gate and source metallization and final packaging steps.

Several other aspects of the MOSFET structure shown in Fig. 21-1 should be noted. First, the source is constructed of many small polygon-shaped areas that are connected in parallel and surrounded by the gate region. The geometric shape of the source regions to some degree influences the on-state resistance of the MOSFET, and some manufacturers even advertise their particular line of MOSFET devices by the

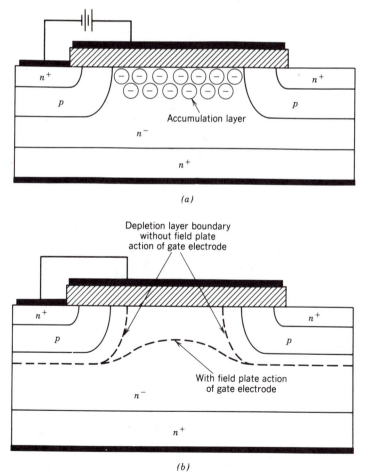

FIGURE 21-2: Gate electrode overlapping the drain drift region (a) to create an accumulation in the on-state, and (b) to act as a field plate in the off-state.

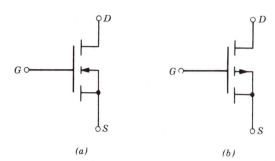

FIGURE 21-3: Circuit symbols for (a) a n-channel and (b) a p-channel MOSFET.

(a)   (b)

shape of the source region (i.e., International Rectifier's HEXFET). The basic reason for the many small source regions is to maximize the width (the lateral dimension perpendicular to the direction of drain current flow) of the gate region compared to its length (the channel length). The gate width W of the MOSFET is the peripheral length of each cell times the number of cells that make up the device. A very large gate width-to-length ratio is desired because this maximizes the gain of the device.

Second, there is a parasitic *npn* BJT between the source and drain contacts as shown in Fig. 21-1 with the *p*-type body region serving as the base of the parasitic BJT. To minimize the possibility that this transistor is ever turned on, the *p*-type body region is shorted to the source region by overlapping the source metallization onto the *p*-type body region, as in Fig. 21-1. As a result of this body-to-source short, there is a parasitic diode connected between the drain and source of the MOSFET, as shown in Fig. 21-1. This integral diode can be used in half-bridge and full-bridge converters.

Third, there is the overlap of the gate metallization across the n^- drift region, where it protrudes to the surface of the wafer. This overlapping of the gate metallization serves two purposes. First, it tends to enhance the conductivity of the drift region at the n^-–SiO_2 interface by forming an accumulation layer (a region of enhanced conductivity to be discussed in later sections), as is shown in Fig. 21-2a, thus helping to minimize the on-state resistance. Second, the metallization tends to act as a field plate when the MOSFET is off that keeps the radius of curvature of the depletion region of the drain–body *pn* junction from getting too small and thus reducing the breakdown voltage of the device. This field plate function is diagrammed in Fig. 21-2b.

The circuit symbol for a *n*-channel MOSFET is shown in Fig. 21-3a and for a *p*-channel MOSFET in Fig. 21-3b. The direction of the arrow on the lead that goes to the body region indicates the direction of current flow if the body–source *pn* junction were forward biased by breaking the short between the two and a forward bias voltage were applied. Thus, a *n*-channel MOSFET that has a *p*-type body region has the arrow pointing into the MOSFET symbol, as is shown in Fig. 21-3a, and the arrow points outwardly for a *p*-channel device.

21-3   *I – V* CHARACTERISTICS

The MOSFET, like the BJT, is a three-terminal device where the input, the gate in the case of the MOSFET, controls the flow of current between the output terminals, the source, and drain. The source terminal is usually common between the input and output of a MOSFET. The output characteristics, drain current i_D as a function of drain-to-source voltage v_{DS} with gate-to-source voltage V_{GS} as a parameter, are shown in Fig. 21-4a for an *n*-channel MOSFET. The output characteristics for a *p*-channel device are the same except that the current and voltage polarities are reversed so that

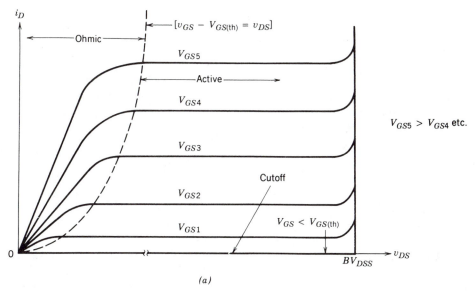

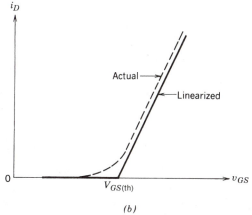

FIGURE 21-4: Current voltage characteristics of a *n*-channel enhancement mode metal-oxide-semiconductor field effect transistor: (*a*) output (i_D vs. v_{DS}) characteristics; (*b*) a transfer curve.

the characteristics for the *p*-channel device would appear in the third quadrant of the $i_D - v_{DS}$ plane rather than the first quadrant, as do the characteristics of Fig. 21-4*a*.

In power electronic applications, the MOSFET is used as a switch to control the flow of power to the load in a manner analogous to the usage of the BJT. In these applications the MOSFET traverses the $i_D - v_{DS}$ characteristics from cutoff through the active region to the ohmic region as the device turns on and back again when it turns off. The cutoff, active, and ohmic regions of the characteristics are shown on Fig. 21-4*a*.

The MOSFET is in cutoff when the gate-source voltage is less than the threshold voltage $V_{GS(th)}$, which is typically a few volts in most power MOSFETs. The device is an open circuit and must hold off the power supply voltage applied to circuit. This means that the drain-source breakdown voltage BV_{DSS} shown on Fig. 21-4 must be larger than the applied drain-source voltage to avoid breakdown and the attendant high power dissipation. When breakdown occurs, it is due to the avalanche breakdown of the drain–body junction.

When the device is driven by a large gate-source voltage, it is driven into the ohmic region (the reason for this designation is twofold, first having to do with the physical mechanisms operative in the MOSFET, which will be discussed in the next section, and second, to avoid confusion with the terminology of saturation, which means one thing when applied to BJTs but another when applied to MOSFETs) where the drain-source voltage $V_{DS(on)}$ is small. In this region the power dissipation can be kept within reasonable bounds by minimizing $V_{DS(on)}$ even if the drain current is fairly large. The MOSFET is in the ohmic region when

$$v_{GS} - V_{GS(th)} > v_{DS} > 0 \qquad (21\text{-}1)$$

In the active region the drain current is independent of the drain-source voltage and depends only on the gate-source voltage. The current is sometimes said to have saturated, and consequently this region is sometimes called the saturation region or the pentode region. We will term this region the active region to avoid the use of the term saturation and the attendant possible confusion with saturation in BJTs. Simple first-order theory predicts that in the active region the drain current is given approximately by

$$i_D = K \left[v_{GS} - V_{GS(th)} \right]^2 \qquad (21\text{-}2)$$

where K is a constant that depends on the device geometry. At the boundary between the ohmic region and active region where $v_{GS} - V_{GS(th)} = v_{DS}$, Eq. 21-2 becomes

$$i_D = K v_{DS}^2 \qquad (21\text{-}3)$$

which is a convenient way of delineating the boundary between the two regions, as is in Fig. 21-4a.

The relationship expressed by Eq. 21-2 is followed reasonably well by logic level MOSFETs. However, a plot of i_D versus v_{GS} (with the MOSFET in the active region) in Fig. 21-4b, usually termed the transfer curve, shows that this equation is followed only at lower values of drain current in power MOSFETs. Overall the transfer curve of a power MOSFET is quite linear in contrast to the parabolic transfer curve of the logic level device. The reasons for the different behavior of the transfer curve between logic level and power MOSFETs will be considered in the next section.

21-4 PHYSICS OF DEVICE OPERATION

21-4-1 Inversion Layers and the Field Effect

The gate portion of the MOSFET structure shown in Fig. 21-1 is the key to understanding how the MOSFET operates The gate region is composed of the gate metallization, the silicon dioxide underneath the gate conductor, which is termed the gate oxide, and the silicon beneath the gate oxide. This region forms a high-quality capacitor, as is shown in Fig. 21-5, and it is sometimes termed an MOS capacitor. Although the capacitor shown is usually composed of aluminum metallization, SiO_2 insulator, and silicon bottom layer, the same basic structure can be fabricated on other semiconductors such as gallium arsenide and other insulators such as aluminum nitride or silicon nitride can be used for the insulator. The top metallization layer can also be other conducting materials. Polysilicon, refractory metals such as tungsten, and other metals have been used on MOSFET devices.

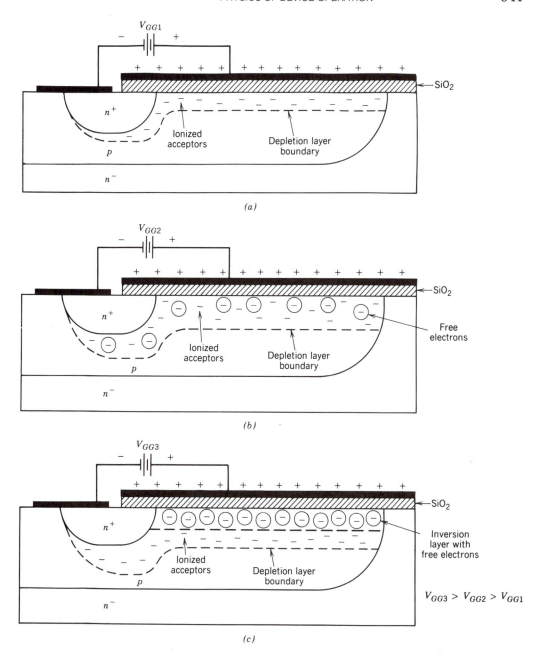

FIGURE 21-5: (a) Formation of the depletion layer, and (b, c) then the inversion layer at the Si–SiO₂ interface as the gate-source voltage is increased.

When a small positive gate-source voltage is applied to the capacitor structure in the simplified diagram of the *n*-channel MOSFET shown in Fig. 21-5a, a depletion region forms at the interface between the SiO₂ and the silicon. The positive charge induced on the upper metallization (the gate side) by the applied voltage requires an equal negative charge on the lower plate, which is the silicon side of the gate oxide. The electric field from the positive charge repels the majority carrier holes from the interface region and thus exposes the negatively charged acceptors, thus creating a depletion region.

Further increases in v_{GS} cause the depletion layer to grow in thickness, as is shown in Fig. 21-5b, to provide the additional negative charge. The growth of this depletion layer can be modeled approximately as a one-sided step junction, such as was considered in Chapter 18. As the voltage is increased, the electric field at the oxide–silicon interface gets larger and begins to attract free electrons as well as repelling free holes. The immediate source of the electrons is electron–hole generation via thermal ionization with the free holes being pushed into the semiconductor bulk ahead of the depletion region. The extra holes are neutralized by electrons that are attracted from the n^+ drain by the positive charge of the holes.

Eventually, as the bias voltage is increased, the density of free electrons at the interface will become equal to the free hole density in the bulk of the body region away from the depletion layer. The layer of free electrons at the interface will be highly conducting and will have all the properties of an n-type semiconductor. At this point the layer of free electrons is termed an inversion layer, as is illustrated in Fig. 21-5c. This n-type layer is a conductive path or channel between the n^+ drain and source regions (thus the terminology channel), which permits the flow of current between source and drain. This ability to modify the conductivity type of the semiconductor immediately beneath the gate insulator by means of an applied voltage or electric field is termed the field effect. The field effect enhances the conductivity of the interface and, hence, the name enhancement mode field effect transistor, which is based on this mechanism.

The value of v_{GS} where the inversion layer is considered to have formed is termed the threshold voltage $V_{GS(th)}$. As v_{GS} is increased beyond $V_{GS(th)}$ the inversion layer gets somewhat thicker and, more important, more conductive as the density of free electrons gets larger as the bias voltage increases. The inversion layer screens the depletion layer adjacent to it from the further bias voltage increases so that the depletion layer thickness now remains constant. The value of the threshold voltage is a function of several factors. A major factor is the oxide capacitance per unit area C_{ox}, which is given by

$$C_{ox} = \frac{\varepsilon_{ox}}{t_{ox}} \qquad (21\text{-}4)$$

where ε_{ox} is the dielectric constant of the silicon dioxide (1.05×10^{-12} farads/cm) and t_{ox} is the thickness of the gate oxide (typically 1000 Å). The threshold voltage is inversely proportional to C_{ox}. Other factors that influence $V_{GS(th)}$ are the work functions of the silicon and the gate metal, any charge bound or trapped in the silicon dioxide, impurities at the interface or in the silicon dioxide, as well as others. In spite of the complex array of factors that influence $V_{GS(th)}$, device manufacturers can adjust its value to whatever value is desired (typically a few volts). Major adjustments are done by the choice of gate metallization, doping density of the body region, and thickness of the gate oxide. Minor adjustments of the threshold voltage during fabrication are done via ion implantation of impurities in the body region just beneath the gate oxide.

21-4-2   Gate Control of Drain Current Flow

We now embed the n-channel MOSFET of Fig. 21-5 into the circuit shown in Fig. 21-6, which has both a gate-source bias supply V_{GS} and a drain-source bias supply V_{DD}. Initially it is assumed that V_{GS} is greater than $V_{GS(th)}$ and that V_{DD} is small. The MOSFET will be in the ohmic region with a relatively small value of I_D, and the inversion layer will have a spatially uniform thickness, as is shown in Fig. 21-6a. Now V_{DD} is slowly increased to ever larger values while V_{GS} is held constant. The drain

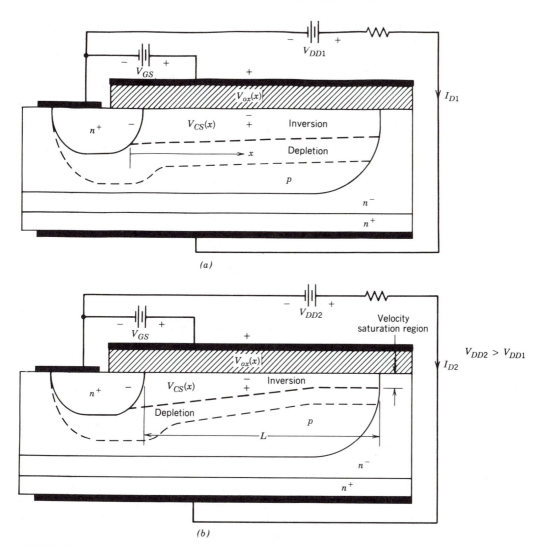

FIGURE 21-6: Change in the inversion layer thickness from being (*a*) spatially uniform at low drain current to being (*b*) spatially nonuniform at larger drain current values.

current will initially increase in proportion to the increase in V_{DD}, since the inversion layer appears as an ohmic resistance connecting the drain to the source. This current increase will cause a voltage drop along the channel, which is shown in Fig. 21-6*a* as $V_{CS}(x)$ (channel-to-source voltage), with x being the distance from the source to the location x in the channel where the voltage is being specified.

In the discussion of the inversion layer formation in the previous section, the gate-to-body voltage, which is the voltage drop across the oxide, was greater than $V_{GS(th)}$. It was also implicitly assumed that this voltage was spatially uniform along the length of the oxide from source to drain so that the inversion layer thickness would also be uniform. But in the structure of Fig. 21-6, the oxide voltage is actually $V_{GS} - V_{CS}(x)$, as indicated in the figure. As I_D increases, $V_{CS}(x)$ increases and the voltage across the oxide at the position x decreases. Since $V_{CS}(x)$ has its largest value, V_{DS}, at $x = L$ (the drain end of the channel), the voltage drop across the oxide, which determines the parameters of the inversion layer, will have its smallest value of $V_{GS} - V_{DS}$. The decrease in the oxide voltage from source to drain when I_D is flowing means that the thickness of the inversion layer must also decrease from source to drain, as indicated in Fig. 21-6b.

As the inversion layer thins out at the drain end of the channel, its resistance increases and the curve of I_D versus V_{DS} for a constant V_{GS} begins to flatten out, as is shown in Fig. 21-4a. This produces the concave curvature in the ohmic region curves shown in Fig. 21-4a. The larger the drain current becomes, the flatter the I_D versus V_{DS} characteristic.

But now an apparent dilemma develops. If I_D is made large enough, it would appear that the voltage drop along the channel would get large enough so that $V_{GS} - V_{DS}$ might get reduced to $V_{GS(\text{th})}$. Then, the inversion layer would essentially disappear at the drain end and no current would be able to flow. In reality the situation is more complicated. As I_D increases, the inversion layer does narrow down in thickness, as indicated. However, since the total current is the same everywhere in the channel, the current density at the drain end is higher, since the inversion layer thickness is less. Since the current is flowing by drift (there is no injection of minority carriers, since the junctions are shorted out by the inversion layer), the electric field parallel to the current flow is also larger at the drain end. (Recall $J = \sigma E$ and σ is constant.)

This larger electric field at the drain end is important in two respects. First, as it becomes larger, the electric field (due to V_{GS}) across the gate oxide at the drain end is becoming too small to maintain the inversion layer. The large electric field due to the constricted area of current flow takes over the maintenance of a minimum thickness of inversion layer at the drain end and thus circumvents the apparent dilemma mentioned in the preceding paragraph. Second, the velocity of the charge carriers is a function of the electric field with the velocity saturating at a constant value as the field is increased, as is indicated in Fig. 21-7. At the point where the field at the drain end is large enough to saturate the carrier velocity, the oxide voltage is approximately at the threshold value so that $V_{GS} - V_{DS} = V_{GS(\text{th})}$ and the device is about to enter the active region. Further increases in V_{DD} will increase the electric field in the narrowest part of the channel and will lead to a growth in the length of the minimum thickness region of the channel toward the source, as diagrammed in Fig. 21-6b. The voltage drop across the oxide will remain fixed and thus so will the inversion layer thickness. Hence, for $V_{DS} > V_{GS} - V_{GS(\text{th})}$, the drain current remains relatively constant, as is indicated in the $i - v$ characteristic of Fig. 21-4.

If V_{GS} is larger, then the inversion layer thickness is larger and a larger total current is required before the electric field at the drain end is large enough to cause carrier velocity saturation. Simple first-order theoretical calculations indicate that in the active region i_D is given by Eq. 21-2, where the constant K is given by

$$K = \mu_n C_{\text{ox}} \frac{W}{(2L)} \tag{21-5}$$

with μ_n being the inversion layer majority carrier mobility and the other parameters are defined below. This equation points out one of the most important design considerations in the fabrication of MOSFETs, that in order to have appreciable gain it is essential that the gate width W be much larger than the channel length L. In modern power MOSFETs the length L is kept to a minimum, consistent with breakdown voltage requirements, to minimize the on-state losses (typical channel lengths are a few microns). The W/L ratio is typically 10^5 or greater and is achieved by using many small source regions, as is indicated in Fig. 21-1. The effective gate width in this structure is the total peripheral distance around all the source regions.

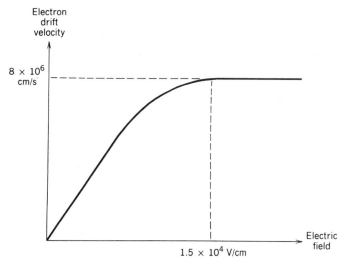

FIGURE 21-7: Electron drift velocity in silicon versus electric field intensity illustrating velocity saturation. The electron mobility is the incremental slope of the velocity versus electric field curve.

The square law $i_D - v_{GS}$ relationship is not maintained at larger values of drain current but instead becomes a linear relationship, as indicated in Fig. 21-4b. The reason for the changeover to a linear relationship is that the mobility in Eq. 21-5 does not remain constant as i_D increases, but instead decreases as the electric field in the inversion layer increases with increasing current. The mobility decreases both because of the velocity-electric field relationship diagrammed in Fig. 21-7 and because larger values of V_{GS} increase the free electron density in the channel. At larger carrier densities the mobility decreases because of what is termed carrier–carrier scattering (see Eq. 19-14a).

21-5 SWITCHING CHARACTERISTICS

21-5-1 MOSFET Circuit Models

MOSFETs are intrinsically faster than bipolar devices because they have no excess minority carriers that must be moved into or out of the device as it turns on or off. The only charges that must be moved are those on the stray capacitances and depletion layer capacitances, which are shown in the cross-sectional view of the MOSFET given in Fig. 21-8. These capacitances can be modeled by the equivalent circuit shown in Fig 21-9a, which is valid when the MOSFET is in cutoff or in the active region. Circuit models such as this are needed for a detailed study of the turn-on and turn-off characteristics of the MOSFET so that appropriate gate drive circuits can be designed.

The drain-source capacitance shown in Fig. 21-8 is not included in the equivalent circuit because it does not materially effect any of the switching characteristics or waveforms. However, it should be considered when designing snubbers, for example, in Chapter 7 where a lossless capacitor snubber is needed in zero-voltage switching converter topologies. There, C_{ds} can be a part of the total snubber capacitance requirement.

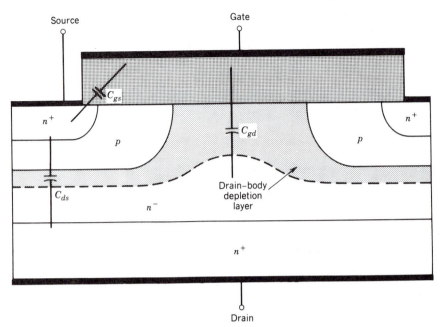

FIGURE 21-8: Cross-sectional view of a n-channel MOSFET showing the approximate origin of the parasitic capacitances that govern the switching speed of the device.

The gate-voltage-controlled current source shown in the equivalent circuit is equal to zero when $v_{GS} < V_{GS(\text{th})}$ and is equal to $g_m[v_{GS} - V_{GS(\text{th})}]$ when the device is in the active region. This method of accounting for the flow of drain current in the active region is suggested by the fact that the transfer characteristic shown in Fig. 21-4b is linear over most of its range. The slope of the transfer characteristic in the active region is the transconductance g_m.

The MOSFET enters the ohmic region when v_{DS} is equal to or less than $v_{GS} - V_{GS(\text{th})}$. In switch-mode power applications $v_{GS} \gg V_{GS(\text{th})}$ when the device is on, so that the criteria for entering the ohmic region can be simplified to $v_{DS} < v_{GS}$. In the ohmic region the dependent current source model is no longer valid because the inversion layer is no longer nearly pinched off at the drain end of the channel but instead has a nearly spatially uniform thickness since v_{DS} is quite small. The inversion layer essentially shorts the drain to the source and so the drain end of C_{gd} is shown in the ohmic region equivalent circuit of Fig. 21-9b as grounded. An on-state resistance $r_{DS(\text{on})}$ is added to the equivalent circuit to account for the ohmic losses, which arise principally from the drain drift region. There are other contributions to the on-state resistance such as ohmic losses in the channel, but they are usually small compared with the drain drift region contribution except for low breakdown voltage devices. These other contributions will be discussed in later sections of this chapter.

Note that the capacitances C_{gs} and C_{gd} are not constant but vary with the voltage across them because part of the capacitance is contributed by depletion layers. For example, the gate-source capacitance is the combination of the electrostatic capacitance of the oxide layer in series with the capacitance of the depletion layer that forms at the Si–SiO$_2$ interface. The most significant change in capacitance occurs in C_{gd} because the voltage change across it, v_{DS}, is much larger than the voltage change across C_{gs}. The change in C_{gd} with v_{DG} ($\approx v_{DS}$), which is diagrammed in Fig. 21-9c, can be as large as a factor of 10. For approximate calculations of switching waveforms, C_{gd} is approximated by the two discrete values C_{gd1} and C_{gd2} shown in Fig.

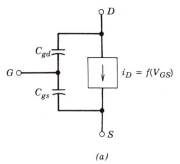

$$i_D = f(V_{GS})$$

(a)

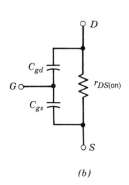

(b)

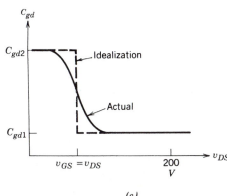

(c)

FIGURE 21-9: Circuit models for MOSFETs for transient analysis: (a) MOSFET equivalent circuit for transient analysis in cutoff and active region; (b) MOSFET equivalent circuit in the ohmic region; (c) variation in gate-drain capacitance with drain-source voltage.

21-9c with the change in value occurring at $v_{DS} = v_{GS}$ where the MOSFET is either entering or leaving the ohmic region. The gate-source capacitance will be assumed to be constant.

21-5-2 Switching Waveforms

The turn-on behavior of the MOSFET embedded in the inductively loaded step-down dc–dc converter, a commonly encountered circuit in power electronics, will be examined. As was done for the analogous BJT circuit, the inductive load is modeled as a constant current source I_o in parallel with a diode D_f as shown in Fig. 21-10. The MOSFET is replaced in Fig. 21-10 by its active region equivalent circuit. The gate is driven by an ideal voltage source, which is assumed to be a step voltage between 0 and V_{GG} in series with an external gate resistance R_G. To keep the explanation simple, we assume that the free-wheeling diode in Fig. 21-10 is ideal with a zero reverse recovery current.

The turn-on waveforms are shown in Fig. 21-11, where the gate drive voltage changes in a step-function manner at $t = 0$ from 0 to V_{GG}, which is well above $V_{GS(th)}$. During the turn-on delay time $t_{d(on)}$ the gate-source voltage v_{GS} rises from 0 to $V_{GS(th)}$ because of the currents flowing through C_{gs} and C_{gd}, as is shown in Fig. 21-12a. The rate of rise of v_{GS} in this region is almost linear, although it is a part of an exponential curve shown dotted in Fig. 21-11, which has a time constant $\tau_1 = R_G(C_{gs} + C_{gd1})$. Beyond $V_{GS(th)}$, v_{GS} continues to rise as before, and the drain current begins to increase according to the linearized transfer curve shown in Fig 21-4b. Therefore, the equiva-

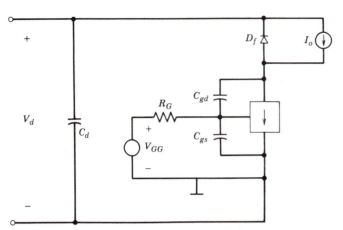

FIGURE 21-10: MOSFET used to switch a diode-clamped inductive
load. The circuit is essentially that of a step-down
dc–dc converter: the equivalent circuit is valid for
cutoff and active region transient analysis.

lent circuit shown in Fig. 21-12b applies. The drain-source voltage remains at V_d as
long as $i_D < I_o$ and the free-wheeling diode D_f is conducting. The time required for
i_D to build up from zero to I_o is the current rise time t_{ri}.

Once the MOSFET is carrying the full load current I_o but is still in the active
region, the gate-source voltage becomes temporarily clamped at V_{GS, I_o}, which is the
gate-source voltage from the transfer curve of Fig. 21-4b needed to maintain $i_D = I_o$.
The entire gate current i_G, which is given by

$$i_G = \frac{\left[V_{GG} - V_{GS, I_o}\right]}{R_G} \tag{21-6}$$

flows through C_{gd}, as is indicated in the equivalent circuit of Fig. 21-12c. This causes
the drain-source voltage to drop at a rate

$$\frac{dv_{DG}}{dt} = \frac{dv_{DS}}{dt} = \frac{i_G}{C_{gd}} = \frac{\left[V_{GG} - V_{GS, I_o}\right]}{R_G C_{gd}} \tag{21-7}$$

(Recall that $v_{GS} = V_{GS, I_o}$ during this interval so $dv_{GS}/dt = 0$.) The decrease in v_{DS}
occurs in two distinct time intervals t_{fv1} and t_{fv2}. The first time interval corresponds to
the traverse through the active region where $C_{gd} = C_{gd1}$. The second time interval
corresponds to the completion of the transient in the ohmic region where the
equivalent circuit shown in Fig. 21-12d applies and $C_{gd} = C_{gd2}$.

Once the drain-source voltage has completed its drop to the on-state value of
$I_o r_{DS(on)}$, the gate-source voltage becomes unclamped and continues its exponential
growth to V_{GG}. This part of the growth occurs with a time constant $\tau_2 = R_G(C_{gs} + C_{gd2})$,
and simultaneously the gate current decays toward zero with the same time constant
as is shown in the waveforms of Fig. 21-11.

If the free-wheeling diode D_f is not ideal, but has a reverse recovery current,
then the switching waveforms are modified, as is shown in Fig. 21-13a. During the
current rise time interval, the drain current increases beyond I_o to $I_o + I_{rr}$ because of
the reverse recovery current of D_f. This causes v_{GS} to increase beyond V_{GS, I_o}, as is
shown in Fig. 21-13a. When the diode current snaps off and recovers to zero, there is a

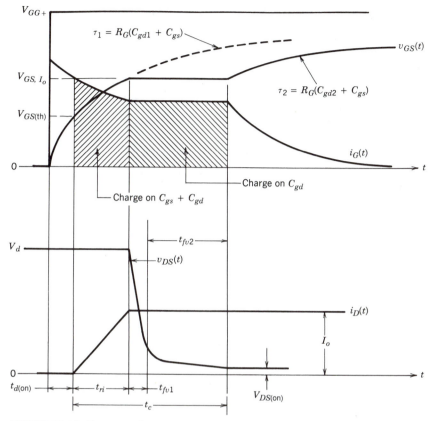

FIGURE 21-11: Turn-on voltage and current waveforms of the MOSFET embedded in a circuit having a diode-clamped inductive load with an ideal (zero reverse recovery current) free-wheeling diode.

rapid decrease in v_{GS} to V_{GS, I_o}, and this rapid decrease provides an additional current to C_{gd} in addition to i_G, as is shown in the equivalent circuit of Fig. 21-13b. This additional current causes v_{DG} and v_{DS} to decrease very rapidly during this recovery interval, as is indicated in the waveforms of Fig. 21-13a. Once the reverse recovery interval is over, the drain current is back to I_o, and the rest of the transient proceeds as in the ideal diode case shown in Fig. 21-11.

The turn-off of the MOSFET involves the inverse sequence of events that occurred during turn-on. The same basic analytical approach used to find the turn-on switching waveforms can be used to find the turn-off waveforms. The turn-off waveforms and associated time intervals are shown in Fig. 21-14 for an assumed step change in the gate drive voltage at $t = 0$ from V_{GG} to zero. The actual values of the switching times will vary depending on whether the gate drive voltage is set to zero or made negative to speed up the transient. Moreover, the value of R_G used during turn-off may be different from that used during turn-on.

During turn-on and turn-off, the instantaneous power loss occurs primarily during the crossover time t_c indicated in Figs. 21-11, 21-13, and 21-14 where $p(t) = v_{DS}i_D$ is high. Since the MOSFET capacitances do not vary with the junction temperature, the switching power losses in the MOSFET are also independent of the junction temperature. However, the on-state resistance does vary with temperature, and thus the conduction loss will vary with junction temperature.

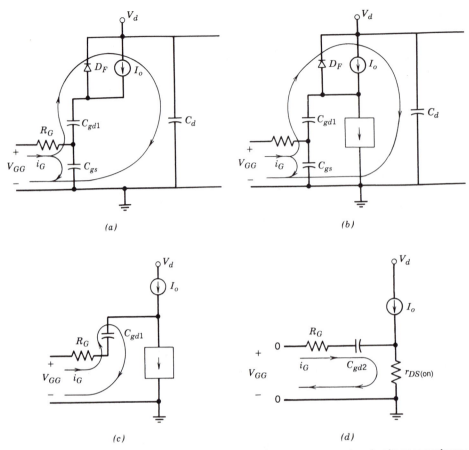

FIGURE 21-12: Equivalent circuits used to estimate the turn-on current and voltage waveforms of the MOSFET used in the diode-clamped inductive load circuit: (a) equivalent circuit during $t_{d(on)}$; (b) equivalent circuit during t_{ri}; (c) equivalent circuit for the t_{fv1} interval; (d) equivalent circuit during t_{fv2}.

21-6   OPERATING LIMITATIONS AND SAFE OPERATING AREAS

21-6-1   Voltage Breakdown

    MOSFETs have two voltage ratings that should not be exceeded: $V_{GS(max)}$ and BV_{DSS}. The maximum allowable gate-source voltage $V_{GS(max)}$ is determined by the requirement that the gate oxide not be broken down by large electric fields. Good-quality thermally grown SiO_2 breaks down at electric field values on the order of 5 to 10 million V/cm. This means that a gate oxide 1000 Å thick can theoretically withstand a gate-source voltage of 50 to 100 V. Typical specifications for $V_{GS(max)}$ are 20 to 30 V, which indicates that device manufacturers put a margin of safety into their ratings. This is done because the breakdown of the gate oxide means permanent failure of the device. Note that even static charge inadvertently put on the gate oxide by careless handling may be sufficient to rupture the oxide. The device user should carefully ground himself before handling any MOSFET to avoid any static charge problems. If transient gate-source voltages in excess of $V_{GS(max)}$ are a possibility, the gate should be protected by a series connection of two zener diodes connected back-to-back between the gate and source terminals. The breakdown voltage of the zeners should be less than $V_{GS(max)}$.

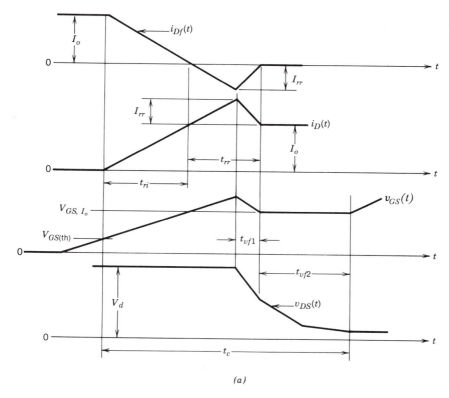

(a)

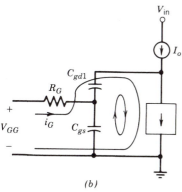

(b)

FIGURE 21-13: Effect of free-wheeling diode reverse recovery current on MOSFET current and voltage waveforms at turn-on: (a) MOSFET turn-on waveforms modified by free-wheeling diode turn-off; (b) equivalent circuit for estimating effect of free-wheeling diode reverse recovery.

The maximum allowable drain-source voltage BV_{DSS} is the largest voltage the MOSFET can hold off without avalanche breakdown of the drain–body pn junction. Large values of breakdown voltage are achieved by the use of the lightly doped drain drift region. The lightly doped drain drift region is used to contain the depletion layer of the reverse-biased drain–body junction. The length of the drift region is determined by the desired breakdown voltage rating and is given roughly by Eq. 19-2. The light doping of the drift region compared to the heavy doping of the body region ensures that the depletion layer of the junction does not extend far into the body toward the source region so that breakdown via reach-through (described in conjunction with the BJT) is avoided.

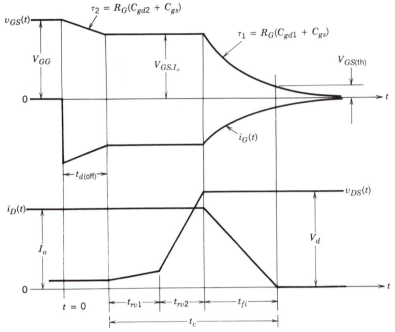

FIGURE 21-14: MOSFET current and voltage waveforms at turn-off in the diode-clamped inductive load circuit. The free-wheeling diode is assumed to be ideal.

The relatively sharp curvature of the diffused p-type body region shown in Fig. 21-1 might lead to a reduction of BV_{DSS} if proper corrective steps are not taken. The extension of the gate metallization over the drain drift region that was pointed out in earlier sections of this chapter acts as a field plate that reduces the curvature of the depletion region. This in turn prevents a severe reduction of the breakdown voltage.

21-6-2   On-State Conduction Losses

Except at higher switching frequencies, nearly all of the power dissipated in a MOSFET in a switched-mode power application occurs when the device is in the on-state The instantaneous power dissipation in the on-state of the MOSFET is given by

$$p_{on} = I_o^2 r_{DS(on)} \tag{21-8}$$

The on-state resistance has several components, as is illustrated in Fig. 21-15. At lower breakdown voltages (a few hundred volts or less), all of these resistance components contribute more or less equally to the total on-state resistance. The device manufacturer attempts to minimize all of the contributions by using the heaviest doping in each region consistent with other requirements such as breakdown voltage requirements. An example of the detailed attention paid to these resistance contributions is given by the extension of the gate metallization over the drain drift region where it protrudes to the silicon surface between the p-type body regions. This overlapping of the gate metal allows the gate-source bias to enhance the conductivity of the drift region at the interface region between the drift region and the gate oxide by attracting additional free electrons to the interface and creating an accumulation layer.

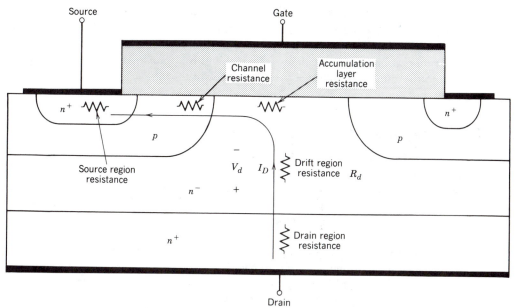

FIGURE 21-15: On-state resistance components in a n-channel enhancement mode MOSFET.

A premium is also placed on dimensional control of the MOSFET features so that the lengths of current paths in highly resistive regions are minimized. Generally speaking, the design of the source cell and its dimensional tolerances have the greatest impact on $r_{DS(on)}$ in low breakdown voltage MOSFETs. The significant progress that has been made in reducing the on-state losses in lower voltage MOSFETs is attested to by the fact that except at perhaps very high current levels, MOSFETs can have lower conduction losses than BJTs for breakdown voltage ratings below a few hundred volts.

Two of the resistance components, the channel resistance and the accumulation layer resistance, are affected by the gate-source bias as well as doping and dimensional considerations. In both of these components larger values of gate-source bias will lower these resistances. Hence, it is desirable to use as large a value of gate-source drive voltage as possible consistent with other considerations such as gate oxide breakdown.

For BV_{DSS} greater than a few hundred volts, the drain drift region resistance R_d dominates the on-state resistance. An optimistic estimate of the specific resistance (ohms-cm^2) of the drift region as a function of breakdown voltage rating can be obtained using Eq. 19-20. The result is

$$\frac{V_d}{J} = R_d A \approx 3 \times 10^{-7} BV_{DSS}^2 \tag{21-9}$$

where A is the cross-sectional area through which the drain current flows. Experimental results and more exact theoretical treatments indicate that the dependence is actually $BV_{DSS}^{2.5-2.7}$. Because of the steep dependence of R_d on the breakdown voltage rating, MOSFETs will have higher on-state losses at the larger blocking voltages compared with BJTs.

The on-state resistance increases significantly with increasing junction temperature. This in turn means that the on-state power dissipation will increase with temperature in most power electronic applications like that illustrated in Fig. 21-10. The positive temperature coefficient of the on-state resistance arises from the decrease of the carrier mobility as the semiconductor temperature increases. This occurs because at higher temperatures the charge carriers undergo more collisions per unit time with the semiconductor lattice because each lattice atom has a larger amplitude of vibration at higher temperatures. The mobility is approximately inversely proportional to the number of collisions per unit time with the lattice, and $r_{DS(on)}$ is inversely proportional to the mobility.

21-6-3   Paralleling of MOSFETs

MOSFETs can be paralleled very easily as are the two shown in Fig. 21-16 because of the positive temperature coefficient of their on-state resistance. For the same junction temperature, if $r_{DS(on)}$ of T_2 exceeds that of T_1, then during the on-state, T_1 will have a higher current and thus higher power loss compared to T_2, since the same voltage appears across both transistors. Therefore, the junction temperature of T_1 will increase along with its on-state resistance. This will cause its share of current to decrease and, hence, there is a thermal stabilization effect.

During switching, the current in each MOSFET is determined by the transfer characteristic, as was explained in previous sections (see Fig. 21-4). The variation in the transfer characteristic from one device to another of the same part number is modest. Hence, it is best to keep the gate-source voltage of the paralleled transistors the same during switching so that they share approximately equal currents during switching. However, the gates cannot be directly connected together, but rather a small resistance or ferrite bead must be used in series with the individual gate connections, as is shown in Fig. 21-16. This is because the gate inputs are highly capacitive with almost no losses, although some stray inductance is always present. The stray inductances in combination with the gate capacitances can result in unwanted high-frequency oscillations in the MOSFETs, which are avoided by the damping resistance shown in Fig. 21-16. Another consideration to keep in mind while paralleling MOSFETs is that their layout should be symmetrical.

21-6-4   Parasitic BJT

The MOSFET has a parasitic BJT as an integral part of its structure, as is shown in Fig. 21-1. The body region of the MOSFET serves as the base of the BJT, the source as the BJT emitter, and the drain as the BJT collector. The beta of this

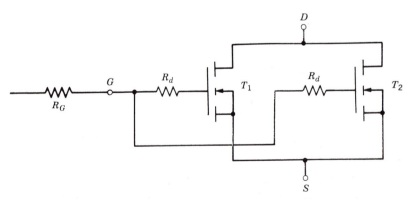

FIGURE 21-16: Parallel connection of MOSFETs. A small damping resistance should be included in series with the gate of each MOSFET to minimize any possible high-frequency oscillation.

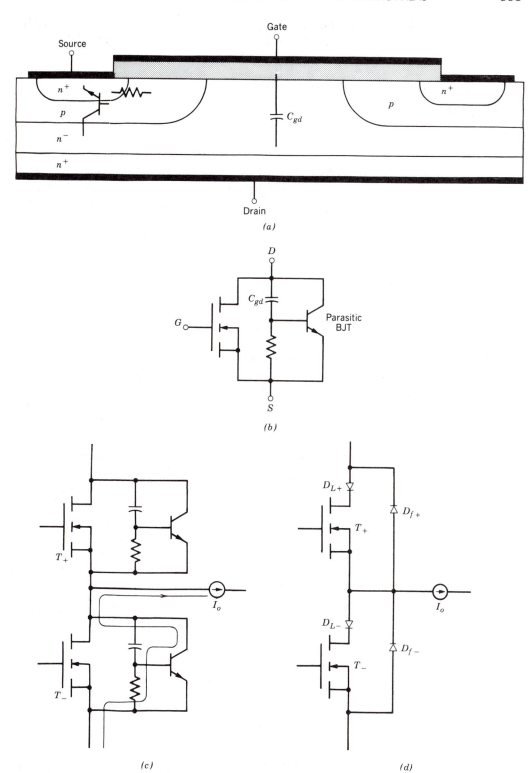

FIGURE 21-17: Potential mechanism for turn-on of the parasitic BJT in a power MOSFET. (*a*) Cross section of the MOSFET showing how the parasitic BJT might be turned on by a fast rising drain-source voltage. (*b*) Equivalent circuit of the MOSFET showing the potential turn-on of the parasitic BJT. (*c*) Mechanism of possible destructive latchup of MOSFETs used in a bridge configuration. This latchup can be prevented using diodes D_{L+} and D_{L-} instead of directly connecting D_{f+} and D_{f-} to the MOSFET drains.

parasitic BJT may be significantly greater than one because the length of the body region where the channel of the MOSFET is formed is kept as short as possible to help minimize the on-state resistance.

It is imperative that this BJT be kept cut off at all times by keeping the potential of the parasitic base as close to the source potential as possible. This is the purpose of the body–source short mentioned previously and shown in Fig. 21-1. If the base of the parasitic BJT were allowed to float, two potential problems would arise. First, the breakdown voltage of the MOSFET would be reduced from $BV_{DSS} = BV_{CBO}$ to BV_{CEO}, a drop that might be as large as 50%. This drop in breakdown voltage could lead to excessive power dissipation because of large breakdown currents if the drain-source voltage exceeded the reduced breakdown voltage.

Second, the base–emitter potential may get large enough for the BJT to turn on and possibly go into saturation, a condition termed latchup. This situation is danger-ous because there will be significant power dissipation and worse yet, the BJT cannot be turned off via the base terminal because the base is not accessible. The only way to turn off the BJT once latchup has occurred is to interrupt the flow of drain current external to the device.

Although the body–source short is quite effective in preventing BJT turn-on from a static viewpoint, it does not guarantee that turn-on cannot occur during high-speed turn-off of the MOSFET. The base of the parasitic BJT is connected to the drain terminal by a portion of the drain-gate capacitance, as is shown in Fig. 21-17a and 21-17b. If the rate of rise of v_{DS} is large enough during turn-off of the MOSFET, the displacement current through the coupling capacitor in Fig. 21-17 to the base of the BJT may be large enough to induce a voltage drop in the parasitic resistance connected between the base and emitter, as is shown in Fig. 21-17a, to turn the BJT on. This parasitic resistance arises from the distributed nature of the base (body) region and the remote location of the body–source short compared to the interior of the body region. This potential BJT turn-on mechanism imposes a maximum rate of rise dv_{DS}/dt on the MOSFET. Fortunately, this potential problem has been recognized by device designers and has been largely eliminated. Modern power MOSFETs have dv_{DS}/dt capabilities in excess of 10,000 V/μs. This potential problem can be easily circumvented by slowing down the switching speed of the MOSFET by using larger values of gate resistance R_G and smaller values of turn-off gate drive.

This unwanted turn-on is most likely to happen in bridge circuits such as shown in Fig 21-17c where the integral or parasitic diode (the B-C diode of the parasitic BJT) mentioned previously is used as the freewheeling diode. When T_+ and T_- are both off, the integral diode of T_- is carrying the load current. At the end of the blanking time, T_+ is turned on and the integral diode of T_- undergoes reverse recovery and the reverse-recovery current flows through the C_{gd} of T_-. If the reverse-recovery (snap-off) of this integral diode is very fast, a large positive dv_{DS}/dt will be imposed on C_{gd} and the combination of this displacement current plus the reverse recovery current may be sufficient to turn-on the parasitic BJT of T_- as discussed in the preceding paragraph. This will result in the destruction of T_- since the parasitic BJT will be conducting a large current while the full voltage between the positive and negative power supply rails is applied across the drain-source terminals of T_-.

This problem can be solved by using the circuit shown in Fig. 21-17d where an external free-wheeling diode is used and another diode D_L is put in series with each MOSFET as is indicated in the figure. The D_L diodes prevent any current flowing through the parasitic diode when the MOSFET is controlled to be off and so all of the current is forced to flow through the external free-wheeling diode. In the off-state the

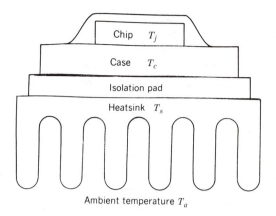

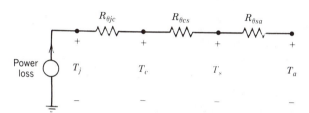

FIGURE 21-18: Estimating heat sink requirements for a MOSFET in order to keep the junction temperature within specifications.

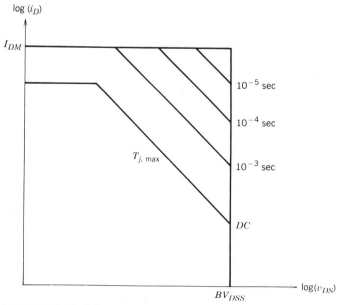

FIGURE 21-19: Safe operating area of a *n*-channel enhancement mode MOSFET. Note the absence of second breakdown.

D_L diodes need only block the on-state voltage of the free-wheeling diodes, which will be small (a few volts) so Schottky diodes could be used for D_{L+} and D_{L-}. Fortunately MOSFETs are now available where the reverse-recovery current of the parasitic diodes is minimized and the parasitic body-source resistance is small enough so that the MOSFETs can be used as shown in Fig. 21-17c without external free-wheeling diodes and series diodes.

21-6-5   Thermal Considerations

In using the MOSFET, like any other semiconductor device, the internal temperature of the semiconductor, the so-called junction temperature T_j, must be kept below a maximum value $T_{j,\max}$, which is generally found on the manufacturer's specification or data sheets for the device. Crudely speaking, this temperature is set by the requirement that the intrinsic carrier density in the most lightly doped portion of the device not exceed the background doping density. If the intrinsic carrier density in the drain drift region exceeds the doping density, then one of the most desirable properties of the MOSFET, the positive temperature coefficient of $r_{DS(\text{on})}$ is lost, since the intrinsic carrier density increases with temperature. If the doping density in the drift region is 10^{14} cm^{-3}, the temperature at which the intrinsic carrier density equals this value is, using Eq. 18-1, about 280°C. Most manufacturers specify a maximum junction temperature between 150 and 225°C.

In using a MOSFET in a demanding application, the control of the junction temperature is an important design consideration. For a worst case design, the maximum junction temperature $T_{j,\max}$, the maximum ambient temperature $T_{a,\max}$, the maximum operating voltage, and maximum on-state current $I_{D,\max}$ are specified. The maximum conduction losses in the MOSFET can be calculated by knowing the maximum duty ratio, $I_{D,\max}$, and $r_{DS(\text{on}),\max}$, which can be obtained from the data sheets corresponding to $T_{j,\max}$ and $I_{D,\max}$. The switching losses can be obtained by integrating the instantaneous power loss with respect to time and averaging it over the switching time period. Therefore, P_{loss}, which is the sum of the on-state losses and the switching losses, can be estimated.

From this information the maximum allowable junction-to-ambient thermal resistance $R_{\theta ja}$ in Fig. 21-18 can be estimated as

$$R_{\theta ja} = \left[T_{j,\max} - T_{a,\max} \right] / P_{\text{loss}} \tag{21-10}$$

Now

$$R_{\theta ja} = R_{\theta jc} + R_{\theta cs} + R_{\theta sa} \tag{21-11}$$

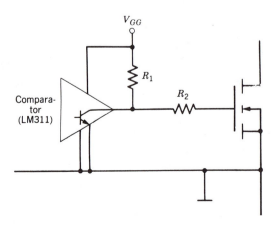

FIGURE 21-20: Simple gate drive circuit suitable for low-speed and low-switching-frequency applications.

where the junction-to-case thermal resistance $R_{\theta jc}$ can be obtained from the data sheets, and the case-to-sink thermal resistance $R_{\theta cs}$ depends on the thermal compound and the voltage insulator (if any) used. Knowing $R_{\theta cs}$ and $R_{\theta jc}$, we can calculate the thermal resistance of the heat sink-to-ambient thermal resistance $R_{\theta sa}$ from Eqs. 21-10 and 21-11. A proper heat sink can then be selected based on the information provided by the heat sink manufacturer's data sheets.

21-6-6 Safe Operating Area

The safe operating area of a power MOSFET is shown in Fig. 21-19. Three factors determine the SOA of the MOSFET, the maximum drain current I_{DM}, the internal junction temperature T_j, which is governed by the power dissipation in the device, and the breakdown voltage BV_{DSS}. These limiting factors have been discussed to some extent already in this chapter and are analogous to those of the BJT and thus need no further elaboration. The MOSFET does not have any second breakdown limitations, as does the BJT, and so none show up on the safe operating area.

For switch-mode applications, the SOA of the MOSFET is square, as is indicated in Fig. 21-19. There is no distinction between forward bias and reverse bias safe operating areas for the MOSFET; they are identical.

21-7 DESIGN OF GATE DRIVE CIRCUITS

21-7-1 Drive Circuit Topologies

As we discussed in the switching characteristics section, the rate of change of v_{DS} and i_D depend on the gate current, which determines how fast the device capacitances are charged and discharged. Therefore, the circuit designer can control the MOSFET switching times by controlling the gate current supplied by the gate drive circuit. The

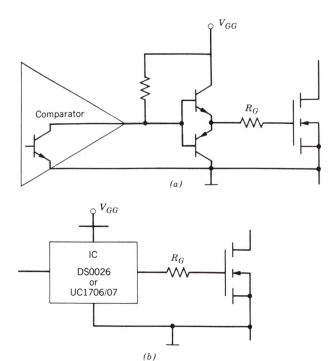

FIGURE 21-21: Gate drive circuit with a totem pole configuration for faster turn-off times: (*a*) discrete totem pole gate drive circuit; (*b*) integrated circuit totem pole gate drive circuit.

advantage of fast switching speed is the reduction in switching power loss due to the reduction of the crossover time t_c shown in Figs. 21-11, 21-13, and 21-14. This is important in high-frequency applications. But there are disadvantages to fast switching, including higher EMI (electromagnetic interference), increased reverse recovery current problems in the free-wheeling diode D_f, and overvoltages due to stray inductances, which were discussed in the previous chapter on BJTs. Hence, it is preferable to switch at slow speeds unless the switching losses become significant.

MOSFETs are used almost exclusively to achieve the high switching frequencies desired in many power electronic applications because of their capability to switch at faster speeds than other devices. In such high-frequency applications, the gate current required by the MOSFETs can be substantial, on the order of 1 A or more, to provide the fast switching speeds.

Since dv_{DS}/dt and di_D/dt are proportional to i_G, the switching times of the MOSFET depend on the operating voltage V_d and the load current I_o. In MOSFET data sheets, the switching times are generally specified at very low gate resistance and fairly low operating voltages and currents. Thus, the switching times should be calculated for the actual operating conditions by using the approach outlined in Section 21-5.

The control signal to switch the MOSFET is usually supplied by a logic circuit consisting of some dedicated ICs or a microprocessor, for example. The signal output stage of such logic circuits cannot drive the MOSFET gate directly because it is often not designed to provide a gate current of appropriate magnitude in both the positive and negative direction. Therefore, a need exists for a gate drive circuit to interface the logic control signal to the gate of the MOSFET.

A simple gate drive circuit with only one switch to control the gate current is shown in Fig. 21-20, where the output transistor of a comparator (eg. LM311)

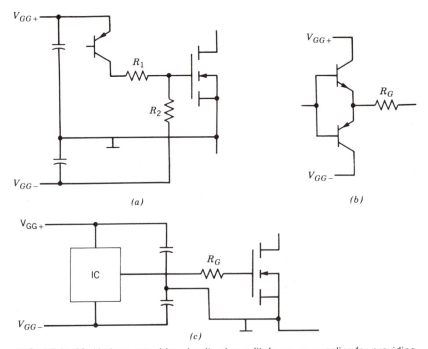

FIGURE 21-22: Various gate drive circuit using split dc power supplies for providing a *n*-channel MOSFET with positive gate-source voltages at turn-on and negative gate drive at turn-off.

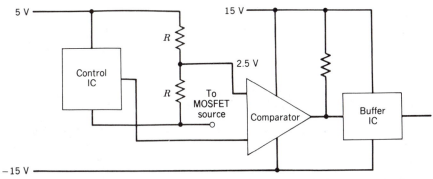

FIGURE 21-23: Circuit configuration for adjusting the reference to the comparator in a gate drive circuit that uses split power supplies.

controls the MOSFET. When the output transistor is off, the MOSFET is on, and vice versa. When the comparator is on, it must sink a current V_{GG}/R_1 and to avoid large losses in the drive circuit, R_1 should be large. This will slow down the MOSFET turn-on time. This means that this drive circuit is suitable only for low switching speed applications.

The inadequacy of this circuit can be overcome by the gate drive circuit shown in Fig. 21-21, where two switches are used in a totem-pole arrangement with the comparator (type LM311) controlling the *npn–pnp* totem pole stack. Here, to turn the MOSFET on, the output transistor of the comparator turns off, thus turning the *npn* BJT on, which provides a positive gate voltage to the MOSFET. At the turn-off of the MOSFET, the gate is shorted to the source through R_G and the *pnp* transistor. Very often, instead of using discrete components, similar performance can be obtained, as is shown in Fig. 21-21*b*, by using buffer ICs such as CMOS 4049 or 4050 if a low gate current is needed, or a DS0026 or UC1707, which can source or sink currents in excess of 1 A.

All of the drive circuits discussed thus far can be made to provide positive gate voltages at turn-on and negative gate voltages at turn-off by means of a split power supply with respect to the MOSFET source, as shown in Fig. 21-22*a* through 21-22*c*.

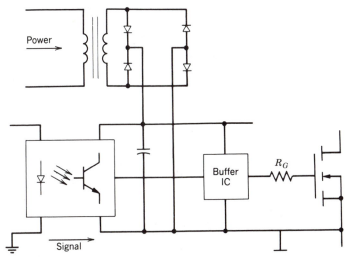

FIGURE 21-24: Electrical isolation of gate drive circuits using an optocoupler.

If the control signal is supplied by a logic circuit connected between V_{GG+} and the source of the MOSFET, then the reference input to the comparator should be shifted to be at the midpotential between V_{GG+} and the source of the MOSFET, as is shown in Fig. 21-23.

21-7-2   Electrical Isolation of Gate Drive Circuits

The need for electrical isolation between the logic level control signal and the MOSFET is the same as that described in the previous chapter for the BJT. Here also, the basic ways to provide electrical isolation is by means of fiber optics, optocouplers, or a transformer.

The optocoupler may be used as is shown in Fig. 21-24, where an isolated dc power supply is required with respect to the source terminal of the MOSFET. Such a design is discussed in the BJT chapter. The optocoupler in Fig. 21-24 can be replaced by fiber optics to provide increased voltage isolation and other advantages discussed in the previous chapter on BJTs. In gate drive circuits with transformer isolation, it is easy to transfer the control signal and the power needed to switch to the MOSFET

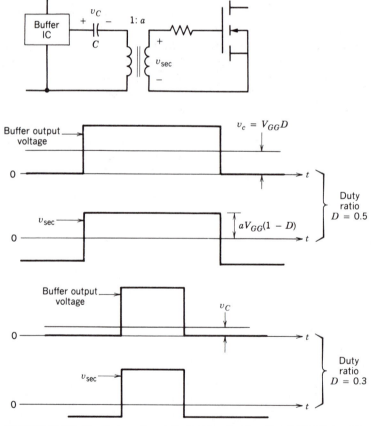

FIGURE 21-25: Simple transformer-isolated gate drive circuit. The duty ratio is limited to about 50% or less. Higher duty ratios result in larger voltages across the capacitor, which means that a fixed buffer output voltage produces a smaller secondary voltage V_{sec} to drive the gate.

through the same transformer, thus eliminating the need for an auxiliary dc power supply with respect to the source of the MOSFET. The design of such a gate drive depends on the application in which the MOSFET is used.

For switching dc power supply applications where the duty ratio is limited to 0 to about 50%, a simple transformer isolated gate drive circuit as shown in Fig. 21-25 is used, where a buffer of the type discussed earlier in this section drives the transformer. The capacitor C is needed to block the dc component in the output voltage of the buffer to prevent transformer saturation, as shown in Fig. 21-25. The reason why the upper limit on the duty ratio is about 50% is that otherwise, the gate supply voltage will not be sufficient for turning the MOSFET on.

A pulse transformer-isolated gate drive that can be used in all applications is shown in Fig. 21-26. It uses two buffers and a Schmitt trigger. When the control voltage goes high, v_A immediately goes low and v_B goes low after some delay based on the R-C time constant. During this delay interval, a voltage pulse is applied across the primary, which turns the MOSFET on. After the voltage pulse, the zener diode

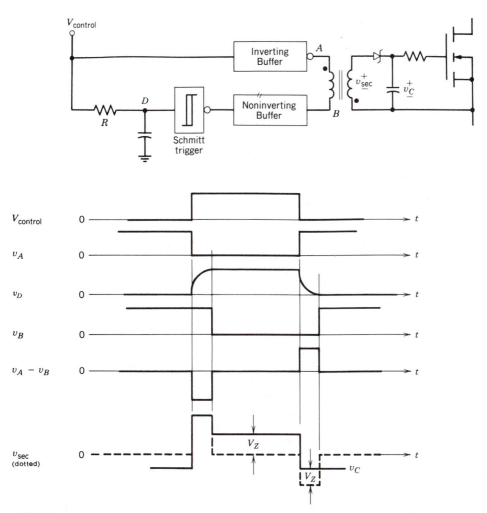

FIGURE 21-26: A pulse-transformer-isolated gate drive circuit that also provides the dc bias voltages. The zener diode breakdown voltage must be less than the negative pulse out of the transformer secondary in order that a negative pulse get to the MOSFET gate to turn it off.

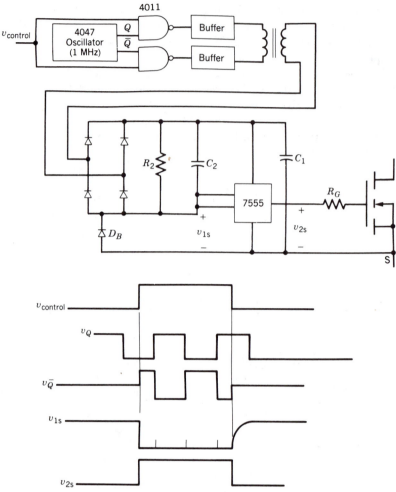

FIGURE 21-27: Transformer-isolated gate drive circuit using a high-frequency modulated carrier so that the MOSFET can be held on for long periods. No auxiliary dc power supplies are needed since both the control signal and bias power come through the transformer.

prevents the gate capacitance from discharging into the transformer secondary during the MOSFET on period and the gate voltage stays at the zener voltage, as is shown in the waveforms in Fig. 21-26. When the control voltage goes low, a voltage pulse of the opposite polarity is applied to the transformer primary. With the zener breakdown voltage less than the secondary voltage, a negative polarity voltage as shown in Fig. 21-26 is applied to the gate, thus turning it off.

If in a given application, the MOSFET to be controlled is to be on for a long time, the gate voltage will go below the desired on-state value due to the leakage current through the zener in the circuit of Fig. 21-26. This limitation is overcome by the circuit shown in Fig. 21-27, where the control voltage is modulated by a high-frequency oscillator output before being applied to the buffer circuits. Now a high-frequency ac signal appears across the transformer primary when the control voltage is high, thus charging the energy storage capacitance C_1 and the capacitance C_2 at the input to the 7555 IC, which is used here as a buffer and a Schmitt trigger because of its low power consumption. With the input to the 7555 low, it provides a positive voltage to the MOSFET gate, thus turning it on as is shown in Fig. 21-27. At turn-off, the control voltage goes low and the voltage across the transformer primary goes to zero. Now C_2 discharges through R_2 and the input voltage to the 7555 goes high, which causes its output voltage to go low, thus turning the MOSFET off. The diode D_B is used to prevent the energy stored in the capacitance C_1 from discharging into the resistance R_2.

In all of the foregoing drive circuits, back-to-back connected zener diodes can be connected across the gate-source terminals to protect the MOSFET gate from overvoltages.

21-8 SNUBBER CIRCUITS

Because of the square safe operating area that the MOSFET has for switch-mode applications, the need for snubbers in MOSFET circuits is greatly minimized compared with BJTs. However, a small R-C turn-off snubber as shown in Fig. 21-28 can be used to prevent voltage spikes and voltage oscillations across a MOSFET during device turn-on. The need for such a turn-off snubber increases with faster switching

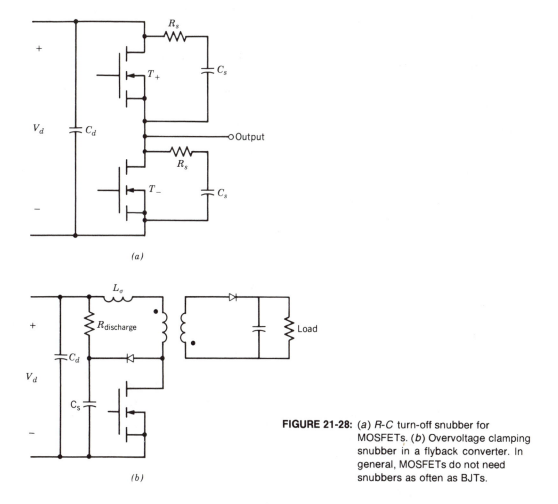

FIGURE 21-28: (a) R-C turn-off snubber for MOSFETs. (b) Overvoltage clamping snubber in a flyback converter. In general, MOSFETs do not need snubbers as often as BJTs.

speeds and higher leakage and stray inductances in the current-commutation loop as was described in the previous chapter on BJTs.

The large peak current handling capability of the MOSFET and the fact that its switching speed can be easily controlled by controlling the gate current eliminates the need for a turn-on snubber in most cases.

21-9   SUMMARY

This chapter has studied the structure and characteristics of MOSFETs designed for power applications. Based on this study, drive circuits for the MOSFET have been explored and the need for snubber circuits ascertained. The important conclusions are summarized:

1. MOSFETs have a vertically oriented structure with a lightly doped drain drift region and a highly interdigitated gate-source structure.

2. The MOSFET is a normally-off device and it is turned on by the application of a sufficiently large gate-source voltage to induce an inversion layer in the MOSFET channel region that shorts the drain to the source.

3. The MOSFET turns on and off very rapidly because it is a majority carrier device and there is no stored charge that must be injected into or removed from it as there is with the BJT.

4. On-state losses in a MOSFET rise much more rapidly with blocking voltage rating than do those in a BJT.

5. Because the MOSFET is a majority carrier device, its on-state resistance has a positive temperature coefficient, which makes it easy to parallel MOSFETs for increased current-handling capability.

6. The safe operating area of a MOSFET for switch-mode applications is large (rectangular) because it is not subject to second breakdown.

7. The MOSFET has a parasitic BJT in its structure that may latch in the on-state in extreme circumstances such as very large rates of increase in the drain-source voltage. Latchup of MOSFETs is most likely to occur in bridge circuits.

8. Drive circuits for the MOSFET are simpler than those for the BJT. However, MOSFET drive circuits still require careful design in order to control the rate of device turn-on and turn-off and to provide electrical isolation between the device and the logic level control circuitry.

9. The large SOA of the MOSFET means that in most situations snubber circuits are not needed for the device.

PROBLEMS

21-1. The small signal gate-source capacitance of a MOSFET decreases as V_{GS} is increased from zero volts. For $V_{GS} > V_{GS(\text{th})}$, the small signal capacitance is constant. Qualitatively explain the reasons for this behavior.

21-2. An n-channel MOSFET is to be used in a step-down converter circuit. The dc voltage $V_d = 300$ V, the load current $I_o = 10$ A, the free-wheeling diode is ideal, and the MOSFET is driven by a 15-V square wave (50% duty cycle and zero dc value) in series with 50 Ω. The MOSFET characteristics are $V_{GS(\text{th})} = 4$ V, $I_D = 10$ A at $V_{GS} = 7$ V, $C_{gs} = 1000$ pF, $C_{gd} = 150$ pF, and $r_{DS(\text{on})} = 0.5$ Ω.
(a) Sketch and dimensions $v_{DS}(t)$ and $i_D(t)$.
(b) Estimate the power dissipation at a switching frequency of 20 kHz.

21-3. The switching times of a MOSFET with $V_{GS(th)} = 4$ V, $g_m = 1$ mho, and $C_{gs} = 1000$ pF are measured in a resistively loaded test circuit having a load resistance $R_D = 25$ Ω and a power supply voltage $V_d = 25$ V. The MOSFET is driven by a unipolar square wave of 15 V in series with 5 Ω. The measured switching times are $t_{fv} = t_{ri} = 30$ ns and $t_{fi} = t_{rv} = 70$ ns. The MOSFET is to be used in a resistively loaded circuit having $R_D = 150$ Ω, $V_d = 300$ V, and the drive circuit is a 15-V unipolar square wave in series with 100 Ω. What will the switching times be in this circuit?

21-4. The MOSFET used in Problem 21-3 has an on-state resistance $r_{DS(on)} = 2$ Ω at a junction temperature $T_j = 25°C$. This resistance increases linearly with increasing T_j, becoming equal to 3 Ω at $T_j = 100°C$. Plot the power dissipated in the MOSFET versus T_j when the MOSFET is used in the circuit of Problem 21-3 ($R_D = 150$ Ω and $V_d = 300$ V). Assume a switching frequency of 10 kHz.

21-5. Three MOSFETS, each rated at 2 A, are to be used in parallel to sink a 5 A load current when they are on. The nominal on-state resistance of the MOSFETs is 2 Ω at $T_j = 25°C$, but measurements indicate that the actual values for each MOSFET are $r_{DS(on)1} = 1.8$ Ω, $r_{DS(on)2} = 2.0$ Ω, and $r_{DS(on)3} = 2.2$ Ω. The on-state resistance increases linearly with temperature and is 1.8 times larger at $T_j = 125°C$. How much power is dissipated in each MOSFET at a junction temperature of 105°C? Assume a 50% duty cycle.

21-6. A hybrid power switch composed of a MOSFET and a BJT connected in parallel is to be used in a switch-mode power application. Explain what the advantages of such a hybrid switch are and what the relative timing of the turn-on and turn-off of the two devices should be.

21-7. A MOSFET is to be used in a step-down converter circuit. The power supply voltage V_d for the circuit is 100 V and the load current $I_o = 100$ A. The totem-pole drive circuit of Fig. 21-22b is used to drive the MOSFET. The BJTs can be considered ideal switches. Complete the design of the drive circuit by specifying the values or a range of values for V_{GG+}, V_{GG-}, and R_G. The rate of rise and fall of v_{DS}, dv_{DS}/dt, must be limited to 500 V/μs or smaller. The MOSFET parameters are listed:

$$C_{gs} = 1000 \text{ pF} \qquad C_{gd} = 400 \text{ pF} \qquad I_{DM} = 200 \text{ A} \qquad V_{GS(max)} = \pm 20 \text{ V}$$

$$BV_{DSS} = 200 \text{ V} \qquad V_{GS(th)} = 4 \text{ V} \qquad I_D = 60 \text{ A at } V_{GS} = 7 \text{ V}$$

21-10 REFERENCES

1. B. JAYANT BALIGA, *Modern Power Devices*, John Wiley & Sons, New York, 1987, Chapter 6.

2. PIERRE ALOISI, *Power Switch*, Motorola Inc., 1986, Section 2.

3. EDWIN S. OXNER, *Power FETs and Their Applications*, Prentice-Hall, Englewood Cliffs, NJ, 1982, Chapters 1–4, 9.

4. B. W. WILLIAMS, *Power Electronics, Devices, Drivers, and Applications*, John Wiley & Sons, New York, 1987, Chapters 3, 4, 7–10.

5. ADOLPH BLICHER, *Field Effect and Bipolar Power Transistor Physics*, Academic Press, New York, 1981, Chapters 12, 13.

6. B. JAYANT BALIGA and DAN Y. CHEN, Editors, *Power Transistors, Device Design and Applications*, IEEE Press, Institute of Electrical and Electronic Engineers, New York, 1984, Part III, *Power MOSFET Field Effect Transistors*, pp. 197–290.

7. MICHAEL S. ADLER, KING W. OWYANG, B. JAYANT BALIGA, and RICHARD A. KOKOSA, "The Evolution of Power Device Technology," *IEEE Transactions on Electron Devices*, Vol. ED-31, No. 11, pp. 1570–1591, Nov. 1984.

22

Thyristors

_____
_____
_____

22-1 INTRODUCTION

Thyristors (sometimes termed SCRs, meaning semiconductor controlled recti-fiers) are one of the oldest (1957 in General Electric Research Laboratories) types of solid-state power device and still have the highest power-handling capability. They have a unique four-layer construction and are a latching switch that can be turned on by the control terminal (gate) but cannot be turned off by the gate. The characteristics of the thyristor (particularly their large power-handling capability) ensures that they will always have important power electronic applications. Thus, the designer and user of power electronic devices and circuits must have a working knowledge of these devices, how to drive them, and the type of snubber circuits that should be used to protect these devices. This chapter addresses these topics.

22-2 BASIC STRUCTURE

The vertical cross section of a generic thyristor is shown in Fig. 22-1a. The approximate thicknesses of each of the four alternating layers of p-type and n-type doping that comprise the structure are indicated as well as the approximate doping densities. In terms of their lateral dimensions, thyristors are among the largest semiconductor devices made. A complete silicon wafer as large as 10 cm may be used to make a single high-power thyristor. Plan views of two different gate and cathode layouts are shown in Fig. 22-1b. The distributed gate structure is for a large-diameter (10 cm) thyristor while the localized gate electrode structure is for a smaller diameter thyristor. In general the layout of the gates and cathodes of thyristors varies greatly depending on the diameter of the thyristor, the intended di/dt capability, and the intended range of switching speeds.

The circuit symbol for the thyristor is shown in Fig. 22-1c. It is essentially the symbol of a diode (or rectifier) with a third control terminal, the gate, added to it. The voltage and current conventions for the thyristor are shown in the figure.

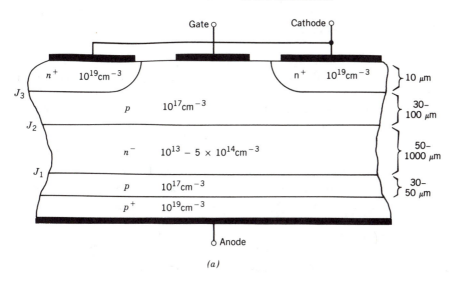

(a)

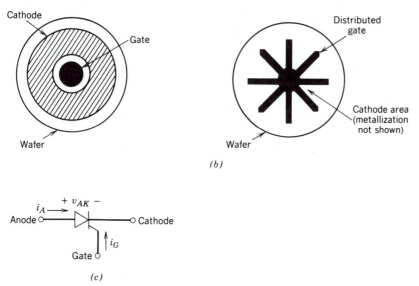

(b)

FIGURE 22-1: Structural details of a generic thyristor: *(a)* vertical cross section; *(b)* gate and cathode layouts; *(c)* circuit symbol.

The vertical cross section of the thyristor appears similar to that of the bipolar junction transistor, including some of the doping densities and layer thicknesses. The cathode is in the same location as the emitter of the BJT and the thyristor gate location is analogous to the base of the BJT. The n^- region of the thyristor absorbs the depletion layer of the junction that blocks the applied voltage when the thyristor is in the off-state, thus performing the same function as the n^- collector drift region of the BJT.

The p layer that forms the anode of the thyristor is a feature of the thyristor structure not found in the BJT. This anode layer causes the thyristor to have characteristics quite different from those of the BJT.

22-3 *I-V* CHARACTERISTICS

The uniqueness of the thyristor lies principally in its i-v characteristic (anode current i_A as a function of the anode-to-cathode voltage v_{AK}), which is shown in Fig. 22-2. In the reverse direction the thyristor appears similar to a reverse-biased diode, which conducts very little current until avalanche breakdown occurs. For a thyristor the maximum reverse working voltage is termed V_{RWM} (devices with values of V_{RWM} as large as 7000 V are available). In the forward direction the thyristor has two stable states or modes of operation that are connected together by an unstable mode that appears as a negative resistance on the i-v characteristic. The low-current, high-voltage region is the forward blocking state or the off-state, and the low-voltage, high-current mode is the on-state. Both of these states are indicated on the i-v characteristic. In the on-state a high-power thyristor can conduct average currents as large as 2000 to 3000 A with on-state voltage drops of only a few volts.

Specific voltage and current values in the forward bias quadrant of the i-v characteristic are of interest to the user and appear on specification sheets. The holding current I_H represents the minimum current that can flow through the thyristor and still maintain the device in the on-state. This current value and the accompanying voltage across the device, termed the holding voltage V_H, represent the lowest possible extension of the on-state portion of the i-v characteristic. For the forward blocking state, the quantities of interest are the forward blocking voltage V_{BO} (sometimes termed the breakover voltage V_{BO} because the i-v curve breaks over and goes to the on-state portion of the characteristic) and the accompanying breakover current I_{BO}.

The breakover voltage and current are defined for zero gate current, that is, the gate is open-circuited. If a positive gate current is applied to the thyristor, then the transition or breakover to the on-state will occur at smaller values of anode-to-cathode voltage, as indicated in Fig. 22-2. As indicated in this figure, the thyristor will switch

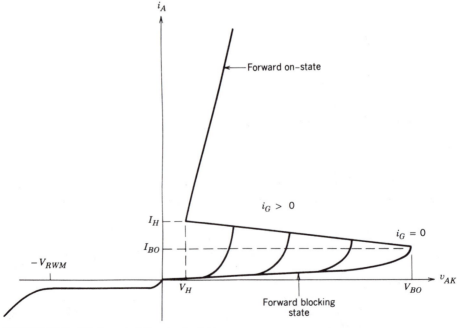

FIGURE 22-2: *I-V* characteristic of a thyristor.

to the on-state at low values of v_{AK} if the gate current is reasonably large. Although not indicated on the *i-v* characteristic, the gate current does not have to be a dc current, but instead can be a pulse of current having some minimum time duration. This ability to switch the thyristor on by means of a current pulse has been the basis of the widespread applications of the device.

However, once the thyristor is in the on-state, the gate cannot be used to turn the device off. The only way to turn off a thyristor is for the external circuit to force the current through the device to be less than the holding current for a minimum specified time period. Special designs of thyristors termed GTOs (gate turn-off devices) have been developed where the gate can be used to turn off the device. These devices will be discussed in the next chapter.

22-4 PHYSICS OF DEVICE OPERATION

22-4-1 Blocking States

In describing how the thyristor operates from a physical point of view, it is convenient to consider the device as the idealized one-dimensional structure shown in Fig. 22-3a. An approximate low-frequency equivalent circuit composed of a *pnp* and a *npn* transistor is shown in Fig. 22-3b, which is easily derived from the one-dimensional model.

In the reverse blocking state, the anode is biased negative with respect to the cathode. Junctions J_1 and J_3 indicated in Fig. 22-3a are reverse biased and J_2 is forward biased. Junction J_1 must support the reverse voltage because J_3 has a low breakdown voltage as a consequence of the heavy doping on both sides of the junction (examine Fig. 22-1). The reverse blocking capability of junction J_1 is usually limited

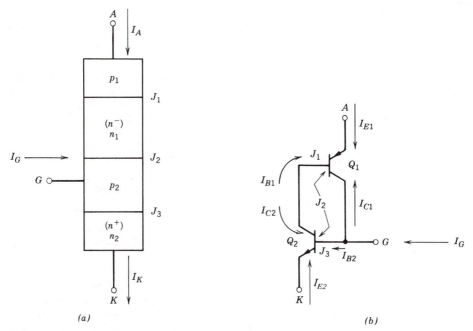

FIGURE 22-3: Simplified models of a thyristor: (a) one-dimensional model of a thyristor, (b) two-transistor equivalent circuit of thyristor.

by the length of the n^- (n_1) region, which is approximately set by the avalanche breakdown limit given by Eq. 19-3.

In the forward blocking state, the junctions J_1 and J_3 are forward biased and J_2 is reverse biased. The doping densities in each of the layers are such that the n^- layer (n_1 layer) is where the depletion region of the reverse-biased J_2 junction appears, and thus this region again determines the blocking voltage capability, this time for the forward blocking state. Generally, the thyristor is designed so that the forward blocking voltage V_{BO} will be about the same as the reverse blocking voltage V_{RWM}.

The values of V_{B0} and V_{RWM} begin to fall rapidly with increasing temperature as the junction temperature rises above roughly 150°C. The detailed explanation for this temperature dependence is beyond the scope of this discussion. Suffice it to say, most device manufacturers specify the maximum junction temperature for their thyristors at 125°C.

On the basis of the bias conditions of the three junctions in the forward blocking state, it can be concluded that both transistors Q_1 and Q_2 in the thyristor equivalent circuit, Fig. 22-3b, are in the active region. Recognition of this fact makes possible a relatively simple, although admittedly qualitative, explanation of the forward bias i-v characteristic of the thyristor. A BJT in the active region can be described, at low frequencies, by the Ebers-Moll equations. For transistors Q_1 and Q_2 in the active region, these equations are (the reader is referred to the references for a detailed derivation of these relations)

$$I_{C1} = -\alpha_1 I_{E1} + I_{CO1} \tag{22-1}$$

and

$$I_{C2} = -\alpha_2 I_{E2} + I_{CO2} \tag{22-2}$$

where the leakage current I_{CO} is given by

$$I_{CO} = I_{CS}\left[1 - \alpha_f \alpha_r\right] \tag{22-3}$$

In Eq. 22-3 I_{CS} is the reverse saturation current of the collector-base diode with the emitter open-circuited and α_f and α_r are the forward active mode and reverse active mode base transport coefficients (or alphas where $\alpha \approx \beta/[1 + \beta]$), respectively, of the transistors. (In the description of how an npn BJT operates, which is given in Chapter 20, the base transport factor is the ratio of the electron current at the collector divided by the injected electron current at the emitter or using the notation of Eqs. 20-2 through 20-4, $\alpha = I_{nc}/I_{ne}$.) If we note that $I_A = I_{E1}$ and that $I_K = -I_{E2} = I_A + I_G$ and setting the sum of all the currents into one of the transistors to zero, it can be shown that I_A is given by

$$I_A = \left[\frac{\alpha_2 I_G + I_{CO1} + I_{CO2}}{1 - (\alpha_1 + \alpha_2)}\right] \tag{22-4}$$

In the blocking state the sum of $\alpha_1 + \alpha_2$ must be much less than unity so that the anode current I_A can be kept quite small, ranging from microamps for low-current devices to a few hundred milliamps for high-current thyristors.

22-4-2   Turn-on Process

Equation 22-4 indicates that if $\alpha_1 + \alpha_2$ approaches unity, the anode current will be arbitrarily large. If this occurs, the thyristor will be at the breakover point where it is about to go into the negative resistance state, where the current gain (β) of both BJTs is greater than one. The negative resistance region is unstable because of the regenerative (positive feedback) connection of the two transistors. Once in this region, the device will quickly carry itself to the stable on-state. The key to the turn-on process is then understanding how the base transport coefficients (alphas) of the BJTs can be increased from the small values required for blocking state operation to the point where their sum is unity.

The mechanism that causes alphas to increase is the growth of the depletion region of junction J_2 into the n_1 layer as the anode–cathode voltage is increased. This causes the effective base thickness of the $p_1 n_1 p_2$ BJT to get smaller and, hence, for α_{pnp} to increase. The extension of the J_2 depletion layer into the p_2 region (the base region of the npn transistor) will likewise cause an increase in α_{npn}.

The combination of the positive feedback connection of the npn and pnp BJTs and the current dependent base transport factors is what makes it possible for the gate terminal to effect a turn-on of the thyristor. If a positive gate current of sufficient magnitude is applied to the thyristor, a significant amount of electron injection across the forward biased J_3 junction into the p_2 base layer of the npn transistor will occur. The electrons will diffuse across the base and be swept across junction J_2 into the n_1 base layer of the pnp transistor.

These extra electrons in the n_1 layer will have two simultaneous effects. First, the depletion layer of junction J_2 will grow in thickness because additional positive space charge from ionized donors is needed to partially compensate for the negative space charge of the electrons. This growth of the depletion layer into the n_1 layer will reduce the effective base thickness of the pnp transistor and thus cause α_{pnp} to get larger. Secondly, the injection of majority carriers (electrons) into the base of the pnp transistor will cause the injection of holes into this base layer (because of space charge neutrality requirements) via injection from the base-emitter junction of the pnp transistor (the $p_1 n_1$ junction). These injected holes will diffuse across the base and be swept across the depletion region of the J_2 junction and into the base region of the npn BJT. These holes will cause a further increase in electron injection (because the positive space charge of the holes attracts the electrons from the n_2 layer) into the base region of the npn BJT. These additional injected electrons will go through the same cycle just described for the initial injection of electrons. This positive reinforcement is thus a regenerative process.

This regenerative action carries the thyristor into the on-state. The large current flow between the anode and cathode injects enough carriers into the base regions to keep the BJTs saturated without any continuous gate current flow. This is the origin of the latching action of the thyristor described earlier.

22-4-3   On-State Operation

In the on-state there is strong minority carrier injection in all four regions of the thyristor structure. The stored charge distribution in the four regions is shown schematically in Fig. 22-4. Junction J_2 is forward biased, and the BJTs in the equivalent circuit are saturated. In this situation large forward currents can flow (dictated by the external circuit), and only a small forward voltage drop occurs

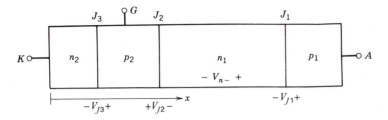

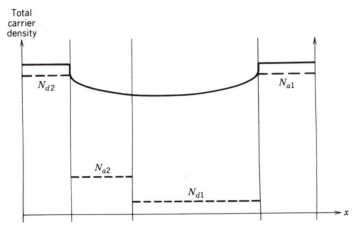

FIGURE 22-4: On-state carrier distributions in a thyristor.

because of the large conductivity modulation represented by the stored charge distribution illustrated in Fig. 22-4. The on-state voltage is given approximately by

$$V_{AK(on)} = V_{J1} - V_{J2} + V_{J3} + V_{n-} \tag{22-5}$$

where the V_Js are the forward bias junction voltages (0.7 to 0.9 V) and V_{n-}, which is on the order of a few tenths of a volt, is relatively independent of the current through the device, and is given approximately by Eq. 19-13. The value of $V_{AK(on)}$ in Eq. 22-5 is quite similar to the expression for the on-state voltage of a diode (Eq. 19-17). At large current densities, the on-state voltage will increase with increasing current because of Auger recombination and reduction in carrier mobilities as described in Chapter 19 as well as parasitic ohmic resistances. The net effect of these factors is an on-state resistance.

22-4-4   Turn-off Process

As we mentioned previously, once the thyristor is latched into the on-state, the gate terminal no longer has any control over the state of the device. In particular, the gate cannot be used to turn the thyristor off. Turn-off can be accomplished only by the external circuit, reducing the anode current below the holding current for a minimum specified period of time. During this time period, the simultaneous action of internal recombination and carrier sweep-out will remove enough stored charge so that the BJTs are pulled out of saturation and into the active region. When this occurs the device will turn off via the regenerative connection of the transistors.

In the standard thyristor, a negative gate current cannot turn off the device because the cathode region is much greater in area than the gate area. When a

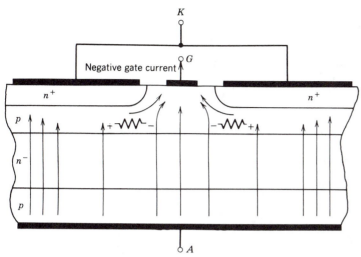

FIGURE 22-5: Current density distribution in a thyristor during attempted turn-off with a negative gate current illustrating current constriction at the center of the cathode.

negative gate current flows, it can only locally reverse bias the gate-cathode (base-emitter of the *npn* transistor) junction as is shown in Fig. 22-5. Lateral voltage drops in the p_2 region caused by the negative i_G will result in current crowding toward the center of the cathode region similar to emitter current crowding during the turn-off of a BJT. The anode current that now flows more in the center of each cathode region keeps the gate–cathode junction forward biased in this center portion and the thyristor stays on in spite of a negative gate current.

22-5 SWITCHING CHARACTERISTICS

22-5-1 Turn-on Transient and *di/dt* Limitations

In describing the turn-on transient of a thyristor, we will assume that the device is embedded in the circuit of Fig. 22-6, which is a simplified diagram of a multiple-phase controlled rectifier. Turn-on of a thyristor is accomplished by applying a pulse of current of specified magnitude and duration to the gate of the device. The gate current is applied at $t = 0$ as is shown in Fig. 22-7 to thyristor T_A in Fig. 22-6 with $t = 0$ corresponding to a time when the voltage in phase A is larger than in the other two phases. The resulting waveforms for the anode current and anode–cathode voltage are shown in the Fig. 22-7. The anode current increases at a fixed rate di_F/dt, which is set by the external circuit because of the switching times of other devices or because of stray inductance in the circuit. Three distinct time intervals can be defined including the turn-on delay time $t_{d(on)}$, the rise time t_r, and the spreading time t_{ps}.

During the turn-on delay time, the thyristor appears to remain in the blocking state. However, the gate current during this time is injecting excess carriers into the p_2 layer (the base of the *npn* transistor in the equivalent circuit) in the vicinity of the gate contact, as is illustrated in Fig. 22-8a. This increase in excess carriers causes the sum of the base transport factors, $\alpha_1 + \alpha_2$, to increase until they equal unity. At this point, the thyristor is at breakover and heavy electron injection into the p_2 layer from the n_2 cathode layer and hole injection from the p_1 layer into the n_1 layer commences in the

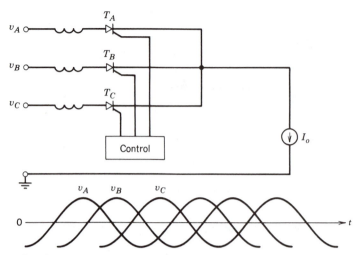

FIGURE 22-6: Multiple-phase controlled rectifier using thyristors.

vicinity of the gate regions, as is illustrated in Fig. 22-8b. The anode current begins to increase, and this marks the end of the turn-on delay time and the start of the rise time interval.

During the rise time interval, a large excess carrier density or plasma is built up in the vicinity of the gate regions, which then spreads laterally across the face of the cathode until the entire cross-sectional area of the thyristor is filled with a high excess carrier density. Simultaneously, there is the commencement and growth of carrier injection from the p_1 anode region into the n_1 layer that forms the base of the pnp transistor. The rate of current rise is usually large enough that the anode current reaches its constant on-state value in significantly shorter times than are required for the excess carrier injection to spread laterally across the entire face of the cathode

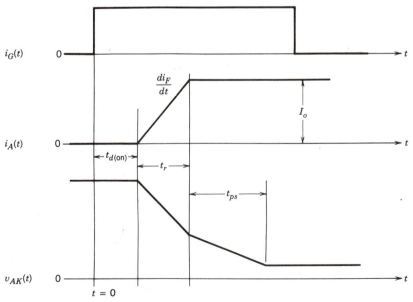

FIGURE 22-7: Thyristor current and voltage waveforms during turn-on.

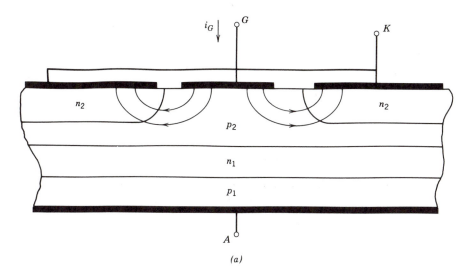

(a)

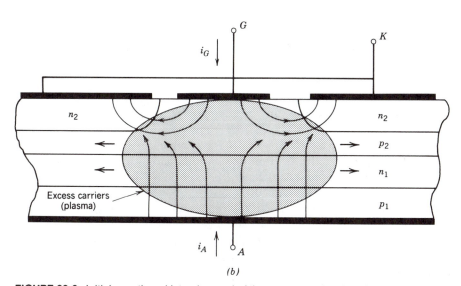

(b)

FIGURE 22-8: Initial growth and lateral spread of the excess carriers in a thyristor at turn-on illustrating the need to limit di_F/dt: (a) injection of minority carriers into the p_2 base region by the gate current during the turn-on delay time that initiates the regenerative switching action; (b) initial turned-on areas of the thyristor in the vicinity of the gate electrodes shortly after the turn-on delay time. The further lateral expansion of this area is also shown.

region. The attainment of the on-state value of the anode current marks the end of the rise time interval.

As the excess carrier density becomes established and grows, as is shown in Fig. 22-8b, the anode–cathode voltage begins to drop. During the rise time interval, the voltage drop is fairly rapid because the localized regions of high excess carrier densities in the vicinity of the gate regions provide a significant reduction in the blocking capability of the thyristor. Even after the rise time interval is over, the plasma still continues to spread over the lateral area of the thyristor, as is shown in

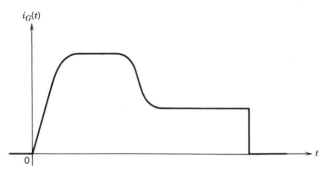

$i_G(t)$

0 t

FIGURE 22-9: A gate current with an initial large value in order
to maximize the initial turned-on areas of the
thyristor. The current is then reduced to smaller
values for a sufficient period time to guarantee
device turn-on.

Fig. 22-8*b*, until the thyristor is completely shorted out by the large excess carrier
densities. The time required for the plasma to spread from the initial regions around
the gate terminals to the entire device cross section is the plasma spreading time t_{ps}.
Typical rates of plasma spreading are given in terms of the plasma spreading velocity,
which has values in the range of 20 to 200 $\mu m/\mu s$. For large-area devices having
diameters of centimeters, it can take several hundred microseconds to completely
turn-on the device if the plasma must spread from a single gate electrode such as is
shown in Fig. 22-1*b*. The rate of voltage drop during the plasma-spreading time is
slower because t_{ps} is larger than t_r and because most of the drop occurs during the t_r
interval.

It is important that the rate of rise of the anode current be kept less than a
maximum value given on the thyristor specification sheet. If di_F/dt exceeds this
maximum rate, the device may be damaged or even fail. Such damage may occur
because large rates of current growth mean that rise time will be short and,
consequently, the turned-on area around the gate region of the thyristor will be quite
small at the end of the rise time interval compared with the cross-sectional area of the
device. A small turned-on area further means that the voltage across the thyristor will
not have fallen very far from the blocking state value during the rise time interval.
Hence, the instantaneous power dissipation during the t_r interval will both be large
and confined to a relatively small volume. In such a situation, the ability to remove
the heat generated by the dissipation will be less than the rate of dissipation and thus
the internal temperature of the region may grow so large that thermal runaway in the
turned-on area will occur and, hence, device damage or failure.

Larger values of gate current during the $t_{d(on)}$ and t_r intervals will increase the
size of the turned-on area by providing a larger amount of excess carriers. A larger
turned-on area will reduce the peak instantaneous power dissipation. For this reason,
gate current applied is often large at the start of the turn-on interval and is gradually
reduced as time proceeds as is shown in Fig. 22-9. There are also structural modifications that can improve the di_F/dt rating, which will be discussed in a later section.

22-5-2   Turn-off Transient

Turn-off of the thyristor requires that it be reverse biased by the external circuit
for a minimum time period. For thyristor T_A in Fig. 22-6, turn-off commences when

thyristor T_B is turned on which would be done when the voltage in phase B is larger than it is in phase A. The larger phase B voltage clearly will reverse bias T_A as soon as T_B is fully on. The turn-on time of a thyristor is appreciably shorter than the turn-off time; hence, as far as the discussion of the turn-off of T_A is concerned, the turn-on of T_B is nearly instantaneous.

However, the commutation of current from T_A to T_B will not be instantaneous, but will occur over an extended time period as is shown in the turn-off waveforms for T_A in Fig. 22-10. The current through T_A starts decreasing at $t = 0$ at a fixed rate di_R/dt, which is governed by the external circuit. The overall turn-off process is quite similar to the turn-off of the power diode described in Chapter 19. As the current decreases, the excess carrier in the four regions of the thyristor are decreasing from the steady-state values shown in the carrier distribution of Fig. 22-4 by a combination of internal recombination and carrier sweep-out.

As time proceeds, the current continues to decrease and soon passes through zero at a time t_1 and then grows toward negative values, as shown in Fig. 22-10. The voltage across the thyristor remains small and positive until either junction J_1 or J_3 start to become reverse biased, an event that does not occur until the excess carrier density at the junction has decayed to zero. Usually J_3 will become reverse biased first, occurring at time t_2 in Fig. 22-10, and then very quickly goes into avalanche breakdown as the anode–cathode voltage goes negative because the heavy doping in the n_2 and p_2 layers means that the reverse blocking capability of this junction is not very large (20 to 30 V typically).

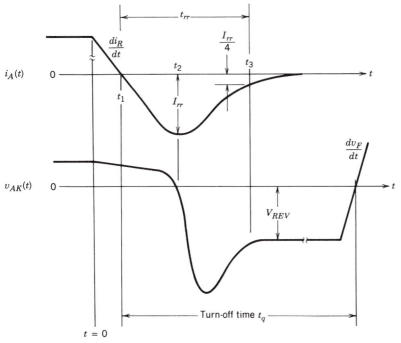

FIGURE 22-10: Thyristor voltage and current waveforms during turn-off. A reapplied forward-blocking voltage must not be impressed on the thyristor until a specified time period, the recovery time t_q, has passed. The rate of rise of the reapplied forward voltage dv_F/dt must be kept below a specified value.

At or shortly after time t_2 the excess carrier distribution in the thyristor is no longer large enough to sustain the ever-growing negative anode current and so the current attains its peak negative value I_{rr} and begins to decay back toward zero. At about the same time the excess carrier density at junction J_3 goes to zero, and it becomes reverse biased. The growth of the negative anode–cathode voltage, which began at t_2, continues and overshoots the value $V_{REV} = V_B - V_A$, which will eventually be imposed on thyristor T_A by the circuit. The voltage overshoot arises from the inductance of the circuit and is governed by how rapidly the anode current decays to zero from its peak reverse value of I_{rr}. This overall process is quite similar to that described for the power diode in Chapter 19.

22-5-3   Turn-off Time and Reapplied dv_F/dt Limitations

In the case of the power diode the reverse recovery transient was defined to be over when the reverse current had decayed to some conveniently small value such as $I_{rr}/4$ or $I_{rr}/10$, which is marked on the anode current waveform in Fig. 22-10 as time t_3. However, such a definition is not suitable for thyristors. Even at such a time as t_3, there are still substantial excess carriers remaining in the n_1 and p_2 regions of the thyristor. If a forward voltage is reapplied to the thyristor at a rate dv_F/dt as is shown in Fig. 22-11, a pulse of decaying forward current, a forward recovery current, would flow as the remaining excess carriers just mentioned continued to simultaneously recombine internally and be swept out by the growing forward voltage.

The pulse of forward recovery current may have the same consequence as a deliberately applied pulse of gate current. If the forward recovery current is large enough, it can turn the thyristor on even though turn-on was not intended. Since the larger the value of dv_F/dt the larger the peak forward recovery current will be, two things must be done to prevent accidental turn-on. First, the time that the thyristor is maintained in the reverse blocking mode must be lengthened beyond the time t_3. Device manufacturers specify a turn-off time t_q for their thyristors, which represents the minimum time their thyristor should remain in the reverse blocking mode before

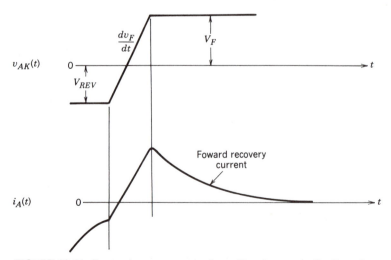

FIGURE 22-11: Forward recovery current resulting from reapplication of a forward voltage across the thyristor before the end of the specified recovery time. If the forward recovery current is too large, inadvertent turn-on of the thyristor may result.

any forward voltage is reapplied. This turn-off time is usually several excess carrier lifetimes in length.

Second, the rate of growth dv_F/dt of the reapplied forward voltage should be kept below a maximum value, which is also specified by the device manufacturer. This maximum value is usually arrived at on the basis of how large of a displacement current a given dv_F/dt can drive through the space charge capacitance of junction J_2. If this displacement current $C_{j_2} \, dv_F/dt$ exceeds the breakover current I_{BO}, device turn-on may result. Hence, the reapplied dv_F/dt should be limited to

$$dv_F/dt|_{max} < I_{BO}/C_{j_2} \qquad (22\text{-}6)$$

Maximum values of dv_F/dt range from perhaps 100 V/μs for slow devices intended for low-frequency phase control applications to several thousand volts per microsecond or larger in devices intended for higher frequency inverter applications and high-voltage dc.

22-6 METHODS OF IMPROVING *di/dt* AND *dv/dt* RATINGS

22-6-1 Improvement in *di/dt*

The key to improving the di/dt rating is to increase the amount of initial area of cathode conduction, since this is the factor that limits the rate of rise of the anode current. One way to increase this area is to increase the gate current as has already been mentioned. But it is desirable to do this without requiring the gate drive circuit to deliver substantially higher gate currents. One way of accomplishing this goal is to use a smaller auxiliary or pilot thyristor to provide large gate currents to the main thyristor, as diagrammed in Fig. 22-12. Furthermore, this pilot thyristor can be integrated onto the same silicon wafer as the main device.

Another improvement that can be made is to modify the gate–cathode geometry so that there are many small cathode and gate regions intermixed together, much the same as the base and emitter regions of a power BJT are constructed. By such intermixing or interdigitating using a variety of complex geometries (for example, some devices use a complicated involute gate structure), the gate–cathode periphery is made large compared with the cathode area. The distributed gate structure shown in Fig. 22-1*b* is a step in this direction. The greater gate–cathode periphery leads to a significant increase in the initial conducting area of the thyristor and hence to a larger di/dt capability.

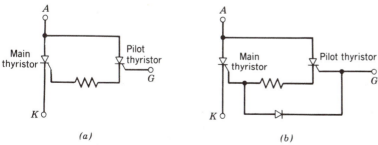

FIGURE 22-12: Thyristor modifications for more rapid turn-on and turn-off: (*a*) thyristor with an auxiliary thyristor to provide large turn-on gate currents, (*b*) gate-assisted turn-off thyristor (GATT).

The use of an interdigitated gate–cathode structure can help to shorten the turn-off time of the thyristor. The large-area cathode and relatively small gate–cathode periphery in a conventional phase control thyristor makes negative gate currents ineffectual in turning off the device, as we explained earlier. However, a highly interdigitated gate–cathode structure where the center of the cathode region is not too far from the gate–cathode boundary makes current crowding toward the center of the cathode much weaker and, thus, allows a negative gate current to be more effective in sweeping out stored charge in the n_2 and p_2 regions, which will in turn shorten the turn-off time.

Since a highly interdigitated gate–cathode structure is usually used in conjunction with a pilot thyristor, a further modification, the addition of a diode as is shown in Fig. 22-12b, is necessary. If this diode is not added, then only the pilot thyristor would benefit from the negative gate current and no negative gate current would be drawn from the main thyristor. The circuit shown in Fig. 22-12b, which combines a pilot thyristor, a diode, and a highly interdigitated gate–cathode structure, is sometimes termed a gate-assisted turn-off thyristor (GATT). Even in this device reverse biasing of the anode–cathode terminals by the external circuit is required to turn off the thyristor. Turn-off times of 10 μs or less have been achieved in devices with forward blocking voltages of 2000 V and on-state currents of 1000 to 2000 A. Such devices can be used at switching frequencies of a few tens of kilohertz.

22-6-2   Cathode Shorts

One useful way to reduce the effects of displacement currents that limit dv_F/dt is by means of cathode shorts illustrated in Fig. 22-13. These shorts, which are realized by overlapping the cathode metallization over portions of the gate region (p_2 region), can partially intercept the displacement current, as diagrammed in the figure. Any portion of the displacement current that is diverted to the cathode short does not pass through the gate–cathode junction and, hence, does not cause carrier injection into the p-type base region. This in turn means that the total displacement current and thus dv_F/dt can be larger without turning on the device. It should also be evident that

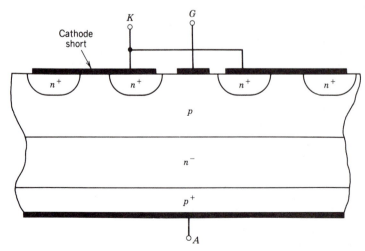

FIGURE 22-13: Thyristor with cathode shorts for enhancing the dv/dt rating of the device.

a highly interdigitated gate–cathode structure will make the use of cathode shorts more effective than in the conventional geometry of a phase control thyristor.

22-7 THYRISTOR GATING CIRCUITS

22-7-1 Gate Current Pulse Requirements

As was discussed in previous sections, a pulse of gate current is needed to turn the thyristor on, and once triggered on, the thyristor continues to conduct without any continuous gate current because of the regenerative action of the device. In estimating how large the pulse of gate current should be to ensure device turn-on, the gate current–voltage characteristic shown in Fig. 22-14a must be used. For a given thyristor type, the range of variation possible in the gate characteristic is specified by maximum and minimum curves in the device data sheets, which are similar to the maximum and minimum curves shown in Fig. 22-14a. The minimum gate current and corresponding gate voltage needed to ensure that the thyristor will be triggered at various operating temperatures is also specified in the device data sheets and is shown in Fig. 22-14a as the dotted line. This minimum gate current curve is sometimes called the locus of minimum firing points.

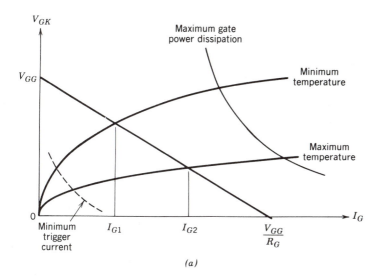

(a)

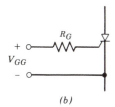

(b)

FIGURE 22-14: Design of thyristor gate drive circuits: (*a*) gate-cathode *i-v* characteristic used for designing gate trigger circuits; (*b*) equivalent circuit of gate-pulse amplifier.

The load line of the gate pulse amplifier should result in a gate current that is greater than that specified in the minimum gate current curve. An equivalent circuit for a gate pulse amplifier is shown in Fig. 22-14b, which consists of an open circuit output voltage of V_{GG} and an output resistance of R_G. This circuit will produce a gate current in the range of I_{G1} at low temperatures to I_{G2} at high temperatures, as shown in the load line construction of Fig. 22-14a. By proper selection of the load line parameters (V_{GG} and R_G), a gate current well in excess of the minimum required current is obtained. The minimum time duration of the current pulse, usually a few tens of microseconds, during which the gate current must flow, is specified on the device data sheets.

The data sheets also specify the maximum allowable gate current and gate power dissipation. A maximum gate power dissipation hyperbola is shown in Fig. 22-14a. These quantities are normally very large in relation to the gate trigger current needed and do not present any design constraint.

To allow a large di/dt during the turn-on of a thyristor, a large gate current pulse is supplied during the initial turn-on phase with a large di_G/dt. After the thyristor turns on, the gate current is then reduced and kept on for some time at a lower value in order to avoid unwanted turn-off of the device. Such a shaped gate current pulse is shown in Fig. 22-9.

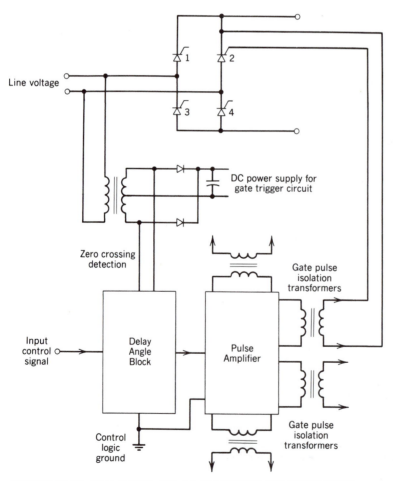

FIGURE 22-15: General block diagram of a thyristor gate trigger circuit.

Thyristors are often used in line frequency converters where the devices are naturally turned off by the line-frequency voltages. To control the dc output voltage of the converter and the magnitude and direction of the power flow through the converter, the thyristors must be turned on at a proper delay angle relative to the zero crossing of the ac line voltages. For load voltage commutated thyristor converters, like those used in very large-power synchronous motor drives and induction heating inverters, the gate trigger times of the thyristors are synchronized with the ac voltages of the load.

A general block diagram of a gate trigger circuit, for example, in a single-phase converter, is shown in Fig. 22-15. The thyristors are at line potential and the trigger circuit must be referenced with respect to a logic ground associated with the control input. Therefore, the zero crossing detection of the line-voltage synchronization and the gate pulse generated within the gate trigger circuit must be isolated from the line potential by means of transformers, as shown in Fig. 22-15. The gate trigger circuit also requires a dc power supply referenced with respect to the logic ground potential. This dc voltage can be supplied by rectifying the output of the line voltage synchronization transformer, as is shown in Fig. 22-15.

In the delay angle block, the ac synchronization voltage is converted into a ramp voltage, which gets synchronized to the zero crossing of the line voltage, as is shown by waveforms in Fig. 22-16. This ramp voltage, which has a constant peak-to-peak amplitude, is compared with a control voltage. During alternate half-cycles when the ramp voltage equals the control voltage, a pulse signal of controllable duration is generated, as is shown in Fig. 22-16. In this manner the delay angle can be varied over nearly the full range between 0° and 180° and the delay angle is proportional to the control voltage. Normally an integrated circuit such as one of the TCA780 family is used to implement this control function. Such integrated circuits also incorporate additional features for start-up, shut down, and so on.

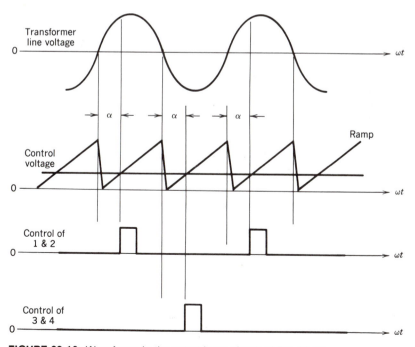

FIGURE 22-16: Waveforms in the gate trigger circuit of Fig. 22-15.

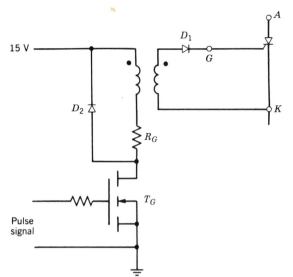

FIGURE 22-17: Pulse amplifier for a thyristor gating circuit.

22-7-2   Gate Pulse Amplifiers

In low-power thyristors used in consumer applications, the trigger current needed by the thyristor is small enough that it can be supplied by ICs without the need for an external pulse amplifier. In high-power thyristors, the trigger current requirement is purposely kept high in order to provide noise immunity. In such thyristors the initial peak gate current at turn-on such as is shown in Fig. 22-9 may be as large as 3 A and then may drop to about 0.5 A for the duration of the pulse.

A pulse amplifier is shown in Fig. 22-17, where the pulse output from the delay angle block turns on the MOSFET that supplies an amplified pulse of gate current to the thyristor through the pulse transformer. The diode D_1 in the secondary side is used to prevent a negative gate current due to the transformer magnetizing current when the MOSFET T_G turns off. The diode D_2 is used to provide a path for the transformer magnetizing current so that the energy in the magnetic core gets dissipated in R_G. Waveshaping to produce a waveform similar to the waveform of Fig. 22-9 can be provided by an R-C network in parallel with R_G.

A similar gate trigger circuit can be built for three-phase full-bridge thyristor converters where the six thyristors are triggered in sequence at $60°$ intervals. However, to get started and with discontinuous load currents, it is necessary to trigger a thyristor pair, one from the top group and one from the bottom group, in sequence at a $60°$ interval. Therefore, a thyristor will receive gate pulses, as shown in Fig. 22-18, at turn-on.

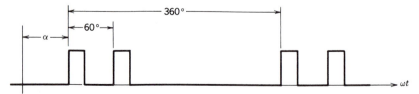

FIGURE 22-18: Gate trigger pulse waveform for a thyristor in a three-phase full-bridge converter.

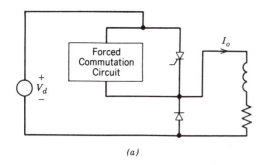

(a)

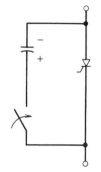

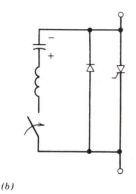

(b)

FIGURE 22-19: Turn-off of a thyristor by forced commutation: (*a*) thyristor switch-mode converter circuit illustrating the need for a commutating circuit, (*b*) idealized examples of commutation circuits. The plus and minus signs represent the polarity of the precharge voltage on the capacitors before the closure of the switches to turn off the thyristors.

22-7-3 Commutation Circuits

In the line-frequency and load-commutated converters, the thyristor current is naturally commutated, and the device turns off when the next thyristor in the sequence is gated on. However, in a switch-mode converter, a commutation circuit such as is shown in Fig. 22-19 is needed to turn off the thyristor. Because of the cost, complexity, and losses associated with the commutation circuits and, more important, because of the evolution of the power-handling capabilities of BJTs and GTOs, the switch-mode thyristor converter is not used in new designs, even at multi-megawatt power ratings for dc and ac motor drives.

These commutation circuits circulate a current through a conducting thyristor in the reverse direction and thus force the total thyristor current to go to zero, turning it off. These circuits often consist of some form of an *L-C* resonant circuit, similar to those discussed in Chapter 7. Such circuits are thoroughly discussed in the literature and so are not considered in this book.

22-8 SNUBBER CIRCUITS FOR THYRISTORS

The reverse recovery current of the thyristor, which was discussed earlier in this chapter, may result in unacceptably large overvoltages because of series inductance if snubbers are not used. In Chapter 19, it was shown that an equivalent circuit for a step-down dc–dc converter could be used to analyze the diode overvoltage snubber in any converter. That equivalent circuit is used here for a three-phase line-frequency thyristor converter of the type discussed in Chapter 4. The ac-side inductances shown in Fig. 22-20*a* are due to line reactances plus any transformer leakage inductance.

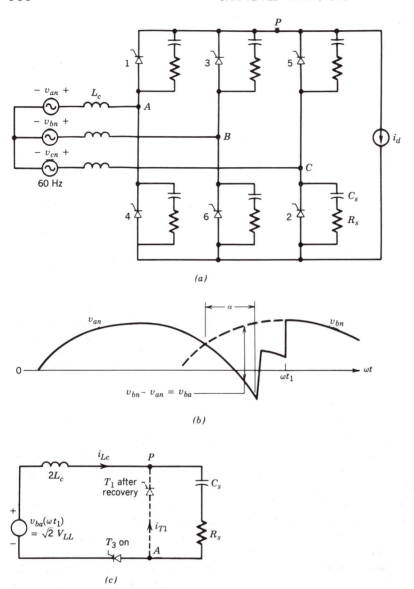

FIGURE 22-20: Turn-off snubbers for thyristors in a three-phase line frequency converter circuit: (*a*) three-phase line frequency converter; (*b*) triggering times; (*c*) equivalent circuit.

The dc-side is represented by a current source where i_d is assumed to flow continuously.

It is assumed that thyristors T_1 and T_2 have been conducting and that thyristor 3 is gated on at a delay angle α, as is shown in Fig. 22-20*b*. The current i_d will commutate from thyristor T_1 (connected to phase *a*) to thyristor T_3 (connected to phase *b*). The voltage v_{ba} is responsible for the commutation of the current. The subcircuit consisting of T_1 and T_3 is shown in Fig. 22-20*c* with T_3 on and T_1 is off and at its reverse recovery at ωt_1, with $i_{Lc} = I_{rr}$. The voltage source in the circuit of Fig. 22-20*c* can be assumed to be a constant dc voltage with a value of v_{ba} at ωt_1, because of the slow variation of 60-Hz voltages compared with the fast voltage and current transients in this circuit. The snubber voltage and current waveforms will be identical to those described in Fig. 19-14, which is in the diode chapter.

To discuss the design of the snubber, a worst case line impedance of 5% is used as explained in Chapter 19, Eq. 19-40, where

$$x_c = \omega L_c = \frac{0.05 V_{LL}}{\sqrt{3}\, I_{a1}} \qquad (22\text{-}7)$$

where V_{LL} is the rms line-to-line voltage and I_{a1} is the rms of the fundamental component of the line current. For a worst case design, the voltage source in Fig. 22-20c will have its maximum value of $\sqrt{2}\, V_{LL}$, which corresponds to $\alpha = 90°$. Thus, during the current commutation, assuming that the commutation voltage has a constant value of $\sqrt{2}\, V_{LL}$, the di/dt through thyristor T_1 is

$$\frac{di}{dt} = \frac{\sqrt{2}\, V_{LL}}{2 L_c} \qquad (22\text{-}8)$$

and, therefore,

$$I_{rr} = \left(\frac{di}{dt}\right) t_{rr} = \frac{\sqrt{6}\, \omega V_{LL} t_{rr} I_{a1}}{0.1 V_{LL}} = 0.09 I_{a1} \qquad (22\text{-}9)$$

where it is assumed $t_{rr} = 10\ \mu s$, that the $t_3 - t_2$ interval of t_{rr}, shown in Fig. 22-10 is zero which is a worst case, and finally $\omega = 377$.

As we discussed in Chapter 19, $C_s = C_{\text{base}}$ is close to an optimum value. If we relate Fig. 22-20c to Fig. 19-13a and Eq. 19-29 of Chapter 19, we get

$$C_{\text{base}} = L_c \left[\frac{I_{rr}}{V_{LL}}\right]^2 \qquad (22\text{-}10)$$

Substituting L_c from Eq. 22-7 at $\omega = 377$ and I_{rr} from Eq. 22-9 into Eq. 22-10 yields

$$C_s = C_{\text{base}}(\mu F) = \frac{0.6 I_{a1}}{V_{LL}} \qquad (22\text{-}11)$$

$R_s = R_{\text{opt}}$ can be obtained from Fig. 19-16 of Chapter 19. Here, assuming the normalized $R_s = R_{\text{opt}} = 1.3$, and using the value of normalizing resistance $\sqrt{2}\, V_{LL}/I_{rr}$, we obtain, using Eq. 22-9,

$$R_s = R_{\text{opt}} = 1.3\sqrt{2}\, \frac{V_{LL}}{I_{rr}} \approx 20 \frac{V_{LL}}{I_{a1}} \qquad (22\text{-}12)$$

To estimate the loss in each snubber, the voltage waveform across a thyristor having a worst case trigger angle of $\alpha = 90°$ is shown in Fig. 22-21. It can be shown that the total energy loss in each snubber equals

$$W_{\text{snubber}} = 3 C_s V_{LL}^2 \qquad (22\text{-}13)$$

or using Eq. 22-11

$$W_{\text{snubber}} = 1.8 \times 10^{-6} I_{a1} V_{LL} \qquad (22\text{-}14)$$

FIGURE 22-21: Voltage waveform across a thyristor triggered at a trigger angle of 90 degrees.

If the three-phase converter kVA is S, then at 60 Hz, each snubber power loss equals

$$P_{\text{snubber}}(\text{in watts}) \approx 0.06S \qquad (22\text{-}15)$$

A similar procedure can be followed for any values of t_{rr} and the ac-line inductance. A conservative design may require C_s to be larger than C_{base} and, therefore, R_s would be smaller than the value found above. In that case, the snubber losses would be higher since they are proportional to C_s.

22-9 SUMMARY

This chapter has considered the structure and physical principles of operation of the thyristor. The unique switching characteristics of the thyristor were studied and from this the requirements for gating circuits to turn the thyristor on were obtained. The basic design of gating circuits was discussed as well as the need for snubber circuits. The important conclusions are listed:

1. The thyristor has a unique four-layer construction of alternating p-type and n-type regions.

2. The forward bias portion of the thyristor's i-v characteristic has two stable operating regions, one being the on-state and the other the off-state. The reverse bias portion of the characteristic is a blocking state.

3. A current pulse applied to the gate will latch the thyristor on but then the gate cannot turn the device off. The external power circuit must reverse bias the thyristor in order to turn it off.

4. The thyristor is a minority carrier device and has the highest blocking voltage capabilities and the largest current conduction capabilities of any of the solid-state switching devices.

5. The thyristor is inherently a slow switching device compared to BJTs or MOSFETs because of the long carrier lifetimes used for low on-state losses and because of the large amount of stored charge. It is therefore normally used at lower switching frequencies.

6. The rate of rise of the on-state current must be kept within bounds because the slow spread of the plasma during the turn-on transient leads to current crowding that could result in device failure if di/dt is too large.

7. The rate of rise of the reapplied forward blocking voltage after turn-off must be limited or the device may be triggered back into the on-state by induced displacement currents. Furthermore, the forward voltage must not be reapplied too soon after turn-off or the device will turn back on.

8. Special structural modifications, such as highly interdigitated gate–cathode layouts and the use of cathode shorts, can substantially improve the di/dt and dv/dt ratings.

9. The requirements on the gating circuits for turning on the thyristor are relatively easy to meet. However, the forced commutation requirement in order to turn off the thyristor usually means that it is mainly used in ac circuits so that natural commutation of the thyristor occurs.

10. The large reverse recovery current in a thyristor necessitates the use of a turn-off snubber in order to prevent large overvoltages because of series inductance.

PROBLEMS

22-1. A thyristor is connected in series with a load resistor R_L and a 60-Hz sinusoidal voltage source with an rms voltage of V_s. A phase control circuit is used to set the trigger angle α so that a specified amount of power is delivered to the load. The on-state voltage of the thyristor is given by $V_{on} = 1.0 + R_{on}i(t)$ where R_{on} is the on-state resistance of the thyristor and $i(t)$ is the current flowing in the circuit through R_L and the thyristor. Develop an expression for the average power dissipated in the thyristor as a function of the trigger angle α.

22-2. In Problem 22-1, $R_L = 1\ \Omega$ and $V_s = 220$ V. The thyristor characteristics are listed below. The thyristor must operate in ambient temperatures as high as $120°$F. How much power can be delivered to the load and what is the trigger angle?

$$R_{on} = 0.002\ \Omega \qquad V_{RWM} = V_{BO} = 800\ \text{V} \qquad I_{A(max)} = 1000\ \text{A}$$

$$T_{j(max)} = 125°\text{C} \qquad R_{\theta j-a} = 0.1°\text{C/W}$$

22-3. A crude estimate of the di/dt limitation in a thyristor can be obtained from the following model. Assume a gate–cathode structure such as is shown in Fig. 22-1 for a phase control thyristor and the radius of the central gate is r_o. When the current begins to rise at turn-on, it starts out in a small conducting area of radius r_o and spreads radially outward as is illustrated in Fig. 22-8 with a velocity u_s. The current rises at a constant rate di/dt for a time t_f and the anode–cathode voltage $v_{AK} = V_{AK}(1 - t/t_f)$ during the current rise interval.

   If it is assumed that all of the power dissipated in the thyristor during the transient goes into raising the temperature T_j of the turned-on area and none gets to the heat sink, then the temperature rise ΔT_j is given by

$$\Delta T_j = \frac{1}{C_v}\int_0^{t_f} P(t)\ dt$$

where C_v is the specific heat of silicon and $P(t)$ is the instantaneous power dissipation in the turned-on area. Assuming the following numerical values, find the maximum allowable di/dt.

$$V_{AK} = 1000\ \text{V} \qquad u_s = 100\ \mu\text{m/}\mu\text{s} \qquad r_o = 0.5\ \text{cm} \qquad t_f = 20\ \mu\text{s} \qquad T_{j(max)} = 125°\text{C}$$

22-4. A modification to the basic structure of a thyristor has been proposed that consists of putting an n^+ layer between the p_1 and n_1 layers shown in Fig. 22-3a. This means that a punch-through structure has been set up for junction J_2 when it is reverse biased. Explain the advantages and disadvantages this structure would have on the thyristor characteristics.

22-5. The turn-off snubber for a thyristor does not include a diode as it did for the BJT and MOSFET. Explain why.

22-6. A thyristor is not completely latched in the on-state until the plasma has spread across the entire cross section of the device. If the thyristor of Problem 22-3 has a diameter of 8 cm, how long does it take to reliably be latched in the on-state?

22-7. The thyristor circuit of Fig. 22-15 is a single-phase line frequency converter. Assume that the line voltage is 230 V rms, 60 Hz and that the stray line inductance L_σ (such as is indicated in Fig. 22-21 for a three-phase converter) has a magnitude given by $\omega L_\sigma = 0.05(V_s/I_{a1})$ where V_s is the rms phase voltage and I_{a1} is the rms value of the fundamental harmonic of the phase current. The thyristors have a reverse recovery time of 10 μs and $I_{a1} = 100\ A$ rms. The load is highly inductive and can be approximated as a dc current source (hence the current flow out of the converter is continuous).

(a) Explain how a single series R-C circuit connected across the line will serve as a turn-off snubber for all four thyristors.

(b) Derive equations for the proper values of snubber capacitance C_s and snubber resistance R_s as functions of V_s, I_{a1}, t_{rr}, and other circuit parameters given above.

(c) Find values for C_s and R_s that will limit the overvoltage to 1.5 times the peak line voltage.

22-10 REFERENCES

1. Sorab K. Ghandhi, *Semiconductor Power Devices*, John Wiley & Sons, New York, 1977, Chapter 5.

2. *SCR Manual* 6th Ed., General Electric Company, Syracuse, NY, 1979.

3. B. M. Bird and K. G. King, *Power Electronics*, John Wiley & Sons, New York, 1983, Chapters 1 and 6.

4. Michael S. Adler, King W. Owyang, B. Jayant Baliga, and Richard A. Kokosa, "The Evolution of Power Device Technology," *IEEE Transactions on Electron Devices*, Vol. ED-31, No. 11, pp. 1570–1591, Nov. 1984.

5. M. H. Rashid, *Power Electronics: Circuits, Devices, and Applications*, Prentice-Hall, Englewood Cliffs, NJ, 1988, Chapters 14–15.

6. B. W. Williams, *Power Electronics, Devices, Drivers, and Applications*, John Wiley & Sons, New York, 1987, Chapters 3, 4, 8–10.

23

Gate Turn-off Thyristors

_____
_____
_____

23-1   INTRODUCTION

In several respects, thyristors are nearly the ideal switch for use in power electronic applications. They can block high voltages (several thousand volts) in the off-state and conduct large currents (several thousand amperes) in the on-state with only a small on-state voltage drop (a few volts). Most useful of all is their capability of being switched on when desired by means of a control signal applied at the gate of the thyristor.

However, the thyristor has a serious deficiency that prevents its use in switch-mode applications; the inability to turn off the device by application of a control signal at the thyristor gate. The inclusion of a turn-off capability in a thyristor requires device modifications and some compromises in the operational capabilities of the device. This chapter describes the structure and operation of thyristors that have a gate turn-off capability, the so-called GTO or gate turn-off thyristor, and the performance compromises required to achieve the turn-off capability. Drive circuits and snubber circuits commonly used with GTOs are also described.

23-2   BASIC STRUCTURE AND *I-V* CHARACTERISTICS

The vertical cross section of a gate turn-off thyristor with its highly interdigitated gate–cathode structure is shown in Fig. 23-1. The GTO retains the basic four-layer structure of the thyristor described in the previous chapter and its doping profile. The width w of the p_2 base layer is generally somewhat smaller in a GTO than in a conventional thyristor.

There are three significant differences between a GTO and a conventional thyristor. First, the gate and cathode structures are highly interdigitated, with various types of geometrical forms being used to lay out the gates and cathodes, including complicated involute structures. The basic goal is to maximize the periphery of the cathode and minimize the distance from the gate to the center of a cathode region.

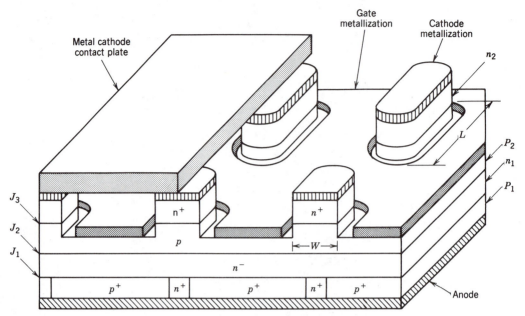

FIGURE 23-1: Vertical cross section and perspective view of a gate turn-off thyristor (GTO).

Second, the cathode areas are usually formed by etching away the silicon surrounding the cathodes so that they appear as islands or mesas, as indicated in Fig. 23-1. When the GTO is packaged, the cathode islands are directly contacted to a metal heat sink, which also forms the cathode connection to the outside world.

A third major difference is noted in the anode region of the GTO. At regular intervals, n^+ regions penetrate the p-type anode (p_1 layer) to make contact with the n^- region that forms the n_1 base layer. The n^+ regions are overlaid with the same metallization that contacts the p-type anode resulting in a so-called anode short, as is shown in Fig. 23-1. The anode short structure is used to speed up the turn-off of the GTO, as will be explained in a later section of this chapter. Some GTOs are made without this anode short so that the device can block reverse voltages.

The i-v characteristic of a GTO in the forward direction is identical to that of a conventional thyristor. However, in the reverse direction, the GTO has virtually no blocking capability because of the anode short structure. The only junction that blocks in the reverse direction is junction J_3, and it has a rather low breakdown voltage (20 to 30 V typically) because of the large doping densities on both sides of the junction. The circuit symbol for the GTO is shown in Fig. 23-2. The two-way arrow convention on the gate lead distinguishes the GTO from the conventional thyristor.

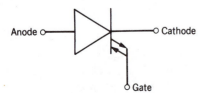

FIGURE 23-2: Circuit symbol for a GTO.

23-3 PHYSICS OF TURN-OFF OPERATION

23-3-1 Turn-off Gain

The basic operation of the GTO is the same as that of the conventional thyristor. The principal differences between the two devices lie in the modifications made in the basic thyristor structure to achieve a gate turn-off capability. Hence, in this chapter we discuss only this aspect of the GTO operation in any detail.

A convenient starting point for appreciating why the GTO structure differs from the conventional thyristor and what performance compromises must be made is to analyze the turn-off conditions in the two-transistor model of the thyristor given in Fig. 22-3b. In the equivalent circuit both Q_1 and Q_2 are saturated in the thyristor on-state. However, if the base current to Q_2 could briefly be made less than the value needed to maintain saturation ($I_{B2} < I_{C2}/\beta_2$), then Q_2 would go active and the thyristor would begin to turn off because of the regenerative action present in the circuit when one or both of the transistors is active.

Using the equivalent circuit of Fig. 22-3b we can write I_{B2} in terms of the thyristor terminal currents as

$$I_{B2} = \alpha_1 I_A - I'_G \qquad (23\text{-}1)$$

where I'_G is the negative of the normal gate current. From the equivalent circuit it is clear that a negative gate current I'_G is the only way that Q_2 can be brought out of saturation. The collector current I_{C2} can be expressed as

$$I_{C2} = (1 - \alpha_1)I_A \qquad (23\text{-}2)$$

Setting up the inequality $I_{B2} < I_{C2}/\beta_2$ with $\beta_2 = \alpha_2/(1 - \alpha_2)$ and using Eqs. 23-1 and 23-2 yields

$$I'_G > \frac{I_A}{\beta_\text{off}} \qquad (23\text{-}3)$$

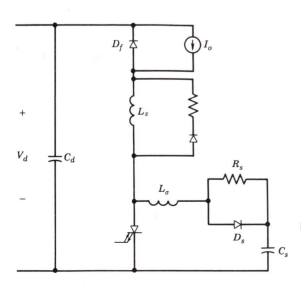

FIGURE 23-3: Step-down converter circuit using a gate turn-off thyristor as the switching device with turn-on and turn-off snubbers.

The parameter β_{off} is the turn-off gain and is given by

$$\beta_{\text{off}} = \frac{\alpha_2}{(\alpha_1 + \alpha_2 - 1)} \tag{23-4}$$

23-3-2   Required Structural Modifications and Performance Compromises

The first step in converting a conventional thyristor into a GTO is to make the turn-off gain as large as feasible so that overlarge values of negative gate current can be avoided. This means that α_2 should be near unity and α_1 should be small. Making α_2 near unity involves the use of a narrow p_2 layer for the npn transistor Q_2 and the use of heavy doping in the n_2 cathode layer (emitter of Q_2). Note that these are the same steps needed to achieve a large value of beta in a conventional BJT, and that they are the normal steps used in the fabrication of a conventional thyristor.

To make α_1 small, the n_1 thyristor layer (the base of transistor Q_1) should be as thick as possible and the carrier lifetime in this layer should be short. A thick n_1 layer is standard in thyristor fabrication because this layer must accommodate the depletion layer of junction J_2 during device operation in the forward blocking state. However, the need for a short carrier lifetime conflicts with the need for a long lifetime to minimize on-state power dissipation in this region. To achieve gate turn-off action, some reduction in carrier lifetimes must be accepted and, consequently, a GTO will have a higher on-state voltage drop at a given current level than a conventional thyristor.

Fortunately, the conflicting requirements on carrier lifetimes are substantially resolved by the anode shorting structure shown in Fig. 23-1. For the GTO to turn off, the excess carriers, especially holes, must be removed from the n_1 layer. The initial distribution of excess carriers at the start of turn-off is shown in Fig. 22-4. Because of the shorted anode structure, there can be no reverse anode–cathode voltages and thus no reverse anode currents to remove the excess carriers by carrier sweep-out. The only way to remove excess carriers is via internal recombination and by diffusion. Unfortunately, the conventional thyristor structure shown in Fig. 22-4 along with the carrier distributions suppresses the diffusion of holes from the n_1 layer because either side of this layer is a heavily doped p-type region where the equilibrium hole densities are larger than the excess hole density in the n_1 region. This suppresses the diffusion of holes out of the n_1 layer into either of two p-type layers (p_1 and p_2). Thus, the only way to remove holes is via internal recombination.

However, the n^+ regions in the GTO anode structure remove the barrier to hole diffusion and provide a sink for excess holes. This permits hole diffusion to occur at a substantial rate so that excess holes in the n_1 layer are removed at least as much by diffusion as by internal recombination. The net result is that the total stored charge is removed much faster during device turn-off, and the GTO thus has both a desirably shorter turn-off and forward recovery times compared with conventional thyristors without severely compromising its on-state losses. This shorted anode structure is so effective in reducing turn-off and recovery times that it is sometimes used in special thyristor structures called reverse conducting thyristors (RCTs). RCTs have short turn-off and recovery times like GTOs but cannot be turned off by a negative gate current because some of the other needed structural modifications are not included.

The essential modification for gate turn-off capability is the use of a highly interdigitated gate and cathode structure that minimizes the lateral voltage drops in the p_2 layer during turn-on and turn-off. Such lateral voltage drops, which were described in the previous chapter (see Fig. 22-5), are especially noticeable in conventional thyristor structures. Such lateral voltage drops lead to current crowding problems and di/dt limitations. However, as has already been mentioned, the use of highly interdigitated gate–cathode structures, which have relatively short distances between the gate contacts and the center of the cathode regions, minimizes these problems. The need for such interdigitation is a two-dimensional consideration and is not predicted by the one-dimensional two-transistor model. To avoid significant voltage drops in the gate metallization with large gate turn-off currents, contacts to the gate metal are uniformly spaced over the wafer surface.

23-4   GTO SWITCHING CHARACTERISTICS

23-4-1   Inclusion of Snubber and Drive Circuits

In describing the switching characteristics of other semiconductor devices, we have embedded the device in a typical switching circuit application without any snubber circuits and with only the most general of driving circuits. The motivation was to describe the switching characteristics of the device without the complications that are added by the presence of the snubber circuits. However, gate turn-off thyristors must normally be used with snubbers, as will be detailed later, and so any realistic description of the GTO switching behavior must include the effects of the snubber circuits. Similarly, we will include the gating circuit in describing the switching waveforms of the GTO.

The step-down converter circuit shown in Fig. 23-3, which uses a GTO as the switching element, will be used to describe the switching waveforms. It is important to realize that GTOs are only used in medium- to high-power applications where not only are the voltage and current levels large, but also the other solid-state components that may be used in conjunction with the GTO are likely to be rather slow. Thus, the free-wheeling diode shown in the circuit of Fig. 23-3 will not be a very fast recovery diode. On the other hand, the GTO will have a fast current rise time at turn-on compared with the diode's reverse recovery time because of the GTOs highly interdigitated gate–cathode structure. The consequence of this is that without protective circuits, very large overcurrents would flow in both the GTO and the diode because the relatively slow reverse recovery of the diode. The generation of such large overcurrents in this type of circuit was described in detail in Chapter 20 in conjunction with BJTs. Hence, the snubber inductor shown in Fig. 23-3 is included in the circuit to act as a turn-on snubber, as discussed in the BJT chapter (Chapter 20) and illustrated in Fig. 20-42.

When the GTO is being turned off, the rate of anode–cathode voltage growth, dv/dt, must be limited to specified levels. Otherwise retriggering of the GTO back into the on-state may occur, as was described in Chapter 22 with conventional thyristors. For this reason, a turn-off snubber is included as a part of the switching circuit, as is illustrated in Fig. 23-3. A gate drive circuit capable of meeting the recommended gating conditions suggested by GTO manufacturers is shown in Fig. 23-4.

23-4-2   GTO Turn-on Transient

When the GTO in Fig. 23-3 is off, the current free-wheels through the diode D_f. Turn-on is initiated by a pulse of gate current shown in Fig. 23-5. The sequence of events occurring inside the GTO during the turn-on process are essentially the same as those described for the conventional thyristor in the previous chapter and so will not

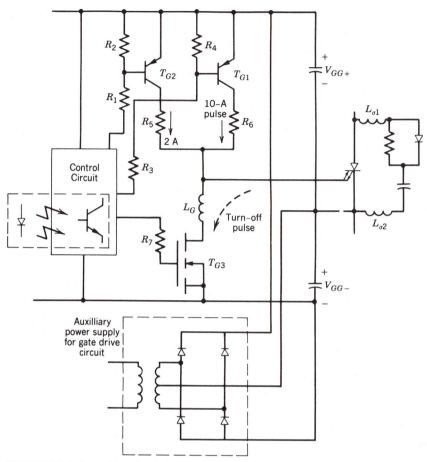

FIGURE 23-4: Gate drive circuit for a gate turn-off thyristor.

be repeated here. During turn-on, both the rate of gate current increase, di_G/dt, and the peak gate current I_{GM} should be large to ensure that all cathode islands begin to conduct and that there is good dynamic sharing of the anode current. Otherwise only a small number of islands might be carrying the total current and localized thermal runaway could occur, resulting in the destruction of the GTO.

A large I_{GM} value is supplied for a long enough time period, for example 10 μs, to ensure that the turn-on process is complete. After completion of turn-on, there must be a minimum continuous gate current I_{GT} flowing during the entire on-state period to prevent unwanted turn-off. The current I_{GT} is sometimes termed the "backporch" current. If the gate current is zero and the anode current gets too low, some of the cathode islands may stop conducting, and if the anode current were to subsequently increase, the remaining conducting islands may not be able to handle the current and the GTO may be destroyed by the ensuing thermal runaway.

The initial large pulse of gate current is provided by the gate drive circuit of Fig. 23-4 by turning on both transistor T_{G1} and T_{G2}. The stray inductance in the positive gate drive circuit should be kept to a minimum to achieve a large di_G/dt value at turn-on. After a time duration t_{w1}, the gate current is reduced from I_{GM} to I_{GT} by turning T_{G1} off.

During the growth of the anode current, the input voltage is shared between the turn-on snubber inductance and the GTO. If the di/dt of the anode current is limited

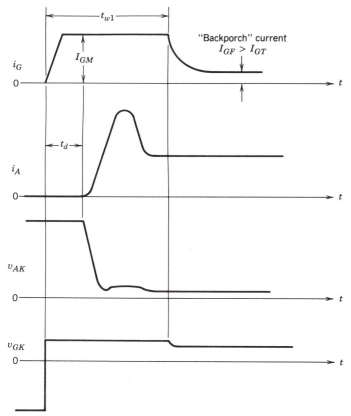

FIGURE 23-5: Turn-on waveforms for a GTO embedded in a step-down converter with turn-on and turn-off snubbers.

by this inductance because of its large value, then the voltage across the GTO will drop quickly to a fairly low value, as is shown in Fig. 23-5. The overshoot in the anode current comes from the reverse recovery of the free-wheeling diode D_f.

23-4-3 GTO Turn-off Transient

The GTO is turned off by applying a large negative gate current, as is shown in Fig. 23-6. The resulting current and voltage waveforms for the GTO in the circuit of Fig. 23-3 are shown in Fig. 23-6. There are several distinct intervals during the turn-off, which are described in the following paragraphs.

The gate drive circuit of Fig. 23-4 supplies the negative gate current by turning on transistor T_{G3}. The gate current must be very large, on the order of $\frac{1}{5}$ to $\frac{1}{3}$ (corresponding to turn-off gains of 3 to 5) of the anode current being turned off, but fortunately this large negative current is required for only a relatively short time. Low voltage MOSFETs are a nearly ideal choice for T_{G3}. The negative di_G/dt must be large in order to have a short storage time and a short anode current fall time and to reduce the gate power dissipation. However, too large a value of negative di_G/dt will result in the anode tail current to be described shortly. Hence, di_G/dt should be kept in the range specified by the device manufacturer.

The negative di_G/dt is controlled by V_{GG-} and L_G of the negative gate drive portion of the circuit of Fig. 23-4. V_{GG-} must be chosen to be less than the

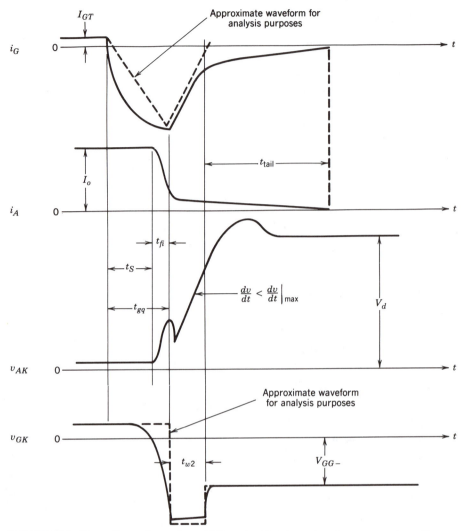

FIGURE 23-6: Turn-off waveforms for a GTO embedded in a step-down converter with turn-on and turn-off snubbers.

gate–cathode junction breakdown voltage. Knowing V_{GG-}, L_G is selected to give the specified di_G/dt. For large GTOs, the stray inductance in the negative gate drive circuit may equal the required L_G.

During the first time interval, the storage time t_s, the growing negative gate current is removing charge stored in the p_2 and n_2 layer at the periphery of the cathode islands, as shown in Fig. 23-7. As the stored charge continues to be removed from the periphery, the size of the plasma-free region grows as it expands in the lateral direction toward the centers of the cathode islands with a so-called squeezing velocity. In essence this removal of the plasma is the inverse of how it was established during turn-on. When a sufficient amount of stored charge has been removed, the regenerative action in the GTO is stopped and the anode current begins to fall. This marks the end of the storage time interval.

Once the regenerative action of the GTO is stopped, the anode current begins to fall rapidly. The current $[I_o - i_A]$ commutates to the turn-off snubber capacitor C_s,

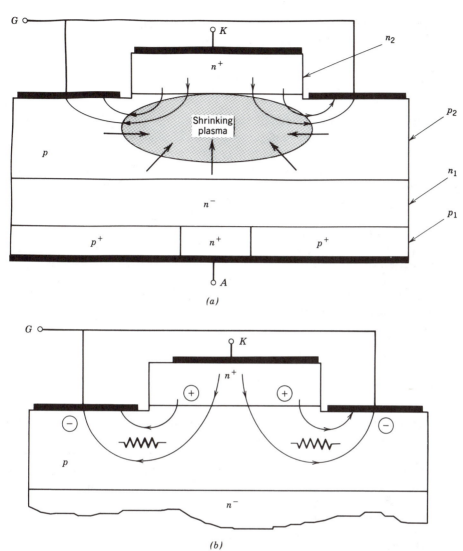

FIGURE 23-7: Mechanisms that determine the maximum anode current that can be turned off by a negative gate current: (a) negative gate current squeezing the excess carrier plasma down to a small volume at the center of the cathode island the farthest possible distance away from the gate electrode; (b) lateral ohmic resistance in the p_2 base layer limiting the maximum gate current.

which is fairly large in GTO applications. There is simultaneously a rapid rise in voltage across the GTO because of stray inductance in the turn-off snubber circuit loop. This stray inductance, (L_σ in Fig. 23-3) should be kept to a minimum to keep the peak of the voltage spike during the anode current fall time interval to a specified value. The anode current fall time t_{fi} interval ends when the excess carriers at the gate–cathode junction have been swept out and the junction recovers its reverse blocking capability.

As the gate–cathode junction recovers its reverse blocking capability, the gate–cathode voltage begins to grow toward negative values and the negative gate current thus begins to decrease rapidly in magnitude, as is shown in Fig. 23-6. The voltage induced in the inductance L_G forces the gate current to keep flowing and the gate–cathode junction goes into avalanche breakdown. The gate–cathode junction now operates as a zener diode and the di_G/dt during this interval is given by

$$\frac{di_G}{dt} = \frac{(V_{GK,\text{breakdown}} - V_{GG-})}{L_G} \tag{23-5}$$

This avalanche breakdown is desirable for a short duration t_{w2} (the gate–cathode junction avalanche breakdown time) to sweep out as much stored charge from the gate and p_2 layer as possible. This interval depends on di_G/dt, which is controlled by the selection of L_G and V_{GG-}. The interval t_{w2} should be kept less than a maximum specified value in order to avoid destruction of the gate–cathode junction.

At the end of the t_{w2} interval there will still be some excess stored charge in the two base regions (n_1 and p_2 layers) of the GTO. A small anode current, usually termed the anode tail current, will continue to flow between the anode and the negatively biased gate, which is due to the sweep-out of this remaining stored charge. This current is driven by the growing voltage difference between the anode and gate. The time interval during which this tail current flows is termed the anode tail-current time t_{tail}. During most of the tail time interval, the gate voltage is at V_{GG-}, the value it will have during the entire off-state interval.

During the tail time interval, the voltage across the GTO grows at a constant rate given by

$$\frac{dv_{AK}}{dt} \approx \frac{I_o}{C_s} \tag{23-6}$$

This interval will contribute a major part of the turn-off losses because this interval is relatively long and the voltage across the GTO during the interval is fairly large.

The overvoltage at turn-off shown in the anode–cathode voltage waveform in Fig. 23-6 is due to stray inductance in the power circuit. It is identical in origin to the overvoltage described in the BJT chapter (Chapter 20) in Fig. 20-36. The anode–cathode overvoltage can be reduced by means of an overvoltage snubber of the type shown in Fig. 20-41 for the BJT.

23-4-4   Minimum On- and Off-State Times

It is strongly recommended that the GTO not be turned on until it has been off for a specified time because of the possibility of poor current sharing between the various cathode islands. Some excess minority carriers will remain in the GTO for fairly long times because of the long lifetimes, and these remaining carriers will cause the few cathode islands in the vicinity of the carriers to have a better conduction characteristic than the rest of the cathode islands. Hence, if turn-on is attempted before all of the carriers have recombined or been swept out, then most of the current will be carried by these few islands (poor current sharing) and device destruction may occur.

Similarly, the GTO should be maintained in the on-state for a specified time period before turn-off is initiated. Again, the reason is because otherwise there may be poor current sharing between the various cathode islands.

The circuit designer should also recognize that the turn-on and turn-off snubbers also require a minimum off-state and minimum on-state time, respectively, to operate properly. This was described in detail in the BJT chapter (Chapter 20), where the design of these snubbers was discussed in detail.

23-4-5 Maximum Controllable Anode Current

The excess carriers, principally those in the p_2 layer, are the source of carriers for the negative gate current. As the negative gate current grows and the plasma-free region grows, as is diagramed in Fig. 23-7, there is a build-up of a substantial voltage across the gate–cathode junction, as is indicated because of the lateral flow of gate current in the p_2 layer. The voltage across the junction is largest at the cathode periphery nearest the gate contact. If this voltage exceeds the breakdown voltage of the junction, then the negative gate current will flow only at the cathode periphery where breakdown has occurred and none of the remaining stored charge will be removed and, hence, the GTO will not be turned off. For this reason, the voltage V_{GG-} must be kept less than the breakdown voltage of the gate–cathode junction.

The limitation on the negative gate–cathode voltage means that there is a maximum gate current that can be pulled out of the GTO. As the removal of stored charge enters its final phase, the region of excess carriers has shrunk to a small area near the center of the cathode island and is the greatest distance from the gate contact. Under these circumstances the reverse voltage across the junction is at its greatest value. The lateral ohmic resistance shown in Fig. 23-7b, which is a function of the device geometry, and the doping level of the p_2 layer along with the junction breakdown voltage determined how large the maximum negative gate current can be. This also means that there is a maximum anode current that can be turned off since by Eq. 23-3 $I_A < \beta_{\text{off}} I_{G,\text{max}}$. The maximum controllable anode current is given on GTO specification sheets by the device manufacturer.

23-5 SNUBBER CIRCUITS

A GTO is capable of turning off a significantly larger current compared to its rms or average current capability. The maximum controllable current for a given GTO in the circuit of Fig. 23-3 depends on the turn-off snubber capacitance C_s. This dependence arises because there is a maximum rate of change in the increase in the anode–cathode voltage at turn-off. Exceeding this maximum $dv_{AK}/dt|_{\text{max}}$ would cause retriggering of the GTO back into the on-state due to large displacement currents. Since, dv_{AK}/dt is inversely proportional to C_s according to Eq. 23-6, so for a given dv_{AK}/dt, the larger C_s is, the larger I_0 can be. This, of course, assumes that the maximum controllable anode current given on the GTO specification sheet is not exceeded. A large C_s, however, results in higher overall switching losses and in a larger current through the GTO at turn-on. Therefore, the capacitance C_s should be just sufficient to turn off the maximum current dictated by the particular application.

The capacitor C_s should have a low internal inductance and a large peak current rating. In practice, this may require the paralleling of many capacitors to achieve these required properties for C_s.

The turn-off snubber diode D_s needs to carry the entire load current for a short time. Its average current is very low but, since its dynamic forward voltage at turn-on must be low, often a diode with a larger average current rating is chosen.

The turn-off snubber resistance R_s must be selected based on trade-offs between maximum additional discharge current into the GTO at turn-on and the requirement on the minimum on-state time of the GTO to discharge C_s so that it can properly operate during the next turn-off, as was described in the BJT chapter. In a GTO, it is much more important to discharge C_s during the on-time interval than it is in a BJT. On the other hand, the peak current conduction capability in a GTO is much larger than in a BJT. Therefore, these factors ought to be considered in the selection of R_s. There is a considerable power loss in R_s and therefore it may require mounting on a heat sink.

It has already been explained why the stray inductance in the turn-off snubber current loop should be as small as possible. To achieve this, the snubber components should be mounted as close to the GTO as possible. Some of the layout and interconnection techniques that should be used are discussed in the last chapter of this book.

The design considerations for a turn-on snubber for the GTO are similar to those described in the diode chapter (Chapter 19).

For GTOs in a bridge configuration, the snubbers similar to those shown in Fig. 23-3 can be used with each GTO. Alternatively, the turn-on snubbers can be combined as shown in Fig. 20-45a of the BJT chapter. The snubber loss recovery, as explained in the BJT chapter, can be significant in GTO applications because of the higher power level of operation and higher snubber losses due to a large C_s.

23-6 OVERCURRENT PROTECTION OF GTOs

In a MOSFET and a BJT, the accidental overcurrent causes the device to go out of saturation and enter the active region. The device itself limits the maximum current, but the voltage across the device becomes large. Thus, the overcurrent condition can be easily detected by measuring the on-state voltage. The overcurrents can also be detected by a current sensor or in the case of a MOSFET, by using a SENSEFET. Once the overcurrent is detected, the BJTs and the MOSFETs can be protected by turning them off within a few microseconds.

The overcurrent protection of GTOs is more complicated. As is shown in Fig. 23-8a, the GTO is designed for an allowable peak operating current that is chosen to

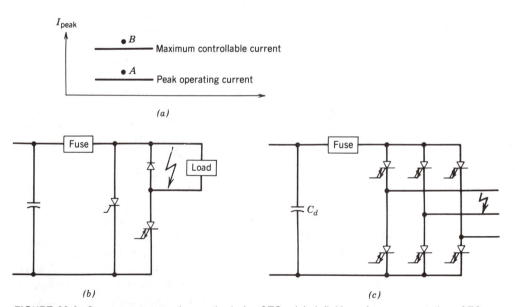

FIGURE 23-8: Overcurrent protection methods for GTOs: (a) definition of overcurrents in a GTO; (b) overcurrent protection of a GTO by "crowbarring," (c) overcurrent protection of a GTO by turning on all the GTOs in the bridge to share the current until the fuse opens up.

be less than the maximum controllable current by a safety factor. The overcurrent in a GTO must be detected by current sensing. If the detected overcurrent is less than the maximum controllable current, for example, at point A in Fig. 23-8a, then the GTO can be turned off by a negative gate current.

However, if the detected overcurrent is greater than the maximum controllable current, for example, at point B in Fig. 23-8a, then an attempt to turn off the GTO by a negative gate current will result in the failure of the GTO. Hence, the GTO is protected by the so-called crowbarring technique, where a thyristor in parallel with the GTO as is shown in Fig. 23-8b is turned on quickly, which then blows the fuse. Without crowbarring, the only way to protect the GTO in the circuit of Fig. 23-8b would be to use a GTO of much larger current rating, which is quite expensive.

In the case of a three-phase configuration, which is shown in Fig. 23-8c, the crowbarring can be obtained by turning on all six GTOs simultaneously. As was discussed in Chapter 6, under normal operation the current through the fuse is in the same range as the current through one GTO. By turning on all three legs simultaneously, the current through the fuse is shared by three legs, which the GTOs will be capable of carrying until the fuse blows.

23-7 SUMMARY

This chapter has examined the structure and characteristics of the gate turn-off thyristor, GTO. The important conclusions are listed:

1. The GTO has the same four-layer structure as the conventional thyristor, but special modifications are made to the structure to enable the gate to turn off the device.
2. The major modifications include a highly interdigitated gate–cathode structure with small cathode and gate widths, the use of anode shorts, and a shorter carrier lifetime in the drift region than is used in a conventional thyristor.
3. The forward bias portion of the GTO i-v characteristic is the same as for the conventional thyristor, but GTOs with anode shorting have very limited reverse blocking capability.
4. The turn-off gain of the GTO is not large (typically 5 or less), so large negative gate current pulses are required to turn off the device.
5. The magnitude of negative gate current that can be applied is limited by current crowding phenomena and, hence, there is a maximum anode current that can be safely turned off.
6. Special gating requirements for the GTO include not only large positive and negative gate current pulses but a continuous on-state gate current to ensure complete turn-on of all the cathode islands.
7. GTOs are used almost exclusively for medium- and high-power applications. Turn-off snubber circuits must be used. The GTO must particularly be protected against overcurrents because the gate cannot turn off currents that exceed a specified maximum value.

PROBLEMS

23-1. Each cathode island in the GTO diagram of Fig. 23-1 has a width W and a length L. The p_2 layer has a thickness t and a resistivity ρ_{p_2}. The GTO has a turn-off gain of β_{off}, N cathode islands connected in parallel, and junction J_3 has a breakdown voltage of BV_{J_3}. Develop an approximate expression for the maximum controllable anode current I_{AM}.

23-2. Consider the step-down converter circuit shown in Fig. 23-3 without the turn-on snubber. The dc input voltage V_d is 500 V, the load current $I_o = 500$ A, and the switching wave form is a 1-kHz square wave. The free-wheeling diode has a reverse recovery time $t_{rr} = 10$ μs. The GTO has a current fall time $t_{fi} = 1$ μs, a maximum reapplied voltage rate $dv/dt = 50$ V/μs and a maximum controllable anode current $I_{AM} = 1000$ A.
(a) Find the appropriate values for resistance R_s and capacitance C_s for the turn-off snubber circuit.
(b) Estimate the power dissipated in the snubber resistance.

23-3. The stray inductance L_σ in the turn-off snubber circuit of Fig. 23-3 will cause an overvoltage across the GTO. Estimate the maximum stray inductance that can be tolerated in the circuit if the overvoltage is not to exceed 1.5 V_d. Express the estimate in terms of the circuit parameters. Assume that the turn-on snubber circuit acts like a constant current source of value I_o during the GTO current fall time t_{fi}.

23-4. The GTO in the circuit of Problem 23-2 is to be protected by a turn-on snubber circuit such as is shown in Fig. 23-3. The maximum rate of rise of the anode current, di_A/dt, is 300 A/μs. Find appropriate values for the inductance and resistance. Ignore the turn-off snubber.

23-5. The drive circuit of Fig. 23-4 is used to turn off the GTO in the circuit of Problem 23-2. The GTO has a gate-cathode breakdown voltage $BV_{J_3} = 25$ V and a turn-off gain $\beta_{\text{off}} = 5$. Assume that the negative voltage V_{GG-} in the drive circuit is 15 V and that the time interval t_{gq} defined in Fig. 23-6 is equal to 5 μs. Estimate the required value of the inductor L_G and the magnitude of the time interval t_{w_2} defined in Fig. 23-6. Use the approximate waveforms for $i_G(t)$ and $v_{AK}(t)$ shown in Fig. 23-6 to simplify the analysis.

23-8 REFERENCES

1. Sorab K. Ghandhi, *Semiconductor Power Devices*, John Wiley & Sons, New York, 1977, Chapter 5.

2. Thyristor Application Notes, "Applying International Rectifier's Gate Turn-off Thyristors," AN-315A, International Rectifier, El Segundo, CA, 1984.

3. *Semiconductor Devices for Power Conditioning*, Roland Sittig and P. Roggwiller (Eds.), Plenum Press, New York, 1982, pp. 91–120.

4. *SCR Manual*, 6th edition, General Electric Company, Syracuse, NY, 1979.

5. *Power Transistors: Device Design and Applications*, B. Jayant Baliga and Dan Y. Chen, Editors, IEEE Press, Institute of Electrical and Electronics Engineers, New York, 1984, Part II. *Gate Turn-off Thyristors/Latching Transistors*, pp. 123–189.

6. Michael S. Adler, King W. Owyank, B. Jayant Baliga, and Richard A. Kokosa, "The Evolution of Power Device Technology," *IEEE Trans. on Electron Devices*, Vol. ED-31, No. 11, pp. 1570–1591, Nov. 1984.

7. M. H. Rashid, *Power Electronics: Circuits, Devices, and Applications*, Prentice-Hall, Englewood Cliffs, NJ, 1988, Chapters 14–15.

8. B. W. Williams, *Power Electronics, Devices, Drivers, and Applications*, John Wiley & Sons, New York, 1987, Chapters 3, 4, 8, 9.

24

Insulated Gate Bipolar
Transistors

24-1 INTRODUCTION

BJTs and MOSFETs have characteristics that complement each other in some respects. BJTs have lower conduction losses in the on-state, especially in devices with larger blocking voltages, but have longer switching times, especially at turn-off. MOSFETs can be turned on and off much faster, but their on-state conduction losses are larger, especially in devices rated for higher blocking voltages (a few hundred volts and greater). These observations have led to attempts to combine BJTs and MOSFETs monolithically on the same silicon wafer to achieve a circuit or even perhaps a new device that combines the best qualities of both types of devices.

These attempts have led to the development of a new device called the insulated gate bipolar transistor (IGBT), which is finding increasingly wide applications. Other names for this device include GEMFET, COMFET (conductivity-modulated field effect transistor), IGT (insulated gate transistor), and bipolar-mode MOSFET or bipolar-MOS transistor. This chapter describes the basic structure and physical operation of the IGBT and the operating limitations that should be observed in using this new device.

24-2 BASIC STRUCTURE

The vertical cross section of a generic n-channel IGBT is shown in Fig. 24-1a. This structure is quite similar to that of the vertical diffused MOSFET shown in Fig. 21-1. The principal difference is the presence of the p^+ layer that forms the drain of the IGBT. This layer forms a pn junction (labeled J_1 in the figure), which injects minority carriers into what would appear to be the drain drift region of the vertical MOSFET. The gate and source of the IGBT are laid out in an interdigitated geometry similar to that used for the vertical MOSFET.

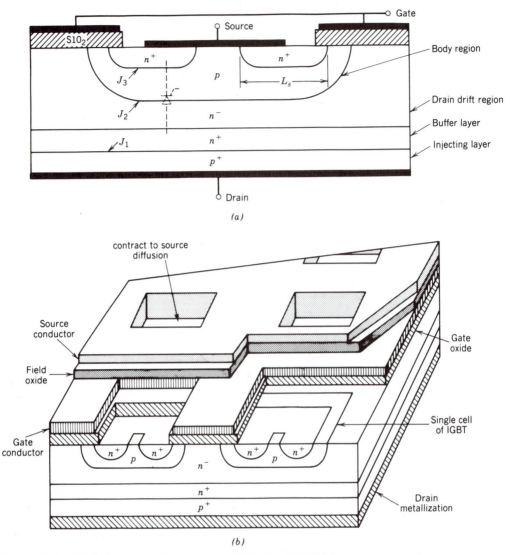

FIGURE 24-1: Vertical cross section and perspective view of an IGBT.

The doping levels used in each of the IGBT layers are similar to those used in the comparable layers of the vertical MOSFET structures except for the body region, as is explained later. It is also feasible to make p-channel IGBTs, and this would be done by changing the doping type in each of the layers of the device.

It is shown in Fig. 24-1a that the IGBT structure has a parasitic thyristor. Turn-on of this thyristor is undesirable and several structural details of a practical IGBT geometry, principally in the p-type body region that forms junctions J_2 and J_3, are different from the simple geometry shown in Fig. 24-1a to minimize the possible activation of this thyristor. These structural changes will be discussed in later sections of this chapter. The IGBT does retain the extension of the source metallization over the body region that is also used in power MOSFETs, such as is illustrated in Fig. 21-1. The body–source short in the IGBT helps to minimize the possible turn-on of the parasitic thyristor, as we explain later.

The n^+ buffer layer between the p^+ drain contact and the n^- drift layer is not essential for the operation of the IGBT, and some IGBTs are made without it (sometimes termed symmetric IGBTs, whereas those with this buffer layer are termed asymmetric IGBTs). If the doping density and thickness of this layer are chosen appropriately, the presence of this layer can significantly improve the operation of the IGBT in two important respects. First, it can lower the on-state voltage drop of the device and, second, shorten the turn-off time. However, the presence of this layer greatly reduces the reverse blocking capability of the IGBT. These effects on the characteristics of the IGBT will be discussed in a later section of this chapter.

The circuit symbol for an n-channel IGBT is shown in Fig. 24-2c. The directions of the arrowheads would be reversed in a p-channel IGBT. This symbol is essentially the same as that used for an n-channel MOSFET, but with the addition of an arrowhead in the drain lead pointing into the body of the device, indicating the injecting contact. There is some disagreement in the engineering community over the proper symbol and nomenclature to use with the IGBT. Some prefer to consider the IGBT as basically a BJT with a MOSFET gate input and, thus, to use a modified BJT symbol for the IGBT. This device has a collector and emitter rather than a drain

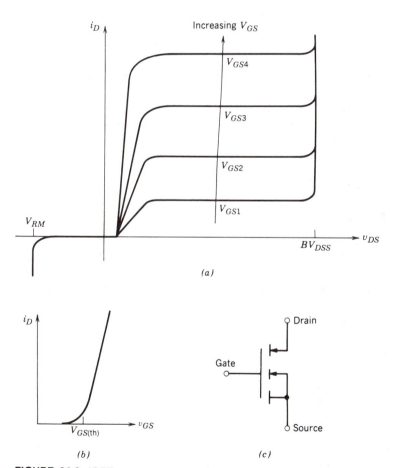

FIGURE 24-2: IGBT current-voltage characteristics and circuit symbol:
(a) output characteristics; (b) transfer characteristics;
(c) N-channel IGBT circuit symbol.

and source. The symbol and nomenclature shown in Fig. 24-2c is more widely used and is the one we have adopted.

24-3 I-V CHARACTERISTICS

The i-v characteristic of an n-channel IGBT is shown in Fig. 24-2a. In the forward direction they appear qualitatively similar to those of a logic level bipolar junction transistor except that the controlling parameter is an input voltage, the gate-source voltage, rather than an input current. The characteristics of a p-channel IGBT would be the same except that the polarities of the voltages and currents would be reversed.

The junction labeled J_2 in Fig. 24-1a blocks any forward voltages when the IGBT is off. The reverse blocking voltage indicated on the i-v characteristic can be made as large as the forward blocking voltage if the device is fabricated without the n^+ buffer layer. Such a reverse blocking capability is useful in some types of ac circuit applications. The junction labeled J_1 in Fig. 24-1a is the reverse blocking junction. However, if the n^+ buffer layer is used in the device construction, the breakdown voltage of this junction is lowered significantly, to a few tens of volts, because of the heavy doping now present on both sides of this junction, and the IGBT no longer has any reverse blocking capability.

The transfer curve i_D-v_{GS} shown in Fig. 24-2b is identical to that of the power MOSFET. The curve is reasonably linear over most of the drain current range, becoming nonlinear only at low drain currents where the gate-source voltage is approaching the threshold. If v_{GS} is less than the threshold voltage $V_{GS(\text{th})}$, then the IGBT is in the off-state. The maximum voltage that should be applied to the gate-source terminals is usually limited by the maximum drain current that should be permitted to flow in the IGBT, as will be discussed in Section 24-7.

24-4 PHYSICS OF DEVICE OPERATION

24-4-1 Blocking State Operation

In simplistic terms, the IGBT is intended to operate as a MOSFET whose drain-drift region is conductivity modulated by the injection of minority carriers (holes in the case of the n-channel IGBT illustrated in Fig. 24-1) into the drift region. The injection is obtained by adding an additional layer to the MOSFET at its drain end so that a forward-biased pn junction (labeled J_1 in Fig. 24-1a) is placed between the drift region and the drain contact. The carrier injection lowers the resistance of the drift region and, hence, its contribution to the on-state voltage. Since the voltage drop in the drift region is what dominates the on-state conduction losses of high-voltage MOSFETs, this conductivity modulation will significantly increase the current-carrying capabilities of high-voltage MOSFETs.

Since the IGBT is basically a MOSFET, the gate-source voltage controls the state of the device. When v_{GS} is less than $V_{GS(\text{th})}$, there is no inversion layer created to connect the drain to the source and, hence, the device is in the off-state. The applied drain-source voltage is dropped across the junction labeled J_2 and only a very small leakage current flows. This blocking state operation is essentially identical to that of the MOSFET.

The depletion region of the J_2 junction extends principally into the n^- drift region, since the p-type body region is purposely doped much more heavily than the drift region. If the thickness of drift region is large enough to accommodate the depletion layer so that the depletion layer boundary does not touch the p^+ injecting layer, then the n^+ buffer layer shown in Fig. 24-1a is not needed. This type of IGBT is sometimes termed a symmetrical IGBT or nonpunch-through IGBT, and it can block reverse voltages as large in magnitude as the forward voltages it is designed to block. As mentioned earlier, this reverse blocking capability is useful in some ac circuit applications.

However, it is possible to reduce the required thickness of the drift region by approximately a factor of two if a so-called punch-through structure similar to that described in Chapter 19 for the power diode and illustrated in Fig. 19-3 is used. In this geometry, the depletion layer is allowed to extend all the way across the drift region at voltages significantly below the desired breakdown voltage limit. The reach-through of the depletion layer to the p^+ layer is prevented by inserting a n^+ buffer layer between the drift region and the p^+ region as is shown in Fig. 24-1a. This type of IGBT structure is sometimes termed an antisymmetric IGBT. The shorter drift region length means lower on-state losses, but the presence of the buffer layer means that the reverse blocking capability of this punch-through geometry will be quite low (a few tens of volts) and therefore nonexistent as far as circuit applications are concerned.

24-4-2 On-state Operation

When the gate-source voltage exceeds the threshold, an inversion layer forms beneath the gate of the IGBT. This inversion layer shorts the n^- drift region to the n^+ source region exactly as in the MOSFET. An electron current flows through this inversion layer as is diagrammed in Fig. 24-3 which in turn causes substantial hole injection from the p^+ drain contact layer into the n^- drift region, as also indicated in the figure. The injected holes move across the drift region by both drift and diffusion taking a variety of paths as is indicated in Fig. 24-3 and reach the p-type body region that surrounds the n^+ source region. As soon as the holes are in the p-type body region, their space charge attracts electrons from the source metallization that contacts the body region, and the excess holes are quickly recombined.

The junction formed by the p-type body region and the n^- drift region is "collecting" the diffusing holes and thus functions as the collector of a thick base pnp transistor. This transistor, diagrammed in Fig. 24-3b has the p^+ drain contact layer as an emitter, a base composed of the n^- drift region, and a collector formed from the p-type body region. From this description an equivalent circuit for modeling the operation of the IGBT can be developed, which is shown in Fig. 24-4a. This circuit models the IGBT as a Darlington circuit with the pnp transistor as the main transistor and the MOSFET as the driver device. The MOSFET portion of the equivalent circuit is also diagrammed in Fig. 24-4a along with the BJT portion. The resistance between the pnp base and the MOSFET drain represents the resistance of the n^- drift region.

Unlike the conventional Darlington circuit, the driver MOSFET in the equivalent circuit of the IGBT carries most of the total terminal current. This unequal division of the total current flow is desirable for reasons having to do with potential turn-on of the parasitic thyristor, a subject that we will discuss shortly. In this situation the on-state voltage $V_{DS(on)}$, using the equivalent circuit of Fig. 24-4a, can be expressed as

$$V_{DS(on)} = V_{J1} + V_{drift} + I_D R_{channel} \qquad (24-1)$$

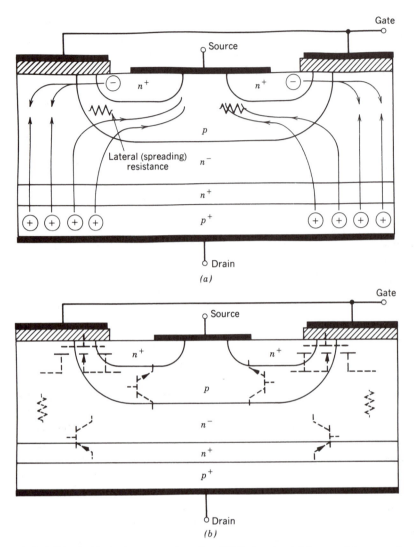

FIGURE 24-3: Vertical cross section of an IGBT showing (a) the on-state current flow paths and (b) the effective MOSFET and BJT operating portions of the structure.

The voltage drop across the injecting junction J_1 is a typical forward bias voltage drop across a *pn* junction, which depends exponentially on the current and to first order has an approximately constant value of 0.7 to 1.0 V. The drop across the drift region is similar to that developed across the drift region in a high-power *pn* junction and is approximately constant and is given approximately by an equation similar to Eq. 19-13 in Chapter 19. The V_{drift} voltage is much less in the IGBT than in the MOSFET because of the conductivity modulation of the drift region and is what makes the overall on-state voltage of the IGBT much less than that of a comparable power MOSFET. The use of the punch-through structure also aids in keeping V_{drift} small. The voltage drop across the channel is due to the ohmic resistance of the channel and is similar to the comparable drop in the power MOSFET discussed in Chapter 21.

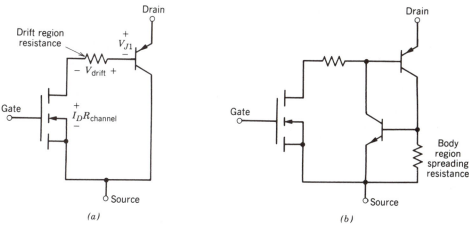

FIGURE 24-4: Equivalent circuits for the IGBT: (*a*) approximate IGBT equivalent circuit valid for normal operating conditions; (*b*) more complete IGBT equivalent circuit showing the transistors comprising the parasitic thyristor.

24-5 LATCHUP IN IGBTs

24-5-1 Causes of Latchup

The paths traveled by the holes injected into the drift region (or *pnp* transistor base) are crucial to the operation of the IGBT. A component of the hole current travels in fairly straight-line paths directly to the source metallization. However, most of the holes are attracted to the vicinity of the inversion layer by the negative charge of the electrons in the layer. This results in a hole current component that travels laterally through the *p*-type body layer, as is diagrammed in Fig. 24-3*a*. This lateral current flow will develop a lateral voltage drop in the ohmic resistance of the body layer (modeled as the spreading resistance in Fig. 24-3*a*), as indicated in the figure. This will tend to forward bias the n^+p junction (labeled J_3 in Fig. 24-1*a*) with the largest voltage across the junction occurring where the inversion layer meets the n^+ source.

If the voltage is large enough, substantial injection of electrons from the source into the body region will occur and the parasitic *npn* transistor diagrammed in Fig. 24-3*b* will be turned on. If this occurs then both the parasitic *npn* and *pnp* transistors will be on and, hence, the parasitic thyristor composed of these transistors will latch on and latchup of the IGBT will have occurred. For a given IGBT with a specified geometry, there is a critical value of drain current that will cause a large enough lateral voltage drop to activate the thyristor. Hence, the device manufacturer specifies the peak allowable drain current I_{DM} that can flow without latchup occurring. There is also a corresponding gate-source voltage that permits this current to flow that should not be exceeded.

Once the IGBT is in latchup, the gate no longer has any control of the drain current. The only way to turn on the IGBT in this situation is by forced commutation of the current, exactly the same as for a conventional thyristor. If latchup is not terminated quickly, the IGBT will be destroyed by the excessive power dissipation. A more complete equivalent circuit for the IGBT, which includes the parasitic *npn* transistor and spreading resistance of the body layer, is shown in Fig. 24-4*b*.

The description of latchup just presented is the so-called static latchup mode because it occurs when the continuous on-state current exceeds a critical value. Unfortunately, under dynamic conditions when the IGBT is switching from on to off, it may latchup at drain current values less than the static current value. Consider the IGBT embedded in a step-down converter circuit. When IGBT is turned off, the MOSFET portion of the device turns off quite rapidly and the portion of the total device current that it carries goes to zero. There is a corresponding rapid buildup of drain-source voltage, as will be described in detail in the next section, which must be supported across the J_2 drift–body junction. This results in a rapid expansion of the depletion layer of this junction into both the body and the drift regions, especially the drift region because of its low doping. This increases the base transport factor α_{pnp} of the pnp transistor, which means that a greater fraction of the holes injected into the drift region will survive the traverse of the drift region and will be collected by the J_2 junction. The magnitude of the lateral hole current flow will then increase and, hence, the lateral voltage will increase. As a consequence, the conditions for latchup will be satisfied even though the on-state current prior to the start of turn-off was below the static value needed for latchup. The value of I_{DM} specified by the device manufacturer usually is given for the dynamic latchup mode.

24-5-2 Avoidance of Latchup

There are several steps that can be taken by the device user to avoid latchup and that the device manufacturer can take to increase the critical current required for the initiation of latchup. The user has the responsibility to design circuits where the possibility of overcurrents that exceed I_{DM} are minimized. However, it is impossible to eliminate this possibility entirely. Another step that can be taken is to slow down the IGBT at turn-off so that the rate of growth of the depletion region into the drift region is slowed down and the holes present in the drift region have a longer time to recombine, thus reducing the lateral current flow in the p-type body region during the turn-off. The increase in the turn-off time is easily accomplished by using a larger value of series gate resistance R_g, as will be explained in the next section.

The device manufacturer seeks to increase the latching current threshold I_{DM} by lowering the body-spreading resistance in the equivalent circuit of Fig. 24-4b. This is done in several ways. First the lateral width of the source regions, labeled L_s in Fig. 24-1a, is kept as small as possible consistent with other requirements. Second, the p-type body region is often partitioned into two separate regions of different levels of acceptor doping density, as is illustrated in Fig. 24-5a. The channel region where the inversion layer is formed is doped at a moderate level, on the order of 10^{16} cm^{-3} and the depth of the p-region is not much deeper than the n^+ source region. The other portion of the body layer beneath the n^+ source regions is doped much more heavily, on the order of 10^{19} cm^{-3} and is made much thicker (or equivalently deeper). This makes the lateral resistance much smaller both because of the larger cross-sectional area and because of the higher conductivity.

Another possible modification to the body layer is shown in Fig. 24-5b, where one of the source regions is eliminated from the basic IGBT cell. This allows the hole current to be collected by the entire side of the cell where the source has been removed. This so-called hole bypass structure in effect provides for an alternate path for the hole current component that does not have to flow laterally beneath a source region. This geometry is quite effective in raising the latchup threshold but it does so at the expense of reducing the transconductance of the IGBT, since the effective width of the gate is reduced by the loss of the second source region in the basic cell.

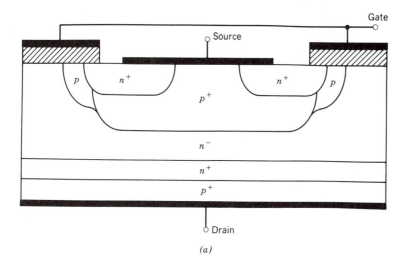

(a)

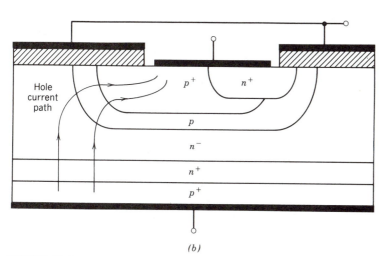

(b)

FIGURE 24-5: IGBT with modified body-source regions to lower the spreading resistance so that the drain current threshold for latchup is increased: (*a*) modification of the body region by heavier doping and greater depth to lower the spreading resistance; (*b*) modified IGBT with a hole current bypass structure to lower the spreading resistance.

By such means as these, the problem of latchup in IGBTs has been greatly minimized. Prototype devices have been demonstrated where it is claimed that the device is latchup proof.

24-6 SWITCHING CHARACTERISTICS

24-6-1 Turn-on Transient

The current and voltage waveforms for the turn-on of an IGBT embedded in a step-down converter circuit similar to that shown in Fig. 21-10 for the MOSFET are

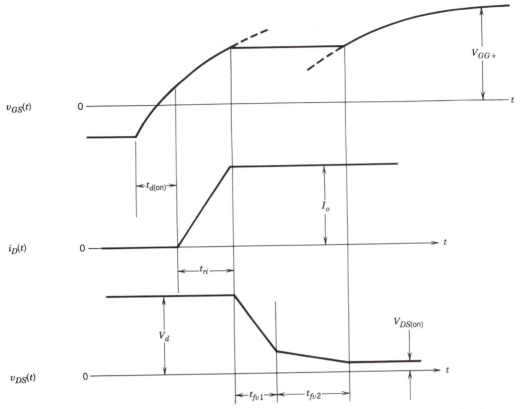

FIGURE 24-6: Turn-on voltage and current waveforms of an IGBT in a step-down converter circuit.

shown in Fig. 24-6. The turn-on portions of the waveforms appear similar to those of the power MOSFET shown in Fig. 21-11 of Chapter 21. The similarity is to be expected, since the IGBT is essentially acting as a MOSFET during most of the turn-on interval. The same equivalent circuits used in Chapter 21 for discussing the MOSFET turn-on waveforms can also be used for calculating the turn-on characteristics of the IGBT.

The t_{fv2} interval observed in the MOSFET drain-source voltage waveform in Fig. 21-11 is usually observed in the IGBT drain source voltage waveform. Two factors will contribute to the t_{fv2} interval in the IGBT waveform. First the gate-drain capacitance C_{gd} will increase in the MOSFET portion of the IGBT at low drain-source voltages in a manner similar to that observed with power MOSFETs. Second, the *pnp* transistor portion of the IGBT traverses the active region to its on-state (hard saturation) more slowly than the MOSFET portion of the IGBT. Until the *pnp* transistor is full on, the full benefit of conductivity modulation of the drain-drift region has not been achieved and thus the voltage across the IGBT has not dropped to its final on-state value.

24-6-2   Turn-off Transient

The turn-off voltage and current waveforms for the IGBT in the step-down converter circuit are shown in Fig. 24-7. The observed sequence of first a rise in the drain-source voltage to its blocking state value before any decrease in the drain current is identical to that observed in all devices used in a step-down converter

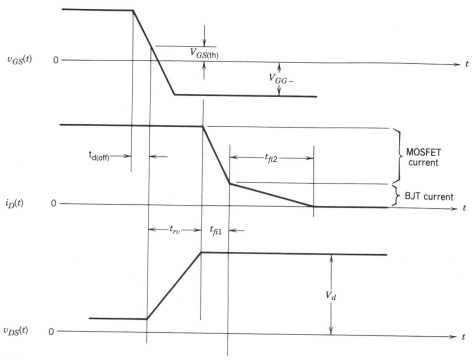

FIGURE 24-7: Turn-off voltage and current waveforms of an IGBT embedded in a step-down converter circuit.

circuit. The initial time intervals, the turn-off delay time $t_{d(\text{off})}$, and the voltage rise time t_{rv} are governed by the MOSFET portion of the IGBT. The equivalent circuits used in Chapter 21 to describe these portions of the power MOSFET turn-off can also be applied to the IGBT. The only modification required is the use of only a single value of gate-drain capacitance C_{gd} rather than two values as used in the power MOSFET calculations. The reasons for this difference are the same as discussed for the turn-on transient of the IGBT.

The major difference between the IGBT turn-off and the power MOSFET turn-off is observed in the drain current waveform where there are two distinct time intervals. The rapid drop that occurs during the t_{fi1} interval corresponds to the turn-off of the MOSFET section of the IGBT. The "tailing" of the drain current during the second interval t_{fi2} is due to the stored charge in the n^- drift region. Since the MOSFET section is off and there is no reverse voltage applied to the IGBT terminals that could generate a negative drain current, there is no possibility for removing the stored charge by carrier sweep-out.

The only way that these excess carriers can be removed, at least in an IGBT without the n^+ buffer layer shown in the IGBT geometry of Fig. 24-1a, is by recombination within the n^- drift region. Since it is desirable that the excess carrier lifetime in this region be large so that the on-state voltage drop is low, then the duration of the t_{fi2} interval at turn-off will be correspondingly long. However, a long t_{fi2} interval is undesirable because the power dissipation in this interval will be large since the drain-source voltage is at its off-state value. This time increases with temperature, as does the tailing time in a power BJT. Thus, a trade-off between on-state losses and faster turn-off times must be made in the IGBT, which are quite similar to those made with minority carrier devices such as BJTs, thyristors, diodes, and the like. Electron irradiation of the IGBT is often used to set the carrier lifetime in the drift region to desired value.

The removal of the stored charge from the drift region by diffusion of the holes to the p^+ layer (so-called back injection) could significantly shorten the t_{fi2} interval if the flux of diffusing holes could be made large. In an IGBT structure without the n^+ buffer layer, such diffusion cannot occur because the gradient of hole distribution is in the wrong direction, that is, the hole density on the p^+ side is greater than the excess hole density in the drift region. Hence, the excess holes are effectively trapped in the drift region. However, the presence of a properly designed n^+ buffer layer modifies this apparently bleak situation markedly. This layer has a much shorter excess carrier lifetime and, thus, acts as a sink for excess holes. The greater recombination rate of holes in the buffer layer sets up a hole density gradient in the drift region during turn-off that causes a large flux of diffusing holes toward the buffer layer. This greatly enhances the removal rate of holes from the drift region and thus shortens the t_{fi2} interval. IGBTs are commercially available with blocking voltages of 1000 V and on-state current capabilities of 200 A that have turn-off times of 1 μs or less. Prototype devices with similar turn-off times but larger blocking voltages and on-state currents have been reported (1800–2000 V). It should be noted that this method of shortening the turn-off time does not require reduction of the carrier lifetime in the drift region, so there is no significant increase in the on-state conduction losses.

24-7 DEVICE LIMITS AND SAFE OPERATING AREAS

The IGBT has robust safe operating areas both during turn-on and turn-off. The forward bias safe operating area shown in Fig.24-8a is square for short switching times, identical to the FBSOA of the power MOSFET shown in Fig. 21-19 for turn-on times shorter than 1 ms. For longer switching times the IGBT is thermally limited, as shown in the FBSOA, and this is also identical to the behavior FBSOA of the power MOSFET.

The reverse bias safe operating area RBSOA is somewhat different than the FBSOA, as is illustrated in Fig. 24-8b. The upper right-hand corner of the RBSOA is progressively cut out and the RBSOA becomes smaller as the rate of change of reapplied drain-to-source voltage dv_{DS}/dt becomes larger. The reason for this restriction on the RBSOA as a function of reapplied dv_{DS}/dt is to avoid latchup. Too large a value of dv_{DS}/dt during turn-off will cause latchup of the IGBT exactly as it can in thyristors and GTOs. Fortunately, this value is quite large, comparing favorably with other power devices. In addition, the device user can easily control the reapplied dv_{DS}/dt by proper choice of V_{GG-} and gate drive resistance.

The maximum drain current I_{DM} is set so that latchup is avoided. It is usually determined on the basis of dynamic latchup conditions. There is also a maximum permissible gate-source voltage $V_{GS(max)}$. As long as this voltage is not exceeded, then an external circuit fault that tries to force the drain current to be larger than I_{DM} will cause the IGBT to leave the on-state and enter the active region where the current becomes a constant independent of the drain-source voltage. Under these conditions the IGBT must be turned off as quickly as possible because of the excessive power dissipation. This behavior is desirable because latchup will not occur and the gate control over the drain current is maintained.

When V_{GS} is 10 to 15 V, drain currents of 4 to 10 times the nominal rated current are obtained. Recent measurements (see Reference 1) indicate that the device can withstand such currents for 5 to 10 μs depending on the value of V_{DS} and can be turned off by V_{GS}.

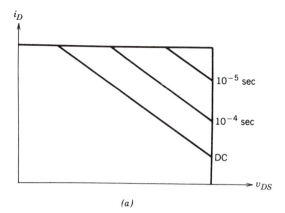

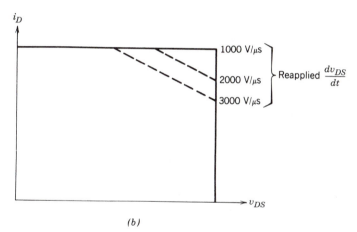

FIGURE 24-8: Safe operating areas of an insulated gate bipolar transistor: (*a*) forward bias safe operating area; (*b*) reverse bias safe operating area.

The maximum permissible drain-source voltage is set by the breakdown voltage of the *pnp* transistor. The beta of the *pnp* transistor is quite low, so its breakdown voltage is essentially BV_{CBO}, the breakdown voltage of the drift–body junction (junction J_2 in Fig. 24-1a). Devices with blocking capabilities as large as 2000 V have been reported.

The maximum permissible junction temperature in commercially available IGBTs is 150°C. A very desirable feature of the IGBT is the fact that the on-state voltage $V_{DS(\text{on})}$ varies very little between room temperature and the maximum junction temperature. In a power MOSFET the on-state voltage increases significantly as the junction temperature increases. The reason for the flat temperature characteristic of the IGBT is the combination of the positive temperature coefficient of the MOSFET voltage drop portion of $V_{DS(\text{on})}$ and the negative temperature coefficient of the voltage drop across the drift region.

24-8   DRIVE AND SNUBBER CIRCUITS

Insulated gate bipolar transistors are similar to power MOSFETs as far as gate–source drive voltage requirements are concerned. The same considerations that govern the design of drive circuits for power MOSFETs thus also apply to the design

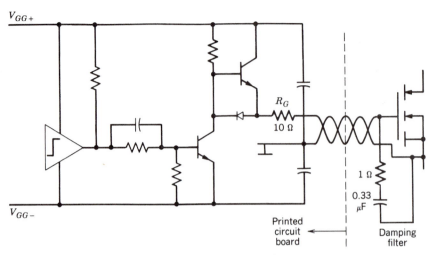

FIGURE 24-9: IGBT drive circuit for larger gate currents.

of drive circuits for IGBTs. This means that the same drive circuits discussed in Chapter 21 for power MOSFETs can be used with IGBTs. If a larger gate current is required, the circuit shown in Fig. 24-9 can be used. A damping filter located near the gate-source terminals can be used to minimize oscillations.

The square safe operating area of the IGBT for switch-mode operation minimizes the need for snubber circuits in most situations. The ability to control the turn-on and turn-off times by controlling the gate current through appropriate sizing of the series gate resistance further minimizes the need for turn-on and turn-off snubbers. The peak current handling capability of the IGBT, which is greater than that of most power MOSFETs, is another factor that makes the use of snubbers unnecessary in most applications. If special circumstances dictate the use of a snubber circuit, the types of snubbers discussed in Chapter 21 for power MOSFETs would also be appropriate for IGBTs.

24-9 SUMMARY

This chapter has examined the structure and characteristics of a new power device, the insulated gate transistor or IGBT. The important conclusions are listed:

1. The IGBT is designed to operate as a MOSFET with an injecting region on its drain side to provide for conductivity modulation of the drain drift region so that on-state losses are reduced.

2. The performance of the IGBT is thus midway between that of a MOSFET and a BJT. It is faster than a comparable BJT but slower than a MOSFET. Its on-state losses are much smaller than those of a MOSFET but not as low as those of a BJT.

3. The IGBT structure contains a parasitic thyristor that must not be allowed to turn on or else the gate will lose the ability to turn off the device.

4. Prevention of the turn-on of the parasitic thyristor involves special structural modifications of the IGBT structure by the device manufacturer and observance of maximum current and voltage ratings of the device by the user. New devices appear to be latchup-proof.

5. The turn-on speed of the IGBT can be controlled by the rate of change of the gate-source voltage.

6. The IGBT has a rectangular safe operating area for switch-mode applications similar to the MOSFET and thus has minimal need for snubber circuits.

PROBLEMS

24-1. *P*-channel MOSFETs require about three times the area on a silicon wafer to achieve a performance comparable to an *n*-channel MOSFET. However, *p*-channel IGBTs have the same area as *n*-channel IGBTs. What are the reasons for the differences between the IGBT behavior and the MOSFET behavior?

24-2. During turn-off, the drain current in an IGBT will exhibit different behaviors depending on whether the carrier lifetime in the drift region is longer or shorter. Qualitatively sketch the drain current versus time during turn-off for a short lifetime IGBT and for long lifetime IGBT and explain the reasons for the differences.

24-3. A punch-through IGBT will have a higher output resistance in the active region (flatter i_D-v_{DS} curves in the active region) than a non-punch-through IGBT. Explain why.

24-4. Estimate the forward and reverse breakdown voltages of the IGBT shown in Fig. 24-1. The doping levels are $p^+ = n^+ = 10^{19}$ cm^{-3}, $p = 10^{17}$ cm^{-3}, and $n^- = 10^{14}$ cm^{-3}. The length (dimension parallel to the current flow direction) of the drift region is 25 μm.

24-5. An IGBT circuit module complete with its own drive circuitry has been made with the following performance specifications:

$$V_{DSM} = 800 \text{ V} \qquad I_{DM} = 150 \text{ A} \qquad dv_{DS}/dt < 800 \text{ V}/\mu\text{s}$$

$$t_{on} = t_{d(on)} + t_{fv} + t_{ri} = 0.3 \ \mu\text{s} \qquad t_{off} = t_{d(off)} + t_{fi} + t_{rv} = 0.75 \ \mu\text{s}$$

This module is to be used in a step-down converter circuit with a diode-clamped inductive load. In this circuit the free-wheeling diode is ideal, the dc supply voltage $V_d = 700$ V, the load current $I_o = 100$ A, and the switching frequency is 50 kHz with a 50% duty cycle.

(a) Show that a turn-off snubber is needed for the IGBT.

(b) Design a turn-off snubber that will provide a factor of safety of two to dv_{DS}/dt.

24-10 REFERENCES

1. Terje Rogne, N. A. Ringheim, J. Eskedal, B. Odegard, and T. M. Undeland, "Short Circuit Capability of IGBT (COMFET) Transistors," 1988 IEEE Industrial Applications Society Meeting, Pittsburg, PA, Oct. 1988.

2. B. Jayant Baliga, *Modern Power Devices*, John Wiley & Sons, New York, 1987, Chapter 7.

3. B. Jayant Baliga, "The Insulated Gate Transistor (IGT)-A New Power Switching Device," *Power Transistors: Device Design and Applications*, B. Jayant Baliga and Dan Y. Chen, (Eds.) IEEE Press, Institute of Electrical and Electronic Engineers, New York, 1984, pp. 354–363.

4. Hamza Yilmaz, John L. Benjamin, Raymond F. Dyer, Jr., Li-shu S. Chen, W. Ron Van Dell, and George C. Pifer, "Comparison of Punch-Through and Non-Punch-Through IGT Structures," *IEEE Transactions on Industrial Applications*, Vol. IA-22, No. 3, pp. 466–470, May/June 1986.

5. Akio Nakagawa, Yoshihiro Yamaguchi, Kiminori Watanabe, and Hiromichi Ohashi, "Safe Operating Area for 1200 V Nonlatchup Bipolar-Mode MOSFETs," *IEEE Transactions on Electron Devices*, Vol. ED-34, No. 2, pp. 351–355, Feb. 1987.

6. Michael S. Adler, King W. Owyang, B. Jayant Baliga, and Richard A. Kokosa, "The Evolution of Power Device Technology," *IEEE Transactions on Electron Devices*, Vol. ED-31, No. 11, pp. 1570–1591, Nov. 1984.

7. M. H. Rashid, *Power Electronics: Circuits, Devices, and Applications*, Prentice-Hall, Englewood Cliffs, NJ, 1988, Chapters 13, 15.

25

Emerging Devices and Circuits

―――――――
―――――――
―――――――

25-1 INTRODUCTION

The number of semiconductor power devices available today is impressively large compared with just a few years ago. The list includes diodes, bipolar junction transistors, monolithic Darlingtons, metal-oxide-semiconductor field effect transistors, thyristors, gate-turn-off thyristors, and insulated gate bipolar transistors. Research efforts will continue to improve these devices, increasing their blocking voltage capabilities, lowering their on-state losses, and increasing their switching speeds. A recent example of this is the development of the ring emitter BJT by Siemens.

Other device and integrated circuit concepts are also currently being explored that show significant potential for future power electronic applications. These concepts that have not yet found general commercial acceptance or are still in the laboratory prototype stage we term emerging devices and circuits. A list of such emerging devices would include power junction field effect transistors (also termed static induction transistors), field controlled thyristors (bipolar static induction transistors), MOS-controlled thyristors (MCTs), high-voltage integrated circuits and so-called smart power circuits and devices. Some of these devices may become widely used in the future, so it is important for designers of power electronic circuits to be aware of these potentially useful devices. In this chapter we briefly summarize the characteristics of these emerging devices and discuss their physical principles of operation and operational limitations.

25-2 POWER JUNCTION FIELD EFFECT TRANSISTORS

25-2-1 Basic Structure and *I-V* Characteristics

The vertical cross section of an n-channel power junction field effect transistor (JFET) is shown in Fig. 25-1a. This particular geometry is utilizing a so-called recessed gate structure, one of the more promising structures for the JFET. The gate and source regions are highly interdigitated in a manner similar to the interdigitation

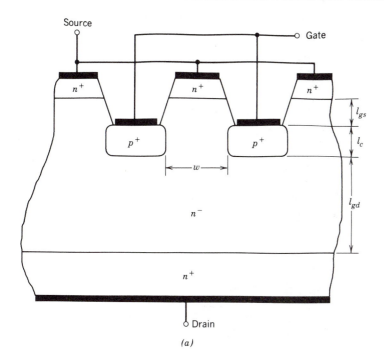

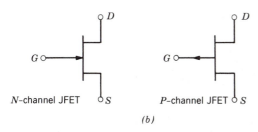

FIGURE 25-1: Structure and circuit symbols for junction field effect transistors: (a) recessed gate JFET cross section; (b) JFET circuit symbols.

used in MOSFETs. Hundreds and even thousands of these basic gate-source cells are connected in parallel to make up a single-power JFET. The doping levels indicated qualitatively in the figure are similar to those used in other power semiconductor devices and designated by the same qualitative symbols. The dimensions shown symbolically in the figure play a major role in determining the device characteristics, as will be discussed shortly.

In a power JFET the channel width w is fairly narrow, typically a few microns to a few tens of microns and the channel length l_c is made smaller than the width w. The device manufacturer also attempts to minimize the dimension l_{gs}. The length of the gate-drain region l_{gd} is dependent on the desired value of blocking voltage capability. This lightly doped region is essentially a drain drift region analogous to that of the power MOSFET.

The circuit symbol for an n-channel JFET is shown in Fig. 25-1b. The arrow pointing into the gate indicates the direction of gate current that would flow if the gate–source junction becomes forward biased. The symbol for a p-channel JFET is also shown in this figure.

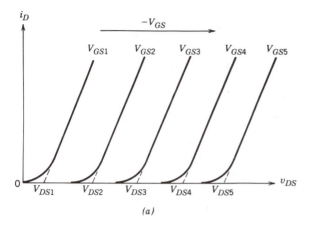

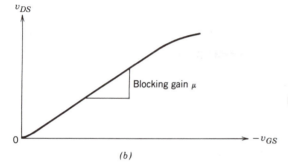

FIGURE 25-2: Current-voltage characteristics of a power JFET: (*a*) output characteristics; (*b*) transfer curve.

The *i-v* characteristics for an *n*-channel power JFET are shown in Fig. 25-2*a*. These characteristics are quite different from those of a MOSFET and are often referred to as triode-like characteristics because of their resemblance to the *i-v* characteristics of vacuum triodes. An approximate transfer curve of drain-source voltage versus gate-source voltage is shown in Fig. 25-2*b*. The slope of this transfer curve is often termed the blocking gain μ because when the curves of i_D vs v_{DS} are extrapolated to zero drain current as indicated in Fig. 25-2*a*, this slope represents the incremental increase in drain-source voltage that can be blocked by the device in the off-state for a given incremental increase in the gate-source voltage.

The most important feature to note about the *i-v* characteristics is that the JFET is a normally-on device, meaning that when the gate is shorted to the source, the device is in the on-state. In contrast, all other semiconductor power devices that we have discussed are normally-off devices. A normally-on characteristic is undesirable in most power electronic applications because the normally-on device may permit unacceptably large transient current flows at system power-up, whereas a normally-off device has a built-in safety feature of being off at system power-up. Because of this drawback, the power JFET has not found any widespread usage even though some are commercially available.

25-2-2   Physics of Device Operation

As we indicated earlier, the JFET is in the on-state when the gate-source voltage is zero. This occurs because there is no hindrance to the flow of current between drain and source in the simplified structure shown in Fig. 25-3*a*. The width of the channel

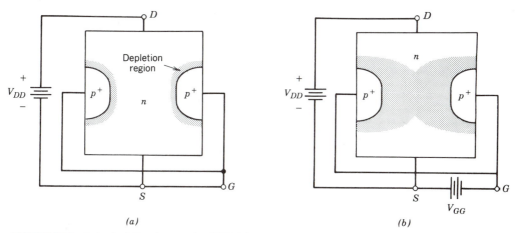

FIGURE 25-3: Gate depletion layers of a JFET (*a*) in the on-state and (*b*) in the blocking state.

region between the p^+ gate regions is large enough so that the depletion regions of the gate–source junction at $v_{GS} = 0$ do not meet in the center of the channel and thus pinch it off. In this on-state, the device designer seeks to minimize the ohmic resistance between the drain and source by making the channel length l_c as short as well as the drain drift region length l_{gd}. The lateral layout of the gate and source regions also impacts the on-state resistance of the device.

When a reverse bias is applied to the gate–source pn junction, the depletion layers grow in width, and at the particular value of v_{GS}, termed the pinch-off voltage V_p, the depletion layers meet in the middle of the channel and pinch it off as is indicated in Fig. 25-3b. In this circumstance there will be no flow of current between drain and source as long as the drain-source voltage is kept small. This is the off-state of the JFET.

If the gate-source voltage is kept at a constant value $|V_{GG}| > |V_p|$, initially no drain current will flow as $v_{DS} = V_{DD}$ is increased from zero. The depletion layers that are blocking the channel have set up a potential barrier to the flow of drain current, as is illustrated in Fig. 25-4. The electric field E_{GS}, diagrammed in Fig. 25-4, which results from this potential barrier, opposes the electric field E_{DS}, which is set by the applied drain-source voltage. As V_{DD} increases, E_{DS} increases while E_{GS} remains fixed because of the fixed V_{GG} and the net electric field in the depletion region in the channel that is blocking the flow of current becomes progressively smaller. The result is that the potential barrier to the flow of drain current gets smaller and smaller, as diagrammed in Fig. 25-4.

When V_{DS} is large enough to suppress the potential barrier set up by V_{GG}, the depletion layer has been effectively eliminated from the center of the channel and current begins to flow. Increases in drain-source voltage beyond this, the threshold value of $\mu |V_{GG}| = V_{DD}$ will cause large increases in the drain current, since the incremental slope of the i-v characteristic in this operating region is essentially the on-state resistance. To maintain the JFET in the off-state for some specified maximum value of $v_{DS} = V_{DSM}$, the gate-source voltage must be larger than V_{DSM}/μ. The length l_{gd} of the drain drift region is set by the requirement that this region accommodate the depletion layer of the gate–drain junction at the maximum value of drain-source voltage V_{DSM} without avalanche breakdown occurring.

It is possible to operate the JFET in a so-called bipolar mode when it is in the on-state. Instead of reducing v_{GS} to zero volts in order to turn on the JFET, the

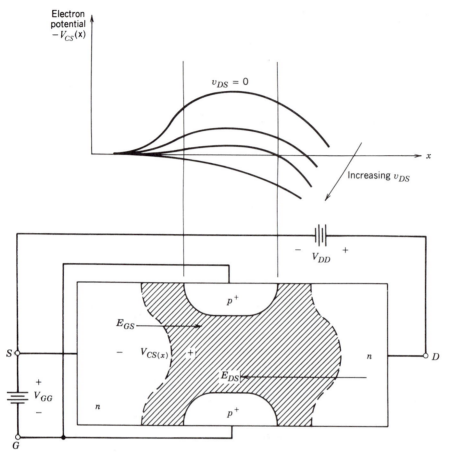

FIGURE 25-4: Potential barrier to current flow in a power JFET because of the applied gate-source voltage. As the applied drain-source is increased, it gradually suppresses this barrier.

gate–source junction is forward biased. This leads to a significant reduction in the on-state resistance because the forward-biased junction injects minority carriers into the channel that conductivity modulates the on-state channel resistance. In this situation, significant gate currents must flow and, from a terminal viewpoint, the device appears similar to a BJT in the on-state. In fact, the drain current that can flow in this mode of operation is proportional to the magnitude of the gate current, and the incremental current gain is typically 100 or more at small drain currents and falls as I_D is increased. If the JFET is designed with narrow channel widths so that the channel is pinched off at zero gate-source bias, the only way to turn on the device is to forward bias the gate–source junction. This produces a JFET with a normally-off characteristic, and it has been called a bipolar static induction transistor (BSIT). The $i\text{-}v$ characteristic of the BSIT looks similar to that of a BJT.

25-2-3 Switching Characteristics

A normally-on JFET is very similar to the MOSFET as far as its switching characteristics are concerned. The equivalent circuit for the normally-on JFET is identical to that of the MOSFET, and the switching waveforms and switching times will be the same as for a comparable MOSFET. The analysis presented in Chapter 21 for the MOSFET can be used with the JFET with only minimal changes. The major difference between the two devices is that the n-channel JFET requires a negative going gate-source voltage to turn off and a positive going v_{GS} to turn on, whereas just the opposite is required for the MOSFET.

If the JFET is operated in the bipolar mode or is fabricated as a normally-off device (BSIT), then it will behave in a switching application more as a bipolar junction transistor than as a MOSFET. This will include some of the phenomena associated with stored charge since the forward-biased gate–source junction injects minority carriers into the channel when the device is in the on-state.

Two principal differences will be noted between the turn-off waveforms of a bipolar mode JFET and the BJT. First, there will be only one current fall time interval with the bipolar-mode JFET because there is no quasi-saturation region such as is observed with the BJT. Second, the turn-off time of the JFET including both storage delay time and current fall time will be considerably shorter for the JFET than for comparable BJT.

The reason is that the JFET has no pn junction in the path of the drain current that can interrupt the sweep-out of excess carriers as the device switches off. When an in-line pn junction, such as the collector-base junction or emitter base junction becomes reverse biased, any stored charge in the device becomes trapped in the device and can be removed only by internal recombination. This considerably slows down the turn-off of the device. Special modifications to the device structure such as anode shorts in GTOs or special drive circuit arrangements such as emitter-open switching with the BJT, are often used to mitigate the effects of this trapped stored charge.

25-3 FIELD-CONTROLLED THYRISTOR

25-3-1 Basic Structure and I-V Characteristic

If the drain of a power JFET structure is modified into an injecting contact by making it into a pn junction, then a new device is produced. This new device, which is variously termed a field-controlled thyristor (FCT), a field-controlled diode, and in Japan, a bipolar static induction thyristor (BSIThy), has the basic structure shown in Fig. 25-5. What would normally be the drain of an n-channel JFET is converted into a pn junction, as is shown, and it becomes the anode of the device. The source of the JFET portion of the new structure is now termed the cathode. The circuit symbol for the FCT is shown in Fig. 25-5b and is essentially a diode symbol with a gate terminal added. The arrow on the gate terminal indicates the direction of forward bias current flowing into the gate–source pn junction.

The i-v characteristic of a normally-on FCT is shown in Fig. 25-6. In the forward bias portion of the characteristic, the FCT appears similar to the power JFET. The differences in the forward bias operation of the two devices is quantitative, with the FCT being able to conduct far larger currents than the JFET for the same on-state voltage. The FCT also blocks in the reverse direction because of the addition of the pn junction at the anode. This reverse blocking is independent of the voltage applied to the gate–source junction.

The FCT can also be made with a normally-off characteristic by using the same approach as was described for the normally-off JFET. The i-v characteristics of a

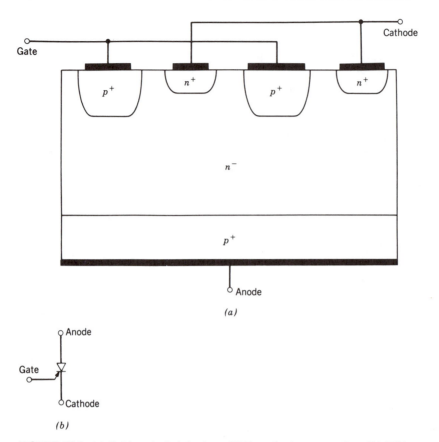

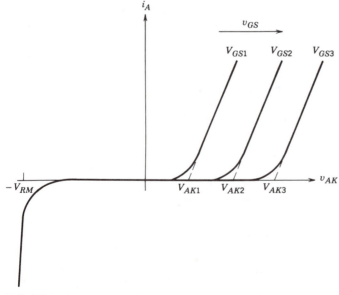

FIGURE 25-5: (*a*) Field-controlled thyristor (FCT) vertical cross section. (*b*) FCT circuit symbol.

FIGURE 25-6: *I-V* characteristic of a normally-on FCT.

normally-off FCT are similar to those of a BSIT except that the current levels are far larger because of the lower on-state resistance of the FCT.

25-3-2 Physical Description of FCT Operation

As was explained in the preceding section, the FCT is basically a power JFET structure with an injecting contact at the anode (JFET drain). The injection of minority carriers from the anode into the anode drift region results in large conductivity modulation of the this region. Consequently, there is a small value of on-state resistance compared with a JFET and a corresponding low value of on-state voltage, even at large values of currents.

The JFET-like gate structure gives the FCT a turn-off capability in addition to a turn-on capability. In the normally-on FCT, turn-off is accomplished by applying a large reverse bias to the gate-cathode terminals so that the gate–cathode junction is reverse biased. The depletion region of this junction then grows and pinches off the channel connecting the anode to the cathode, thus preventing any current flow. The negative bias on the gate draws the excess carriers out of the device as a large negative gate current, similar to the action in a GTO. If the FCT is fabricated as a normally-off device, then its turn-on and turn-off are similar to the turn-on and turn-off of a normally-off JFET.

Even though the FCT is termed a field-controlled thyristor, it is important to note that the device does not have any regenerative turn-on or turn-off as does a gate-turn-off thyristor. The FCT does not latch on or off. If the gate drive holding the normally-on FCT in the blocking state is removed, the FCT will turn on. Similarly, removal of the gate drive holding the normally-off FCT in the on-state will cause the device to turn off.

25-3-3 Switching Characteristics

The switching times of the FCT will be considerably slower than those of the normally-on JFET. This is because of the large amount of stored charge in the drift region and the channel region of the device. The switching waveforms will be qualitatively similar to those of a normally-off JFET, although the turn-off times will be significantly longer because of the large amount of drift region stored charge. As explained earlier, turn-off of the FCT requires a large negative gate current pulse, similar to a GTO. The normally-off JFET does not have nearly the amount of drift region stored charge, since it does not have the injection of charge from the anode.

The FCT would be expected to have large reapplied dv_{AK}/dt ratings, since it does not have any regenerative turn-on mechanism such as limits the conventional thyristor. This expectation has been verified in laboratory prototypes of FCT structures.

Laboratory prototypes of FCTs have also demonstrated that limits exist to the allowed values of di_A/dt, especially during device turn-on. Several factors lead to turn-on of the FCT in localized areas and then expansion of these localized areas to encompass the entire active region of the device as time proceeds. The localized turn-on leads to small regions of high-power dissipation, which will be too large if the rate of growth of the anode current di_A/dt exceeds some maximum rate. The excessive power dissipation in this small area can then cause device failure just as it can in comparable situations in BJTs or thyristors. Improvements in the di_A/dt rating have been made in recent years so that FCTs now have ratings in excess of 1000 A/μs.

25-4   JFET-BASED DEVICES VERSUS OTHER POWER DEVICES

The blocking voltage capability of the JFET that has been achieved to date compares reasonably well with the BJT or the MOSFET. The limiting factor in the JFET is not avalanche breakdown across the depletion region in the drift region but rather the achievable value of blocking gain. The maximum voltage that can be blocked between anode and cathode or drain and source is crudely modeled as the product of the blocking gain and the breakdown voltage of the gate–source junction.

The on-state losses in a JFET are larger than in a comparable MOSFET. The reasons are largely technological rather than fundamental and further research will narrow the gap between the two types of devices. Intuitively one would expect that JFETs and MOSFETs made to block the same voltages in the off-state would have the same on-state losses. If the JFET is operated in the bipolar mode, then the on-state losses of the JFET lie between those of the MOSFET and the BJT.

The switching speeds of normally-on JFETs are presently somewhat slower than those of comparable MOSFETs. This is basically a technological limitation rather than a fundamental one. The normally-off JFET (BSIT) has switching speeds that are currently comparable or somewhat better than those of a comparable BJT. In principle a BSIT should have faster turn-off times than a BJT because of the lack of an in-line *pn* junction that can lead to an open-base turn-off problem such as can slow down the BJT. The FCT is as fast as a GTO, the device against which it is normally compared.

Normally-on JFETs, like MOSFETs, are majority carrier devices and have no serious propensities to second breakdown as do BJTs and other minority carrier devices. Bipolar-mode JFETs and FCTs do have nonuniform turn-on of the active regions and, hence, some likelihood of second breakdown under the right conditions. This potential problem of current constrictions in bipolar-mode JFETs and FCTs also means that limits on *di/dt* and *dv/dt* may also exist. The seriousness of these potential problems is somewhat unclear and further research will be needed to address this issue.

Probably more than any other reason, the normally-on characteristic of the JFET is most responsible for the lag in its use in switch-mode applications compared with other devices. The disadvantages of this normally-on characteristic have already been detailed. Although it is possible to fabricate normally-off JFET-based devices, they have not yet matched the capabilities of other normally-off devices. Further research and development with the JFET-based structures will be needed to rectify this situation. It is likely that it will not happen on a broad front, since the other devices discussed in the previous chapters have already found wider acceptance and their capabilities are continuing to be developed. JFET-based devices will most likely find niche applications where their unique properties offer advantages that other devices cannot match.

25-5   MOS-CONTROLLED THYRISTORS

The MOS-controlled thyristor or MCT is a new device that is in an early stage of development. It is basically a thyristor with one or more MOSFETs built into the gate structure. In one version, whose equivalent circuit is shown in Fig. 25-7a, the built-in MOSFETs both turn on and turn off the thyristor. In a second version whose equivalent circuit is shown in Fig. 25-7b, the device is turned on by a gate current like a conventional thyristor, but it is turned off by the built-in MOSFET. Both of these

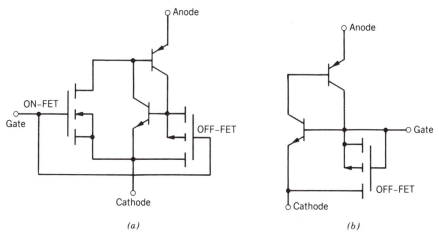

FIGURE 25-7: Equivalent circuit for a MOS-controlled thyristor: (*a*) MCT with MOSFET turn-on and turn-off; (*b*) MCT with conventional gate turn-on and MOSFET turn-off.

devices would have applications similar to those of the GTO, but the MCT would have simpler gating requirements, particularly in terms of the magnitude of the needed gating signals.

25-5-1 MOSFET-Controlled Turn-on and Turn-off

Turn-off is accomplished by turning on the OFF-FET, which in turn shunts base current away from the *npn* BJT in the thyristor pair. This causes the *npn* transistor to begin to turn off as the stored charge in the *p*-base of the transistor disappears and its current gain falls to a low value where the latch-on condition of the thyristor is no longer satisfied. Once this occurs, the thyristor turns itself off by regenerative action. During the turn-off of the thyristor, the other MOSFET in the circuit, the ON-FET, must be kept in the blocking state. As explained later, a zero or negative value gate-anode voltage will guarantee that the ON-FET is in the blocking gate.

Turn on is accomplished by driving the ON-FET into the conducting state. Since the ON-FET is an *n*-channel MOSFET and the OFF-FET is a *p*-channel device, a positive going gate voltage will simultaneously turn on the ON-FET and turn off the OFF-FET. Turn on of the ON-FET permits the flow of base current out of the *pnp* transistor in the thyristor pair that thus activates the *pnp* transistor. The collector current from the *pnp* transistor then flows as base current to the *npn* transistor and turns it on. Once both transistors are on, the regenerative action of the connection will cause the thyristor to turn on. The better conduction characteristic of the *npn* transistor ensures that it will carry most of the base current of the *pnp* transistor in the on-state rather than in the ON-FET that is in parallel with the *npn* BJT. During the on-state, the OFF-FET must be kept in the blocking state, which is ensured by a positive gate voltage.

The conceptual cross section of the unit cell of this version of the MCT is shown in Fig. 25-8. A complete MCT would be composed of many of these cells connected in parallel in order to achieve the desired current capability. This complete device has not yet been fabricated. However, simpler derivatives of it, containing either just the ON-FET or the OFF-FET have been fabricated and tested. Structures with just the ON-FET incorporated in the device have been termed FET-controlled thyristors or

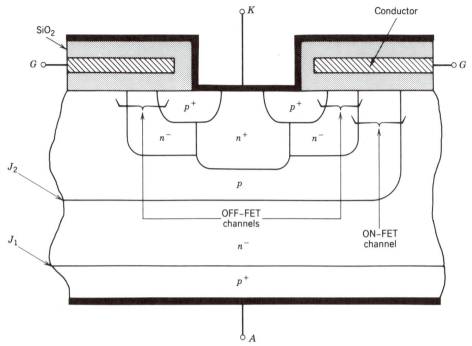

FIGURE 25-8: Conceptual cross section of a MOS-controlled thyristor.

MOS-gated SCRs. Experimental results achieved to date indicate that turn-on and turn-off times for the MCT are faster than those realized with a GTO.

25-5-2   Conventional Gate Turn-on and MOSFET Turn-off

The unit cell of the MCT shown in Fig. 25-8 indicates one potentially significant disadvantage to the structure, the curvature of the $n^- p$ boundary (which forms the J_2 junction of the thyristor). This curvature will lead to reductions in the blocking voltage capabilities of the device. The use of field plates or guard rings to partially overcome this problem is not a viable option because the area required for them would not be available for carrying current and, thus, the large current capabilities expected for a thyristor would be compromised.

A possible solution is to remove the ON-FET, as is indicated in the equivalent circuit of Fig. 25-7a, and to rely on the conventional gate turn-on of the thyristor. Only the OFF-FET is retained. Removal of the ON-FET permits the $n^- p$ boundary to remain straight, as is shown in the unit cell shown in Fig. 25-9, and thus the reduction in breakdown voltage because of field crowding is avoided. Turn-off of this version of the MCT would be accomplished in the same manner as described previously for the other version of the MCT.

Both versions of the MCT will require more research and development to realize a practical device. If its full potential can be realized, it will become a desirable alternative for the GTO.

25-6   HIGH-VOLTAGE INTEGRATED CIRCUITS

Modern semiconductor power control circuits have a considerable amount of control and trigger circuitry in addition to the power device itself. Several examples of

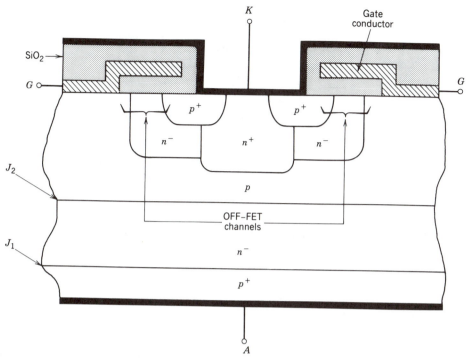

FIGURE 25-9: Conceptual cross section of a MOS-controlled thyristor with a conventional gate for turn-on and a MOSFET for turn-off.

this are presented in earlier chapters. The control circuitry often includes logic circuitry controlled by microprocessors. The inclusion of such control/trigger circuitry on the same chip or wafer as the power device would greatly simplify the overall circuit design and broaden the range of potential applications. A cheaper and more reliable power control system would result from this integration. Some such integration has already been demonstrated including so-called smart switches. Schemes for sensing overvoltage and overcurrent conditions have even been incorporated in such integration demonstrations.

For such integration to become widely used, the IC fabrication process must meet four basic requirements. First, it must isolate high-voltage devices, primarily the power devices, from the low-voltage components, which are primarily the control and trigger circuits. Second, it must make provision for on-chip interconnection of high-voltage components with conductor runs over low-voltage regions. Third, the fabrication process must include the full range of standard IC components including n-channel and p-channel MOSFETs, npn and pnp BJTs, diodes, resistors, and capacitors. Finally, circuit techniques compatible with IC fabrication methods must be found for interconnecting the low-voltage circuits with the high-voltage power devices without leaving the chip or wafer.

The first requirement, the isolation of low-voltage devices from high-voltage elements, can be accomplished by either dielectric isolation or pn junction isolation. Dielectric isolation basically consists of etching a pocket in the chip or wafer and then growing a layer of silicon dioxide in it, as is shown in Fig. 25-10. Next a layer of silicon is deposited over the SiO_2. After annealing the deposited silicon at a high temperature, it becomes recrystallized and can then be used for fabricating the low-voltage devices. Dielectric isolation is free of parasitic devices, such as diodes, that

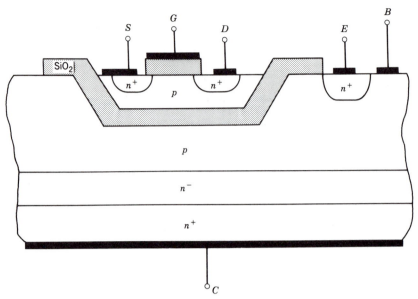

FIGURE 25-10: High-voltage integrated circuit composed of a power BJT with dielectric isolation of a MOSFET.

could become activated in certain circumstances and negate the isolation. However, dielectric isolation is relatively expensive to implement and results in lower yields.

Junction isolation, on the other hand, is much cheaper and easier to implement. A *pn* junction is fabricated so as to completely surround the area to be isolated, as is diagrammed in Fig. 25-11. This junction is then reverse biased at all times, thus achieving the desired isolation. The junction isolation of 500 V has been demonstrated. The principal disadvantage of this isolation method is the parasitic diode that comes with it. Potential problems with this diode include possible turn-on and temperature-dependent leakage currents.

A completely effective means of high-voltage interconnections on the chip or wafer has yet to be devised. The basic difficulty is that wherever an interconnect must pass over either an isolation region or some other heavily doped region that is at a significantly different potential than the interconnect, the equipotential lines between

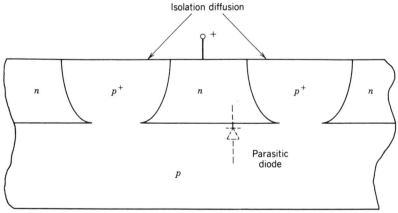

FIGURE 25-11: *PN* junction isolation.

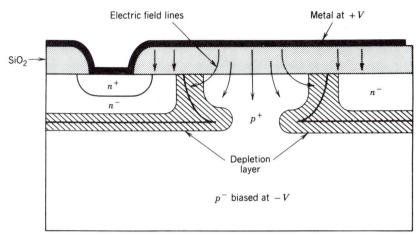

FIGURE 25-12: Electric field crowding where high-voltage interconnects cross over an isolation diffusion.

the low- and high-voltage regions have appreciable curvature. This results in a considerable amount of "field crowding," as is illustrated in Fig. 25-12, for the case of an isolation region. This field crowding will lead to premature breakdown of the isolation region and thus to shorting of the high-voltage to ground.

Two possible approaches to this problem include the use of thick insulators and the use of field shields. The use of thicker insulators will require further research and development. Questions concerning what is the most appropriate insulator and how to prevent the inevitable strain (which increases in the deposited material as it gets thicker) from delaminating the film from the substrate remain to be answered. In the second approach, the field in the insulator can be made more uniform by the use of field shields, which are illustrated in Fig. 25-13. The rationale for the field shields is the same as for field plates discussed in previous chapters. In addition, the use of field shields would mean that the insulating film would not have to be as thick.

The problem of how to connect the low-voltage control circuits to the high-voltage power devices without leaving the chip has yet to be completely answered. For

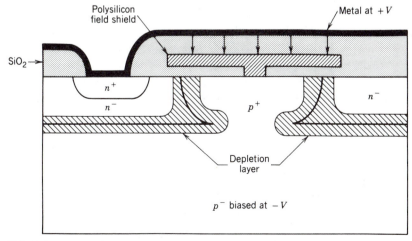

FIGURE 25-13: Use of a field shield to minimize field crowding in the case of an interconnect over an isolation region.

discrete circuits, special methods, including the use of transformers or optocouplers, were described in previous chapters for achieving this interconnection. The use of transformers is precluded if the interconnection is to be made on the chip or wafer. Currently it is not feasible to fabricate optocoupler circuits on the silicon wafer where the power device and low-voltage control circuitry reside. The basic difficulty is that the light-emitting portion of the optocoupler must be made from gallium arsenide GaAs, and the integration of gallium arsenide devices on silicon is only in an early research phase at the present time.

The most feasible method of connecting low-voltage circuits to high-voltage devices is with the use of level-shifting circuitry. Such circuitry maintains a relatively fixed voltage drop between two nodes while permitting the flow of any desired current. A zener diode in breakdown is an example of a device that could be used for level shifting. Other circuits are known that could be used. The problem of how to maintain the voltage difference without excessive power dissipation is the basic issue that will require further research before a commonly accepted method of electrical connection will be available.

Several prototype fabrication processes have been demonstrated recently that include the full range of devices mentioned previously. But because of the problems mentioned in this section, none of these processes can yet be considered a "standard" process. The primary driving forces that will ultimately determine which, if any, process becomes "standard" are yield and cost. Based on what has already been demonstrated, it is clear the high-voltage ICs have a promising future and will have wide applications. Some have even predicted that integrated power electronic devices and circuits could trigger a second electronic revolution that will surpass the present so-called IC-based revolution.

25-7 NEW SEMICONDUCTOR MATERIALS

Silicon is presently the only semiconductor material used in making commercially available power devices. This is because silicon can be grown in single-crystal form with larger diameters and the greatest purity of any available semiconductor. There are, however, other materials that have superior properties compared to silicon for power device applications. Unfortunately, they are not available in high enough purity or large enough sizes to be considered for device manufacturing.

Gallium arsenide would be a highly desirable material for device fabrication if the problems of purity and crystal size could be overcome. It has a larger energy gap than silicon, which means that devices made in GaAs could be used to higher temperatures than silicon devices (larger energy gap translates into smaller intrinsic carrier densities in GaAs than in Si at the same temperature). The carrier mobilities in GaAs are larger than in silicon, which means that the on-state resistances, especially in majority carrier devices, would be smaller in gallium arsenide than in silicon.

There is currently a significant amount of research being centered on GaAs devices for high-speed digital circuitry and microwave devices. This has led to advances in the quality of single-crystal GaAs material. If this level of interest is maintained, the quality of single-crystal GaAs technology may become sufficiently advanced so as to make GaAs power devices practical. There is currently some efforts underway to fabricate power GaAs Schottky diodes. Presently achievable carrier lifetimes in GaAs are too short for good conductivity modulation in high-voltage devices. Thus, BJTs and thyristors fabricated in GaAs are not currently feasible.

Other materials with desirable properties that are of potential interest include silicon carbide and diamond. Both of these materials have larger band gaps than silicon or gallium arsenide. Diamond has carrier mobilities that are comparable to silicon and a higher value of thermal conductivity. However, the state of the technology for these materials is primitive compared with silicon and will require a significant research investment over many years before the material quality will be suitable for power device studies.

25-8 SUMMARY

This chapter has examined the structure and characteristics of several devices and circuits that currently are in an early state of development (hence, our classification of emerging devices) but that appear to have potentially useful properties for power electronic applications. Power junction field effect transistors are included in this classification even though such devices are commercially available because they have not yet found widespread usage. The major conclusions are listed as follows:

1. JFETs are a normally-on majority carrier device with a triode-like i-v characteristic. They are similar to MOSFETs in their switching characteristics and have somewhat higher on-state losses.

2. Field-controlled thyristors have a JFET structure with an injecting drain–channel pn junction that leads to heavy conductivity modulation of the drain drift region and the channel region and, hence, lower on-state losses.

3. The FCT has both gate-controlled turn-on and turn-off and so is a potential supplement to the GTO. It appears to have faster switching speeds than the GTO.

4. The normally-on characteristic of the JFET and the FCT has limited their usage in power electronic applications. They can be made with normally-off characteristic, but this presently leads to higher on-state losses.

5. The MOS-controlled thyristor is essentially a GTO with integrated MOS-driven gates controlling both turn-on and turn-off that potentially will significantly simplify the design of circuits using GTOs.

6. High-voltage integrated circuits offer the promise of drive circuits and logic level control circuits even perhaps sensing elements for overcurrent and thermal protection fabricated on the same chip as the power device. This would significantly lower costs and increase the reliability of power devices and systems.

7. Other semiconductor materials such as gallium arsenide have properties such as larger carrier mobilities than silicon that are highly desirable for the fabrication of power devices. However, the quality and availability of these other materials is currently inferior to silicon and, hence, cannot be used for commercial power device products.

PROBLEMS

25-1. Construct an equivalent circuit for the power JFET (normally-on SIT) similar to the equivalent circuit of the MOSFET that can be used for estimating switching times in circuit applications. Assume that the JFET is working in the triode mode.

25-2. Design a simple drive circuit for the normally-on field controlled thyristor. Consider an arrangement of a MOSFET in series with the cathode of the FCT. Qualitatively describe the operation of the circuit and comment on the required characteristics of the MOSFET.

25-3. Consider a power diode such as is shown in Fig. 19-1. The diode is to be fabricated in gallium arsenide and silicon and both diodes are to have the same on-state voltage drop and reverse-blocking capability. Which material can have the shorter carrier lifetime and by how much?

25-4. Approximately how thick does an insulating layer of silicon dioxide used in a high-voltage IC have to be in order to hold off 1000 V?

25-9 REFERENCES

1. B. JAYANT BALIGA, *Modern Power Devices*, John Wiley & Sons, New York, 1987, Chapters 7–9.

2. J. NISHIZAWA, "Junction Field-Effect Devices," *Semiconductor Devices for Power Conditioning*, Rolland Sittig and P. Roggwiller (Eds.), Plenum Press, New York, 1982, pp. 241–270.

3. VICTOR A. K. TEMPLE, "MOS-Controlled Thyristors—A New Class of Power Devices," *IEEE Transactions on Electron Devices*, Vol. ED-33, No. 10, pp. 1609–1618, Oct. 1986.

4. MICHAEL S. ADLER, KING W. OWYANG, B. JAYANT BALIGA, and RICHARD A. KOKOSA, "The Evolution of Power Device Technology," *IEEE Transactions on Electron Devices*, Vol. ED-31, No. 11, pp. 1570–1591, Nov. 1984.

5. BRIAN R. PELLY, "Power Semiconductor Devices—A Status Review," 1982 International Power Semiconductor Converter Conference Proceedings, CH1682-4182/000-001$00.75 © 1982 IEEE.

6. *Power Transistors: Device Design and Applications*, B. Jayant Baliga and Dan Y. Chen, Editors, IEEE Press, Institute of Electrical and Electronics Engineers, New York, 1984, Part IV; *Emerging Transistors Technology*, pp. 291–374.

26

Passive Components and
Practical Converter
Design Considerations

26-1 INTRODUCTION

The large currents and voltages found in power electronic circuits place severe stresses on passive components as well as on the semiconductor devices. Therefore, passive devices must be selected as carefully as the solid-state switching element. This chapter presents the fundamental differences found among the various types of capacitors so that the user can select components with the appropriate properties from the manufacturers' catalogs. Inductors present a special problem, since they are not available with a wide range of properties but instead are usually designed and built specifically for the given application. Some simple design fundamentals for building inductors and transformers are presented in this chapter. Cooling considerations and design fundamentals for heat sinking are also presented. As we discussed in the previous chapters, it is important to minimize leakage inductances; hence, some general rules for the layouts of both power and logic circuits are presented.

26-2 DESIGN OF INDUCTORS

Generally inductors must be designed and built specifically for the given application. Magnetic cores on which inductors are wound are available in a wide variety of shapes, sizes, and magnetic materials. Off-the-shelf cores are generally used in most applications. An iterative step-by-step inductor design procedure is given below.

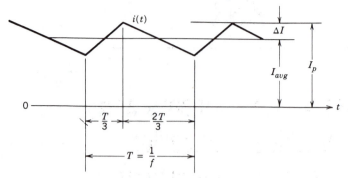

FIGURE 26-1: Current through the filter inductor of a dc–dc converter.

26-2-1   Design Inputs

The information required to build an inductor includes the inductance L, the peak current I_p, the rms current I_{rms}, and the frequency f. These values are found via the detailed design calculations for the specific converter of interest as described in Chapters 3 to 7.

26-2-2   Core Material and B_{max}

In an inductor, the flux ϕ and the flux density B are proportional to the current i. For example, the current through the filter inductor in a dc–dc converter is shown in Fig. 26-1, where

$$\frac{\Delta I}{I_p} = \frac{\Delta B}{B_{max}} \tag{26-1}$$

and the ripple frequency f is also given. Based on an initial selection of the material, B_{max} can initially be assumed to be equal to the saturation flux density B_{sat}. This allows ΔB to be estimated from Eq. 26-1. Knowing ΔB and f, the losses are obtained from loss curves like that shown in Fig. 26-2 for a widely used ferrite material. If the core losses are too high, either a smaller value of B_{max} can be used in Eq. 26-1 or a better core material must be selected. If the core losses are low, a material with a higher saturation flux density could be chosen. A commonly used upper limit for core losses is 10 mW/cm^3.

Cores are available in a variety of materials including ferrites, iron laminations, iron powder, and so on, which have different saturation flux densities, frequency-dependent losses, Curie temperatures, and other properties such as brittleness. The iron laminations are available in a variety of thicknesses, including amorphous materials (metallic glasses) that are useful for high-frequency applications.

26-2-3   Core Size and Shape

Knowing B_{max} and the material, a core size and shape can be selected from the manufacturer's catalog. A variety of shapes are available to suit the given application. This is particularly true of ferrite cores that are available as toroids, pot cores with an air gap, and in U, E, and I shapes. Laminated materials are available as tape-wound

CORE LOSS vs. FLUX DENSITY

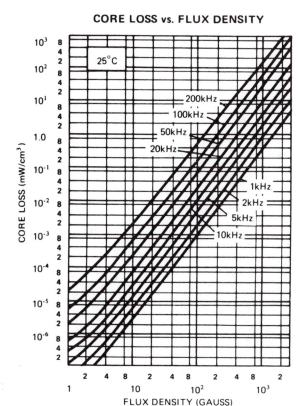

FIGURE 26-2: Core losses versus frequency in a widely used ferrite material. (*Source*: Ferroxcube Division of Amperex Elecronic Corporation.)

toroids and C-cores. In the following description, a double E-core is selected as an example (shown in Fig. 26-3), though the procedure applies to any other shape. An initial choice of the core size can be made using Eq. 26-8*a*, which is derived in a later section. The parameters on the left-hand side of this equation are the design inputs listed above. For the initial estimate assume $k_{cu} = 0.6$ and $J = 2$ A/mm². B_{max} is known from Eq. 26-1. The area product $A_w A_{core}$ allows the core size to be picked using the information in Reference 5.

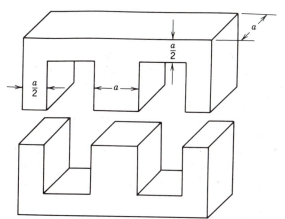

FIGURE 26-3: A magnetic core in the shape of an E, often termed a double-E core.

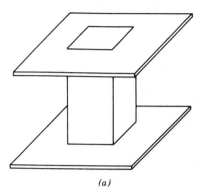

(a)

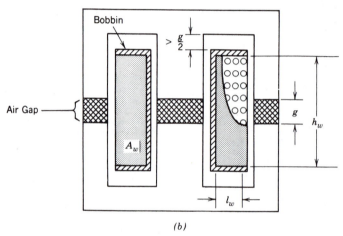

(b)

FIGURE 26-4: Bobbin and an assembled inductor. The air-gap spacing is supported by an insulating material, which is indicated shaded: (a) bobbin; (b) assembled inductor.

26-2-4 Bobbin

A bobbin is provided with most cores and the effective cross-sectional area $A_w = h_w l_w$ for winding on the bobbin is given, as shown in Fig. 26-4. These bobbins are also available in a wide variety of sizes and shapes.

26-2-5 Copper Wire

Knowing I_{rms}, the cross-sectional area of the copper wire can be chosen from thermal considerations so that the highest temperature in the middle of the winding is less than the specified maximum temperature for the insulating material (lacquer) on the wire. The surface-to-volume ratio of the winding decreases with the increasing dimension l_w defined in Fig. 26-4. Therefore, from thermal considerations, a larger inductor must be designed for lower power loss per unit volume.

The maximum permissible current density decreases with increasing bobbin size, as is indicated in Fig. 26-5. Using this figure, the current density J in the copper wire can be obtained for a natural convection-cooled inductor. The cross-sectional area of the copper wire in terms of I_{rms} and J is

$$A_{cu} = I_{rms}/J \tag{26-2}$$

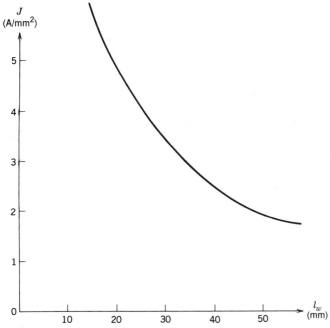

FIGURE 26-5: Permissible current density in copper wire as a function of bobbin dimension l_w with $h_w = 2l_w$.

From wire tables, the nearest wire with a cross-sectional area A'_{cu} equal to or greater than A_{cu} is selected.

Assuming a fill-factor k_{cu} for the winding to be in a range of 0.6 to 0.8, the number of turns N is given by

$$A_w k_{cu} = N A'_{cu} \qquad (26\text{-}3)$$

Assuming $A_{cu} = A'_{cu}$, Eqs. 26-2 and 26-3 yield

$$N = A_w \frac{k_{cu} J}{I_{rms}} \qquad (26\text{-}4)$$

where N is the maximum number of turns possible for the given I_{rms}, the current density optimized from Fig. 26-5, and the selected core.

26-2-6   Maximum *L* for the Selected Core

Having selected B_{max} in Section 26-2-2 and knowing the core cross-sectional area A_{core}, we can determine that the maximum flux in the core is given by

$$\phi_{max} = B_{max} A_{core} \qquad (26\text{-}5)$$

To calculate the maximum possible inductance L_{max} of the inductor, the air gap must be chosen such that at I_p, the flux is ϕ_{max} (known from Eq. 26-5). Using the definition of inductance in terms of current and flux linkage yields

$$L_{max} I_p = N \phi_{max} \qquad (26\text{-}6)$$

then

$$L_{\max} = \frac{N\phi_{\max}}{I_p} = \frac{NB_{\max}A_{\text{core}}}{I_p} \tag{26-7a}$$

The inductance given in Eq. 26-7a is the maximum inductance possible with the selected core and the given I_{rms} and I_p. If the inductance given by Eq. 26-7a is larger than the value desired, then a smaller core size should be chosen. The opposite is true if the inductance is too small. If the next smaller core size gives too low a value of the inductance, then the previous core size is kept with a fewer number of turns N' in Eq. 26-7a to save on copper weight, cost, and losses.

$$L = \frac{N'B_{\max}A_{\text{core}}}{I_p} \tag{26-7b}$$

Combining Eqs. 26-4 and 26-7a yields

$$L_{\max}I_pI_{\text{rms}} = \left\{ k_{\text{cu}}JB_{\max} \right\} A_wA_{\text{core}} \tag{26-8a}$$

For the selected number of turns N', combining Eqs. 26-7a and 26-7b in Eq. 26-8a results in

$$LI_pI_{\text{rms}} = \frac{N'}{N}k_{\text{cu}}JB_{\max}A_wA_{\text{core}} \tag{26-8b}$$

Equation 26-8a shows that the $L_{\max}I_pI_{\text{rms}}$ product results in the product of two areas associated with the core multiplied by the term within the brackets, which contains quantities that are material and slightly size dependent. The area product A_wA_{core} is an indicator of core size and, therefore, of cost.

26-2-7   Air Gap

The air gap is calculated on the basis of the maximum assumed flux density $B_{\max}$ being generated by the peak current I_p. Because of the fringing of the flux as is indicated in Fig. 26-6, the effective cross-sectional area of the air gap A_g can be estimated as is shown in Fig. 26-6b. The flux density in the air gap can be estimated as

$$B_{g,\max} = \frac{A_{\text{core}}B_{\max}}{A_g} \tag{26-9}$$

and

$$H_g = \frac{B_g}{\mu_0} \tag{26-10}$$

From Ampere's law, at the peak current

$$\oint H_{g,\max}\, dl = N'I_p \tag{26-11}$$

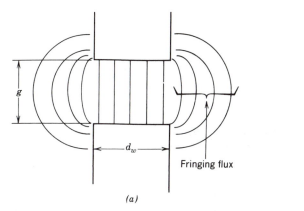
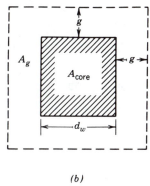

(a)   (b)

FIGURE 26-6: Fringing flux and an estimation of the area of the air gap A_g: (a) fringing flux in the air gap; (b) effective cross-sectional area of the air gap.

If a high relative permeability is assumed for the core material in Fig. 26-4 (so that all of the mmf is dropped in the air gap), then Eq. 26-11 becomes

$$H_{g,\max} 2g = N' I_p \qquad (26\text{-}12)$$

The air-gap length g can be estimated from Eqs. 26-9, 26-10, and 26-12 as

$$g = \frac{\mu_0 A_g N' I_p}{2 A_{\text{core}} B_{\max}} \qquad (26\text{-}13)$$

For the specific air gap given by the foregoing equation, I_p will produce exactly $\phi_{\max}$, thus satisfying Eq. 26-7b.

Fringing flux shown in Fig. 26-6a "cuts" the copper wire of the winding and thus produces eddy current losses in the winding. These losses can be reduced by using distributed air gaps or powder cores.

26-2-8 Loss Estimation

The core losses can be obtained from a graph provided by the core manufacturer that is similar to that in Fig. 26-2 if ΔB, ripple frequency, and core volume or weight are known. The winding losses can be estimated by calculating the winding resistance. Knowing the cross-sectional area and the length of the copper wire and the temperature, the dc resistance of the winding at the operating temperature can be estimated. There are additional eddy current losses in the winding, especially if ΔB is large. These losses are further discussed in the transformer design section.

26-2-9 General Comments

In the foregoing design procedure, both ΔB and the copper area A_{cu} are chosen based on the maximum operating temperature limitation. This may result in unacceptably large losses, which can be remedied by selecting a copper wire with a larger cross-sectional area and designing the inductor for a smaller $B_{\max}$ and, hence, a smaller ΔB. This will undoubtedly require a larger core than the previous design.

26-3   TRANSFORMER DESIGN

The design of small, natural convection-cooled transformers is described in this section. An iterative step-by-step design procedure is given in the next sections.

26-3-1   Design Inputs

The required design input information consists of the primary voltage v_1, the primary rms current I_1, the frequency f, and the turns ratio a.

26-3-2   Core Material and B_{max}

In a transformer with bidirectional ac excitation, $\Delta B = B_{max}$. Since the core losses depend on ΔB as is shown in Fig. 26-2, B_{max} may be chosen to be much smaller than B_{sat}, depending on the losses of the initially selected core material and the operating frequency. In converters such as forward converters with unidirectional core excitation, $\Delta B = B_{max}/2$. If the core losses are too high, either a smaller value of B_{max} can be used or a better core material must be selected.

26-3-3   Core Sizes and Shapes

Knowing the voltage waveform, the flux ϕ can be obtained as

$$v_1 = N_1 \frac{d\phi}{dt} \tag{26-14}$$

where it is recognized that the average voltage is zero. For a square-wave voltage input with $D = 0.5$ as shown in Fig. 26-7, Eq. 26-14 becomes

$$N_1 = \frac{V_1}{4 f \phi_{max}} \tag{26-15}$$

where $f = 1/T$ is the operating frequency. An initial selection of the core size can be made using Eq. 26-27 (to be derived in Section 26-3-6), where S and the frequency f are known from the design inputs. Approximate estimates of k_{cu}, J, and B_{max} can be

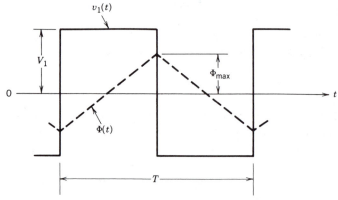

FIGURE 26-7: Transformer flux waveform when $v_1(t)$ is a square wave.

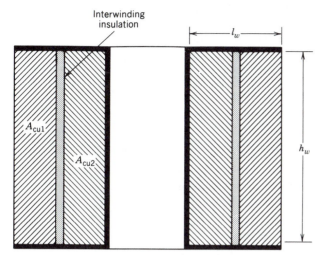

= Secondary winding

= Primary winding

FIGURE 26-8: Transformer bobbin with windings.

made in the same manner as was described in Section 26-2-3 for the inductor. This allows the product $A_W A_{core}$ to be estimated and the core size can be selected using the information given in Reference 5. Now that the core area A_{core} is known, B_{max} is selected per the discussion in Section 26-3-2. Therefore, $\phi_{max} = A_{core} B_{max}$ can be substituted into Eq. 26-15 to yield

$$N_1 = \frac{V_1}{4 f A_{core} B_{max}} \tag{26-16}$$

and therefore

$$N_2 = \frac{N_1}{a} \tag{26-17}$$

26-3-4 Bobbin

The bobbin for a transformer wound on a double E core is shown in Fig. 26-8. Based on the initial core selection, we know the two dimensions l_w and h_w and, therefore, the effective area A_w of the winding cross section in the bobbin.

26-3-5 Copper Wire

From thermal considerations similar to those discussed in the design of inductors, the current density J in the copper wires can be obtained from Fig. 26-5. Assuming that the magnetizing current is negligible,

$$N_1 I_1 = N_2 I_2 \tag{26-18}$$

Both windings are designed for the same current density so that

$$I_1 = A_{cu1} J \tag{26-19}$$

and

$$I_2 = A_{cu2}J \tag{26-20}$$

Combining Eqs. 26-18, 26-19, and 26-20 yields

$$N_1 A_{cu1} = N_2 A_{cu2} \tag{26-21}$$

Assuming that the same winding fill factor k_{cu} for the primary and secondary windings yields

$$A_w k_{cu} = N_1 A_{cu1} + N_2 A_{cu2} \tag{26-22}$$

Combining Eqs. 26-21 and 26-22, the area of the copper wire can be estimated as

$$A_{cu1} = \frac{A_w k_{cu}}{2N_1} \tag{26-23}$$

and

$$A_{cu2} = \frac{A_w k_{cu}}{2N_2} \tag{26-24}$$

26-3-6   Maximum Currents for the Selected Core

With A_{cu1} selected per Eq. 26-23, the maximum allowable current $I_{1,max}$ at the current density J is

$$I_{1,max} = \frac{k_{cu}JA_w}{2N_1} \tag{26-25}$$

similarly

$$I_{2,max} = \frac{k_{cu}JA_w}{2N_2} \tag{26-26}$$

If $I_{1,max}$ and $I_{2,max}$ are smaller than the currents required by the design specifications, then a larger core must be selected. The opposite is true if the maximum currents possible are too large.

The volt–ampere rating can be obtained by combining Eqs. 26-15 and 26-25 and assuming $I_1 = I_{1,max}$

$$S = V_1 I_1 = \left[2k_{cu}JB_{max} \right] A_{core} A_w f \tag{26-27}$$

This equation shows that the maximum volt-ampere rating for the selected core is proportional to the product of the two areas associated with the core and the operating frequency. The quantities within the square brackets are material and slightly size dependent. Another important conclusion to be drawn is that for a given VA rating, the core size varies inversely proportional to the frequency as long as B_{max} is kept constant.

26-3-7   Eddy Current Considerations

In the foregoing design, the eddy current losses in the copper wire have not been included. If the diameter of the wire is larger than the penetration depth δ, which is given by

$$\delta = \sqrt{\frac{2}{\omega\mu\sigma}} \tag{26-28}$$

(where $\omega = 2\pi f$, μ is the permeability of the material, and σ is the conductivity), then there will be substantial eddy current losses. In that case, a multistrand or Litz wire should be used, with the diameter of each strand being less than the penetration depth. Another solution is to use a foil winding with the foil thickness being less than the penetration depth.

26-3-8   Leakage Inductance

Because of the electrical isolation requirement, it is not possible to intertwine the primary and secondary winding conductors. The flux density in the window depends inversely on the window height, and therefore the leakage inductance varies inversely with the (window height)^2. For example, the transformer in Fig. 26-9a will have sixteen times the leakage inductance of the transformer shown in Fig. 26-9b. Toroidal cores are a good choice for minimizing the leakage inductance in a transformer.

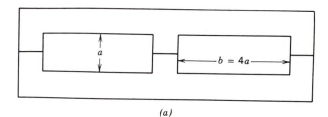

(a)

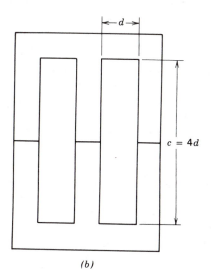

(b)

FIGURE 26-9: The leakage inductance of transformer (a) is at least sixteen times larger than that of transformer (b). The eddy current losses are also larger in transformer (a).

26-3-9   Comparison of Inductor and Transformer Size

With the same operating frequency and the same value of B_{max}, a comparison of Eq. 26-8b with $N' = N$ for the inductor with Eq. 26-27 for the transformer yields

$$S = 2f\left(LI_pI_{rms}\right) \qquad (26\text{-}29)$$

Given the inductance and inductor currents, it is possible to equate the inductor size to that of a transformer at a frequency f whose volt-ampere rating S can be calculated from Eq. 26-29.

26-4   SELECTION OF CAPACITORS

Capacitors must be selected based on the required capacitance, operating voltage, rms current, and frequency. In power electronics, there are basically three types of capacitors used: electrolytic, metallized polypropylene, and ceramic.

26-4-1   Aluminum Electrolytic Capacitors

Electrolytic capacitors offer a larger capacitance per unit volume and are polarized. The large capacitance is due to the aluminum foil connected to the positive terminal being etched so that its surface is porous like a sponge. This results in an increase in its surface area by as much as a factor of 100 compared to its original unetched area. On this etched foil, an insulating layer of aluminum oxide is formed electrochemically. The negative terminal of the capacitor is connected to another aluminum foil that is in electrical contact with the liquid electrolyte, which is an electrically conducting material.

Because of the resistance of the electrolyte, these capacitors have a significant equivalent series resistance (ESR). These capacitors should not be used at temperatures below the specified minimum temperature, since the tendency of the electrolyte to crystallize results in a larger resistance. As discussed in previous chapters, ESR in the output filter capacitors must be low to minimize the ripple in the output voltage.

The capacitor package or can is sealed at the top with an insulating layer that surrounds the electrical terminals. The rate of evaporation of the electrolyte through the seal increases significantly with temperature. Therefore, the capacitor lifetime decreases significantly with temperature. It should be noted that the electrolytic capacitors have by far the shortest lifetime of any element, active or passive, used in power electronic converters. The temperature within the capacitor depends on the power loss. This loss increases with the rms current because of ohmic losses in the capacitor. For a given current, the ripple voltage across the dielectric decreases with increasing frequency. Therefore, the dielectric power loss decreases with increasing frequency. For a given lifetime of the capacitor, its current-carrying capacity increases with increasing frequency and decreasing ambient temperature. Information about these factors are specified in the data sheets.

26-4-2   Metallized Polypropylene Capacitors and Ceramic Capacitors

In snubbers and thyristor commutation circuits, capacitors must handle large currents, but the capacitance value required is small. Metallized polypropylene capacitors are a good choice for such applications because of a very small loss coefficient of the polypropylene dielectric material. The dielectric losses are propor-

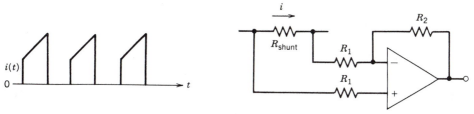

FIGURE 26-10: Current measurement with a resistive shunt and an op-amp.

tional to the square of the voltage and frequency. Since the voltage across the dielectric is proportional to the current and inversely proportional to the frequency, the dielectric power loss is proportional to the square of the current and inversely proportional to the frequency. Therefore, for a specified temperature, the current-handling capability increases slightly with frequency.

Ceramic capacitors have extremely low series inductance. They are used as filters, for example, on printed circuit boards to reduce ripple in the supply voltage.

26-5 RESISTORS

At rated power, the temperature of a resistor is in the range of 300° to 400°C. To avoid such high temperatures, as a general rule, the power loss in a resistor should be limited to approximately one-half its rated power. Resistors for handling higher power are generally wire-wound. Some of these are designed to be mounted on heatsinks.

26-6 CURRENT MEASUREMENTS

In many applications, electrical isolation is not needed between the current measurement circuit and the control electronics circuit. In such cases, the current can be measured by means of a shunt resistor and an op-amp such as is shown in Fig. 26-10. This circuit avoids common-mode voltage jumps in the measured output.

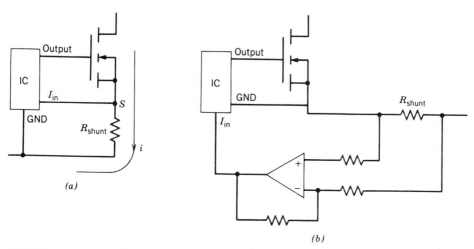

FIGURE 26-11: Use of a control IC in conjunction with (*a*) a current-measuring resistor and (*b*) an improved current-measuring method.

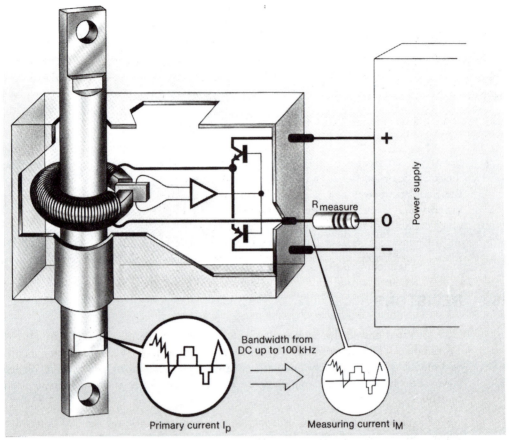

FIGURE 26-12: Hall-effect compensated current sensor with electrical isolation. (Courtesy of LEM, Geneva, Switzerland.)

Moreover, the slew-rate of the op-amp provides excellent filtering of noise, which may be introduced since the current to be measured often has a step waveform. The power loss in the resistor can be minimized by amplifying the voltage by the op-amp. Normally the voltage across the resistor can be as low as 50 to 500 mV without unacceptable noise problems.

Figure 26-11 shows two alternatives for current measurement in conjunction with control and driver ICs. In Fig. 26-11*a*, the voltage across R_{shunt} must be fairly large, and this may cause noise problems. Furthermore, the inductance in the source-to-ground wire can result in oscillations in the gate signal due to the di/dt of the current though the MOSFET. In the circuit of Fig. 26-11*b*, these problems are reduced.

Often in high-power applications, electrical isolation is required between the current to be measured and the control electronics. For ac currents, current transformers can be used. If there is a dc bias in the current, the current transformer will not work. Often the instantaneous current must be measured, which includes the dc bias. In such cases, the current can be measured by a Hall-effect current sensor such as that shown in Fig. 26-12. In this current sensor, a compensating current is applied to the secondary winding such that the field in the toroid is kept at zero. In this way, current from dc up to 100 kHz bandwidth can be measured.

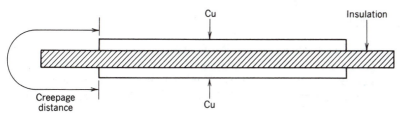

FIGURE 26-13: Transmission-line-like structure composed of copper strips for a low-inductance electrical interconnection.

26-7 HEATSINKING

Extruded aluminum heatsinks of various shapes are used for cooling of the power semiconductor devices. If the heatsinks are cooled by natural convection, the distance between each fin should be at least 10 to 15 mm. A coating of black oxide results in a reduction of the thermal resistance by 25%, but the cost may be higher by the same factor. Thermal time constants of natural convection-cooled heatsinks are in the range of 4 to 15 minutes. Forced-air cooled heatsinks can be made much more compactly. Their thermal time constants may be less than one minute. In higher power ratings, water or oil cooling is used to further improve the thermal conduction.

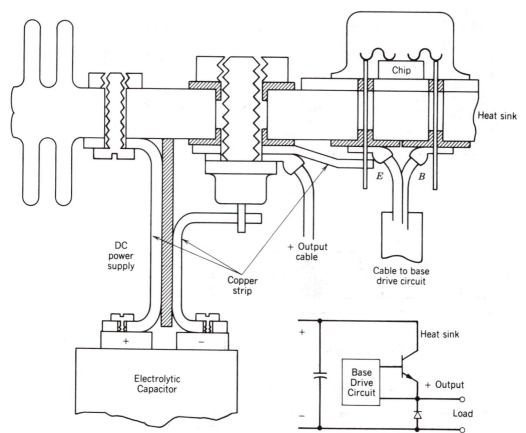

FIGURE 26-14: Use of low-inductance copper strips in the interconnection of an electrolytic capacitor, power transistor, and a free-wheeling diode.

26-8   CIRCUIT LAYOUT

Leakage or stray inductance in a circuit that experiences a large di/dt should be kept as low as possible to minimize overvoltages. Any attempt to minimize stray inductance also reduces stray fluxes and hence EMI. Faster switching devices and the desirability to operate at higher switching frequencies combine to result in larger di/dt.

Copper strips, with a thin insulator sandwiched between them, as is shown in Fig. 26-13, comprise a transmission line and provide an excellent means of reducing the stray inductance. The creepage distance can be increased as is shown in the figure. To use these strips may be cumbersome in practice, but such a design establishes the upper limit in reducing stray inductance. The actual design then can be adapted to meet the manufacturing and cost constraints. As an example, a step-down dc–dc converter layout is presented in Fig. 26-14 where a transistor in a TO-3 case and a stud-mounted free-wheeling diode are used.

These copper strips can also be used to make contact with high-power semiconductor modules as shown in Fig. 26-15. For a three-phase inverter, a parallel combination of electrolytic capacitors can be connected in a similar manner.

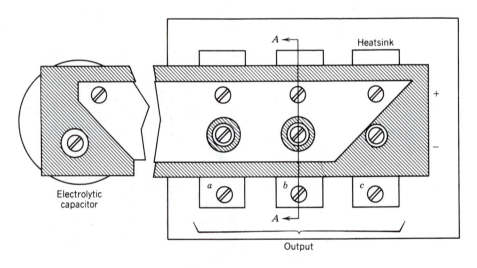

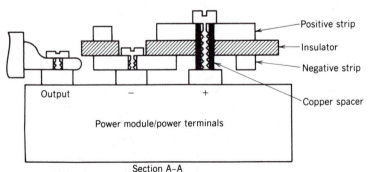

FIGURE 26-15: Example of the use of low-inductance copper strips to interconnect transistor modules of a three-phase inverter and an electrolytic capacitor. Several electrolytic capacitors may be connected in parallel.

26-9 SUMMARY

This chapter has examined some of the practical converter design considerations. The considerations have included passive component selection, cooling considerations, and circuit layout. The important conclusions are as follows:

1. A step-by-step inductor design procedure is presented. B_{max} in the core is selected based on ΔI, f, the core loss, and saturation considerations. For an initial choice of core size, the maximum permissible current density in the copper wires is estimated. The number of turns N is calculated based on current density and I_{rms}. As a last step, the air gap is adjusted such that I_p results in the previously calculated B_{max}.

2. A step-by-step transformer design procedure is presented.

3. Selection of capacitors is based on the capacitance, breakdown voltage, and thermal considerations. The allowable current depends on the capacitor lifetime, cooling, and ambient temperature.

4. Resistors should be operated at less than one-half their rated power.

5. Currents measurements with and without electrical isolation are discussed.

6. Heatsinks are selected based on power loss in the semiconductor device and compactness of design. Forced cooling results in a more compact design and a lower thermal time constant that decreases the capability of the converter to provide peak power above rated values for any significant time duration.

7. Circuit layout to reduce stray inductance by the use of copper strips forming a transmission line is presented.

PROBLEMS

26-1. Design an inductor with an inductance $L_s = 10$ mH for the current waveform shown in Fig. 26-1 where $I_p = 10$ A and $\Delta I = 1.5$ A. A ferrite core material is to be used due to a high ripple frequency of 100 kHz. Because of loss and saturation considerations $B_{max} = 0.3$ T. The core is a double-E core as shown in Fig. 26-3 and the bobbin is shown in Fig. 26-4. Normally only a limited selection of core sizes is available, but in this problem it is assumed that the dimensions can vary continuously. The core area is a^2 and for the bobbin $l_w = 0.5a$ and $h_w = a$. Assume $A_g = 1.2 A_{core}$ and $k_{cu} = 0.7$.
   (a) Find I_{rms}.
   (b) Choose a reasonable current density J and calculate a.
   (c) Knowing $l_w = 0.5a$, find the allowable current density J from Fig. 26-5. Do, if necessary, a few iterations to find the final value of a and J.
   (d) What is the number of turns N?
   (e) Calculate the air gap g.

26-2. An inductor is to be used for improving the line current into a 1-kW single-phase diode rectifier. A cut C-core is to be used with the same relative size as is shown in Fig. 26-3. As in Problem 26-1, it is assumed that the dimensions can vary continuously and that $A_{core} = a^2$ and $A_g = 3a^2$. The ac input voltage V_s is 220 V rms at 60 Hz. The maximum flux density is limited to 1.5 T. Assume $A_g = 1.2$ A_{core}, $k_{cu} = 0.7$, and $J = 2 A/\text{mm}^2$.
   (a) Find I_d, I_p, I_{rms}, and L_s for $(L_s I_d)/V_s = 0.01, 0.04$, and 0.12 [mH A/V]. (*Hint*: See Figs. 3-6 and 3-7 of Chapter 3.)
   (b) Find the area product $A_{core} A_w$ for the three cases in part (a) and sketch it as a function of $(L_s I_d)/V_s$.

26-10 REFERENCES

1. R. E. TARTER, *Principle of Solid-State Power Conversion*, SAMS, 1985.
2. K. THORBORG, *Power Electronics*, Prentice-Hall, New York, 1988.
3. D. LANCASTER, *CMOS Cookbook*, SAMS, 1977.
4. C. W. T. McLYMAN, *Transformer and Inductor Design Handbook*, Dekker, New York, 1978.
5. C. W. T. McLYMAN, *Magnetic Core Selection for Transformers and Inductors*, Dekker, New York, 1982.

INDEX